T0135385

Advances in Intelligent Systems and Computing

Volume 953

Series Editor

Janusz Kacprzyk, Systems Research Institute, Polish Academy of Sciences, Warsaw, Poland

Advisory Editors

Nikhil R. Pal, Indian Statistical Institute, Kolkata, India
Rafael Bello Perez, Faculty of Mathematics, Physics and Computing, Universidad Central de Las Villas, Santa Clara, Cuba
Emilio S. Corchado, University of Salamanca, Salamanca, Spain
Hani Hagras, School of Computer Science & Electronic Engineering, University of Essex, Colchester, UK
László T. Kóczy, Department of Automation, Széchenyi István University, Gyor, Hungary
Vladik Kreinovich, Department of Computer Science, University of Texas at El Paso, El Paso, TX, USA
Chin-Teng Lin, Department of Electrical Engineering, National Chiao Tung University, Hsinchu, Taiwan
Jie Lu, Faculty of Engineering and Information Technology, University of Technology Sydney, Sydney, NSW, Australia
Patricia Melin, Graduate Program of Computer Science, Tijuana Institute of Technology, Tijuana, Mexico
Nadia Nedjah, Department of Electronics Engineering, University of Rio de Janeiro, Rio de Janeiro, Brazil
Ngoc Thanh Nguyen, Faculty of Computer Science and Management, Wrocław University of Technology, Wrocław, Poland
Jun Wang, Department of Mechanical and Automation Engineering, The Chinese University of Hong Kong, Shatin, Hong Kong

The series "Advances in Intelligent Systems and Computing" contains publications on theory, applications, and design methods of Intelligent Systems and Intelligent Computing. Virtually all disciplines such as engineering, natural sciences, computer and information science, ICT, economics, business, e-commerce, environment, healthcare, life science are covered. The list of topics spans all the areas of modern intelligent systems and computing such as: computational intelligence, soft computing including neural networks, fuzzy systems, evolutionary computing and the fusion of these paradigms, social intelligence, ambient intelligence, computational neuroscience, artificial life, virtual worlds and society, cognitive science and systems, Perception and Vision, DNA and immune based systems, self-organizing and adaptive systems, e-Learning and teaching, human-centered and human-centric computing, recommender systems, intelligent control, robotics and mechatronics including human-machine teaming, knowledge-based paradigms, learning paradigms, machine ethics, intelligent data analysis, knowledge management, intelligent agents, intelligent decision making and support, intelligent network security, trust management, interactive entertainment, Web intelligence and multimedia.

The publications within "Advances in Intelligent Systems and Computing" are primarily proceedings of important conferences, symposia and congresses. They cover significant recent developments in the field, both of a foundational and applicable character. An important characteristic feature of the series is the short publication time and world-wide distribution. This permits a rapid and broad dissemination of research results.

** **Indexing: The books of this series are submitted to ISI Proceedings, EI-Compendex, DBLP, SCOPUS, Google Scholar and Springerlink** **

More information about this series at http://www.springer.com/series/11156

Hasan Ayaz
Editor

Advances in Neuroergonomics and Cognitive Engineering

Proceedings of the AHFE 2019 International
Conference on Neuroergonomics
and Cognitive Engineering, and the AHFE
International Conference on Industrial
Cognitive Ergonomics and Engineering
Psychology, July 24–28, 2019,
Washington D.C., USA

 Springer

Editor
Hasan Ayaz
Drexel University
Philadelphia, PA, USA

ISSN 2194-5357   ISSN 2194-5365 (electronic)
Advances in Intelligent Systems and Computing
ISBN 978-3-030-20472-3   ISBN 978-3-030-20473-0 (eBook)
https://doi.org/10.1007/978-3-030-20473-0

© Springer Nature Switzerland AG 2020
This work is subject to copyright. All rights are reserved by the Publisher, whether the whole or part of the material is concerned, specifically the rights of translation, reprinting, reuse of illustrations, recitation, broadcasting, reproduction on microfilms or in any other physical way, and transmission or information storage and retrieval, electronic adaptation, computer software, or by similar or dissimilar methodology now known or hereafter developed.
The use of general descriptive names, registered names, trademarks, service marks, etc. in this publication does not imply, even in the absence of a specific statement, that such names are exempt from the relevant protective laws and regulations and therefore free for general use.
The publisher, the authors and the editors are safe to assume that the advice and information in this book are believed to be true and accurate at the date of publication. Neither the publisher nor the authors or the editors give a warranty, expressed or implied, with respect to the material contained herein or for any errors or omissions that may have been made. The publisher remains neutral with regard to jurisdictional claims in published maps and institutional affiliations.

This Springer imprint is published by the registered company Springer Nature Switzerland AG
The registered company address is: Gewerbestrasse 11, 6330 Cham, Switzerland

Advances in Human Factors
and Ergonomics 2019

AHFE 2019 Series Editors

Tareq Ahram, Florida, USA
Waldemar Karwowski, Florida, USA

10th International Conference on Applied Human Factors and Ergonomics and the
Affiliated Conferences

Proceedings of the AHFE 2019 International Conference on Neuroergonomics and
Cognitive Engineering, and the AHFE International Conference on Industrial
Cognitive Ergonomics and Engineering Psychology, held on July 24–28, 2019, in
Washington D.C., USA

Advances in Affective and Pleasurable Design	Shuichi Fukuda
Advances in Neuroergonomics and Cognitive Engineering	Hasan Ayaz
Advances in Design for Inclusion	Giuseppe Di Bucchianico
Advances in Ergonomics in Design	Francisco Rebelo and Marcelo M. Soares
Advances in Human Error, Reliability, Resilience, and Performance	Ronald L. Boring
Advances in Human Factors and Ergonomics in Healthcare and Medical Devices	Nancy J. Lightner and Jay Kalra
Advances in Human Factors and Simulation	Daniel N. Cassenti
Advances in Human Factors and Systems Interaction	Isabel L. Nunes
Advances in Human Factors in Cybersecurity	Tareq Ahram and Waldemar Karwowski
Advances in Human Factors, Business Management and Leadership	Jussi Ilari Kantola and Salman Nazir
Advances in Human Factors in Robots and Unmanned Systems	Jessie Chen
Advances in Human Factors in Training, Education, and Learning Sciences	Waldemar Karwowski, Tareq Ahram and Salman Nazir
Advances in Human Factors of Transportation	Neville Stanton

(continued)

(continued)

Advances in Artificial Intelligence, Software and Systems Engineering	Tareq Ahram
Advances in Human Factors in Architecture, Sustainable Urban Planning and Infrastructure	Jerzy Charytonowicz and Christianne Falcão
Advances in Physical Ergonomics and Human Factors	Ravindra S. Goonetilleke and Waldemar Karwowski
Advances in Interdisciplinary Practice in Industrial Design	Cliff Sungsoo Shin
Advances in Safety Management and Human Factors	Pedro M. Arezes
Advances in Social and Occupational Ergonomics	Richard H. M. Goossens and Atsuo Murata
Advances in Manufacturing, Production Management and Process Control	Waldemar Karwowski, Stefan Trzcielinski and Beata Mrugalska
Advances in Usability and User Experience	Tareq Ahram and Christianne Falcão
Advances in Human Factors in Wearable Technologies and Game Design	Tareq Ahram
Advances in Human Factors in Communication of Design	Amic G. Ho
Advances in Additive Manufacturing, Modeling Systems and 3D Prototyping	Massimo Di Nicolantonio, Emilio Rossi and Thomas Alexander

Preface

This book brings together a set of contributed articles on emerging practices and future trends in cognitive engineering and neuroergonomics–both aiming at harmoniously integrating human operator and computational system, the former through a tighter cognitive fit and the latter through a more effective neural fit with the system. The chapters in this book uncover novel discoveries and communicate new understanding and the most recent advances in the areas of workload and stress, activity theory, human error and risk, mental state, and systemic-structural activity theory. Further topics include neuroergonomic measures, cognitive computing, and associated applications.

The book is organized into nine parts:

1. Health and Neuroergonomics
2. Neurobusiness Applications
3. Mental State Assessment
4. Virtual Reality and Machine Learning for Neuroergonomics
5. Systemic-Structural Activity Theory
6. Cognitive Computing
7. Brain–Machine Interface and Artificial Intelligence Systems
8. Cognitive Neuroscience and Health Psychology
9. Applications in Human Interaction and Design.

Part 1 to 6 includes contributions to the 10th International Conference on Neuroergonomics and Cognitive Engineering. Part 7 to 9 includes contributions to the 1st International Conference on Industrial Cognitive Ergonomics and Engineering Psychology (ICEEP). This conference examines the cognitive ergonomic aspects of a workplace to understand a working task and solve a problem, thus making human–system interaction compatible with human cognitive abilities and limitations at work. It discusses optimal human-work parameters, such as mental workload, decision-making, skilled performance, human reliability, human–system design, human–computer interaction, work stress and training, as these may relate to worker's ability to properly construe the task, in order to avoid hazard, errors, misperception, frustration, and mental work overload.

Special thanks go to Dr. Umer Asgher from the National University of Sciences and Technology for his valuable contribution and for chairing the novel conference track on Industrial Cognitive Ergonomics and Engineering Psychology. Collectively, the chapters in this book have the goal of offering a deeper understanding of the couplings between external behavioral and internal mental actions, which can be used to design harmonious work and play environments that seamlessly integrate human, technical, and social systems.

Each chapter of this book was either reviewed or contributed by members of the Scientific Advisory Board. For this, our sincere thanks and appreciation goes to the board members listed below:

Neuroergonomics, Cognitive Engineering

H. Adeli, USA
Carryl Baldwin, USA
Gregory Bedny, USA
Winston "Wink" Bennett, USA
Alexander Burov, Ukraine
P. Choe, Qatar
M. Cummings, USA
Frederic Dehais, France
Chris Forsythe, USA
X. Fang, USA
Qin Gao, China
Klaus Gramann, Germany
Y. Guo, USA
Peter Hancock, USA
David Kaber, USA
Kentaro Kotani, Japan
Ben Lawson, USA
S.-Y. Lee, Korea
Harry Liao, USA
Y. Liu, USA
Ryan McKendrick, USA
John Murray, USA
A. Ozok, USA
O. Parlangeli, Italy
Stephane Perrey, France
Robert Proctor, USA
A. Savoy, USA
K. Vu, USA
Thomas Waldmann, Ireland
Tomas Ward, Ireland
Brent Winslow, USA
G. Zacharias, USA

L. Zeng, USA
Matthias Ziegler, USA

Cognitive Computing and Internet of Things

Hanan A. Alnizami, USA
Thomas Alexander, Germany
Carryl Baldwin, USA
O. Bouhali, Qatar
Henry Broodney, Israel
Frederic Dehais, France
Klaus Gramann, Germany
Ryan McKendrick, USA
Stephane Perrey, France
Stefan Pickl, Germany
S. Ramakrishnan, USA
Duncan Speight, UK
Martin Stenkilde, Sweden
Ari Visa, Finland
Tomas Ward, Ireland
Matthias Ziegler, USA

Industrial Cognitive Ergonomics and Engineering Psychology

Ellie Abdi, USA
José Arzola-Ruiz, Cuba
Yasar Ayaz, Pakistan
Vania Vieira Estrela, Brazil
Muhammad Jawad Khan, Pakistan
Roland Iosif Moraru, Romania
Noman Naseer, Pakistan
Hoang-Dung Nguyen, Vietnam
Noriyuki Oka, Japan
Aleksandra Przegalinska, USA
Hendrik Santosa, USA
Silvia Serino, Italy
Susumu Shirayama, Japan
Stéphanie Stankovic, France
Redha Taiar, France
Jinhui Wang, China

We hope that this book will offer an informative and valuable resource to professionals, researchers, and students alike and that it helps them understand innovative concepts, theories, and applications in the areas of cognitive engineering and neuroergonomics. Beyond basic understanding, the contributions are meant to inspire critical thinking for future research that further establish the fledgling field of neuroergonomics and sharpen the more seasoned practice of cognitive engineering. While we don't know where the confluence of these two fields will lead, they will certainly transform the very nature of human–system interaction, resulting in a yet to be envisioned design that improve form, function, efficiency, and the overall user experience for all.

This book is dedicated to Gregory Bedny for his pioneering work in the systemic-structural activity theory and for the scientific contribution to the field of neuroergonomics.

July 2019   Hasan Ayaz

Contents

Cognitive Neuroscience and Health Psychology

Applications in Human Interaction and Design

Health and Neuroergonomics

Novel Contributions of Neuroergonomics and Cognitive Engineering to Population Health

Peter A. Hall[(✉)]

School of Public Health and Health Systems, University of Waterloo,
200 University Avenue West, Waterloo, ON N2L 3G1, Canada
pahall@uwaterloo.ca

Abstract. This chapter describes important ways in which brain imaging and brain stimulation technologies are poised to contribute to disease prevention at the level of whole populations, and how neuroergonomics and cognitive engineering have set the conditions for this to occur. The historically limited influence of neuroscience on population health is discussed with reference to logistics, conceptual barriers, and epistemic considerations. With respect to the latter, the brain is typically viewed as an outcome variable, rather than its more nuanced role as a predictor, mediator or moderator. Yet these later roles potentiate a number of important functions for neuroscience research within disease prevention with wide ranging implications. Using examples from multiple laboratories, I highlight several examples of how neuroimaging and neuromodulation technologies can be used to generate new knowledge to shape disease prevention programs and optimize health communication strategies.

Keywords: Neuroergonomics · rTMS · fNIRS · Brain · Health · Population

1 Introduction

Most of the diseases responsible for premature mortality worldwide are chronic rather than infectious, and are fundamentally preventable in nature [1]. Within the preventive medicine perspective, secondary prevention is intervention intended to reduce disease susceptibility among at-risk populations (e.g., reducing binge drinking among college students), whereas primary prevention involves intervening at the whole population level to reduce risk of initial occurrence of disease cases within the population (e.g., reducing everyday consumption of alcohol in the general population). Primary disease prevention is often approached by addressing lifestyle determinants that give rise to new cases or the critical exposure variable of interest. For instance, smoking can be reduced in the general population for the sake of decreasing the prevalence of lung cancer; physical activity can be encouraged for the sake of preventing obesity, heart disease, and type 2 diabetes; safer sexual practices within the general population (e.g., use of condoms) can reduce the spread of sexually transmitted diseases, including HIV. Within population health, surprisingly little attention has been paid to the brain itself as a foundation for primary prevention of chronic disease, despite the central involvement

© Springer Nature Switzerland AG 2020
H. Ayaz (Ed.): AHFE 2019, AISC 953, pp. 3–13, 2020.
https://doi.org/10.1007/978-3-030-20473-0_1

of the brain in processing information about health risks communicated through social networks and the media, and despite the brain's role as the nexus of translating thoughts about risk into risk-avoidance behaviors. This, however, is rapidly changing.

Technological and conceptual developments from subdisciplines of neuroscience and human factors engineering have come together to provide important new avenues for disease prevention. Human factors involves consideration of the interface between humans and other aspects of a given system and/or technology; neuroergonomics is a subfield of the former that focusses specifically on systems that interface with the human brain. With respect to neuroscience, understanding the harmonious integration of brain imaging technology and the individual has motivated and been motivated by miniaturization of brain imaging technology in the interests of enhancing portability. Whereas technologies for imaging the brain at rest or during a functional task were the provenience of laboratory science, portable imaging devices such as functional near infrared spectroscopy (fNIRS) are sufficiently portable to be deployed within the everyday living environment, making human interface of central interest.

2 Historical Challenges

There are several challenges to incorporating brain imaging into population level disease prevention research: logistical, epistemic and conceptual. Some of these have been solved by recent technological innovation in neuroergonomics. I will review each of these in the sections that follow.

Logistical. Sample sizes in neuroimaging studies are often relatively small, comprising dozens (most common) or hundreds of participants (least common). However, population health research studies—particularly longitudinal studies—typically involve thousands or tens of thousands of participants, with some stretching to more than one million. Such large sample sizes are sought in population health research for very good reasons. When paired with random sampling techniques, larger sample sizes are more likely to result in findings that are generalizable to the target population, which is a necessary precondition for the use of such research for informing public policy. The latter is a major objective of population health research, and so sample size maximization is valued because of it. Beyond generalizability, larger sample sizes produce parameter estimates that are more precise (i.e., measured with less error) than smaller samples, all else being equal. Again, policy drives the push for larger samples, in that highly consequential provision of budget funding (in the millions or tens of millions of dollars) may be tied directly to a population estimate of disease/risk factor incidence, prevalence, or estimate of relationship magnitude. Having an imprecise estimate could lead to over funding or under funding of health services, preventative activities or educational initiatives by millions of dollars in a single calendar year. This provides a strong fiscal incentive to seek larger samples in population health studies.

Finally, many population health studies are longitudinal in nature. Longitudinal studies differ from prospective studies in that they follow people over time to examine prospective relationships among variables. However, longitudinal studies differ from garden variety prospective studies in two primary ways: (1) they tend to be longer in duration, in an absolute sense, and (2) they always have two or more follow-up measurements. Thus, to be considered a prospective study, only a baseline and single follow-up measurement is required; however a longitudinal study is one that has several follow-ups, and by convention may stretch for years or even decades. Although these con-ventions are not universal, this differentiation approximates the most common usage of the term in population health research. The longitudinal study design in practice gen-erates more instances of missing data than a simple prospective or cross-sectional design, because of the multiple measurement waves. Such missingness is sufficiently prob-lematic that it has become the singular focus of many academic statisticians. It can be dealt with in many cases using statistical imputation techniques, but it also necessitates more cases of data in order to proactively offset its effects. For this reason, larger sample sizes are preferred in longitudinal research, in order to ensure that expected missing data does not compromise generalizability of the results or precision of parameter estimates derived. Indeed, all imputation methods work better—or indeed may be essentially unnecessary—when the ratio between complete and missing data is favorable (i.e., few missing observations per 100 valid observations, for example). However, even the best imputation methods are doomed to fail if the same ratio is unfavorable.

These reasons—generalizability, precision of estimation, and dataset integrity—all drive sample sizes up in population health research. This has traditionally made it difficult to incorporate neuroimaging methods into large scale population health ini-tiatives, despite a longstanding recognition that the brain is an important determinant and outcome of population health. Instead, cognitive testing has mostly stood in place of imaging, given the better fit with data collection protocols already in place in human longitudinal research studies. Functional imaging in the form of functional magnetic resonance imaging (fMRI) or positron emission tomography (PET) both rely on static equipment with limited accessibility, and no option for deployment to field settings where many population health research teams work. Likewise, brain structural imaging procedures such as magnetic resonance imaging (MRI) and computerized tomography (CT) suffer from the same limitation. Until relatively recently electro encephalogram (EEG) has been similarly hampered by equipment, but less so than the above modalities; however, ongoing challenges in terms of spatial resolution and motion artifact have limited the utility of EEG in large scale data collection initiatives.

Epistemic. Traditionally, population health studies have disproportionately been conduced by researchers in epidemiology and public health. These fields include training in statistics, observational research, as well as measurement. However, few if any programs in either field include core training in neuroscience methods, concepts, or data management approaches (e.g., signal processing). Yet all of the above are nec-essary in order to conduct research involving neuroscience methods such as brain imaging. The few studies that have nonetheless involved imaging (most commonly fMRI) have sidestepped the issue by limiting data collection to urban research centers and incorporating neuroimaging expertise as separate team members. However, it is generally the case that little overlap in expertise exists between population health researchers and those who are involved in the brain health measurements. this lack of common conceptual understanding, lexicon, and skill set has made it difficult to have a

Brain-as-Outcome:

Brain-as-Predictor:

Brain-as-Mediator:

Brain-as-Moderator:

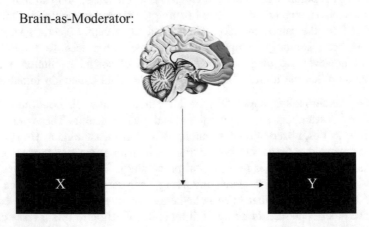

Fig. 1. Four prototypes linking the brain and health variables within the population health perspective. A *mediator* is a variable purported to be part of the causal chain linking two other variables; a *moderator* is a variable that modifies the strength of relationship between two other variables.

full and complete integration of neuroscience methods into existing population health research teams, which requires transdisciplinary familiarity among all team members.

Conceptual. For reasons that are not fully clear, when considered at all in population health contexts the brain has commonly been thought of as an outcome variable. For instance, diabetes, hypertension, obesity or other diseases processes would be expected to be linked to the brain primarily because it may manifest some consequence of the disease process. Indeed, within all of these examples—diabetes, hypertension, and obesity—there is very little consideration of the brain (and the cognitive operations that it supports) as being anything other than another organ that diabetes, hypertension or obesity might effect the function of. For this reason, there might have traditionally been less incentive to include logistically challenging measurement of brain-related parameters in population health research studies already burdened with formidable logistical and measurement challenges in more central ways. Most large-scale population studies involving cognition examine the effect of disease processes on cognitive performance. However, there are many possibilities for how to conceptualize the role of the brain in disease prevention [2]. Beyond the brain-as-outcome perspective, the brain can be thought of also as a predictor of outcomes (the "brain-as-predictor" prototype), a mediator of relationships between an established variable and a health outcome (the "brain-as-mediator" prototype), or a moderator of the effects of such a variable on the outcome of interest (the "brain-as-moderator" prototype). In turn these standard prototypes can be elaborated into more complex models including moderated mediation and mediated moderation models involving the brain. Each of these prototypes are depicted in Fig. 1.

As discussed below, some examples of brain-as-outcome and brain-as-predictor prototypes highlight new avenues of progress in disease prevention at the population level. These have all been facilitated by technological development in neuromodulation (active perturbation of brain function) and neuroimaging (passive measurement of brain function or structure).

3 New Avenues in Disease Prevention

Health Behavior Research: Knowledge Generation. New knowledge about how the brain regulates eating has been gleaned from studies involving neuromodulation technologies, which have been increasingly refined in recent years. Neuromodulation includes any technology that induces change in the excitability of neuron populations in a temporary or long-lasting sense. Neuromodulation has been used both for laboratory experimentation to triangulate causal effects of brain systems on eating behavior, and as a clinical intervention to improve symptoms of disorders of indulgent eating [3, 4]. Given that many public health threats are attributable to (or exacerbated by) indulgent eating (e.g., obesity, Type 2 diabetes, hypertension), the role of neuromodulation technology in knowledge generation has been extremely important.

The two technologies used in neuromodulation research involving eating are repetitive transcranial magnetic stimulation (rTMS) and transcranial electrical stimulation (tES)[1]. In our laboratory we have used rTMS technologies to examine the role of the dorso-lateral prefrontal cortex (dlPFC) in modulation of eating indulgence in the laboratory by using it to temporarily suppress excitability of neuron populations within this structure. Our findings and those of others have suggested that suppression of the dlPFC results in increased consumption, whereas excitation leads to better self-restraint [5, 6], an effect that appears to be amplified when environmental cues encourage indulgence [7]. This research has relied on the use of a special form of rTMS called theta burst stimulation, involving rapid delivery of magnetic pulses (50 Hz), rendered possible only in the past decade with advancements in rTMS hardware [8]. Continued progress in this area will depend on the integration of rTMS systems with robotics to localize brain structures with more accuracy, and possibly even more rapid pulse delivery to induce temporary or long-lasting states of cortical excitation or inhibition. Combining rTMS with neuroimaging has been a challenge in the past, but fMRI, fNIRS and EEG have all now been successfully used in concert with rTMS and tES variants, in part because of the co-evolution of the two technologies.

Another area of research that has made important use of neuroscience methods—and will benefit from continued innovation in portable brain imaging technology—is exercise neuroscience. Many studies over the past number of years have revealed beneficial effects of exercise for the brain, using a brain-as-outcome perspective; the benefits of exercise are particularly evident in relation to brain systems that support cognitive control, attention and memory [9–12]. Likewise, the brain-as-predictor per-spective has also yielded fundamental insights in relation to the role of cognitive control networks in facilitating adherence to this behavior that is notoriously subject to distraction and competing alternative behaviors [13–15]. Most of the existing studies have utilized fMRI yet would benefit from a nuanced understanding of exercise effects on the brain during the course of movement execution; something only possible with mobile fNIRS and EEG. Previously mentioned miniaturization efforts and enhanced portability will enable more ecologically valid assessments of exercise effects on the brain. Further development of everyday wearable versions of fNIR sensors will further expand the potential of population level examination of bi-directional relationships between exercise and brain health. This will be particularly true when brain health data is integrated with accelerometer data from existing mobile devices already widely available in the general population.

Health Communication Research: Knowledge Application. The brain-as-predictor approach is the most quintessential example of how different prototypes beyond the brain-as-outcome default might be important to consider. The work of Emily Falk and

---

[1] tES and rTMS are the two major categories of non-invasive brain stimulation technology currently in use for both experimental and clinical treatment applications. Both technologies purport to increase and/or decrease excitability of target neuron populations within the cortex. Although the proposed effect may be similar in terms of neuron population excitability, the mechanism of action is very different: while tES uses a constant ("direct") current to induce changes in excitability using electrodes places on the scalp, rTMS uses magnetic pulses delivered via an external coil placed against the scalp to induce changes in neuroelectric activity within targeted brain regions.

A.

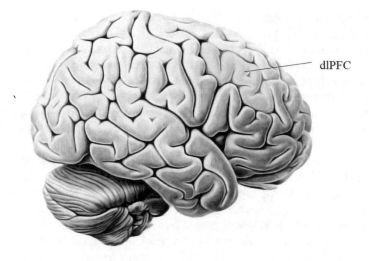

B.

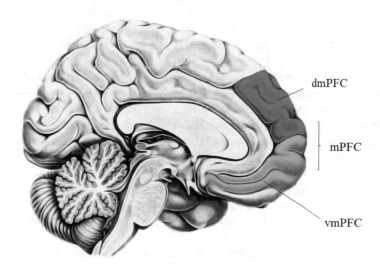

Fig. 2. Brain regions implicated in health behavior and health communication; dlPFC = dorsolateral prefrontal cortex; dmPFC = dorsomedial prefrontal cortex; vlPFC = ventrolateral prefrontal cortex; mPFC = medial prefrontal cortex. (A) lateral view, (B) sagittal view.

colleagues on the neural predictors of health communication success is particularly compelling [16–18]. In a series of studies, Falk and colleagues have demonstrated that health communications that activate the medial prefrontal cortex (mPFC; Fig. 2)—an area involved in self-referential and evaluative processing—are more likely to be spontaneously shared with others within a social network, and are more predictive of post-communication changes in behavior in the general population.

In an initial series of studies involving smokers, the investigators examined self-reported impressions of a smoking cessation ad as well as neural activation in the mPFC as predictors of subsequent smoking; when examining 1-month changes in smoking level using exhaled CO_2. Although self-reported impressions of the media ads predicted significant variability in smoking cessation outcomes at the 1 month mark, neural responses to the ads in the mPFC did as well, and predicted a significant increment in variability beyond the self-reported impressions [19]. A subsequent study examined the extent to which neural responses vs self-reported impressions of ad efficacy predicted real-world uptake of a smoking cessation hotline at the level of the whole population; in this study three ad campaigns were compared by participants, and a-priori rank orderings were created based on self-reported impressions among existing smokers, as well as via magnitude of neural activity during viewing (using the same participants). The rank orderings were then compared to real world change in usage rates of the advertised quit line when the three ad campaigns were actually rolled out in three different states [17]. Findings indicated that mPFC activation-derived rank orderings correctly predicted quit line usage increase in the population. Strikingly, subjective impression-derived rank orderings were negatively predictive of quit line usage, such that the most effective campaign was predicted least effective and the least effective campaign predicted most effective. Similar findings have been reported with other outcome metrics, media modalities and target behaviors [20–22].

These findings illuminate a new path to pre-screening health communications that utilize media transmission. In contrast with the common approach of using traditional focus groups to pre-screen alternative communication options and select solely based on the self-reported impressions, it may be possible to replace or supplement such information with neural response data. Higher degrees of engagement of the mPFC (and/or other brain regions shown to predict relevant communication success) might be used to select from alternative campaigns or media samples, to ensure that advertising dollars are invested appropriately. Taking this approach requires the embedding of neuroscience equipment and expertise (measurement, signal processing and analyses) within population health research teams, and increasing the level of literacy of public health experts in relation to the interpretation of the outputs from such multidisciplinary efforts.

The majority of the studies above make use of fMRI for the pre-screening phase. However, the mPFC can also be imaged using functional near infrared spectroscopy (fNIRS), a technique that also relies on a blood oxygenation dependent (BOLD) response to infer neuronal activity within specific regions. fNIRS has some advantages over fMRI, including the lower equipment cost and ability to be used in highly variable settings, as mentioned earlier. These parameters allow for a good fit with settings that wish to employ neural screening of messages without direct access to fMRI facilities. Recent miniaturization of fNIRS technology has enabled the equipment to be deployed with more flexibility, and health communication initiatives may increasingly make use of it [23–25]. Initial studies have confirmed the mPFC effects described above using fNIRS [20].

Longitudinal Research. Even within the traditional brain-as-outcome approach, there is much to be gained from adding brain imaging to large scale (and large sample) longitudinal studies. Although fMRI is likely not the technology of choice, miniaturization of fNIRS allows for use in field settings where most of the data collection for longitudinal research takes place. To date no such studies have been attempted, but many existing studies have cognitive testing protocols, which readily allows for the integration of fNIRS, given that it can be employed to image cortical activations that are task dependent. The integration of this metric allows for brain health to be assessed as an outcome of any other predictor variables already included in the study, and also enables tracking of brain development over time using growth curve modelling. The other prototypes depicted in Fig. 2 can also be estimated when using fNIRS measurements. For example, brain responses to cognitive test protocols can be used to predict onset of diseases, consistency in health protective behaviors, susceptibility to risk behaviors, and even mortality. Likewise, the brain's role as a moderator and mediator of other effects can be probed.

4   Conclusion

In conclusion, significant advances in neuroimaging have enabled more flexibility in how and when the brain is examined, whether its role is conceptualized as an outcome, predictor, mediator or moderator. Recent studies have demonstrated its power to play an important role in predicting the efficacy of health communications, and ongoing attempts are being made to integrate it into longitudinal research studies. The latter will enable the brain to be examined with respect to any of its putative roles, and with more flexibility than ever before possible. Continued technological advancement in mobile brain imaging and neuromodulation will greatly augment our ability to prevent diseases on the population level. Neuroergonomics and cognitive engineering are two fields of inquiry that are ideally positioned to facilitate this.

Acknowledgments. This work was supported by an operating grant to P. Hall (435-2017-0027) from the Social Sciences and Humanities Research Council of Canada.

References

1. World Health Organization: Global status report on noncommunicable diseases (2014)
2. Erickson, K.I., Creswell, J.D., Verstynen, T.D., Gianaros, P.J.: Health neuroscience: defining a new field. Curr. Dir. Psychol. Sci. **23**, 446–453 (2014)
3. Hall, P.A.: Brain stimulation as a method for understanding, treating and preventing disorders of indulgent food consumption. Curr. Addict. Rep. (2019)
4. Hall, P.A.: Executive-control processes in high-calorie food consumption. Curr. Dir. Psych. Sci. **25**, 91–98 (2016)

5. Lowe, C.J., Staines, W.R., Manocchio, F., Hall, P.A.: The neurocognitive mechanisms underlying food cravings and snack food consumption. a combined continuous theta burst stimulation (cTBS) and EEG study. Neuroimage **177**, 45–58 (2018)

6. Hall, P.A., Lowe, C., Vincent, C.: Brain stimulation effects on food cravings and consumption: an update on Lowe et al. (2017) and a response to Generoso et al. (2017). Psychosom. Med. **79**, 839–842 (2017)

7. Safati, A.B., Hall, P.A.: Contextual cues as modifiers of cTBS effects on indulgent eating (Manuscript under review)

8. Suppa, A., Huang, Y.Z., Funke, K., Ridding, M.C., Cheeran, B., Di Lazzaro, V., Ziemann, U., Rothwell, J.C.: Ten years of theta burst stimulation in humans: established knowledge, unknowns and prospects. Brain. Stim. **9**, 323–335 (2016)

9. Erickson, K.I., Prakash, R.S., Voss, M.W., Chaddock, L., Hu, L., Morris, K.S., White, S.M., Wójcicki, T.R., McAuley, E., Kramer, A.F.: Aerobic fitness is associated with hippocampal volume in elderly humans. Hippocampus **19**, 1030–1039 (2009)

10. Erickson, K.I., Voss, M.W., Prakash, R.S., Basak, C., Szabo, A., Chaddock, L., Kim, J.S., Heo, S., Alves, H., White, S.M., Wojcicki, T.R.: Exercise training increases size of hippocampus and improves memory. P. Natl. Acad. Sci. USA **108**, 3017–3022 (2011)

11. Hillman, C.H., Erickson, K.I., Kramer, A.F.: Be smart, exercise your heart: exercise effects on brain and cognition. Nat. Rev. Neurosci. **9**, 58 (2008)

12. Stillman, C.M., Erickson, K.I.: Physical activity as a model for health neuroscience. Ann. NY. Acad. Sci. **1428**, 103–111 (2018)

13. Best, J.R., Chiu, B.K., Hall, P.A., Liu-Ambrose, T.: Larger lateral prefrontal cortex volume predicts better exercise adherence among older women: evidence from two exercise training studies. J. Gerontol. A-Bio. **72**, 804–810 (2017)

14. Gujral, S., McAuley, E., Oberlin, L.E., Kramer, A.F., Erickson, K.I.: Role of brain structure in predicting adherence to a physical activity regimen. Psychosom. Med. **80**, 69–77 (2018)

15. Hall, P.A., Fong, G.T.: Temporal self-regulation theory: a neurobiologically informed model for physical activity behavior. Front. Hum. Neurosci. **25**, 117 (2015)

16. Falk, E., Scholz, C.: Persuasion, influence, and value: perspectives from communication and social neuroscience. Ann. Rev. Psychol. **4**, 69 (2018)

17. Falk, E.B., Berkman, E.T., Lieberman, M.D.: From neural responses to population behavior: neural focus group predicts population-level media effects. Psychol. Sci. **23**, 439–445 (2012)

18. Falk, E.B., Morelli, S.A., Welborn, B.L., Dambacher, K., Lieberman, M.D.: Creating buzz: the neural correlates of effective message propagation. Psychol. Sci. **24**, 1234–1242 (2013)

19. Falk, E.B., Berkman, E.T., Whalen, D., Lieberman, M.D.: Neural activity during health messaging predicts reductions in smoking above and beyond self-report. Health Psychol. **30**, 177 (2011)

20. Burns, S.M., Barnes, L., Katzman, P.L., Ames, D.L., Falk, E.B., Lieberman, M.D.: A functional near infrared spectroscopy (fNIRS) replication of the sunscreen persuasion paradigm. Soc. Cogn. Affect. Neur. **1**, 9 (2018)

21. Falk, E.B., O'Donnell, M.B., Tompson, S., Gonzalez, R., Dal Cin, S., Strecher, V., Cummings, K.M., An, L.: Functional brain imaging predicts public health campaign success. Soc. Cogn. Affect. Neur. **11**, 204–214 (2015)

22. Doré, B.P., Tompson, S.H., O'Donnell, M.B., An, L., Strecher, V., Falk, E.B.: Neural mechanisms of emotion regulation moderate the predictive value of affective and value-related brain responses to persuasive messages. J. Neurosci. **39**, 1293–1300 (2019)

23. Ayaz, H., Izzetoglu, M., Izzetoglu, K., Onaral, B.: The use of functional near-infrared spectroscopy in neuroergonomics. In: Neuroergonomics, pp. 17–25. Academic Press, London (2019)

24. Curtin, A., Ayaz, H.: The age of neuroergonomics: towards ubiquitous and continuous measurement of brain function with fNIRS. Jpn. Psychol. Res. **60**(4), 374–386 (2018)
25. Ferrari, M., Quaresima, V.: A brief review on the history of human functional near-infrared spectroscopy (fNIRS) development and fields of application. Neuroimage **63**, 921–935 (2012)

A Cross-Sectional Study Using Wireless Electrocardiogram to Investigate Physical Workload of Wheelchair Control in Real World Environments

Shawn Joshi[1,2,3,4,5(✉)], Roxana Ramirez Herrera[6],
Daniella Nicole Springett[3,4,5], Benjamin David Weedon[3,4,5],
Dafne Zuleima Morgado Ramirez[6,7], Catherine Holloway[6,7],
Hasan Ayaz[1,8,9,10], and Helen Dawes[3,4,5]

[1] School of Biomedical Engineering, Science & Health Systems,
Drexel University, Philadelphia, PA, USA
{sj633,ayaz}@drexel.edu
[2] College of Medicine, Drexel University, Philadelphia, PA, USA
[3] Movement Science Group, Oxford Brookes University, Oxford, UK
hdawes@brookes.ac.uk
[4] Oxford Institute of Nursing, Midwifery, and Allied Health Research,
Oxford, UK
[5] Nuffield Department of Clinical Neurosciences,
Oxford University, Oxford, UK
[6] UCL Interaction Centre, University College London, London, UK
[7] Global Disability Innovation Hub, London, UK
[8] Department of Family and Community Health, University of Pennsylvania,
Philadelphia, PA, USA
[9] Center for Injury Research and Prevention,
Children's Hospital of Philadelphia, Phialdelphia, PA, USA
[10] Drexel Business Solution Institute, Drexel University, Philadelphia, PA, USA

Abstract. The wheelchair is a key invention that provides individuals with limitations in mobility increased independence and participation in society. However, wheelchair control is a complicated motor task that increases physical and mental workload. New wheelchair interfaces, including power-assisted devices can further enable users by reducing the required effort especially in more demanding environments. The protocol engaged novice wheelchair users to push a wheelchair with and without power assist in a simple and complex environment using wireless Electrocardiogram (ECG) to approximate heart rate (HR). Results indicated that HR determined from ECG data, decreased with use of the power-assist. The use of power-assist however did reduce behavioral performance, particularly within obstacles that required more control.

Keywords: Wheelchair · Power-assist · Heart rate · Wireless · Real-world · Cognitive workload · Disability

© Springer Nature Switzerland AG 2020
H. Ayaz (Ed.): AHFE 2019, AISC 953, pp. 14–25, 2020.
https://doi.org/10.1007/978-3-030-20473-0_2

1 Introduction

The wheelchair is a tool for equality for individuals with limitations in mobility; it increases independence and opportunities to actively engage in their environment [1, 2]. Typical manual wheelchair propulsion can lead to a variety of negative health outcomes. Wheelchair users (WU) are prone to upper arm injuries related to continuous or excessive use, including damage to rotator cuff muscles – 42–66% of WU often report shoulder pain [3–5] and may suffer from bilateral carpal tunnel syndrome [6]. Reduced mobility can lead to a more sedentary lifestyle often reducing individual physical capacity for many WU [7, 8].

This is particularly problematic, as independent manual wheelchair propulsion requires adept physical capacity and cardiorespiratory fitness [9, 10]. Many musculoskeletal problems manual WU face can be prevented by reducing the use of the wheelchair (however this is wholly impractical as it would limit equality/autonomy), or altering factors related to reducing the physical load (demands of the environment) [11, 12], or increasing power (human characteristics) [13]. Electric remove the need to self-propel and therefore reduce strain injuries, while reducing metabolic demand to allow further travel and in more variable locations [14]. However they encourage an even less physically active lifestyle, predisposing users to long term health problems related to inactivity (obesity, cardiovascular disease, etc.) [15, 16].

Power Assisted Devices (PADs) are new generation mobility interfaces, that offer a middle ground solution to the problems of both manual and electric wheelchairs. They can allow users to reduce physical strain, but not at the cost of removing all the cardiovascular (CV) beneficial physical activity [13]. They are propelled in the same manner as a manual wheelchair but are fitted with small electric motors (either in the wheels or behind the wheelchair) to augment the user's physical power and allow for the social, and mobile benefits of an electric wheelchair while partially retaining the exercise component from a manual wheelchair. While PADs have intrinsic design problems, they are being increasingly considered among manual WU [17].

Wheelchair control is a complicated motor task that increases both the cognitive (or mental) and physical workload of an individual [18]. Cognitive workload (CW) refers to the limited information processing capacity of the brain demanded by a task or environment [19]. When environmental demands increase, subsequent increases in CW are generated. However if environmental demands exceed this capacity for information processing, task performance inevitably decreases [20]. Accidents or errors are a result of decreased or poor task performance [21]. Measuring CW is complex as it represents the interplay between the environmental demands (input), human characteristics (capacities), and task performance (output) on the operator [22, 23]. The association between CW and physical workload is an essential component of physical neuroergonomics, the study of the brain in relation to the control and design of physical tasks incorporating evaluations of brain and body measurements in natural environments as opposed to artificial laboratory settings and simplified tasks [24–32].

Understanding the factors in reducing/optimizing cognitive and physical workload in in order to improve task performance is important, particularly in the context of operating complex machinery such as manual wheelchairs. Excessive workload can lead to serious injuries, increased economic burden, and other maladies to and from the user [33, 34] and can further impact mobility, resulting in activity restriction, affecting social participation, health and wellbeing and quality of life [35]. Physical workload can be measured by a variety of mechanisms, however one of the more common, practical and valid measures includes heart rate (HR) [36].

In the United States alone, there are 3.6 million active WU above the age of 15, and due to our aging population, there are an additional 2 million new WU every year [37]. Therefore, it is imperative that newer generation wheelchair designs, such as PAD's, consider both physical and mental effort implications to optimize control ergonomics to improve safety and better community engagement. The objective of this study was to understand the interplay of expertise, environment, and interface during real-world wheelchair control. Therefore, this paper set out to evaluate the cognitive and physical workload as measured by behavioral task performance and HR for novice WU during manual and power-assisted wheelchair propulsion in both simple and complex environments.

2 Methods

30 novice participants (12 males) were recruited aged 31.8 ± 9yrs. Only those physically able to propel a manual wheelchair for an extended period of time, and without cognitive disability or recent physical injury were recruited. Each participant completed a Physical Activity Readiness Questionnaire (PAR-Q) [38], and had biometric measurements of height, weight, age, skin color, hair color/type, grip strength (left and right), blood pressure, maximal speeds, seat height, and arm lengths. All participants also reported having normal or corrected-to-normal vision and were self-described to be able to control a wheelchair for up to one hour, including difficult terrain.

The study was conducted at the Oxford Brookes Sports Hall located in Oxford, UK with approval obtained from the University Research Ethics Committee with reference number UCLIC/1617/024/StaffHolloway/Herrera between 12/17 and 09/18.

2.1 Measurments and Devices

All participants used the manual wheelchair frame (QUICKIE LIFE R) weighing 10.5 kg to traverse two environments detailed below. The wheelchair had a seat width of 45 cm and fitted with the M24 Alber Twion (*Alber GmbH, Albstadt, Germany*) power assist wheels (additional 6 kg each). The power assist was set in the ECO mode to allow for a maximal propulsion speed of 10 km/hr. Participants wore a portable ECG sensor known as the EcgMove 3 (*Movisens GmbH, Karlsruhe, Germany)* across the chess below T4. Experiments were video recorded using a GoPro Hero Action Digital.

The behavioral performance during the experiment was manually recorded by two research assistants (reduced inter-rater variability) and retrospectively corrected when reviewing the video of each experiment. Total number of errors per obstacle was recorded and converted into percentages based on the maximum error count recorded by all participants. Obstacle percentage scores were averaged to give a total performance percentage (each obstacle given the equal weight/importance), where higher scores indicated better performance/fewer errors.

2.2   Environmental Design

Two environments (simple and complex) were created. The simple environment (flat terrain and free of obstacles) formed the outer rectangle of 13 m × 14 m for a total propulsion distance of 54 m, and the complex environment (four separate obstacles) nested within a 36 m inner square. Each obstacle was approximately 7 m, with 1 m of free space between the start and end of the obstacle to allow for clearance and preparation for the next obstacle. All obstacles were designed to mimic common conditions WU encounter, two required more power (rough terrain and incline slope), and two of which required more motor planning (cones and side slopes). The overall design of the environments and order of the obstacles are depicted in Fig. 1, and further described below. The environments were set in the Oxford Brookes Sports Hall and guiding lines were provided for participants to follow.

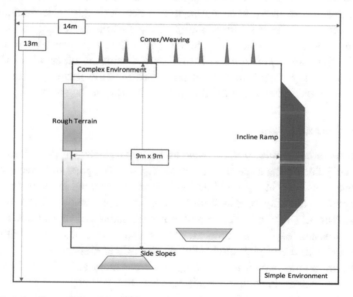

Fig. 1. Sketch of outer simple and nested inner complex (with 4 obstacles) environments.

The obstacle labelled as "Rough Terrain" in Fig. 1 mimicked a high friction environment requiring more power from the user, and was created using foam noodles, a common material used in obstacle designs for children with disabilities [39]. As specified by the wheelchair manufacturer manual, the height of the rough terrain was roughly set at 3 cm repeated every 3–5 cm, under the 5 cm safety limit for the caster wheels and at a total width of 80 cm. Errors while traversing rough terrain including shifting off the obstacle path, hesitating or abruptly stopping while traversing.

The incline ramp also required more power, created from 1.8 cm thick plywood and set 1 m wide with safety barriers to prevent participants from falling off the ramp. The incline ramp was set to American Disability Association (ADA) 2010 guidelines [37] of having a maximum of a 5° gradient at a straight on approach. The ramp was 3 m long, climbing to a horizontal surface at a height of 26 cm and 50 cm in length that continued to an additional 3 m decline at the same 5° gradient to reach back to the flat path. Physical errors while traversing the incline ramp included hitting the boundary lip of the ramp as well as hesitating/stopping during the entire obstacle.

Cones/weaving required more upper limb coordination and control. Cones were set on the guiding line of the path at 92 cm apart (cone edge to cone edge) to mimic ADA guidelines [37] of acceptable wheelchair accessible door width. The start and final cone were set 1.1 m from the ends of the length. The participants were asked to approach the first cone from the outside, and weave back and forth until reaching the end. Errors while traversing the cones included hitting or ignoring a cone.

The side slopes also required more upper limb coordination and control. Each slope was 2.4 m in length and set at a 10° gradient to a maximum of 20 cm, a height tested to be safely balanced and not lead to tipping over. They were set 1.5 m from the ends of the length, and 70 cm away from the path at a parallel angle. The participants were instructed to approach the first side slope at an angle using one wheel on the side slope while keeping the other wheel along the flat path. The participants were requested to exit the side slope and approach the 2nd side slope with the other wheel while maintaining the remaining wheel on the flat path. Errors while traversing the side slopes included hesitations and not maintaining a level height on the slope.

2.3 Experimental Setup

All circuits were completed in clockwise and counterclockwise directions alternating every 4 circuits during the experiment. This setup was designed to prevent fatigue. All circuits were completed in a pseudorandomized predetermined order per participant to reduce a repetitive learning effect. Ultimately, participants completed 16 circuits - 4 in a simple environment (no obstacles) without power assistance, 4 in a simple environment with power assistance, 4 in a complex environment (with obstacles) without power assistance, and 4 in a complex environment with power assistance.

To standardize the experiment participants were fixed at self-paced speeds. Each participant completed their first circuit through the complex environment without power assistance, at self-selected speeds (encouraged to make the fewest errors), to control for inter-individual differences in fitness, and recorded their first circuit completion time. All remaining circuits were attempted to be completed within that specific time ($\pm$ 5 s) regardless of interface by using a research assistant who walked beside the participant at that designated pace. The $\pm$ 5 s accounted for fatigue and learning. Participants were given rests before the start of each circuit (30–50 s) to allow for a more stable physiological baseline of HR and other measures. Total times to complete each circuit were recorded.

2.4   Statistical Analysis

Statistical analysis of behavioral performance (percentages based on errors) and HR information during the experimental procedure employed the use of linear mixed modeling implemented in NCSS (*NCSS, LLC. Kaysville, Utah, USA*). Linear mixed-effects estimates were computed with restricted maximum likelihood.

3   Results

All 30 novice behavioral performance data revealed a significant effect for the type of interface (manual or power-assist) ($F_{1,209}$ = 38.3877, p < 0.001) depicted in Fig. 2. Use of the power-assist interface led to an overall decreased performance of 5.6179% for the complex environment. Post-hoc analysis with Bonferroni adjustment revealed that the power-assist interface significantly decreased performance for only 2 of the 4 obstacles (Fig. 3) - 10.6667% performance reduction for navigating the cones ($F_{1,209}$ = 32.1538, p < 0.001), and 9.1667% reduction for navigating the side slopes ($F_{1,209}$ = 15.3360, p < 0.001), but no significant difference in the obstacles that required more power, the rough terrain or the incline ramp.

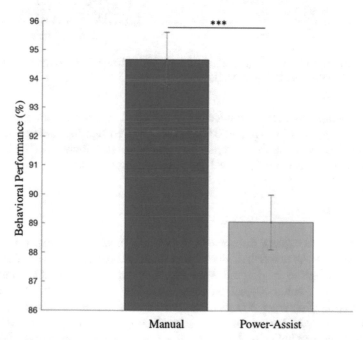

Fig. 2. Overall behavioral performance comparison between the manual and power-assist. interface for the entire complex environment. (***p < 0.001)

ECG data of 23 out of 30 novice participants was processed using a customized MATLAB script to calculate mean HR per circuit. Post-hoc analysis with Bonferroni

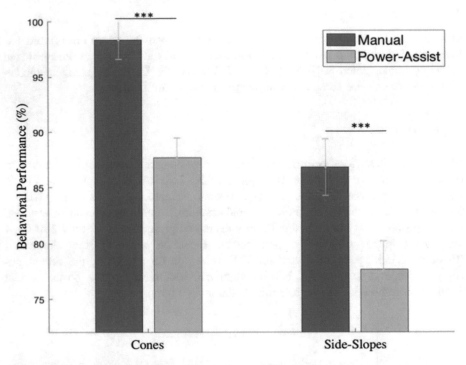

Fig. 3. Behavioral performance comparison of the cones and the side slopes between the manual and power-assist interface. (***p < 0.001)

adjustment revealed that HR of novice users decreased by 4.4767 bpm with the use of power-assist ($F_{1,328}$ = 13.0175, p < 0.001). There was no significant difference in heart rate information between the different environments and there was no significant interaction between the interfaces and the environment (Fig. 4).

4   Discussion

This study set out to explore the concept of PADs in physical and cognitive workload in realistic settings. As newer assistive device interfaces reach users with disabilities, it becomes paramount to begin understanding how these devices affect mental and physical workload within typical environments. We designed tasks that mimicked some of everyday situations WU would encounter. The results of this study observed that HR decreased with the use of power-assist for people learning to use wheelchairs, however at the cost of behavioral performance, particularly with particularly significant increases in errors for obstacles that require more skill and control. Therefore, this methodology along with HR measurements may aid in characterizing physical work-load impact of power-assist interfaces. It appeared that new users were able to take advantage of the power assist to reduce physical workload as measured by HR while

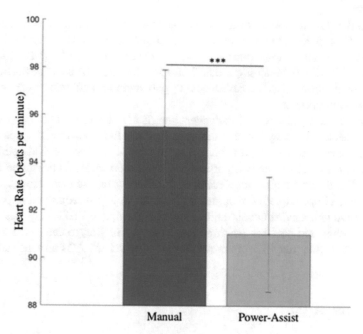

Fig. 4. Heart rate comparison of the interface. (***p < 0.001)

maintaining performance in tasks that require more power, but decreased performance
in obstacles needing more control.

4.1   Environment

Each subject completed 8 simple, and 8 complex circuits, where the complex circuits
were predicted to increase the physical workload. However, HR was not a reliable
predictor for determining environmental difficulty. Furthermore, no interactions
between the interface and environment were found. WU develop skills over time to
negotiate inaccessible environments [40, 41], however increased environmental difficulty did not induce measurable changes in HR as expected from the literature [42].
Environmental complexity [43], as related to task demands, are directly correlated with
increased CW which has been determined via evaluation of task performance [44] and
cardiac responses [45], however this was not the case for measuring HR within this
study.

4.2   Interface

PAD's are designed to augment the physical power of the user, to reduce metabolic
effort and allow the user to expend less energy than typical manual propulsion. This
design intention was indeed reflected with a statistically significant, yet small decrease
in HR of 4.48 bpm. Several studies have explored exercise ergonomics, and even some
in wheelchair ergonomics with similar findings [46], however none to our knowledge

have looked at CV responses to the newer generation wheelchair interface of power-assistance. Champagne et al. reported similar cardiorespiratory reductions with the use of mobility assistance dogs during a natural environment [47]. This reduction in physical workload may be an important factor in allowing new users to further engage in their communities for increased social participation and equality even within nonoptimal environments.

Behavioral performance did however marginally decrease with the use of the power-assistance. This may indicate that as a new interface, power assistance may lead to more accidents specifically for those new to wheelchair control. This performance reduction/increased error rate was particularly true for obstacles that required more fine control and skill (weaving through cones and balancing on the side slopes). However, PAD's may not impact quality of control in environments that require more power (like high friction/rough environments and steep incline ramps). This may lead to more informed, customized decision making on the part of the user, to determine what types of environments they may face more regularly and whether PADs are optimal for their use.

5   Conclusion

In summary, HR is a reliable measurement for assessing the potential for physical workload reduction for power-assist devices/or new mobility interfaces for new users. Power-assistance is an important factor in reducing physical workload for people learning to use wheelchairs, but perhaps at the cost of increasing minor accidents/errors. Portable non-invasive ECG is a safe and reliable measure that can be used in any simple, or physically demanding environment. Measuring HR variability may be an important factor in future studies, along with more robust measures for behavioral performance including smoothness of control. Ultimately, portable physiological measures of WU in natural environments can provide more insight for personalization of mobility devices or improved guided skill training in wheelchair control.

Acknowledgments. We would like to thank Jamie Whitty and Joel Chappell of the School of Architecture from Oxford Brookes University, for constructing our ramps, Ian Allen, the Oxford Brookes sports booking coordinator, for helping us with numerous appointments, and our research assistants Cyrus Goodger, Jessica Andrich, and JoJo Dawes. This research was funded through the Adaptive Assistive Rehabilitative Technologies – Beyond the Clinic grant by the Engineering and Physical Sciences Research Council (EP/M025543/1). SJ is additionally supported by the Fulbright US-UK Commission. HD is supported by the Elizabeth Casson Trust and received support from the NIHR Oxford health Biomedical Research Centre. Additional support provided by CONACYT (National Council of Science and Technology in Mexico).

References

1. McClure, L.A., Boninger, M.L., Oyster, M.L., Williams, S., Houlihan, B., Lieberman, J.A., Cooper, R.A.: Wheelchair repairs, breakdown, and adverse consequences for people with traumatic spinal cord injury. Arch. Phys. Med. Rehabil. **90**, 2034–2038 (2009)
2. Smith, C., McCreadie, M., Unsworth, J., Wickings, H.I., Harrison, A.: Patient satisfaction: an indicator of quality in disablement services centres. Qual. Health Care **44**, 31–3631 (1995)
3. Dalyan, M., Cardenas, D.D., Gerard, B.: Upper extremity pain after spinal cord injury. Spinal Cord **37**, 191–195 (1999)
4. Fullerton, H.D., Borckardt, J.J., Alfano, A.P.: Shoulder pain: a comparison of wheelchair athletes and nonathletic wheelchair users. Med. Sci. Sports Exerc. **35**, 1958–1961 (2003)
5. Holloway, C.S., Symonds, A., Suzuki, T., Gall, A., Smitham, P., Taylor, S.: Linking wheelchair kinetics to glenohumeral joint demand during everyday accessibility activities. In: 2015 37th Annual International Conference of the IEEE Engineering in Medicine and Biology Society (EMBC), pp. 2478–2481 (2015)
6. Asheghan, M., Hollisaz, M.T., Taheri, T., Kazemi, H., Aghda, A.K.: The prevalence of carpal tunnel syndrome among long-term manual wheelchair users with spinal cord injury: a cross-sectional study. J. Spinal Cord Med. **39**, 265–271 (2016)
7. Tawashy, A.E., Eng, J.J., Krassioukov, A.V., Miller, W.C., Sproule, S.: Aerobic exercise during early rehabilitation for cervical spinal cord injury. Phys. Ther. **90**, 427–437 (2010)
8. Van Den Berg-Emons, R.J., Bussmann, J.B., Haisma, J.A., Sluis, T.A., Van Der Woude, L. H.V., Bergen, M.P., Stam, H.J.: Prospective study on physical activity levels after spinal cord injury during inpatient rehabilitation and the year after discharge. Assistive Technol. Res. Ser. **26**, 134–136 (2010)
9. Gauthier, C., Arel, J., Brosseau, R., Hicks, A.L., Gagnon, D.H.: Reliability and minimal detectable change of a new treadmill-based progressive workload incremental test to measure cardiorespiratory fitness in manual wheelchair users. J. Spinal Cord Med. **40**, 759–767 (2017)
10. Van Der Scheer, J.W., De Groot, S., Tepper, M., Gobets, D., Veeger, D.J.H.E.J., Van Der Woude, L.H.V., Woldring, F., Valent, L., Slootman, H., Faber, W.: Wheelchair-specific fitness of inactive people with long-term spinal cord injury. J. Rehabil. Med. **47**, 330–337 (2015)
11. Holloway, C., Tyler, N.: A micro-level approach to measuring the accessibility of footways for wheelchair users using the Capability Model. Transp. Planning Technol. **36**, 636–649 (2013)
12. Holloway, C.S.: The Effect of Footway Crossfall Gradient on Wheelchair Accessibility. Department of Civil, Environmental, & Geomatic Engineering. University College London, London (2011)
13. Kloosterman, M.G.M., Snoek, G.J., Van Der Woude, L.H.V., Buurke, J.H., Rietman, J.S.: A systematic review on the pros and cons of using a pushrim-activated power-assisted wheelchair. Clin. Rehabil. **27**, 299–313 (2013)
14. Cooper, R.A., Boninger, M.L., Spaeth, D.M., Ding, D., Guo, S., Koontz, A.M., Fitzgerald, S.G., Cooper, R., Kelleher, A., Collins, D.M.: Engineering better wheelchairs to enhance community participation. In: IEEE Transactions on Neural Systems and Rehabilitation Engineering 14, pp. 438–455 (2006)
15. van der Woude, L.H.V., de Groot, S., Janssen, T.W.J.: Manual wheelchairs: research and innovation in rehabilitation, sports, daily life, and health. Med. Eng. Phys. **28**, 905–915 (2006)

16. Consortium for Spinal Cord Medicine: Preservation of upper limb function following spinal cord injury: what you should know (2008)
17. Morgado Ramirez, D.Z., Holloway, C.: "But, i don't want/need a power wheelchair": toward accessible power assistance for manual wheelchairs. In: Proceedings of the 19th International ACM SIGACCESS Conference on Computers and Accessibility, pp. 120–129 (2017)
18. Zhao, Y., Tang, J., Cao, Y., Jiao, X., Xu, M., Zhou, P., Ming, D., Qi, H.: Effects of distracting task with different mental workload on steady-state visual evoked potential based brain computer interfaces - an offline study. Front. Neurosci. **12**, 1–11 (2018)
19. Parasuraman, R., Sheridan, T.B., Wickens, C.D.: Situation awareness, mental workload, and trust in automation: viable, empirically supported cognitive engineering constructs. J. Cogn. Eng. Decis. Making **2**, 140–160 (2008)
20. Hancock, P.A., Parasuraman, R.: Human factors and safety in the design of intelligent vehicle-highway systems. J. Saf. Res. **23**, 181–198 (1992)
21. Oyster, M.L., Smith, I.J., Kirby, R.L., Cooper, T.A., Groah, S.L., Pedersen, J.P., Boninger, M.L.: Wheelchair skill performance of manual wheelchair users with spinal cord injury. Top. Spinal Cord Inj. Rehabil. **18**, 138–139 (2012)
22. Causse, M., Chua, Z., Peysakhovich, V., Del Campo, N., Matton, N.: Mental workload and neural efficiency quantified in the prefrontal cortex using fNIRS. Sci. Rep. **7**, 1–15 (2017)
23. Curtin, A., Ayaz, H.: The age of neuroergonomics: towards ubiquitous and continuous measurement of brain function with fNIRS. Jap. Psychol. Res. **60**, 374–386 (2018)
24. Ayaz, H., Dehais, F.: Neuroergonomics: The Brain at Work and in Everyday Life. Elsevier, Academic Press, London (2018)
25. Clark, V.P., Parasuraman, R.: Neuroenhancement: enhancing brain and mind in health and in disease. NeuroImage **85**, 889–894 (2014)
26. Gramann, K., Fairclough, S.H., Zander, T.O., Ayaz, H.: Editorial: trends in neuroergonomics. Front. Hum. Neurosci. **11**, 11–14 (2017)
27. Karwowski, W., Siemionow, W., Gielo-Perczak, K.: Physical neuroergonomics: the human brain in control of physical work activities. Theor. Issues Ergon. Sci. **4**, 175–199 (2003)
28. Mehta, R.K., Parasuraman, R.: Neuroergonomics: a review of applications to physical and cognitive work. Front. Hum. Neurosci. **7**, 1–10 (2013)
29. Parasuraman, R.: Neuroergonomics: research and practice. Theor. Issues Ergon. Sci. **4**, 5–20 (2003)
30. Parasuraman, R.: Neuroergonomics: brain, cognition, and performance at work. Curr. Dir. Psychol. Sci. **20**, 181–186 (2011)
31. Parasuraman, R., Christensen, J., Grafton, S.: Neuroergonomics: the brain in action and at work. NeuroImage **59**, 1–3 (2012)
32. Parasuraman, R., Rizzo, M.: Neuroergonomics: The Brain at Work. Oxford Univesity Press, New York (2007)
33. Fallahi, M., Motamedzade, M., Heidarimoghadam, R., Soltanian, A.R., Miyake, S.: Assessment of operators' mental workload using physiological and subjective measures in cement, city traffic and power plant controlcenters. Health Promo. Perspect. **6**, 96–103 (2016)
34. Sauer, J., Nickel, P., Wastell, D.: Designing automation for complex work environments under different levels of stress. Appl. Ergon. **44**, 119–127 (2013)
35. Chen, W.Y., Jang, Y., Wang, J.D., Huang, W.N., Chang, C.C., Mao, H.F., Wang, Y.H.: Wheelchair-related accidents: relationship with wheelchair-using behavior in active community wheelchair users. Arch. Phys. Med. Rehabil. **92**, 892–898 (2011)
36. Roscoe, A.H.: Assessing pilot workload. Why measure heart rate, HRV and respiration? Biol. Psychol. **34**, 259–287 (1992)
37. ADA Standards for Accessible Design. Title II 279–279 (2010)

38. Thomas, S., Reading, J., Shephard, R.J.: Revision of the physical activity readiness questionnaire (PAR-Q). Can. J. Sport Sci. = J. Can. des sciences du sport **17**, 338–345 (1992)
39. https://www.slideshare.net/chessarose/wheelchair-accessible-obstacle-course
40. Meyers, A.R., Anderson, J.J., Miller, D.R., Shipp, K., Hoenig, H.: Barriers, facilitators, and access for wheelchair users: substantive and methodologic lessons from a pilot study of environmental effects. Soc. Sci. Med. **55**, 1435–1446 (2002)
41. Rimmer, J.H., Riley, B., Wang, E., Rauworth, A., Jurkowski, J.: Physical activity participation among persons with disabilities: barriers and facilitators. Am. J. Prev. Med. **26**, 419–425 (2004)
42. Light, K.C., Obrist, P.A.: Task difficulty, heart rate reactivity, and cardiovascular responses to an appetitive reaction time task. Psychophysiology **20**, 301–312 (1983)
43. Faure, V., Lobjois, R., Benguigui, N.: The effects of driving environment complexity and dual tasking on drivers' mental workload and eye blink behavior. Transp. Res. Part F: Traffic Psychol. Behav. **40**, 78–90 (2016)
44. Lyu, N., Xie, L., Wu, C., Fu, Q., Deng, C.: Driver's cognitive workload and driving performance under traffic sign information exposure in complex environments: a case study of the highways in China. Int. J. Environ. Res. Public Health **14**, 1–25 (2017)
45. Stikic, M., Berka, C., Levendowski, D.J., Rubio, R., Tan, V., Korszen, S., Barba, D., Wurzel, D.: Modeling temporal sequences of cognitive state changes based on a combination of EEG-engagement, EEG-workload, and heart rate metrics. Front. Neurosci. **8**, 1–14 (2014)
46. Hilbers, P.A., White, T.P.: Effects of wheelchair design on metabolic and heart rate responses during propulsion by persons with paraplegia. Phys. Ther. **67**, 1355–1358 (1987)
47. Champagne, A., Gagnon, D.H., Vincent, C.: Comparison of cardiorespiratory demand and rate of perceived exertion during propulsion in a natural environment with and without the use of a mobility assistance dog in manual wheelchair users, vol. 95, p. 685. Published for the AAP by Lippincott Williams & Wilkins, Baltimore, MD (2016)
48. Li, W.-C., Chiu, F.-C., Kuo, Y.-s., Wu, K.-J.: The investigation of visual attention and workload by experts and novices in the cockpit BT - engineering psychology and cognitive ergonomics. In: Applications and Services, pp. 167–176. Springer, Berlin Heidelberg (2013)

Association Between Physicians' Burden and Performance During Interactions with Electronic Health Records (EHRs)

Natalie Grace Castellano[1], Prithima Mosaly[1,2,3],
and Lukasz Mazur[1,2,3($\boxtimes$)]

[1] School of Information and Library Science, University of North Carolina,
Chapel Hill, NC, USA
lmazur@med.unc.edu
[2] Carolina Health Informatics Program, University of North Carolina,
Chapel Hill, NC, USA
[3] Division of Healthcare Engineering, Department of Radiation Oncology,
University of North Carolina, Chapel Hill, NC, USA

Abstract. Suboptimal usability within electronic health records (EHRs) can pose risks for patient safety. This study uses data collected in a simulated environment in which providers interacted with 'current' and 'enhanced' Epic EHR interfaces to manage patients' test results and missed appointments. Interactions were quantified and categorized by high or low burden in terms of displayed behavioral and physiological data. Using recorded video data, providers' workflow and performance was analyzed. Suboptimal performance was found to be associated with high burden levels.

Keywords: Usability · Electronic health record (EHR) · Burden ·
Patient safety

1 Introduction

1.1 Background & Rationale

Adoption and use of electronic health records (EHRs) has dramatically increased over the past decade, and as of 2015, 84% of non-federal acute care hospitals reported adopting at least a basic EHR. This is more than a 900% increase in adoption from 2008, when reported adoption was only at nine percent [1]. EHR adoption has undoubtedly increased since 2015, and it is this increasingly widespread adoption that makes EHR systems so important to consider. EHRs provide opportunities for improved patient care such as more convenient access to patient data, data integration, and clinical decision support, and they should therefore be embraced [2, 3, 4]. According to the Institute of Medicine (IOM), EHR systems provide a means to accumulate health information about individuals over time in an electronic format that is only available to authorized users. They are intended to support knowledge, decisions, and efficiency in clinical settings in order to provide patient care that is higher in quality, efficiency, and safety [5].

© Springer Nature Switzerland AG 2020
H. Ayaz (Ed.): AHFE 2019, AISC 953, pp. 26–35, 2020.
https://doi.org/10.1007/978-3-030-20473-0_3

Some of the greatest benefits of EHR systems are that they allow for interoperability within medical facilities, improve the quality of patient care, and facilitate clinical workflow [6]. In one case study, a pediatric medical practice utilized an EHR system to facilitate improved vaccinations in children by tracking immunization records and alerting providers of missing vaccinations, also allowing for quicker documentation of administered vaccines and electronic ordering of additional vaccines [7]. In addition to improving workflows and providing more quality care, EHRs have also been shown to improve patient safety. A study that looked at patient safety event data in Pennsylvania over a period of ten years found that the adoption of advanced EHRs reduced negative patient safety events by 19%. Additionally, medication errors were reduced by 24% and complications went down by 18% [8].

Unfortunately, suboptimal levels of adoption and integration of EHRs in hospitals in the United States as well as the need for providers to use systems that are poorly designed raise serious challenges that can hinder quality care [9, 10]. Specifically, poor usability of EHR interfaces can "create new hazards in the already complex delivery of care," which in turn negatively affects providers' performance and patient safety [11]. Studies have found that suboptimal usability can contribute to increased burden of interaction such as wasteful clicks and suboptimal information processing, which decreases efficiency and increases the potential for human errors [12].

In the healthcare domain, errors could be considered to hold higher stakes than other disciplines because errors put human lives at risk. Errors associated with poor EHR usability can range from minor to severe, but because of the human consequences involved, the aim should be to reduce all errors, including those categorized as minor. A qualitative study focused on health information system-related errors by Ash, Berg, and Coiera [13] identified errors related to the processes of entering in and retrieving information from the system as well as communication and coordination processes. Poorly designed systems make it easy to accidentally order an incorrect medication or assign a medication to the wrong patient simply because an incorrect selection is in close proximity to the correct selection on the screen. Some systems cause issues because they do not support the typical multitasking environment of providers. Coordination-based issues can occur due to overreliance on the computer system to communicate with other providers and assuming that entry of information into the system is a sufficient means of communication with another provider. This assumption can result in missed critical information because the information transfer requires the provider on the receiving end to proactively enter the system in search of new information [13].

Therefore, considering the usability of EHR interfaces used by healthcare providers during care delivery is incredibly important because of the possibility of high risks due to human errors. These EHRs should be evaluated and improved with users in mind in order to reduce devastating medical-related errors stemming from human mistakes.

1.2 Project Background

Researchers in the Human Factors Laboratory in the department of Radiation Oncology at UNC Hospitals study the usability of interfaces embedded into EHRs, such as Epic, with the specific aim of improving providers' performance while reducing interaction burden. Research investigation has focused on topics such as assessing the effects of task demands and cognitive or mental workload on providers' performance during interactions with EHRs [14]. Investigations have involved various evaluation methods, including measuring EHR interaction task difficulties with task flows, mouse clicks, number of searches, repeated visits to screens, physiological measures for mental effort such as pupillary response and blink frequencies, and electroencephalography (EEG) [14, 15]. Other evaluation methods included performance metrics based on errors as well as subjective measures like subjective workload measured with the NASA Task Load Index (NASA-TLX) [14]. Overall, their findings suggest that improvements to EHR interfaces will reduce providers' burden and improve performance, therefore increasing patient safety.

One study asked providers to use the Epic EHR system to manage patient test results in the In Basket, and data was collected on these interactions with the EHR. The In Basket is an interface of the EHR where new patient results are compiled for providers to view and take appropriate actions, much like an email inbox. Providers were asked to manage patient test results, some of which were abnormal, and some of which were abnormal and had a patient fail to show up to their related follow-up appointment, referred to as a "no-show" appointment. In the first session, all providers interacted with a normal Epic interface, referred to as the "current" EHR, but in the second session, half of providers interacted with an "enhanced" EHR. This enhanced EHR introduced changes into the In Basket interface by placing all patient test results related to a no-show appointment in a separate folder within the In Basket labeled "All Reminders". All screens of the current EHR interface remained the same in the enhanced EHR, but the additional "All Reminders" folder was added to the In Basket to draw attention to no-show appointments in the enhanced EHR interface.

The next phase of this study was conducted for this presented project. This phase focused on investigating how burden is associated with providers' performance in the test data from the second session.

1.3   Definitions

- *Provider burden* is a composite measure made up of two components that can each be individually quantified: mental effort and task difficulty. Literature suggests that mental effort and task difficulty can significantly influence performance, so these components are important to include to assess overall provider burden [15].
- *Mental effort* is the amount of cognitive resources supplied to perform a task and was quantified based on two physiological measures of eye behavior: pupillary dilation and blink rates. Pupillary dilation and blink rates have been strongly correlated to represent mental effort due to task demands as well as task difficulty.
- *Task difficulty* is a measure of the adaptive behaviors used to cope with task demands in order to perform the task, and it is measured by time on task, total clicks, number of searches, and number of revisits to the Chart Review screen. Greater time on task, numbers of clicks, numbers of searches, and numbers of revisits to the Chart Review screen reflect greater task difficulty.

- *Performance* is based on whether the provider documented a patient's no-show appointment status. Optimal performance was displayed if providers used the phrases "no show," "missed appointment," "reschedule appointment," or "will re-refer patient" in their documentation or reason for call, as well as if they added the pre-populated comment that was accessible in the enhanced EHR, reading "No show to follow-up appointment." If these phrases or comments were not included in the provider's workflow, their performance was considered to be suboptimal.

2 Methods

2.1 Original Data Set

The original data collection came from 38 resident physicians from Internal Medicine, Family Medicine, Pediatrics, Gynecology, Oncology, Psychiatry, and Surgery departments (Post-Graduate Year range: 1 to 5) who participated in the study and were incentivized with a $100 gift card for their participation. Participants were recruited through an email sent to the UNC Hospitals resident physician email list. Recruitment criteria included at least six months of experience using the Epic EHR system. Epic is one of the leading manufacturers of EHRs and thus is a widely-used interface among healthcare systems. The 2016 Medscape EHR report found that Epic was the most widely used EHR with 28% of respondents reporting using an Epic system [16]. Therefore, it was important to investigate providers' burden when using this EHR system and use these findings to propose improvements in order to reduce providers' burden and increase patient safety. Data of providers completing simulated patient scenarios in the Epic environment were collected, including video data of the inter-actions, clicks, time on tasks, and eye-tracking data of blinks and pupillary dilation during the tasks. This data was also coded in terms of high and low burden based on displayed mental effort and task difficulty.

2.2 High Vs. Low Burden Grouping Criteria

Mental effort was measured using task-evoked pupillary response (TEPR) and blink rate data (Table 1). Larger pupillary dilations were considered to reflect higher burden and lower dilations were considered to reflect lower burden. To determine indicators of high and low burden, TEPR data was plotted sequentially and values of change in pupillary dilations below the first quartile were considered to reflect low burden while values of change greater than the third quartile were coded as high burden in terms of mental effort. Blink rate was evaluated such that greater numbers of blinks per minute indicated lower burden while fewer blinks per minute indicated higher burden. When plotted sequentially, numbers of blinks below the first quartile were considered to reflect high burden while numbers above the third quartile were considered to reflect low burden.

Task difficulty was evaluated based on time on task, total number of clicks, number of searches, and number of revisits to the Chart Review interface (Table 1). Based on the time taken interacting with each patient in the In Basket until finishing the encounter, high time on task was considered to reflect high burden and low time on task was considered to reflect low burden. In order to determine high and low time on task, the raw data was plotted sequentially and first and third quartiles were identified. Values below the first quartile were considered low time on task and therefore reflected low burden and values above the third quartile were considered high time on task and therefore reflected high burden. High numbers of clicks reflected high burden in terms of task difficulty and low numbers of clicks reflected low burden. As with other data, click counts were plotted sequentially and numbers of clicks below the first quartile were considered to be low and numbers of clicks above the third quartile were considered to be high. Higher numbers of searches per session were considered to reflect high burden while lower numbers of searches reflected low burden. After sequentially plotting the raw data, values below the first quartile were considered to be low numbers of searches while values above the third quartile were considered to be high numbers of searches. Greater numbers of revisits to the Chart Review interface were considered to reflect higher burden. When plotted sequentially, data below the first quartile were considered to be low numbers of revisits while data above the third quartile were considered to be high numbers of revisits.

Table 1. Descriptive statistics of data samples reflecting low and high burden for mental effort and task difficulty.

Burden component	Burden indicator	Statistic	Low burden (N = 8)	High burden (N = 8)
Mental effort	TEPR (mm)	Mean [SD]	−0.1 [.1]	0.3 [.1]
	Blink rate (blinks/min)		22 [5]	11 [4]
Task difficulty	Time on task (sec)		77 [18]	183 [45]
	Clicks (count)		14 [6]	43 [10]
	Number of searches	Count	≤ 3	>3
	Number of re-visits to patient information		1	>1

2.3   Sample

Videos representing either high or low levels of burden for each component of burden (mental effort: pupillary dilation and blink rate; task difficulty: time on task, number of clicks, searches, and revisits) were purposively selected for analysis so that high and low levels of burden for each component were represented. Eight representative interactions were chosen for each burden indicator, one representing each of the eight patient scenarios, in order to equally represent all patient scenarios. Interactions from 35 out of the original 38 participants were represented in this sample. Participants were represented in an average of 2.5 interactions each, with the greatest number of inter-actions from a single participant being 7 interactions. This sampling resulted in a total of 96 interactions: 8 for high burden and 8 for low burden for each of the 6 burden indicators. Figure 1 shows the breakdown of the sample sizes of interactions for each indicator.

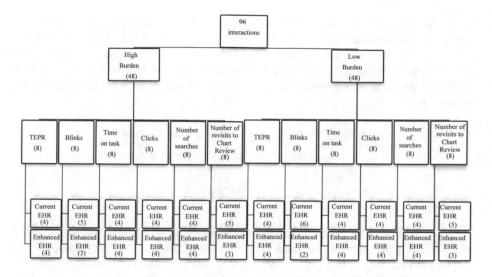

Fig. 1. Grouping of interactions and variables represented in each sample.

2.4 Data Analysis

A total of 96 interactions were analyzed. Of those, 43 used the enhanced EHR interface (22 in high burden and 21 in low burden) and 53 used the current EHR interface (26 in high burden and 27 in low burden). Next, interactions were categorized according to their performance (optimal vs. suboptimal). Two chi-square analyses were conducted. One compared performance across burden level (high burden vs. low burden), and the other compared performance across EHR type (current vs. enhanced). Fisher's Exact Tests were conducted as a follow-up to both chi-square analyses.

3 Results

Data was grouped based on optimal or suboptimal performance, measured by whether individuals documented patients' no-show status. This performance was then compared across burden level (low vs. high) as well as by EHR type (enhanced vs. current). Table 2 shows the counts of interactions by performance and EHR type across burden level.

Table 2. Counts of interactions by performance and EHR type across burden level.

	Optimal performance		Suboptimal performance	
	Current	Enhanced	Current	Enhanced
High burden	2	8	24	14
Low burden	7	13	20	8
Total	9	21	44	22

A chi-square test was conducted comparing performance across burden level and found significant differences, X^2 (1, $N = 96$) = 4.848, p = 0.0277. A one-tailed Fisher's Exact Test indicated that the probability for suboptimal performance is greater for a provider experiencing high burden as compared to low burden (p = 0.0233). Table 3 displays the counts of interactions displaying optimal and suboptimal performance by burden level.

Table 3. Counts by performance and burden level.

	Optimal performance	Suboptimal performance
Low burden	20	28
High burden	10	38

As secondary analysis, a chi-square test was conducted to compare performance across EHR type (current vs. enhanced) and found significant differences, X^2 (1, $N = 96$) = 11.213, p = 0.0008. A one-tailed Fisher's Exact Test indicated that the probability for suboptimal performance is greater for a provider using the current EHR as compared to the enhanced EHR (p = 0.0008). Table 4 displays the counts of interactions displaying optimal and suboptimal performance by EHR type.

Table 4. Counts by performance and EHR type.

	Optimal performance	Suboptimal performance
Enhanced EHR	21	22
Current EHR	9	44

Therefore, providers that experienced high burden or were working in the current EHR were more likely to have suboptimal performance, or were more likely to not explicitly document the patient's no-show status.

4 Discussion

Results suggest associations between performance and burden level as well as between performance and EHR type. More specifically, it was found that the probability of suboptimal performance was greater for providers experiencing high burden. Though the definition of performance was very specific for this project as documenting a patient's failure to show up at an appointment was considered optimal performance, these results demonstrate an association between displayed burden and taking appropriate actions. It is interesting to consider why there was an association between high burden and failing to write a note indicating "no show." In this study, provider burden is a composite measure made up of mental effort, or cognitive resources, and task difficulty, or behaviors used to cope with task demands. Because burden is a composite measure, it is plausible that multiple explanations exist for the association between burden and performance. However, the most salient possible reason that the two indicators appear together is due to providers' previous knowledge of the optimal workflow, or lack thereof, when following up on patients that did not show up to their appointment. When providers do not know what actions to take, perhaps from a lack of prior training or experience, they may fail to take the appropriate action and therefore have suboptimal performance. Further, they also exert more cognitive resources and find the task more difficult due to a lack of familiarity and therefore demonstrate high burden. Although the association between high burden and suboptimal performance was significant, 10 out of 30 providers that had optimal performance also displayed high burden, so burden level is not exclusively associated with performance. It is likely that performance is influenced by external factors such as previous experience with the task being completed, as providers could have familiarity with the task workflow and complete it optimally while still displaying high or low burden depending on their personal mental effort and difficulty with the task.

Results also suggest that the probability of suboptimal performance was greater for providers interacting with the current EHR as compared to the enhanced EHR. The enhanced EHR design drew attention to patients that missed their appointment by placing them in a separate "All Reminders" folder in the In Basket. The addition of the new folder interface changed providers' workflow and directed them to the "All Reminders" folder to interact with no-show patients instead of accessing them from the In Basket. After accessing the patient in the "All Reminders" folder, providers using the enhanced EHR used the same interfaces as the current EHR to take actions, including the "Chart Review," "Telephone Call," and "Medications and Orders" screens. Because all subsequent interfaces used were the same as the current EHR, the use of the "All Reminders" folder in the enhanced EHR could have lowered the likelihood of providers displaying suboptimal performance because it drew attention to no-show patients.

4.1   Limitations

This study was conducted with a small sample size in order to give equal representation to all burden indicators in the sample. Because only a sample of all provider interactions were analyzed, findings from this study may not be comprehensive to all provider behaviors. Also, using a larger sample in the original study or recruiting other medical providers besides resident physicians could have expanded findings.

The sample for this study was chosen to be representative after categorizing the videos based on their display of burden. The criteria for high or low burden in each of the burden categories was chosen for the original study, and the chosen criteria for categorization could have inaccurately represented high or low burden. There also could have been errors in recording data or categorizing interactions that would have put interactions into incorrect groups. Also, burden was only quantified by a few measures, and these may not have been the most representative of a provider's burden. Burden was quantified using pupillary dilation, blink rate, time on task, clicks, number of searches, and number of revisits to the Chart Review interface, and some of these burden indicators may have appeared due to a factor besides burden caused by the interface.

A very specific performance definition was used, which may have misrepresented the success of providers. Providers could have taken the appropriate actions to follow up on a patient's no-show without explicitly documenting "no show" in their notes, but using the performance definition used in this study their performance would be considered suboptimal.

4.2   Future Research

Future research in this area could assess provider burden when interacting with EHRs by using different measures, such as supplementing data with qualitative interviews with providers to gain insight into their experiences or self-perceptions of burden. The addition of a first-hand account from participants may be a more reliable way to determine burden as compared to observations and physiological data alone. Also, allowing providers to explain their decision-making processes may provide new information about how the EHR interface affects their workflow.

Using only burden indicators that can be visually assessed by researchers may help to make connections between interface issues and burden. Measures such as mouse movements, clicks, and gaze patterns are able to be viewed visually in conjunction with participants' interface navigation while completing tasks. The ability to visualize burden during interactions may allow researchers to pinpoint exactly where high burden occurred and what section of the interface participants were interacting with, and they can connect high burden to the issues identified as occurring in the section of the interface that was interacted with.

4.3   Conclusions

This study established associations between performance and burden as well as between performance and EHR design during providers' management of abnormal test results and missed follow-up appointments. The association between performance and burden reinforces the idea that providers experiencing higher burden have a greater chance of suboptimal performance. The association between performance and EHR design provides evidence that the process of redesigning interfaces with a focus on users' awareness of patients' abnormal test results and follow-up status could be able to produce more desirable performance. Findings from this study can be incorporated into future design of Epic EHR systems in order to reduce providers' burden and improve patient safety. An EHR system that burdens providers less will ultimately help to improve patient safety in healthcare.

References

1. Office of the National Coordinator for Health Information Technology: Non-federal acute care hospital electronic health record adoption. https://dashboard.healthit.gov/quickstats/pages/FIG-Hospital-EHR-Adoption.php (May 2016)
2. Bates, D.W., Leape, L.L., Cullen, D.J., Laird, N., Petersen, L.A., Teich, J.M., Burdick, E., Hickey, M., Kleefield, S., Shea, B., Vander Vliet, M., Seger, D.L.: Effect of computerized physician order entry and a team intervention on prevention of serious medication errors. J. Am. Med. Assoc. **280**(15), 1311–1316 (1998). https://doi.org/10.1001/jama.280.15.1311

3. Elnahal, S.M., Joynt, K.E., Bristol, S.J., Jha, A.K.: Electronic health record functions differ between best and worst hospitals. Am. J. Manag. Care **17**(4), e121–e147 (2011)
4. Blumenthal, D., Glaser, J.P.: Information technology comes to medicine. N. Engl. J. Med. **356**(24), 2527–2534 (2007). https://doi.org/10.1056/NEJMhpr066212
5. Institute of Medicine (US) Committee on Data Standards for Patient Safety: Key capabilities of an electronic health record system: letter report. Natl. Acad. Sci., 1–35 (2003). https://www.ncbi.nlm.nih.gov/books/NBK221800/
6. Thakkar, M., Davis, D.C.: Risks, barriers, and benefits of EHR systems: a comparative study based on size of hospital. Perspect. Health Inf. Manag. **3**, 5 (2006)
7. Au, L., Oster, A., Yeh, G.H., Magno, J., Paek, H.M.: Utilizing an electronic health record system to improve vaccination coverage in children. Appl. Clin. Inform. **1**(3), 221–231 (2010). https://doi.org/10.4338/ACI-2009-12-CR-0028
8. Hydari, M.Z., Telang, R., Marella, W.M.: Electronic health records and patient safety. Commun. ACM **58**(11), 30–32 (2015)
9. Jha, A.K., DesRoches, C.M., Campbell, E.G., Donelan, K., Rao, S.R., Ferris, T.G., Shields, A., Rosenbaum, S., Blumenthal, D.: Use of electronic health records in US hospitals. N. Engl. J. Med. **360**(16), 1628–1638 (2009). https://doi.org/10.1056/nejmsa0900592
10. Leape, L.L.: Errors in medicine. Clin. Chim. Acta **404**(1), 2–5 (2009). https://doi.org/10.1016/j.cca.2009.03.020
11. Institute of Medicine: Health IT and patient safety: building safer systems for better care. http://www.nationalacademies.org/hmd/ ~ /media/Files/Report%20Files/2011/Health-IT/HealthITandPatientSafetyreportbrieffinal_new.pdf (November 2011)
12. Rose, A.F., Schnipper, J.L., Park, E.R., Poon, E.G., Li, Q., Middleton, B.: Using qualitative studies to improve the usability of an EMR. J. Biomed. Inform. **38**(1), 51–60 (2005). https://doi.org/10.1016/j.jbi.2004.11.006
13. Ash, J.S., Berg, M., Coiera, E.: Some unintended consequences of information technology in health care: the nature of patient care information system-related errors. J. Am. Medical Inf. Assoc.: JAMIA **11**(2), 104–112 (2004). https://doi.org/10.1197/jamia.M1471
14. Mazur, L.M., Mosaly, P.R., Moore, C., Comitz, E., Yu, F., Falchook, A.D., Eblan, M.J., Hoyle, L.M., Tracton, G., Chera, B.S., Marks, L.B.: Toward a better understanding of task demands, workload, and performance during physician-computer interactions. J. Am. Med. Inf. Assoc. **23**(6), 1113–1120 (2016). https://doi.org/10.1093/jamia/ocw016
15. Mosaly, P.R., Mazur, L.M., Yu, F., Guo, H., Merck, D., Laidlaw, D.H., Moore, C., Marks, L.B., Mostafa, J.: Relating task demand, mental effort and task difficulty with physicians' performance during interactions with electronic health records (EHRs). Int. J. Hum Comput Interact. **34**(5), 467–475 (2018). https://doi.org/10.1080/10447318.2017.1365459
16. Peckham, C., Kane, L.: Medscape EHR report 2016: physicians rate top EHRs. In: Rosensteel, S. (ed.). https://www.medscape.com/features/slideshow/public/ehr2016#page=5. 25 Aug 2016

Beyond Physical Domain, Understanding Workers Cognitive and Emotional Status to Enhance Worker Performance and Wellbeing

Juan-Manuel Belda-Lois, Carlos Planells Palop, Andrés Soler Valero,
Nicolás Palomares Olivares, Purificación Castelló Merce,
Consuelo Latorre-Sánchez, and José Laparra-Hernández[✉]

Instituto de Biomecánica, Universidad Politécnica de Valencia, Camino de Vera,
46022 Valencia, Spain
{Juanma.Belda, Carlos.Planells, Andres.Soler,
Nicolas.Palomares, Puri.Castello, Consuelo.Latorre,
Jose.Laparra}@ibv.org

Abstract. A methodology is presented to obtain measurements of the emotional states of workers from the measurement of Heart Rate Variability. Two methodologies have been used, one based on logistic regression and another using fuzzy trees. The results show promising results to have a single model for using through different persons to obtain an estimation of their internal arousal and valence. This estimation will be validated in a second stage with a measurement of the cognitive load of the worker.

Keywords: Human factors · Emotional · Cognitive · Model

1 Introduction

Nowadays, mainly due to new forms of work organization and technological advances, the concept of mental workload is becoming increasingly important. The profile of the workload of many jobs is changing dramatically due to the existence of higher mental and intellectual demands or situations that require fast response or critical decisions. This contrasts with the physical demands of more traditional jobs. An example is the introduction of collaborative robots in the industry [1].

Besides, mental and physical health problems are closely related [2]: any characteristic of work having an effect on mental health could have a side effect on physical health. These conditions are associated with occupational injuries [3] and absenteeism. Besides, a hostile environment, a physically demanding work and a high exposure to stress are related with early retirement [4].

There are a number of general methods for the assessment of psychosocial factors such as LEST or RNUR. These methods allow the evaluation of mental load associated with work. However, these methods are intended for low qualified jobs. Therefore, its use is well suited for monotonous, repetitive works such as the work in a supply chain. Besides, these methods rely on questionnaires and rating systems based on observation.

© Springer Nature Switzerland AG 2020
H. Ayaz (Ed.): AHFE 2019, AISC 953, pp. 36–44, 2020.
https://doi.org/10.1007/978-3-030-20473-0_4

Subsequently, the assessment is influenced by the subjective perception of the evaluator and the worker.

An alternative for the assessment of cognitive workload is the study of the variations of related physiological indicators. Many physiological signals have been shown to correlate with cognitive and emotional states: brain signals (EEG) [5], heart rate variability (HRV) [5–7], respiration [6], galvanic skin response (GSR) [8, 9], blood pressure [10] and eye movement [7].

Among the main advantages of using physiological signals for cognitive load assessment are the possibility to obtain reliable objective measures related with the cognitive load, and being able to record these measurements in real time. However, often the instrumentation required to obtain these signals interferes with the normal development of the work [11].

For these reasons, HRV is a very well suited physiological signal for these purposes as there are an increasing number of commercial devices providing reliable measurements of heart rate as bracelets or smart watches. In addition, the relationship between HRV and the cognitive load it is widely described in the scientific literature [12].

The goal of this paper is to propose a new approach for a cognitive-emotional model based on the analysis of HRV. We aim that a reliable, real-time objective estimation of the cognitive load of workers is able to maximize well-being while minimizing possible errors.

2 Materials and Methods

2.1 Experimental Protocol

Ten healthy volunteers were recruited, gender balanced and age between 25–40, without cardiac or neurological pathologies. In addition, they were asked not to drink coffee or stimulants 2 h before the test.

The experiment consisted on a set of images with different emotional content. The experiment was carried out in a room with controlled temperature, light and humidity (23 °C, 60% RH, 425 lx). The user sat in a comfortable chair in front of a computer screen. During the entire test, three electrodes placed on the chest collect the user's ECG signal. Biosignalplux® was used to record the ECG, with sampling frequency 200 Hz.

The test consists in the generation of emotions through images previously cataloged in arousal and valence, extracted from the IAPS [13] database. The pictures were shown in the computer screen. The battery of images consists of 4 groups of arousal: low (arousal less than 4), medium-low (arousal between 4 and 5), medium-high (arousal between 5 and 6) and high (arousal greater than 6) where each group had 20 images with increasing valence. Before each group of images, 6 neutral images with very low arousals (less than 3) were presented. In total, 104 images, at 10 s per image, a total of 17 min and 20 s (Fig. 1).

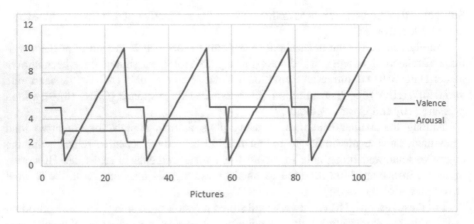

Fig. 1. Experimental protocol.

2.2 Methodology

Heart rate variability (HRV) is defined as changes in heart rate over time. To determine the heart rate, interval between beats must be found. The Pan-Tompkins algorithm [14] has been used to find the R point of the ECG wave. The difference between the point R_n and R_{n+1} determines the heart rate. Finally, the R vector contains all the R points of the ECG signal and was used to analyze HRV. Since the R vector is a vector with a non-uniform sampling frequency, it is resampled at 4 Hz in order to apply different techniques during processing.

The extraction of characteristics has been carried out using the wavelet packet decomposition (WPD) technique [15]. The WPD is an extension of the discrete wavelet transform (DWT) technique. In the DWT a high pass and low pass filter is performed on the original signal. Then, another high and low pass filter is performed on the previous low-passed signal, etc. In each level the original signal is divided into two octave bands. On the other hand, the WPD works both on the output of the high pass and the low pass filters. At each level we have a decomposition of 2^j signals (Fig. 2).

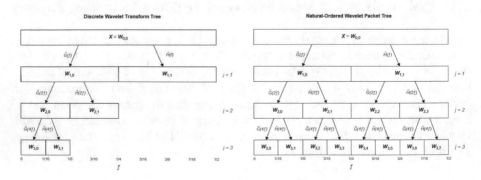

Fig. 2. DWT vs WPD

A 5 levels decomposition has been chosen in order to have the highest decomposition with a time window that fits in the image exposure time. The output signals contains 1 sample per band every 8 s. In a 5 levels Wavelet Packet Decomposition there are 32 bands. The input of the classifiers have been the coefficients of the 32 bands.

Machine learning models were used for the creation of a system that predict the subject's emotional state. A leave-one-out method was carried out to validate the model. The model was trained with 9 subjects and validated with the last one. The extracted characteristics and the sex of the subject were introduced as inputs. The output was the prediction of emotional state in 3 groups: low arousal, high arousal with high valence and high arousal with low valence (Fig. 3).

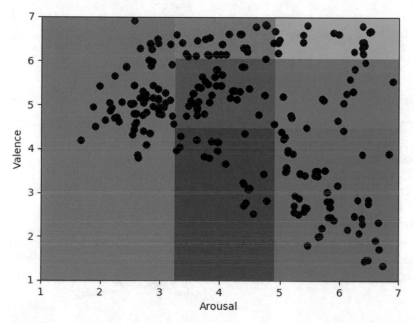

Fig. 3. The three emotional regions for the classifiers. Low arousal (in red), high arousal high valence (green), high arousal low valence (blue). The black dots are the situation in the plane of the images used for the classification. The figure also show the regions where multiple memberships occurs as mixed color regions. Original colors are the prediction regions, while mixed color regions are the difficult cases for each region.

The first model used was a logistic regression model. Logistic regression is a type of regression analysis used to predict the outcome of a categorical variable based on independent or predictive variables. It is useful to model the probability of an event occurring as a function of other factors. The probabilities that describe the possible outcome of a single trial are modeled, as a function of explanatory variables, using a logistic function.

The second model used was a Fuzzy Decision Tree [16]. Decision Trees are a prediction model where, given a set of data, diagrams of logical constructions are made, very similar to the rule-based prediction systems, which serve to represent and categorize a series of conditions that occur successively, to the resolution or classification of a problem.

3 Results

The score obtained in terms of accuracy using the logistic regression model is shown in Table 1:

Table 1. Logistic regression score

	Women	Men	All
Mean score	0,4322	0,4823	0,4015
Best score	0,4605	0,5466	0,5657

After the training with the leave-one-out method, changing the subject of validation, without distinction of sex, the mean accuracy of the model was 40.15%. In the best case, the accuracy was 56.57%.

With distinction of sex, the model for men had an average accuracy of 48.23% with a better case of 54.66%. For women, the average was 43.22% and the best case was 46.05%.

In Fig. 4 the Fuzzy Decision Tree is shown.

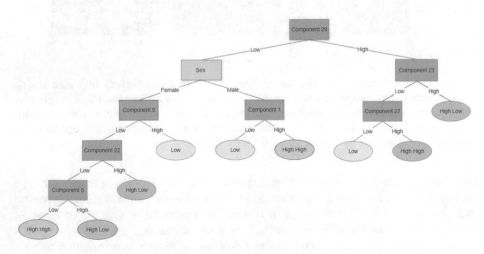

Fig. 4. Fuzzy Decision Tree

In the Decision Tree, 7 components were used, as well as sex, to distinguish between the 3 states. If component 29 was high, the tree could decide with one or two more components if the state is high arousal with negative valence or low arousal. On the other hand, if component 29 was low, the tree created a new branch where it depends on the sex to continue deciding. High arousal with negative valence was the group that depends on fewer components.

Confusion matrix of the Decision Tree is shown in Table 2.

Table 2. Confusion matrix for the Fuzzy Tree classifier. Beta = 0.4, Alfa = 0.8

	Low	High positive	High negative
Low	**122**	72	68
High Positive	107	**119**	67
High Negative	169	182	**217**

The reliability of the classifier was 40.78%. For low arousals, the Fuzzy Tree classifier had a score of 46.56%, for high positive a score of 40.61%, and 38.20% for high negative.

4 Discussion

In this work we have proposed the creation of a predictive model of emotions based on the variability of heart rhythm. An experiment was carried out with 10 subjects showing them a group of images of the IAPS, labeled with valence and arousal, while their ECG was recorded. After extracting features using the wavelet packet decomposition, two machine learning models were trained and validated, giving the previously mentioned results.

During the experiment, 104 images labeled by the IAPS were showed where all the possible combinations of arousal and valence were covered. This experimentation has been carried out in several investigations [17]. Other experiments with a smaller group of images, with the values previously selected for the most common states could be explored. Shorter experiences can help the subject to not get tired [18]

The choice of subjects had as a single exclusion criteria heart disease or neurological disorders, but it is possible that some subject has a low emotional response to stimuli [19], introducing greater variability and worsening the predictive model. A test prior to the experimentation would be necessary to ensure that the selected subjects are suitable for the study [20].

For the study, the possible outcomes predicted in 3 categories were grouped, low arousal, corresponding to a standard situation, high arousal with high valence, corresponding to a happy situation, and a high arousal with low valence, corresponding to an unhappy situation. In other studies, a different number of categories have been proposed [21, 22]. For user-dependent models, more types of emotion could be predicted, with more narrow bands of arousal and valence [23]. Since emotions are very subjective, user-independent models should have less types of emotion, coded with standard ways of elicitation.

In the methodology, the wavelet packet decomposition was used. WPD is a technique that allows finding components with an exponential level according to the number of decomposition, providing a lot of information that can be used to create a predictive model. The results found are a good start to continue working, being able to incorporate other features such as temporal or non-linear HRV parameters into future works, used in numerous studies [21–23].

Since we are converting emotions from a linear space (IAPS) to a discrete one, categorical machine learning models have been used. Other models such as support vector machines have also been used in other studies [24]. In this type of classification of emotions in labels, the worst case of the predictive model is that a predicted emotion fails to an opposite one. The confusion matrix must be calculated to see this error and try to minimize it.

For future work, additional tests that complete this type of experimentation should be studied. Only the emotional part of the subjects is being taken into account, however, the cognitive part has a significant weight [25]. Being able to have a model that is not only emotional, but also capable of detecting situations of high cognitive demand, such as stress, would be very useful. The incorporation of more information through non-invasive techniques will allow obtaining more robust models. These signals can come from cameras that analyze facial features [26] or from the cell phone for movement recording.

Finally, tests should be performed in real environments outside the laboratory after having a predictive model with better results. 24-h real experiment would provide the model with significant information for training and validation in real daily living situations. The applications that can have a cognitive-emotional model are very wide and in all the fields, being able to improve the well-being of the people.

References

1. Peruzzini, M., Grandi, F., Pellicciari, M.: Benchmarking of tools for user experience analysis in industry 4.0. Procedia Manuf. **11**, 806–813 (2017)
2. Scott, K.M., Lim, C., Al-Hamzawi, A., Alonso, J., Bruffaerts, R., Caldas-de-Almeida, J.M., Florescu, S., de Girolamo, G., Hu, C., de Jonge, P., Kawakami, N., Medina-Mora, M.E., Moskalewicz, J., Navarro-Mateu, F., O'Neill, S., Piazza, M., Posada-Villa, J., Torres, Y., Kessler, R.C.: Association of mental disorders with subsequent chronic physical conditions: world mental health surveys from 17 countries. J. Am. Med. Assoc. psychiatry **73**(2), 150–158 (2016)
3. Gillen, M., Yen, I.H., Trupin, L., Swig, L., Rugulies, R., Mullen, K., Font, A., Burian, D., Ryan, G., Janowitz, I., Quinlan, P.A., Frank, J., Blanc, P.: The association of socioeconomic status and psychosocial and physical workplace factors with musculoskeletal injury in hospital workers. Am. J. Ind. Med. **50**(4), 245–260 (2007)
4. De Wind, A., Geuskens, G.A., Reeuwijk, K.G., Westerman, M.J., Ybema, J.F., Burdorf, A., Bongers, P.M., Van der Beek, A.J.: Pathways through which health influences early retirement: a qualitative study. BMC Public Health **13**(1), 292 (2013)

5. Engström, J., Johansson, E., Östlund, J.: Effects of visual and cognitive load in real and simulated motorway driving. Transp. Res. Part F: Traffic psychol. Behav. **8**(2), 97–120 (2005)

6. Fairclough, S.H., Venables, L., Tattersall, A.: The influence of task demand and learning on the psychophysiological response. Int. J. Psychophysiol. **56**(2), 171–184 (2005)

7. Fairclough, S.H., Venables, L.: Prediction of subjective states from psychophysiology: a multivariate approach. Biol. Psychol. **71**(1), 100–110 (2006)

8. Cohen, R.A., Waters, W.F.: Psychophysiological correlates of levels and stages of cognitive processing. Neuropsychologia **23**(2), 243–256 (1985)

9. Scheirer, J., Fernandez, R., Klein, J., Picard, R.W.: Frustrating the user on purpose: a step toward building an affective computer. Interact. Comput. **14**(2), 93–118 (2002)

10. Jorgensen, R.S., Johnson, B.T., Kolodziej, M.E., Schreer, G.E.: Elevated blood pressure and personality: a meta-analytic review. Psychol. Bull. **120**(2), 293 (1996)

11. Wagner, J., Kim, J., Andre, E.: From physiological signals to emotions: implementing and comparing selected methods for feature extraction and classification. In: 2005 IEEE International Conference on Multimedia and Expo, Amsterdam, pp. 940–943 (2005)

12. Appelhans, B., Luecken, L.: Heart rate variability as an index of regulated emotional responding. Rev. Gen. Psychol. **10**, 229–240 (2006). https://doi.org/10.1037/1089-2680.10.3.229

13. Lang, P.J., Bradley, M.M., Cuthbert, B.N.: International affective picture system (iaps): affective ratings of pictures and instruction manual. Technical Report A-8 (2008)

14. Pan, J., Tompkins, W.J.: A real-time qrs detection algorithm. IEEE Trans. Biomed. Eng. **3**, 230–236 (1985)

15. MATLAB. https://es.mathworks.com/help/wavelet/examples/wavelet-packets-decomposing-the-details.html

16. Yuan, Y., Shaw, M.J.: Induction of fuzzy decision trees. Fuzzy Sets Syst. **69**(2), 125–39 (1995). https://doi.org/10.1016/0165-0114(94)00229-Z

17. Nardelli, M., Greco, A., Valenza, G., Lanata, A., Bailón, R., Scilingo, E.P.: A multiclass arousal recognition using HRV nonlinear analysis and affective images. In: 2018 40th Annual International Conference of the IEEE Engineering in Medicine and Biology Society (EMBC), pp. 392–395, Honolulu, HI (2018)

18. Lench, H.C., Flores, S.A., Bench, S.W.: Discrete emotions predict changes in cognition, judgment, experience, behavior, and physiology: a meta-analysis of experimental emotion elicitations. Psychol. Bull. **137**, 834–855 (2011)

19. Amstadter, A.: Emotion regulation and anxiety disorders. J. Anxiety. Disord. **22**, 211–221 (2008)

20. Kroenke, K., Spitzer, R.L., Williams, J.B.: The patient health questionnaire-2: validity of a two-item depression screener. Medicalcare **41**(11), 1284–1292 (2003)

21. Lee, C.K., Yoo, S.K., Park, Y.J., Kim, N.H.: Using neural network to recognize human emotions from heart rate variability and skin resistance. In: 27th Annual International Conference of the Engineering in Medicine and Biology Society, pp. 5523–5525 (2005)

22. Kim, K.H., Bang, S.W., Kim, S.R.: Emotion recognition system using short-term monitoring of physiological signals. Med. Biol. Eng. Compute. **42**, 419–427 (2004)

23. Picard, R.W., Vyzas, E., Healey, J.: Toward machine emotional intelligence: analysis of affective physiological state. IEEE Trans. Pattern Anal. Mach. Intell. **23**(10), 1175–1191 (2001)

24. Yu, S.N., Chen, S.F.: Emotion state identification based on heart rate variability and genetic algorithm (2015)

25. Luque-Casado, A., Perales, J.C., Cárdenas, D., Sanabria, D.: Heart rate variability and cognitive processing: the autonomic response to task demands. Biol. Psychol. **113**, 83–90 (2016)
26. Fan, Y., Lu, X., Li, D., & Liu, Y.: Video-based emotion recognition using CNN-RNN and C3D hybrid networks. In: Proceedings of the 18th ACM International Conference on Multimodal Interaction, pp. 445–450. ACM, October 2016

Performance and Brain Activity During a Spatial Working Memory Task: Application to Pilot Candidate Selection

Mickaël Causse[1(✉)], Zarrin Chua[1], and Nadine Matton[2]

[1] ISAE-SUPAERO, Université de Toulouse, Toulouse, France
mickael.causse@isae.fr
[2] ENAC and CLLE, Université de Toulouse, Toulouse, France
nadine.matton@enac.fr

Abstract. For 18 ab initio airline pilots, we assessed the possibility of predicting flight simulator performance with the performance and the prefrontal activity measured during a spatial working memory (SWM) task. Behavioral results revealed that a better control of the aircraft altitude in the flight simulator was correlated with better strategy during the SWM task. In addition, neuroimaging results suggested that participants that recruited more neural resources during the SWM task were more likely to accurately control their aircraft. Taken together, our results emphasized that spatial working memory and the underlying neural circuitries are important for piloting. Ultimately, SWM tasks may be included in pilot selection tests as it seems to be a good predictor of flight performance.

Keywords: Flight performance · Cognitive performance ·
Functional Near Infrared Spectroscopy (fNIRS) · Pilots selection and training

1 Introduction

Human error is a major contributing factor to accidents in aviation [1]. Common causes encompass the aircrews not initiating the appropriate maneuvers, failing to notice visual and auditory alerts, being unable to maintain an appropriate situation awareness (SA), or making poor decisions. Some errors may appear particularly surprising considering the hard-to-achieve selection criteria that pilots have to meet to enter some training programs or to obtain their flight license. The notion that specific aspects of cognition play a crucial role in the chain of events leading up to aircraft accidents suggests the potential benefice of implementing more efficient cognitive screening procedures for ab initio pilots. A required first step in this direction involves characterizing the human cognitive limitations in such complex environments [2] and identifying which cognitive functions and underlying neural circuitries are predictive of pilots' performance. In the past, a set of studies investigated the associations between cognition and flight proficiency in ab initio pilots. They measured the relationship between students' performance on the selection tests (for admission into the training program) and the training outcome, as indexed by flying performance [3–6].

© Springer Nature Switzerland AG 2020
H. Ayaz (Ed.): AHFE 2019, AISC 953, pp. 45–55, 2020.
https://doi.org/10.1007/978-3-030-20473-0_5

Correlation scores between selection tests and training outcome tend to be weak, the highest ranging between r = .20 and r = .25. The most robust predictor was a composite score encompassing both cognitive and psychomotor variables (e.g., r = .31 [6]), followed by previous training experience (.30). The personality, general intelligence, and academic tests yielded lowest mean validities (.13, .13, and .15, respectively). Damos [5] questioned the low predictive validity of selection test scores and called for the introduction of tests based on theoretical arguments. In other words, a way of improvement could come from using tests that target more precisely the cognitive functions solicited during piloting. Tests assessing working memory seem to be promising, as revealed by several studies showing significant links with piloting performance [7–10] or training success [11]. Moreover, as visual-spatial abilities are known to be essential for pilots [12], an assessment of spatial working memory seems particularly relevant in the context of initial pilot selection.

2 Objectives

The primary aim of the current study was to assess the possibility of predicting flying performance in ab initio airline pilots by measuring their spatial working memory performance. In order to gain a deeper understanding of the neural mechanisms engaged during the two types of tasks, we measured the prefrontal activity with Functional Near-Infrared Spectroscopy (fNIRS). fNIRS is a field-deployable non-invasive optical brain monitoring technology that provides a measure of cerebral hemodynamics within the cortex in response to sensory, motor, or cognitive activation. Its application in the aforementioned contexts allowed us monitoring and localizing the hemodynamic changes associated with the two tasks (piloting and the neuropsychological task) see Fig. 1.

3 Method

3.1 Participants

Eighteen ab initio airline pilots (élèves pilotes de ligne, EPL) from the Ecole Nationale de l'Aviation Civile (ENAC) (mean age: 20.6, SD = 1.1, all males) were recruited to perform the flight simulation and the spatial working memory (SWM) task. All participants gave written informed consent in accordance with the local ethical board committee. The study complied with the Declaration of Helsinki for human experimentation and was approved by the medical Committee (CPP du Sud-Ouest et Outre-Mer IV, n°CPP15-010b/2015-A00458-41). All participants were ab initio airline pilots without professional flying experience. However, their very low experience with light aircraft or flight simulators varied, with a mean total flying experience of 48.88 h (SD = 57.00 h) and a mean recent flying experience (past 2 years) of 24.83 h (SD = 30.35 h).

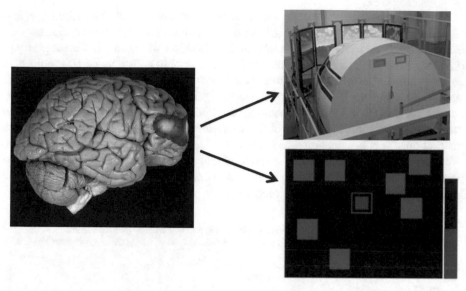

Fig. 1. Are there any communalities in performance and brain activity between piloting (top right) and performing a working memory task such as SWM (bottom right)?

3.2 Flight Simulator

The experiment took place aboard the PEGASE flight simulator. It simulates a twin-engine aircraft flight model, see Fig. 2.

Fig. 2. Illustration of the cockpit of the ISAE-SUPAERO flight simulator ("PEGASE").

Participants sat in the pilot's seat (front-left) of the aircraft. The flight scenario started in-flight, vertically above Tarbes-Lourdes airport (LFBT). The participants were asked to head as straight as possible in the direction of Agen La Garenne airport (LFBA). After this point, they were asked to turn right to a heading of 100° during 10 nautical miles. During the navigation, participants were required to adopt five different speed and altitudes. Pilots could use the navigation display, the auto-pilot was disconnected, and the aircraft was manually controlled. The flight plan was delivered to the participants before the starting of the scenario, under the form of a map with written instructions. This information was made available during the entire scenario duration. We used four different metrics to assess flight performance, as illustrated for one participant in Fig. 3: altitude deviation (feet), speed deviation (knots), and longitudinal/latitudinal deviations (degree).

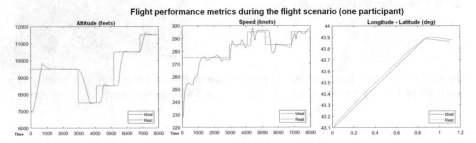

Fig. 3. Ideal performance vs. actual performance during the flight scenario (illustration for one participant). The 5 required changes in speed/altitude and the turn to a heading of 100° after "Agen" are visible.

3.3   Spatial Working Memory (SWM) Task

The SWM test from the CANTAB neuropsychological battery is designed to recruit and assess the ability to maintain and update spatial information in working memory. The goal of the task is to find tokens which are hidden one at a time within a random arrangement of boxes. Once a token has been found within a box, the participant does not need to inspect the same box again, as another token would never yet again appear there. For full details about this task, see De Luca et al. [13]. After practice trials with 3 boxes, assessed trials were administered with increasing difficulty (6, 8, 10 and 12 boxes). Administration time was approximately 8 min. Performance was measured both with the mean number of errors within each level of task difficulty (i.e. number of boxes), defined as the number of times the participant revisited a box in which a token had previously been found; and with a strategy score. An efficient strategy for completing this task is to follow a predetermined search sequence, returning to the same initial box after a token has been found. The strategy score consisted in counting the number of times the participant began a new search with a different box.

3.4 Prefrontal Cortex Activity Measurements

During the entire duration of the flight scenario and the neuropsychological task, the hemodynamics of the prefrontal cortex was recorded using the fNIR100 stand-alone functional brain imaging system (Biopac™, see Fig. 4). Sixteen optodes recorded the hemodynamics at a frequency of 2 Hz with a fixed 2.5 cm source-detector separation. COBI Studio software version 1.2.0.111 (Biopac™ systems) was used for data acquisition and visualization, and the fNIRS raw data were pre-processed using fnirSoft version 1.3.2.3 (Biopac™ systems). For each participant, the variations in light absorption at two different peak wavelengths (730 nm and 850 nm) were used to calculate changes of HbO2 and HHb concentrations (both in μmol/L) using the modified Beer–Lambert Law (MBLL). Before starting the flight scenario and the SWM task, participants were asked to relax for approximately two minutes, and a ten-second baseline measurement was then performed. To remove long-term drift, higher-frequency cardiac or respiratory activity and other noise with other frequencies than the target signal [14–16], we used a band-pass FIR filter with an order of 20 (0.02–0.40 Hz) on this raw time series of HbO2 and HHb signal changes. After this process, a correlation-based signal improvement (CBSI) algorithm [17] was used to filter out spikes and to improve signal quality based on the assumed negative correlation between HbO2 and HHb. Changes in HbO2 concentration from the ten-second rest period baseline was computed over the entire time course of each task [18–20]: for the flight scenario, we calculated the average HbO2 concentrations changes from the ten-second rest period baseline during the entire flight. Concerning SWM, HbO2 concentrations changes were averaged across all trials (entire trial duration) for each condition. More specifically, changes in HbO2 concentrations were computed over the average concentration of the combined trials of each condition. For simplicity, statistical analyzes were focused on HbO2 signal only.

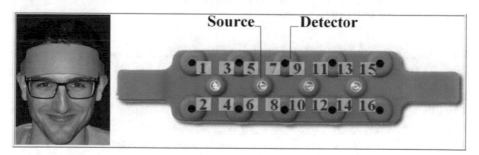

Fig. 4. fNIR100 headband and associated optodes numbering. Roughly, optodes #1 to #6 correspond to the left prefrontal cortex; optodes #7 to #10 correspond to the medial prefrontal cortex; optodes #11 to #16 correspond to the right prefrontal cortex.

4   Results

4.1   Behavioral Results

4.1.1   Association Between Flight Performance and Flight Experience

Even if very low in our *ab initio* pilots sample, recent flight experience (past two years) with light aircraft and flight simulator was associated with one of the flight performance metrics, namely speed deviation ($r(18) = -0.48$, $p = 0.042$), see Fig. 5. There was no significant correlation with the three other flight performance metrics ($r < 0.40$ in all cases).

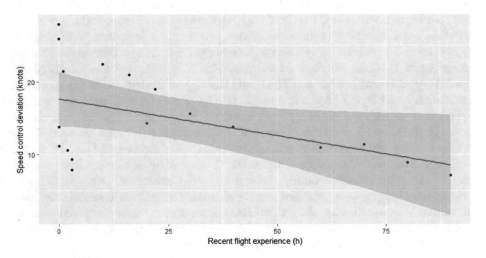

Fig. 5. Correlation between flight experience and aircraft speed deviation (a lower speed deviation indicates better performance).

4.2   Association Between Flight Performance and Spatial Working Memory (SWM) Efficiency

In order to reduce the number of correlation analysis, we focused on the highest level of difficulty of the SWM task (12 boxes). A better strategy during the highest level of difficulty of the SWM task was associated with a better control of the aircraft altitude ($r(18) = 0.57$, $p = 0.013$), see Fig. 6. There was no significant correlation with the three other flight performance metrics ($r < 0.40$ in all cases).

4.3   Neuroimaging (FNIRS) Results

Effects of the Difficulty of the Spatial Working Memory (SWM) Task on Prefrontal Activity As a first step, we checked whether increased SWM task difficulty was associated with an increase of the prefrontal HbO2 concentration.

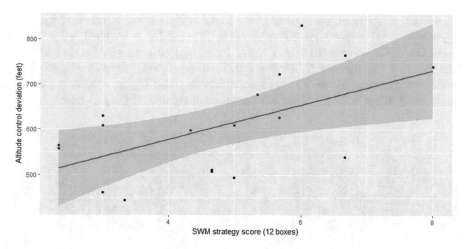

Fig. 6. Correlation between the strategy score (a lower value indicates a better strategy) during the highest level of difficulty of SWM and aircraft altitude deviation (a lower value represents a better performance).

Figure 7 illustrates HbO2 concentration changes for one participant during the entire SWM task performance. Peaks of HbO2 concentration can be seen at the beginning of most trials, in particular during the first occurrence of the highest level of difficulty (12 boxes).

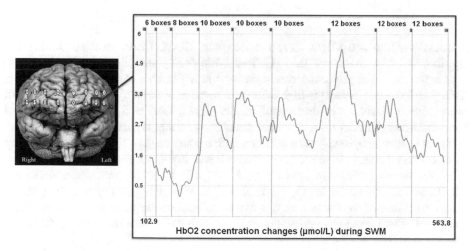

Fig. 7. Illustration of the HbO2 concentration changes in the left prefrontal cortex (optode #4) during the four levels of difficulty of the SWM task for one participant.

At the group level, as expected, we found a significant increase of the HbO2 concentration with increased difficulty ($F(3, 51) = 4.16$, $p < 0.010$, $\eta_p^2 = .20$). In particular, HbO2 concentration was higher in the highest level of difficulty, 12 boxes, than during 6 and 8 boxes (LSD: $p = 0.003$ and $p = 0.046$, respectively), see Fig. 8. HbO2 concentration in the 10 boxes condition was also higher than in the 6 boxes condition (LSD: $p < 0.05$). Other effects were not significant (all $p > 0.05$).

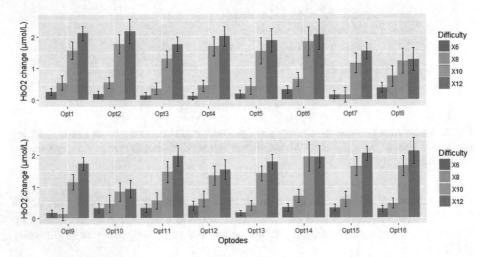

Fig. 8. HbO2 concentration changes in the 16 fNIRS optodes according to the 4 levels of difficulty of the SWM task (6, 8, 10 and 12 boxes).

Association between Flight Performance and HbO2 Concentration during the Spatial Working Memory (SWM) Task. We finally investigated whether HbO2 concentration in the prefrontal cortex during the SWM task could be correlated with flying performance during the flight simulator session. In order to reduce the number of correlation analysis, we again focused on the highest level of difficulty (12 boxes) of the SWM task. Results from 16 correlations showed that higher HbO2 concentration changes in several optodes was associated with a better control of the aircraft trajectory (a lower longitudinal deviation represents a better flight performance). Negative correlations were significant in optodes #3, #5, #7, #9, #10, #12, #13, and #14 (all $r < -0.40$, and all $p < 0.05$), see Fig. 9. Higher PFC activity in this region during the 12 boxes condition of the SWM task was associated with better performance in the flight simulator. There was no significant correlation with the three other flight performance metrics ($r < 0.40$ in all cases).

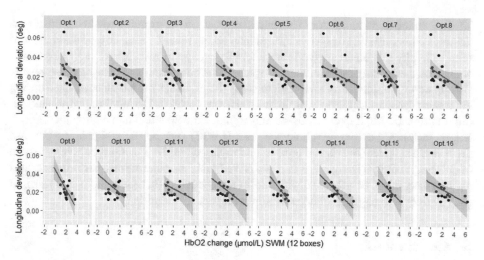

Fig. 9. Correlations between the aircraft longitudinal deviation during the flight scenario and HbO2 concentration changes in the 16 fNIRS optodes during the highest level of difficulty of SWM.

5 Discussion

We submitted a population of *ab initio* airline pilots to a navigation scenario in flight simulator and a neuropsychological task assessing spatial working memory. Our main objective was to investigate the possible behavioral and neurological communalities between these two different environments. The idea was to assess whether flying performance could be predicted by performance and related neurological activity during a well-controlled neuropsychological task such the SWM. We found rather convincing behavioral and neurological communalities between flying and performing the SWM task. First, an important aspect of the flying performance (altitude control) was correlated with the strategy score of SWM. An efficient strategy score indicated that participants employed a pre-determined sequence strategy of searching, an indication on the participant's ability to use heuristic strategy [21]. Thus, participants who were able to deploy an efficient strategy that simplified SWM task accomplishment were also more likely to be more efficient for controlling the altitude of the aircraft. In addition to that, we also found a significant correlation between flight experience (with light aircraft or flight simulator) and the ability to control aircraft speed. This result is somewhat surprising given the extremely low participants' experience, but it suggests that at the initial stage of the pilot's training, a few hours of experience can produce significant improvement in piloting performance.

Our neuroimaging results confirmed that task difficulty modulates fNIRS activity [18, 22–24], more specifically, provokes an increase of the HbO2 concentration when task difficulty is more important. This increased HbO2 concentration likely reflects the fact that mentally demanding tasks require resources in prefrontal-cortex-dependent functions. More importantly, fNIRS results also revealed that a higher fNIRS activity during the highest level of difficulty of SWM was associated with a better control of the aircraft's heading in the flight simulator. This latter result can be interpreted by the fact that participants who were still engaged in the spatial working memory task, despite the context of high difficulty (12 boxes), were more likely to engage in a sustained and fine-grained control of the heading of the aircraft. Together, our behavioral and neuroimaging results are apparently contradictory to a previous study [25] in which performance attained during the neuropsychological tasks was not conclusively correlated to landing performance in flight simulator. Most likely, here we find significant behavioral and neurological correlations between the neuropsychological task (SWM) and the flight simulation because there was a sufficient overlap in terms of underlying cognitive functions between the type of piloting activity (navigation) and the SWM task. In other terms, performance to a realistic task might be predicted with a less ecological one (neuropsychological task), as long as they both engages sufficiently similar aspects of cognition.

6 Conclusions

This paper describes a viable methodology to determine which cognitive functions predict pilots' abilities to fly an aircraft. The aim is to guide the development of an optimized cognitive screening tool well-suited for the successful selection of pilot candidates. We contributed to establish that spatial working memory is a reliable means to predict some aspects of piloting performance. Behavioral and neurological measurements performed during a SWM task significantly predicted some performance metrics in *ab initio* pilots during a flight simulator scenario. Such experiment paves the way for the development of dedicated software designed for the selection and certification of pilots capable of reducing the risk of accidents. Of course, more data is needed to confirm the potential of improvement of current selection test batteries with task like SWM as well as predictive validity studies investigating the association of such test performance with pilot training outcome.

References

1. Wiegmann, D.A., Shappell, S.A.: Human error and crew resource management failures in naval aviation mishaps: a review of U.S. Naval Safety Center data, 1990-96. Aviat. Space Env. Med **70**(12), 1147–1151 (1999)
2. Parasuraman, R.: Neuroergonomics: research and practice. Theor. Issues Ergon. Sci. **4**(1), 5–20 (2003)
3. Burke, E., Hobson, C., Linsky, C.: Large sample validations of three general predictors of pilot training success. Int. J. Aviat. Psychol. **7**(3), 225–234 (1997)
4. Carretta, T.R.: Pilot candidate selection method: still an effective predictor of US air force pilot training performance. Aviat. Psychol. Appl. Hum. Factors **1**, 3–8 (2011)
5. Damos, D.L.: Pilot selection batteries: shortcomings and perspectives. Int. J. Aviat. Psychol. **6**(2), 199–209 (1996)
6. Martinussen, M.: Psychological measures as predictors of pilot performance: a meta-analysis. Int. J. Aviat. Psychol. **6**(1), 1–20 (1996)
7. Causse, M., Dehais, F., Arexis, M., Pastor, J.: Cognitive aging and flight performances in general aviation pilots. Aging Neuropsychol. Cogn. **18**(5), 544–561 (2011)

8. Causse, M., Dehais, F., Pastor, J.: Executive functions and pilot characteristics predict flight simulator performance in general aviation pilots. Int. J. Aviat. Psychol. **21**(3), 217–234 (2011)
9. Taylor, J., O'Hara, R., Mumenthaler, M., Yesavage, J.: Relationship of CogScreen-AE to flight simulator performance and pilot age. Aviat. Space Environ. Med. **71**(4), 373 (2000)
10. Benthem, K.V., Herdman, C.M.: Cognitive factors mediate the relation between age and flight path maintenance in general aviation. Aviat. Psychol. Appl. Hum. Factors **6**, 81–90 (2016). https://doi.org/10.1027/2192-0923/a000102
11. Wang, H., Su, Y., Shang, S., Pei, M., Wang, X., Jin, F.: Working memory: a criterion of potential practicality for pilot candidate selection. Int. J. Aerosp. Psychol. **28**, 1–12 (2019)
12. Dror, I.E., Kosslyn, S.M., Waag, W.L.: Visual-spatial abilities of pilots. J. Appl. Psychol. **78**(5), 763 (1993)
13. De Luca, C.R., et al.: Normative data from the Cantab. I: development of executive function over the lifespan. J. Clin. Exp. Neuropsychol. **25**(2), 242–254 (2003)
14. Lu, C.-M., Zhang, Y.-J., Biswal, B.B., Zang, Y.-F., Peng, D.-L., Zhu, C.-Z.: Use of fNIRS to assess resting state functional connectivity. J. Neurosci. Methods **186**(2), 242–249 (2010)
15. Roche-Labarbe, N., Zaaimi, B., Berquin, P., Nehlig, A., Grebe, R., Wallois, F.: NIRS-measured oxy- and deoxyhemoglobin changes associated with EEG spike-and-wave discharges in children. Epilepsia **49**(11), 1871–1880 (2008)
16. White, B.R., et al.: Resting-state functional connectivity in the human brain revealed with diffuse optical tomography. NeuroImage **47**(1), 148–156 (2009)
17. Cui, X., Bray, S., Reiss, A.L.: Functional near infrared spectroscopy (NIRS) signal improvement based on negative correlation between oxygenated and deoxygenated hemoglobin dynamics. NeuroImage **49**(4), 3039–3046 (2010)
18. Ayaz, H., Shewokis, P., Bunce, S., Izzetoglu, K., Willems, B., Onaral, B.: Optical brain monitoring for operator training and mental workload assessment. Neuroimage **59**(1), 36–47 (2012)
19. Durantin, G., Gagnon, J.-F., Tremblay, S., Dehais, F.: Using near infrared spectroscopy and heart rate variability to detect mental overload. Behav. Brain Res. **259**, 16–23 (2014)
20. Foy, H.J., Runham, P., Chapman, P.: Prefrontal cortex activation and young driver behaviour: a fNIRS study. PLoS ONE **11**(5), e0156512 (2016)
21. Lee, A., Archer, J., Wong, C.K.Y., Chen, S.-H.A., Qiu, A.: Age-related decline in associative learning in healthy Chinese adults. PLoS ONE **8**(11), e80648 (2013)
22. Gateau, T., Durantin, G., Lancelot, F., Scannella, S., Dehais, F.: Real-time state estimation in a flight simulator using fNIRS. PLoS ONE **10**(3), e0121279 (2015)
23. Mandrick, K., Peysakhovich, V., Rémy, F., Lepron, E., Causse, M.: Neural and psychophysiological correlates of human performance under stress and high mental workload. Biol. Psychol. **121**, 62–73 (2016)
24. Takeuchi, Y.: Change in blood volume in the brain during a simulated aircraft landing task. J. Occup. Health **42**(2), 60–65 (2000)
25. Causse, M., Chua, Z., Peysakhovich, V., Del Campo, N., Matton, N.: Mental workload and neural efficiency quantified in the prefrontal cortex using fNIRS. Sci. Rep. **7**, 5222 (2017)

Neurobusiness Applications

Using fNIRS and EDA to Investigate the Effects of Messaging Related to a Dimensional Theory of Emotion

Jan Watson[1]([✉]), Amanda Sargent[1], Yigit Topoglu[1], Hongjun Ye[2],
Wenting Zhong[2], Rajneesh Suri[2,3], and Hasan Ayaz[1,3,4,5]

[1] School of Biomedical Engineering, Science and Health Systems,
Drexel University, Philadelphia, PA, USA
{jlw437,as3625,yt422,ayaz}@drexel.edu
[2] LeBow College of Business, Drexel University, Philadelphia, PA, USA
{hy368,wz326,surir}@drexel.edu
[3] Drexel Business Solutions Institute, Drexel University, Philadelphia, PA, USA
[4] Department of Family and Community Health, University of Pennsylvania,
Philadelphia, PA, USA
[5] Center for Injury Research and Prevention,
Children's Hospital of Philadelphia, Philadelphia, PA 19104, USA

Abstract. Effective techniques for the analysis of messaging strategies is critical as targeted messaging is ubiquitous in society and a major research interest in the fields of psychology, business, marketing and communications. In this study, we investigated the effect of audiovisual messaging on participants' affective state using a two-dimensional theory of emotion with orthogonal valence and arousal axes. Twenty-four participants were recruited and presented with either a positively framed or negatively framed environmental conservation messaging video. We monitored participants' finger based electrodermal activity (EDA) as well as prefrontal hemodynamic activity using functional near infrared spectroscopy (fNIRS) during message viewing to attain measures related to neural activity and arousal. Consistent with our expectations, combined results from EDA and prefrontal asymmetry from fNIRS indicate positively framed messaging stimuli was related to higher arousal and higher valence compared to the negatively framed messaging stimuli. Combined brain and body imaging provides a comprehensive assessment and fNIRS + EDA can be used in the future for the neuroergonomic assessment of cognitive and affective state of individuals in real-world environments via wearable sensors.

Keywords: Functional near infrared spectroscopy · Electrodermal activity · Affective state assessment · Neuroergonomics

1 Introduction

Targeted messaging has been widely researched in psychology, communications, business and marketing specifically within the areas of health promotion, disease prevention, environmental awareness, politics and advertising [1]. Such messaging

© Springer Nature Switzerland AG 2020
H. Ayaz (Ed.): AHFE 2019, AISC 953, pp. 59–67, 2020.
https://doi.org/10.1007/978-3-030-20473-0_6

research seeks to identify a population target by identifying the characteristics and motivations of behavior that are most effectively influenced by a particular messaging approach [1, 2].

Neuroergonomics is an emerging science that aims to understand neural and physiological factors as they contribute to human performance in everyday settings and activities [3–7]. To date, various theories of behavioral change have relied on the self-reported measurement of variables for their support [8]. Although such measures are related (e.g. reports on intentions to change a behavior or reports of self-efficacy), they in themselves are weak predictors of future behavioral change [9–11]. Additional information from sources such as neuroimaging and physiological measures could improve our understanding of the relationship between messaging exposure and actual behavioral modification [12, 13].

Emotions are biologically based and play an important role in the determination of behavior [14, 15]. A two-dimensional (2D) theory of emotion holds that all emotions can be identified as coordinates of valence and arousal [15] where the arousal dimension reflects the level of activation related to the subjective experience and the valence dimension indicates the degree of to which the experience elicits an approach (caused by positive stimuli) or avoidance (caused by negative stimuli) response [16, 17]. Additionally, the valence hypothesis [18] posits that both the expression and experience of emotional valence is lateralized across the left and right frontotemporal hemispheres of the brain where approach related emotions are thought to be lateralized to the right. The valence hypothesis originally arose from Electroencephalography (EEG) data [18–20] to investigate the relationship between affect, engagement and brain activation in frontal brain activity [21, 22] and assumes that emotions contain approach/withdrawal components [23]. Therefore, emotion will be associated with a right or left asymmetry depending on the extent to which it is accompanied by approach or withdrawal behavior [23]. In terms of previous messaging and marketing research using EEG, evidence has shown that there is a correlation of frontal alpha asymmetry with affective states, specifically where a positive perception towards and advertisement correlates with higher frontal alpha asymmetry [24–26] and differences in frontal asymmetry activation has been demonstrated in both fMRI and other (e.g. functional near infrared spectroscopy) neuroimaging studies related to emotional visual, auditory and social cues [20, 27–31].

Electrodermal activity (EDA) is a well verified psychophysiological index of emotional arousal and refers to the variation of the electrical properties of the skin in response to sweat secretion [21]. It is an optimal way to measure the peripheral autonomic nervous system [32, 33]. By applying a low constant voltage, a change in skin conductance can be measured non-invasively [34]. The time series of skin conductance can be characterized by slowly varying tonic activity (i.e., skin conductance level; SCL) and fast varying phasic activity (i.e., skin conductance responses; SCRs) [32]. Higher electrodermal reactions have also been observed in response to acoustic stimuli (pleasant or unpleasant) compared to neutral stimuli [35].

Functional near infrared spectroscopy (fNIRS) uses near infrared light to monitor oxygenated and deoxygenated hemoglobin changes at the outer cortex of the brain [36, 37]. fNIRS measures hemodynamic activity in cortical regions such as the prefrontal cortex (PFC) in relation to a cognitive or emotional task of interest. Although fNIRS cannot observe deep brain regions like fMRI, fNIRS is able to provide comparable information regarding the modulation of emotion from both auditory [38] and visual [39, 40] messaging cues. Additionally, an fNIRS study by [41] has observed activation of the rostromedial prefrontal cortex (rPFC) during the experience of positive emotion in the context of esthetic experience and various studies have examined emotional constructs by measures of hemispheric asymmetry in the PFC using fNIRS [42].

The objective of this study is to verify if brain and body measures can be used together to provide a reliable assessment of the affective state of individuals. We utilized wearable sensors, fNIRS for prefrontal hemodynamics and EDA for finger based galvanic skin response, to compare two stimuli: a positively-framed and a negatively-framed audio-video narrative of the same story. Both stimuli addressed environmental conservation but from different perspectives. We expected that these stimuli locate in opposite directions diagonally within the 2D emotion model. In the study, we used fNIRS to capture frontal asymmetry (difference in activation of left and right hemisphere) to estimate valence level along with the EDA-based skin conductance response to estimate arousal level.

2 Methods

2.1 Participants

Twenty-four participants (16 female, mean age = 29 years) volunteered for the study. All confirmed that they met the eligibility requirements of being right-handed with vision correctable to 20/20, did not have a history of brain injury or psychological disorder and were not on medication affecting brain activity. Prior to the study all participants signed consent forms approved by the Institutional Review Board of Drexel University. Due to equipment malfunction, one participant's EDA data was not available for analysis.

2.2 Experimental Procedure

The experiment was performed over one 10 min session. Participants were brought into a room and sat in front of a single computer monitor. After giving consent, participants were fitted with a sixteen optode continuous wave fNIRS system model 1200 (fNIR Devices LLC, www.fnirdevices.com) to measure prefrontal cortex hemodynamic activity along with a miniaturized electrodermal activity (EDA) sensor, Shimmer 3 GSR + Unit (Shimmer Sensing, www.shimmersensing.com) system that was placed on the participant's left index and middle fingers to assess physiological measures from skin conductance.

After device set up, participants were placed into two groups based on the framing of the environmental resource conservation video they would view, either positively framed or negatively framed. Participants were instructed to look at a fixation cross on a computer monitor and relax for 30 s before being presented with one of two videos previously described. At the conclusion of the messaging intervention, participants were once again instructed to look at a fixation cross on a computer monitor and relax for 30 s.

2.3   EDA Acquisition and Analysis

Raw EDA data was measured in micro Siemens at a rate of 128 Hz using Shimmer Capture Software. Pulse was also measured through an optical pulse ear clip connected to the Shimmer GSR device. Both accelerometer and gyroscope data were collected for the device to monitor for signal artifacts that can be caused by excessive motion. Time synchronized blocks for rest periods and video stimuli were processed using the MATLAB Toolbox Ledalab [32]. Time synchronization markers were not saved with EDA for one participant, and hence, this participant's data was eliminated from analysis, keeping n = 23 for EDA results.

For the preprocessing of data, an infinite impulse response (IIR) Butterworth – low pass filter (zero phase) was applied with a filter order of 2 and a cutoff frequency of 0.5 Hz. After preprocessing the data, continuous decomposition analysis (CDA) was applied using Ledalab. Linear mixed models were used for statistical analysis.

2.4   fNIRS Acquisition and Analysis

Prefrontal cortex hemodynamics were measured using a continuous wave fNIRS 1200 system (fNIR Devices LLC, fnirdevices.com) that was described previously [43]. Light intensity at two near-infrared wavelengths of 730 and 850 nm was recorded at 2 Hz using COBI Studio software [43]. The fNIRS headband contains 4 light-emitting diodes (LEDs) and 10 photodetectors for a total of sixteen optodes. Light intensity data was passed through a finite impulse response (FIR) hamming filter with a filter order of 20 and a cutoff frequency of 0.1 Hz. Time synchronized blocks for video stimuli were processed with the Modified Beer-Lambert Law and baseline-corrected to calculate oxygenation for each optode. Linear mixed models were used for statistical analysis [44, 45].

3   Results

The average phasic component in electrodermal activity is shown below. Figure 1 shows the mean SCR throughout video viewing for the two different message viewing groups (Positive Messaging and Negative Messaging). There is a significant difference between the two groups ($F_{1,21} = 4.93$, $p < .05$).

Next, prefrontal cortex hemodynamics results are summarized in Fig. 2 that show a comparison of frontal asymmetry, the difference between average oxygenation changes of left hemisphere (spatiotemporal averaging of optodes 1–8) and the right hemisphere (spatiotemporal averaging of optodes 9–16) during video viewing for the two different message viewing groups (Positive Messaging and Negative Messaging). There is a significant difference between the two groups ($F_{1,22} = 4.25$, $p = .05$).

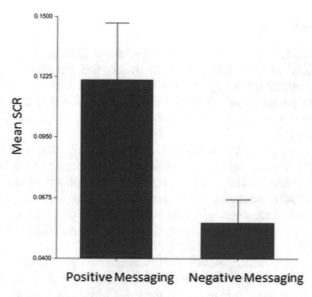

Fig. 1. EDA results: comparison of mean SCR for positively and negatively framed environmental conservation messaging groups. Whiskers are standard error of the mean (SEM).

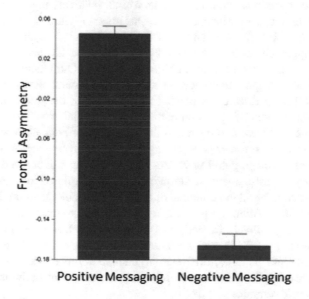

Fig. 2. FNIRS results: comparison of frontal asymmetry for positively and negatively framed environmental conservation messaging groups. Whiskers are standard error of the mean (SEM).

4 Discussion

In this study, our aim was to verify if both fNIRS and EDA measures can be used to assess individuals' affective state. We have investigated the effect of positively and negatively framed messaging (audio-visual stimuli) on viewers. Participants were placed into two groups based on the framing of the environmental resource conservation video they viewed while both their electrodermal and hemodynamic activity were continuously monitored.

EDA is commonly used to estimate arousal [21, 32, 33, 35], one axis of the 2D emotional model. We expected to see a difference between the two stimuli as the video stimuli narrative intensity differed. EDA results indicated that participants that viewed the positively framed messaging video exhibited a higher level of arousal than participants that viewed the negatively framed messaging video. This is consistent with previous EDA results [46] demonstrating differentiating sensitivity of skin conductance measures related to dynamic stimuli.

Prefrontal cortex hemodynamics results indicated that participants viewing the positively framed messaging video exhibited increased frontal asymmetric activation compared to the participants that viewed the negatively framed messaging video. The frontal asymmetry metric in our study is the linear difference between average oxygenation changes of left hemisphere and the right hemisphere. This demonstrates that a higher frontal asymmetry value represents an overall relatively higher activation in the left hemisphere of anterior prefrontal cortex. Our frontal asymmetry findings are in line with the previous EEG [24–26] and fNIRS [47, 48] frontal asymmetry studies that demonstrated approach related emotions lateralized to the left hemisphere and withdraw related emotions lateralized to the right hemisphere. Our results suggest fNIRS is a promising neuroimaging approach to capture the frontal asymmetry related to affective states. Future research is needed to capture both the role and sensitivity of motivation in the context of approach and avoidance [31].

In conclusion, we have demonstrated the capability of both EDA and fNIRS data independently to make inferences regarding affective state by way of a 2D theory of emotion. Both neuroimaging and physiological monitoring can be used together for a more comprehensive assessment of the individual. We explored the mobile brain and body imaging approach and the potential of utilizing a 2D emotional model empowered by the two modalities. Although preliminary, results are promising and suggested that an approach based (positively framed) messaging strategy has both an increased effect on viewer arousal and PFC frontal asymmetric activation. Considering that both sensors are wearable and portable, this neuroergonomic approach can be used in future studies for diverse applications for a continuous assessment of individuals with complex and real-world stimuli.

References

1. Pelletier, E., Sharp, L.G.: Persuasive communication and proenvironmental behaviours: how message tailoring and message framing can improve the integration of behaviours through self-determined motivation. Can. Psychol. **49**(3), 210–217 (2008)

2. Schmid, K.L., Rivers, S.E., Latimer, A.E., Salovey, P.: Targeting or tailoring? Mark. Health Serv. **28**(1), 32–37 (2008)

3. Ayaz, H., Onaral, B., Izzetoglu, K., Shewokis, P.A., McKendrick, R., Parasuraman, R.: Continuous monitoring of brain dynamics with functional near infrared spectroscopy as a tool for neuroergonomic research: empirical examples and a technological development. Front. Hum. Neurosci. **7**(December), 1–13 (2013)

4. Mehta, R.K., Parasuraman, R.: Neuroergonomics: a review of applications to physical and cognitive work. Front. Hum. Neurosci. **7**(December), 1–10 (2013)

5. Ayaz, H., Dehais, F.: Neuroergonomics: The Brain at Work and in Everyday Life. Academic Press, London (2018)

6. Parasuraman, R., Rizzo, M.: Neuroergonomics: The Brain at Work. Oxford University Press, Incorporated, Cary (2006)

7. Parasuraman, R., Jiang, Y.: Individual differences in cognition, affect, and performance: behavioral, neuroimaging, and molecular genetic approaches. Neuroimage **59**(1), 70–82 (2012)

8. Villegas, J., Matyas, C., Srinivasan, S., Cahyanto, I., Thapa, B., Pennington-Gray, L.: Cognitive and affective responses of Florida tourists after exposure to hurricane warning messages. Nat. Hazards **66**(1), 97–116 (2013)

9. Armitage, C.J., Conner, M.: Efficacy of the theory of planned behaviour: a meta-analytic review. Br. J. Soc. Psychol. **40**, 471–499 (2001)

10. Webb, T.L., Sheeran, P.: Does changing behavioral intentions engender behavior change? A meta-analysis of the experimental evidence. Psychol. Bull. **132**(2), 249–268 (2006)

11. Cooper, N., Tompson, S., O'Donnell, M.B., Emily, B.F.: Brain activity in self- and value-related regions in response to online antismoking messages predicts behavior change. J. Media Psychol. **27**(3), 93–109 (2015)

12. Berkman, E.T., Falk, E.B.: Beyond brain mapping. Curr. Dir. Psychol. Sci. **22**(1), 45–50 (2013)

13. Leger, P.-M., Riedl, R., vom Brocke, J.: Emotions and ERP information sourcing: the moderating role of expertise. Ind. Manag. Data Syst. **114**(3), 456–471 (2014)

14. Miller, G.A.: How we think about cognition, emotion and biology in psychopathology. In: Psychophysiology, vol. 46, pp. S12–S12. Blackwell Publishers, Malden (2009)

15. Lang, P.J.: The emotion probe: studies of motivation and attention. Am. Psychol. **50**(5), 372–385 (1995)

16. Ravaja, N., Saari, T., Salminen, M., Laarni, J., Kallinen, K.: Phasic emotional reactions to video game events: a psychophysiological investigation. Media Psychol. **8**(4), 343–367 (2006)

17. Carver, C.S.: Approach, avoidance, and the self-regulation of affect and action. Motiv. Emot. **30**(2), 105–110 (2006)

18. Davidson, R.J.: Cerebral asymmetry, emotion and affective style. In: Davidson, R.J., Hugdahl, K. (eds.) Brain Asymmetry. The MIT Press, Cambridge (1995)

19. Palmiero, M., Piccardi, L.: Frontal EEG asymmetry of mood: a mini-review. Front. Behav. Neurosci. **11**(November), 1–8 (2017)

20. Zhao, Z., Glover, G.: Hemispheric asymmetry for emotional stimuli. NeuroReport **9**(14), 3233–3239 (1998)

21. Lang, P.J., Greenwald, M.K., Bradley, M.M., Hamm, A.O.: Looking at pictures: affective, facial, visceral, and behavioral reactions. Psychophysiology **30**(3), 261–273 (1993)

22. Di Domenico, S.I., Le, A., Liu, Y., Ayaz, H., Fournier, M.A.: Basic psychological needs and neurophysiological responsiveness to decisional conflict: an event-related potential study of integrative self processes. Cogn. Affect. Behav. Neurosci. **16**(5), 848–865 (2016)

23. Davidson, R.J.: What does the prefrontal cortex 'do' in affect: perspectives on frontal EEG asymmetry research. Biol. psychol. **67**(1–2), 219 (2004)
24. Vecchiato, G., et al.: Spectral EEG frontal asymmetries correlate with the experienced pleasantness of TV commercial advertisements. Med. Biol. Eng. Comput. **49**(5), 579–583 (2011)
25. Ohme, R., Reykowska, D., Wiener, D., Choromanska, A.: Application of frontal EEG asymmetry to advertising research. J. Econ. Psychol. **31**(5), 785–793 (2010)
26. Vecchiato, G., et al.: Changes in brain activity during the observation of TV commercials by using EEG, GSR and HR measurements. Brain Topogr. **23**(2), 165–179 (2010)
27. Balconi, M., Molteni, E.: Past and future of near-infrared spectroscopy in studies of emotion and social neuroscience. J. Cogn. Psychol. **28**(2), 129–146 (2016)
28. Aday, J., Rizer, W., Carlson, J.M.: Neural mechanisms of emotions and affect. In: Emotions and Affect in Human Factors and Human-Computer Interaction, pp. 27–87. Elsevier, London (2017)
29. Yuksel, B.F., Oleson, K.B., Chang, R., Jacob, R.J.K.: Detecting and adapting to users' cognitive and affective state to develop intelligent musical interfaces. In: New Directions in Music and Human-Computer Interaction, pp. 163–177. Springer, Cham (2019)
30. Herrmann, M.J., Ehlis, A.-C., Fallgatter, A.J.: Prefrontal activation through task requirements of emotional induction measured with NIRS. Biol. Psychol. **64**(3), 255–263 (2003)
31. Harmon-Jones, E., Gable, P.A., Peterson, C.K.: The role of asymmetric frontal cortical activity in emotion-related phenomena: a review and update. Biol. Psychol. **84**(3), 451–462 (2010)
32. Benedek, M., Kaernbach, C.: A continuous measure of phasic electrodermal activity. J. Neurosci. Methods **190**(1), 80–91 (2010)
33. Greco, A., Valenza, G., Citi, L., Scilingo, E.P.: Arousal and valence recognition of affective sounds based on electrodermal activity. IEEE Sens. J. **17**(3), 716–725 (2017)
34. Poh, M., Member, S., Swenson, N.C., Picard, R.W.: A wearable sensor for unobtrusive, long-term assessment of electrodermal activity. IEEE Trans. Biomed. Eng. **57**(5), 1243–1252 (2010)
35. Bradley, M.M., Lang, P.J.: Affective reactions to acoustic stimuli. Psychophysiology **37**(2), S0048577200990012 (2000)
36. Franceschini, M.A., Boas, D.A.: Noninvasive measurement of neuronal activity with near-infrared optical imaging. Neuroimage **21**(1), 372–386 (2004)
37. Izzetoglu, K., Bunce, S., Izzetoglu, M., Onaral, B., Pourrezaei, K.: Functional near-infrared neuroimaging. IEEE Trans. Neural Syst. Rehabil. Eng. **13**(2), 5333–5336 (2004)
38. Plichta, M.M., et al.: Auditory cortex activation is modulated by emotion: a functional near-infrared spectroscopy (fNIRS) study. Neuroimage **55**(3), 1200–1207 (2011)
39. Yang, H., et al.: Gender difference in hemodynamic responses of prefrontal area to emotional stress by near-infrared spectroscopy. Behav. Brain Res. **178**(1), 172–176 (2007)
40. Leon-Carrion, J., et al.: Differential time course and intensity of PFC activation for men and women in response to emotional stimuli: a functional near-infrared spectroscopy (fNIRS) study. Neurosci. Lett. **403**(1–2), 90–95 (2006)
41. Kreplin, U., Fairclough, S.H.: Activation of the rostromedial prefrontal cortex during the experience of positive emotion in the context of esthetic experience. An fNIRS study. Front. Hum. Neurosci. **7**(December), 1–7 (2013)
42. Di Domenico, S.I., Rodrigo, A.H., Ayaz, H., Fournier, M.A., Ruocco, A.C.: Decision-making conflict and the neural efficiency hypothesis of intelligence: a functional near-infrared spectroscopy investigation. Neuroimage **109**, 307–317 (2015)

43. Ayaz, H., Shewokis, P.A., Curtin, A., Izzetoglu, M., Izzetoglu, K., Onaral, B.: Using MazeSuite and functional near infrared spectroscopy to study learning in spatial navigation. J. Vis. Exp. **56**, 1–13 (2011)
44. McKendrick, R., Ayaz, H., Olmstead, R., Parasuraman, R.: Enhancing dual-task performance with verbal and spatial working memory training: continuous monitoring of cerebral hemodynamics with NIRS. Neuroimage **85**, 1014–1026 (2014)
45. Liu, Y., et al.: Measuring speaker–listener neural coupling with functional near infrared spectroscopy. Sci. Rep. **7**(1), 43293 (2017)
46. Khalfa, S., Isabelle, P., Jean-Pierre, B., Manon, R.: Event-related skin conductance responses to musical emotions in humans. Neurosci. Lett. **328**(2), 145–149 (2002)
47. Balconi, M., Grippa, E., Vanutelli, M.E.: Resting lateralized activity predicts the cortical response and appraisal of emotions: an fNIRS study. Soc. Cogn. Affect. Neurosci. **10**(12), 1607–1614 (2015)
48. Tuscan, L.-A., Herbert, J.D., Forman, E.M., Juarascio, A.S., Izzetoglu, M., Schultheis, M.: Exploring frontal asymmetry using functional near-infrared spectroscopy: a preliminary study of the effects of social anxiety during interaction and performance tasks. Brain Imaging Behav. **7**(2), 140–153 (2013)

Brain Based Assessment of Consumer Preferences for Cognition Enhancing Hot Beverages

Amanda Sargent[1]([⊠]), Jan Watson[1], Yigit Topoglu[1], Hongjun Ye[2],
Wenting Zhong[2], Rajneesh Suri[2,3], and Hasan Ayaz[1,3,4,5]

[1] School of Biomedical Engineering, Science and Health Systems,
Drexel University, Philadelphia, PA 19104, USA
{as3625,ljw437,yt422,ayaz}@drexel.edu
[2] LeBow College of Business, Drexel University, Philadelphia, PA 19104, USA
{hy368,wz326,surir}@drexel.edu
[3] Drexel Business Solutions Institute, Drexel University, Philadelphia,
PA 19104, USA
[4] Department of Family and Community Health, University of Pennsylvania,
Philadelphia, PA 19104, USA
[5] Center for Injury Research and Prevention,
Children's Hospital of Philadelphia, Philadelphia, PA 19104, USA

Abstract. A current trend in food advertising is to emphasize the benefits provided by such products. However, the impact of such messaging on consumer decision-making and cognition is not well understood. Using a new generation of wearable and portable neuroimaging sensors, brain activity can be monitored and used to analyze how consumers respond to product communications and engage with products. In a controlled multi-day study, we explored the effects of product promotions and how they influence consumer preference and cognition. We measured brain activity using functional near infrared spectroscopy (fNIRS) that monitored the anterior prefrontal cortex of participants during behavioral tasks as well as hot beverage drinking. Results indicated fNIRS-based brain activity is related to the varying task load levels consistent with neural efficiency hypothesis. Moreover, promotional material and the cognition enhancing beverage appear to be influencing the participants and requiring less mental effort during the tasks when compared to the control.

Keywords: Functional near infrared spectroscopy · Cognition ·
Consumer behavior · Promotions · Neural efficiency

1 Introduction

Neuroscience and neuroengineering has a growing role in the field of marketing to gain insights into consumer behavior [1]. Marketing guides product design and presentation to make products more compatible with consumer preferences [2, 3]. The combination of neuroscience and marketing presents many new opportunities to improve our understanding of consumer decision making, individual differences in brand

© Springer Nature Switzerland AG 2020
H. Ayaz (Ed.): AHFE 2019, AISC 953, pp. 68–77, 2020.
https://doi.org/10.1007/978-3-030-20473-0_7

preferences, shopping environments, and interpersonal influences in purchasing situations [4]. Consumers make decisions every day, especially when deciding what foods and drinks to consume, yet little is known about how product promotions affect such decisions [5]. Knowing the factors that influence consumers' food and drink choices provides important information for better understanding the factors that affect consumers' preferences and purchase decisions [6]. The use of neuroimaging and physiological data along with traditional measures will provide rich and high-resolution information that can be used to better understand these preferences [2].

A current trend in food advertising is to emphasize the advantages of foods to the consumer by highlighting the benefits provided by such products [5, 7]. Advertising has been shown to influence product preferences and purchasing of products being advertised [8]. Information labels and claims have become an established way of communicating the health benefits of a product to consumers [9]. Moreover, product claims and benefits are widely used to market the product and to create a product differentiation [10]. However, the impact of such messaging on consumer decision-making and cognition is not well understood [5]. Past research is still ambiguous as to whether the labeling of items as healthy or merely having a benefit influences consumer preference [5, 11–13].

In this study, we measured brain activity using functional near infrared spectroscopy (fNIRS) that monitored the anterior prefrontal cortex of participants during task performance. fNIRS has emerged with the dawn of 21st century as a new technique to measure brain activity non-invasively [14–16]. fNIRS uses near infrared light to monitor oxygenated and deoxygenated hemoglobin changes at the outer cortex [17, 18]. fNIRS has the potential to enable diverse studies in the field of consumer neuroscience as it allows for the study of participants in real world, naturalistic settings [19–21]. Studying consumer behavior in a laboratory setting may cause a discrepancy between the measured brain activity and associated tasks [22, 23]. Most studies typically focus on one product or advertisement in isolation however, products are rarely featured alone (e.g. in a grocery store) [22]. Therefore, being able to study consumers in real-world environments could generate better understanding behind consumer preferences.

The goal of the present study is to identify how promotional materials affect consumer perceptions as well as their cognitive ability. We hypothesize that product-messaging accompanying a commercially available cognitive function enhancing hot beverage line will affect consumer's evaluation and cognition more than that for the traditional hot beverage. To test this, we designed a study that investigated differences between the effects of messaging on consumer preferences and cognitive. This study demonstrates a new approach to assessing product performance in naturalistic settings consistent with neuroergonomics approach [24]. Using new generation of wearable and portable optical neuroimaging sensors, brain activity of participants was monitored in ecologically valid realistic office environments [25] and assessed how consumers respond to product communications (e.g. ads, packaging) and engage with the products (e.g. consume), and how these might affect consumer preferences [26, 27]. With traditional self-reported measures, consumers may not be able to fully articulate their preferences. Human behavior can be driven by processes operating below conscious awareness and by measuring neural activity we can better understand their preferences

[28, 29]. Therefore, objective measures complement self-reported measures to provide a richer, more in depth analysis of consumer preferences [2].

2 Method

2.1 Participants

Thirty-nine participants between the ages of 18 and 38 (26 females, mean age = 22.5 ± 4.8 years) have been recruited for the study thus far. All confirmed that they met the eligibility requirements of being right-handed with vision correctable to 20/20, did not have a history of brain injury or psychological disorder, and were not on medication affecting brain activity. Prior to the study all participants signed consent forms approved by the Institutional Review Board of Drexel University.

2.2 Experimental Procedure

The experiment was performed over two one-hour sessions. After giving consent, participants were fitted with sensor devices as described in the next section. After device set up all participants performed three cognitive tasks presented in random order in a task battery for three trials and then were asked to consume a tea (cognition enhancing tea or a control tea) after the first and second task trials. Participants consumed a different beverage in each session but the same beverage (cognition enhancing tea, control tea) was used for all trials during a session.

The first task that participants completed was a mental arithmetic task. Here, one-digit or two-digit numbers were presented to the participant. Participants were instructed to add or multiply numbers mentally and enter the correct response on the screen. The computations included multiplied single digit numbers (2 + 4 or 2×4) or double-digit numbers (12 + 45 or 11×13) to increase task difficulty.

The next cognitive task was a rapid visual processing task (RVP) where digits were presented serially one at a time. The digits, from 1 to 9, appeared in pseudo-random order on a screen at the rate of 100 digits per minute. Participants were asked to detect target sequences of digits (for example, 3-5-7) that appeared. When the participant saw the target sequence, they were asked to respond by pressing enter as quickly as possible. There were two different levels of difficulty that were manipulated by using different lengths of digits (single digit and triple digit).

The Stroop paradigm was used for the final task in this battery. Participants were asked to look at color words (i.e. red, blue, green) in two different conditions: *congruent*, where the word was displayed in the color that its name denoted (e.g. "yellow" displayed in yellow font) and *incongruent*, where the word was displayed in a different color font than its name denoted (e.g. "yellow" displayed in red font).

After completing the task battery at the end of trial 1 and 2, participants were provided an 8 oz cup of the assigned beverage (cognition or control tea) to consume. Participants were asked to consume the tea unsweetened and black (without sugar). They were given three minutes to consume the beverage in this manner and then given the opportunity to add sugar to their preference. They were given four additional minutes to consume the beverage. Participants were provided a total of seven minutes to drink the beverage and then were asked to complete a survey regarding their attitude about the beverage. The beverage was weighed before consumption, after any sugar is added, and after the drinking period to determine how much of the beverage was consumed by the participant. Participants completed a short survey to assess their mood as well their attitude about the beverage from the consumption session after each trial. All surveys were presented on the computer. Participants then proceeded to complete the cognitive tasks a second time. After completion of the cognitive task battery, participants then completed the 7 min beverage consumption task that was previously described before completing the cognitive task battery a third and final time. A diagrammatic version of the protocol can be seen in Fig. 1.

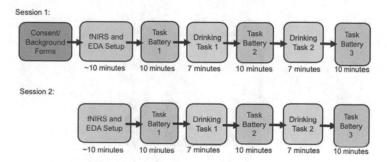

Fig. 1. Diagrammatic version of protocol

2.3 Signal Acquisition and Analysis

Prefrontal cortex hemodynamics were measured using a continuous wave fNIRS system (fNIR Devices LLC, fnirdevices.com) that was described previously [30]. Light intensity at two near-infrared wavelengths of 730 and 850 nm was recorded at 2 Hz using COBI Studio software [30]. The headband system contains four LEDs and ten photodetectors for a total of sixteen optodes. Data was passed through a finite impulse response hamming filter of order 20 and cutoff frequency 0.1 Hz. Time synchronized blocks for each trial were processed with the Modified Beer-Lambert Law to calculate oxygenation for each optode. Linear mixed models were used for statistical analysis.

3 Results

Prefrontal cortex hemodynamics results are summarized in the following figures for the three different cognitive tasks.

Arithmetic Task: The Fig. 2 depicts the comparison between the cognition enhancing tea and the control. The difference between the two groups was significant ($F_{1,183} = 5.064$, p = .025).

However, there was no significant interaction (see Fig. 3) between trials and the type of beverage consumed ($F_{2,183} = 0.82$, p = .44).

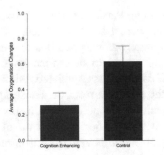

Fig. 2. Comparison of drink types during arithmetic task. Whiskers are standard error of the means (SEM).

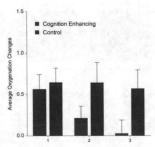

Fig. 3. Comparison of beverage types (cognition enhancing vs. control) across trials during arithmetic task. Whiskers are SEM.

Rapid Visual Processing (RVP) Task: A comparison between the cognition enhancing beverage (in Fig. 4) showed no significant difference between the two tea groups on this task ($F_{1,165} = 1.28$, p = .25).

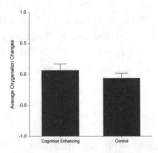

Fig. 4. Comparison of beverage types during the RVP task. Whiskers are SEM.

The differences between the beverage consumed (see Fig. 5 below) showed significant differences across trials ($F_{2,145.5} = 3.00$, p = .05).

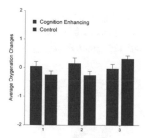

Fig. 5. Comparison of beverage types across trials during the RVP task. Whiskers are SEM.

Stroop Task: The comparison between the cognition enhancing tea and the control (Fig. 6) showed a significant difference between the two tea groups ($F_{1,187} = 6.11$, $p = .01$).

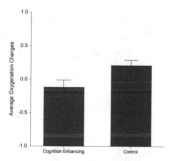

Fig. 6. Comparison of beverage types during the Stroop paradigm. Whiskers are SEM.

A comparison between the cognition enhancing and control tea across trials showed no significant difference between the beverages ($F_{2,187} = 0.34$, $p = .70$) (Fig. 7).

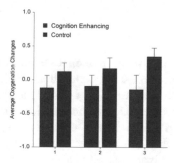

Fig. 7. Comparison of beverage types during the Stroop paradigm across trials. Whiskers are SEM.

Cognitive Activity During Consumption: A comparison of the prefrontal cortex hemodynamics during the consumption of the beverage itself (Fig. 8) showed a marginal difference between the two tea groups ($F_{1,123} = 7.35$, p = .007).

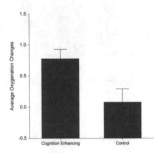

Fig. 8. Comparison of tea groups during the consumption period. Whiskers are SEM.

We also compared differences between cognition enhancing tea and the control across both consumption periods (Fig. 9). There was no significant difference between the type of beverage across the consumption periods ($F_{2,123} = 1.23$, p = .27).

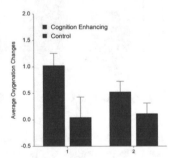

Fig. 9. Comparison of beverage type across both consumption periods. Whiskers are SEM.

4   Discussion

In this study, we investigated how promotional materials accompanying a cognition enhancing beverage may affect consumer perceptions of the beverage as well as their cognitive abilities. We developed a multi-day study to test the effects of the messaging that accompanies such beverages. During one session participants consumed a cognition enhancing beverage in the presence of a display of promotional materials during consumption. During the other session participants consumed a similar control hot beverage with no promotional materials with balanced type. To determine the effects on cognition, participants performed three cognitive tasks and consumed a hot beverage while their prefrontal cortex was monitored with fNIRS.

Pre-frontal cortex hemodynamics results indicated that when consuming the cognition enhancing beverage participants required less mental effort to complete the tasks as compared with the control tea. During the arithmetic task, participants consuming the cognition enhancing tea required less mental effort across trials whereas for the control tea the amount of effort remained consistent. For the RVP task, there was no significant difference between the cognition and control tea overall. However, across trials during the RVP we see that effort for the cognition tea remains consistent whereas for the control during trial 3 more mental effort is required than trials 1 and 2. Next, during the Stroop task we again see a significant difference between the cognition enhancing beverage and control tea overall however there is no difference across trials. Finally, during the consumption period, there was a higher activity for the cognition enhancing tea suggesting that they were thinking more about the beverage more than the control tea.

In conclusion, this study demonstrated it is possible to detect a difference in brain activation between a cognition enhancing tea and a control tea. The study described here provides important albeit preliminary information about fNIRS measures of the PFC hemodynamic response and its relationship to the effects of promotional materials on cognition. The promotional material and the cognition enhancing beverage appear to be influencing the cognition of the participants and requiring less mental effort during the tasks when compared to the control. Since fNIRS technology for wearable and portable sensors, it has the potential to be utilized in real world environments to provide better understanding of consumer preference which lead to eventual purchase decisions and the factors underlying these decisions.

References

1. Agarwal, S., Dutta, T.: Neuromarketing and consumer neuroscience: current understanding and the way forward. Decision 42(4), 457–462 (2015)
2. Ariely, D., Berns, G.: Neuromarketing: the hope and hype of neuroimaging in business. Nat. Rev. Neurosci. 11(4), 284–292 (2010)
3. Karmarkar, U.R., Yoon, C.: Consumer neuroscience: advances in understanding consumer psychology. Curr. Opin. Psychol. 10, 160–165 (2016)
4. Harris, J.M., Ciorciari, J., Gountas, J.: Consumer neuroscience for marketing researchers. J. Consum. Behav. 17(3), 239–252 (2018)
5. Van der Laan, L.N., De Ridder, D.T., Viergever, M.A., Smeets, P.A.: Appearance matters: neural correlates of food choice and packaging aesthetics. PLoS ONE 7(7), e41738 (2012)
6. Carrillo, E., Varela, P., Salvador, A., Fiszman, S.: Main factors underlying consumers' food choice: a first step for the understanding of attitudes toward "healthy eating". J. Sens. Stud. 26(2), 85–95 (2011)
7. Esch, F.-R., Möll, T., Schmitt, B., Elger, C.E., Neuhaus, C., Weber, B.: Brands on the brain: do consumers use declarative information or experienced emotions to evaluate brands? J. Consum. Psychol. 22(1), 75–85 (2012)
8. Harris, J.L., Bargh, J.A., Brownell, K.D.: Priming effects of television food advertising on eating behavior. Health Psychol. 28(4), 404–413 (2009)

9. Sabbe, S., Verbeke, W., Deliza, R., Matta, V., Van Damme, P.: Effect of a health claim and personal characteristics on consumer acceptance of fruit juices with different concentrations of açaí (Euterpe oleracea Mart.). Appetite **53**(1), 84–92 (2009)
10. Leathwood, P.D., Richardson, D.P., Strater, P., Todd, P.M., van Trijp, H.C.: Consumer understanding of nutrition and health claims: sources of evidence. Br. J. Nutr. **98**(3), 474–484 (2007)
11. Raghunathan, R., Naylor, R.W., Hoyer, W.: The unhealthy = tasty intuiton and its effects on taste inferences, enjoyment, and choice of food products. J. Mark. **70**, 170–184 (2006)
12. Borgmeier, I., Westenhoefer, J.: Impact of different food label formats on healthiness evaluation and food choice of consumers: a randomized-controlled study. BMC Pub. Health **9**(1), 184 (2009)
13. Provencher, V., Polivy, J., Herman, C.P.: Perceived healthiness of food. If it's healthy, you can eat more! Appetite **52**(2), 340–344 (2009)
14. Ferrari, M., Quaresima, V.: A brief review on the history of human functional near-infrared spectroscopy (fNIRS) development and fields of application. Neuroimage **63**(2), 921–935 (2012)
15. Villringer, A., Chance, B.: Non-invasive optical spectroscopy and imaging of human brain function. Trends Neurosci. **20**(10), 435–442 (1997)
16. Villringer, A., Planck, J., Hock, C., Schleinkofer, L., Dirnagl, U.: Near infrared spectroscropy (NIRS): a new tool to study hemodynamic changes during activation of brain function in human adults. Neurosci. Lett. **154**(1–2), 101–104 (1993)
17. Franceschini, M.A., Boas, D.A.: Noninvasive measurement of neuronal activity with near-infrared optical imaging. NeuroImage **21**(1), 372–386 (2004)
18. Izzetoglu, M., Izzetoglu, K., Bunce, S., Ayaz, H., Devaraj, A., Onaral, B., Pourezzaei, K.: Functional near-infrared neuroimaging. IEEE Trans. Neural Syst. Rehabil. **12**(2), 153–159 (2005)
19. McKendrick, R., et al.: Into the wild: neuroergonomic differentiation of hand-held and augmented reality wearable displays during outdoor navigation with functional near infrared spectroscopy. Front Hum. Neurosci. **10**, 216 (2016)
20. Curtin, A., Ayaz, H.: The age of neuroergonomics: towards ubiquitous and continuous measurement of brain function with fNIRS. Jpn. Psychol. Res. **60**, 374–386 (2018)
21. Quaresima, V., Ferrari, M.: Functional near-infrared spectroscopy (fNIRS) for assessing cerebral cortex function during human behavior in natural/social situations: a concise review. Organ. Res. Methods **22**(1), 46–68 (2016)
22. Krampe, C., Strelow, E., Haas, A., Kenning, P.: The application of mobile fNIRS to "shopper neuroscience" – first insights from a merchandising communication study. Eur. J. Mark. **52**(1/2), 244–259 (2018)
23. Boksem, M., Smidts, A.: Brain responses to movie trailers predict individual preferences for movies and their population-wide commerical success. J. Mark. Res. **LII**, 482–492 (2015)
24. Ayaz, H., Dehais, F.: Neuroergonomics: The Brain at Work and Everyday Life, 1st edn. Elsevier Academic Press, London (2018)
25. Ayaz, H., Onaral, B., Izzetoglu, K., Shewokis, P.A., McKendrick, R., Parasuraman, R.: Continuous monitoring of brain dynamics with functional near infrared spectroscopy as a tool for neuroergonomic research: empirical examples and a technological development. Front Hum. Neurosci. **7**, 871 (2013)
26. Hsu, M., Yoon, C.: The neuroscience of consumer choice. Curr. Opin. Behav. Sci. **5**, 116–121 (2015)
27. Barnett, S.B., Cerf, M.: A ticket for your thoughts: method for predicting movie trailer recall and future ticket sales using neural similarity among moviegoers. J. Consum. Res. **44**, 160–181 (2017)

28. Khushaba, R.N., Wise, C., Kodagoda, S., Louviere, J., Kahn, B.E., Townsend, C.: Consumer neuroscience: assessing the brain response to marketing stimuli using electroencephalogram (EEG) and eye tracking. Expert Syst. Appl. **40**(9), 3803–3812 (2013)
29. Calvert, G.A., Brammer, M.J.: Predicting consumer behavior: using novel mind-reading approaches. IEEE Pulse **3**(3), 38–41 (2012)
30. Ayaz, H., Shewokis, P., Curtin, A., Izzetoglu, M., Izzetoglu, K., Onaral, B.: Using mazesuite and functional near infrared spectroscopy to study learning in spatial navigation. J. Vis. Exp. **56**, e3443 (2011)

The Effects of Advertising
on Cognitive Performance

Hongjun Ye[1](✉), Siddharth Bhatt[1], Wenting Zhong[1], Jan Watson[1],
Amanda Sargent[1], Yigit Topoglu[1], Hasan Ayaz[1,2],
and Rajneesh Suri[1,2]

[1] Drexel University, 3141 Chestnut Street, Philadelphia, PA 19104, USA
{hy368,shb56,wz326,jlw437,as3625,yt422,
ayaz,surir}@drexel.edu
[2] Drexel Business Solutions Institute, 3220 Market Street,
Philadelphia, PA 19104, USA

Abstract. The market for beverages providing a cognitive boost is growing. Marketers often highlight the benefits of such drinks in their advertisements. Can such communications influence consumers' actual cognitive performance? Our results of cognitive task battery in a controlled experiment show that consumers' cognitive performance was negatively impacted in the presence of an advertisement communicating the benefits of a cognitive drink.

Keywords: Cognitive drinks · Consumer behavior · Advertisement

1 Introduction

The market for health and wellness has grown rapidly in the recent past. With the increase of cognitively-demanding tasks and a continuous engagement of people with the internet, beverages that provide cognitive boost have gained popularity especially with the younger generation.

To highlight the benefits of such beverages, marketing communications have also begun to emphasize the improvement of cognitive performance associated with such beverages. While such "cognitive boost" beverages might be effective in cognitive enhancement, the effects of communicating the benefits alone through advertisements could also have its own influence. Determining how and if the advertisement of such beverages contributes to consumers' cognitive performance is important to understand when assessing the effectiveness of such cognitive beverages.

Although prior research has demonstrated that advertising influences consumers' cognition, affect, experience, and brand choice [1–3], to our knowledge there is no research that has investigated the effects of advertising on cognitive performance.

The present research investigates the influence of communications (advertisements) accompanying a cognitive boost beverage on consumers' cognitive performance. In this study, self-reported measures accompanied the measurement of brain dynamics and affective states. Our results are counter intuitive and show that in the presence of an advertisement for a cognitive beverage, the participants performed less efficiently on cognitive tasks.

© Springer Nature Switzerland AG 2020
H. Ayaz (Ed.): AHFE 2019, AISC 953, pp. 78–83, 2020.
https://doi.org/10.1007/978-3-030-20473-0_8

2 Method

2.1 Participants and Product Stimuli

Eighteen hot tea drinkers (11 females and 7 males) were recruited to participate in this study that was conducted at an East Coast U.S. university. The average age of the participants was 21 years. Cognitive boost tea and regular tea with a similar flavor were used in the study.

2.2 Procedure and Experimental Design

Prior to the lab visit, participants completed a pre-screen survey. In this survey, in addition to demographic information (age, gender, and education), participants' preferences for hot beverages and their daily beverage consumption was collected.

The participants completed this study in two lab visits, each session lasting an hour. In each lab session, participants consumed only one type of hot beverage – a cognitive boost tea or a regular tea. There was at least a 24-h break between the two lab visits by each participant. Although each participant consumed both drinks, and the order of the type of drink (cognition, regular) consumed in each session was randomized.

During the consumption of the cognitive boost tea the participants were exposed to two advertisements that communicated the benefits of the cognitive boost tea. A table top merchandising display (8 inches × 12 inches) and a floor mounted retractable banner display (82 inches × 34 inches) communicated the benefit (improves performance at work) and the ingredients of the cognitive boost tea.

Each participant completed a battery of three tasks – arithmetic task, rapid visual information processing task (RVP) and Stroop task (Appendix A, B, C). These tasks were selected as they represented a microcosm of tasks a person usually accomplished at work (i.e., math problem-solving, attention, and working memory). Further, each task consisted of two levels of difficulty (easy and hard) and each participant completed three trials (length: approximate 10 min/trial) consisting of the same battery of tasks in each session but presented in a random order.

The first trial was used to create a baseline before a participant consumed a beverage. Upon completion of the first trial, participants were provided the hot beverage (12 oz cup) and had 7 min to consume as much beverage as they wanted. Following the consumption period, participants evaluated the beverage followed by the second trial of the task battery. Accuracy and reaction times (RT) were measured to assess task performance. Self-reported items were also used to assess their perceptions of the drink consumed during each session (Appendix D).

3 Results

The analyses were run using SAS software and the default level of significance was set at $p < 0.05$.

3.1   Baseline Cognitive Task Performance

Participants' baseline cognitive task performance was evaluated by comparing the efficiency on the three tasks. Behavioral efficiency is defined as the accuracy in unit time and was computed by using the total number of correct responses over the response time. Across all three tasks, participants showed higher efficiency when the advertisements were absent. Significant differences were found for arithmetic tasks and Stroop tasks (Table 1) but not for the RVP task.

Table 1. Behavioral tasks baseline performance comparison (efficiency).

	Arithmetic (Easy)	Arithmetic (Hard)	RVP (Easy)	RVP (Hard)	Stroop (Easy)	Stroop (Hard)
Regular tea	0.41 ± 0.15	0.11 ± 0.06	2.24 ± 0.21	2.87 ± 0.31	1.35 ± 0.26	1.21 ± 0.33
Cognitive boost tea	0.33 ± 0.61***	0.08 ± 0.05***	2.22 ± 0.20	2.75 ± 0.28	1.20 ± 0.26***	1.07 ± 0.37**

*P < 0.05; **P < 0.01; ***P < 0.001 (pair t-test)

3.2   Changes in Behavioral Efficiency

Although the participants' baseline performance in the cognitive boost tea session was not as strong as the performance in the regular tea session, cognitive boost tea outperformed the regular tea in terms of enhancing the efficiency over trials.

The changes in efficiency after each consumption of beverage were calculated and compared. These changes were in reference to the baseline that was established prior to consumption of the beverage. The changes in efficiency were higher for the cognitive boost tea for both arithmetic tasks and Stroop tasks. For RVP tasks, participants' efficiency on tasks decreased over time, however, the rate of decrease was slower for the cognitive boost tea (Table 2).

Table 2. Changes in efficiency (vis-à-vis baseline).

	Arithmetic After the 1st drink	Arithmetic After the 2nd drink	RVP After the 1st drink	Rvp After the 2nd drink	Stroop After the 1st drink	Stroop After the 2nd drink
Regular tea	10.9%	23.3%	−2.8%	0.1%	1.1%	3.7%
Cognitive boost tea	19.2%	42.8%	−0.4%	1.8%	10.4%	11.1%

3.3   Drink Consumption and Drink Evaluation

On average, participants consumed 134.28 milligrams of cognitive boost tea and 108.78 milligrams of the regular tea in the study. For drink evaluation, we asked the participants to rate the flavor and the perceived benefits of the tea (Appendix E). No difference was found in the self-reported measures for the two types of tea.

4 Discussion

The differences in performance observed in participants' baseline indicated that the advertisements accompanying the cognitive boost tea might have influenced the participants' cognitive performance. Notable changes in cognitive efficiency in the following trials suggests that the cognitive boost beverage did indeed provide the increase in cognitive performance. However, the performance during the baseline condition suggests that the impact of the advertisement that accompanies such products should not be neglected.

Given the results of this study, there seems to be a dual effect of both cognitive boost beverages and advertisements on consumers' cognitive performance. However, the current study is limited as the interaction effect between advertisements and consumption of cognitive boost beverage was not tested. Future research needs to examine the interaction between these variables by conducting controlled experiments. For example, a two-factor between-subject design would be ideal to elicit the interaction between the two variables (advertisements and consumption of cognitive boost beverages). Further, neuro-imaging devices such as functional near infrared spectroscopy (fNIRS) can help understand the neural correlates of changes in cognitive efficiency attributable to consumption of such beverages both in the presence and the absences of advertisements [4, 5].

Appendix A: Example of Arithmetic Task

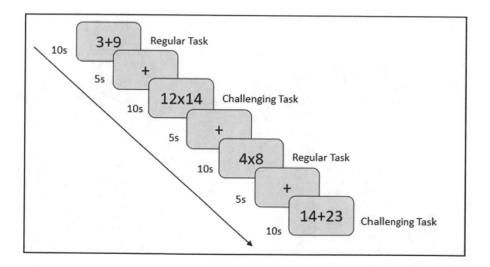

Appendix B: Example of Rapid Visual Processing (RVP) Task

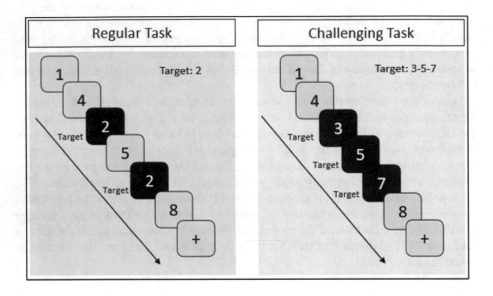

Appendix C: Example of Stroop Task

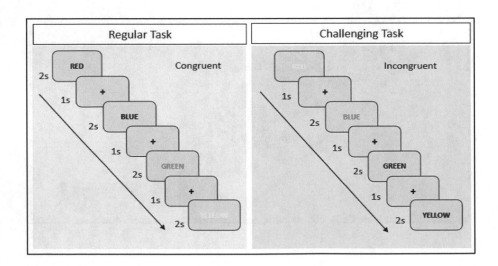

Appendix D: Experiment Procedure

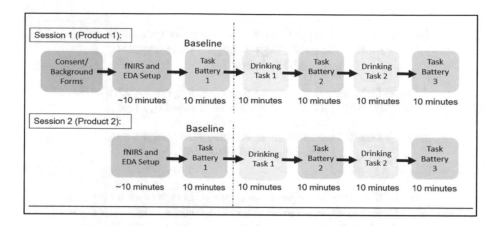

Appendix E: Drink Evaluation Survey Sample Questions

Item no.	Description
1	The hot beverage activated me
2	The hot beverage was strong
3	The hot beverage was good
4	I am feeling active
5	I feel confident at my ability this time
6	I feel confident at my performance on the tasks

Note: items were rated on a nine-point scale (1 = not at all; 9 = extremely)

References

1. Vakratsas, D., Ambler, T.: How advertising works: what do we really know? J. Mark. **63**(1), 26–43 (1999)
2. Edell, J.A., Burke, M.C.: The power of feelings in understanding advertising effects. J. Consum. Res. **14**(3), 421–433 (1987)
3. Mela, C.F., Gupta, S., Lehmann, D.R.: The long-term impact of promotion and advertising on consumer brand choice. J. Mark. Res. **34**(2), 248–261 (1997)
4. Ayaz, H., Shewokis, P.A., Bunce, S., Izzetoglu, K., Willems, B., Onaral, B.: Optical brain monitoring for operator training and mental workload assessment. Neuroimage **59**(1), 36–47 (2012)
5. Ayaz, H., Onaral, B., Izzetoglu, K., Shewokis, P.A., McKendrick, R., Parasuraman, R.: Continuous monitoring of brain dynamics with functional near infrared spectroscopy as a tool for neuroergonomic research: empirical examples and a technological development. Front. Hum. Neurosci. **7**, 871 (2013)

The Effects of Incentives in a Choice-Based Conjoint Pricing Study

Hongjun Ye[1]([⊠]), Siddharth Bhatt[1], Wenting Zhong[1], Jan Watson[1],
Amanda Sargent[1], Yigit Topoglu[1], Hasan Ayaz[1,2],
and Rajneesh Suri[1,2]

[1] Drexel University, 3141 Chestnut Street, Philadelphia, PA 19104, USA
{hy368, shb56, wz326, jlw437, as3625, yt422
ayaz, surir}@drexel.edu
[2] Drexel Business Solutions Institute, 3220 Market Street,
Philadelphia, PA 19104, USA

Abstract. This study investigates the effects of incentives on a consumer's choice of a utility plan. In our study that utilizes conjoint analysis, our research aims to understand the impact of incentives on consumers' choice of a Time-of-Use (TOU) utility pricing plan. The results indicate that underlining the consequences and offering incentives influence the valuation of the attributes in competing choices.

Keywords: Conjoint analysis · Consumer behavior · Price perception · Utility

1 Introduction

The effect of incentives is important to consider for any research [1]. Researchers offer different types of incentives to attract, recruit, or motivate participants. Monetary compensation upon completion of a study is a commonly used method. Additionally, researchers often offer monetary bonuses depending on the participant's performance [2]. However, incentives do not have to be monetary. For instance, prompts such as motivational messages and texts that are designated to elicit greater efforts are also used in experiments. In fact, it is a common belief that intrinsic motivation can generate adequate efforts in the absence of monetary rewards under certain circumstances [1].

The nature and the purpose of the experiments determine whether to offer incentives and what to offer as incentives. In the present study, we were interested in the effects of incentives in a choice-based conjoint pricing study. The study presented choice of competing utility pricing plans that varied on the time of use pricing for electricity.

Choice-based conjoint study is a methodology that has been frequently used in the industry [3]. The purpose of a conjoint study is to determine how consumers are likely to choose between alternative choices which vary on the features of product attributes. In our research, we examined consumers' preference for different time-of-use (TOU) utility pricing plans using a conjoint analysis that accompanied three different incentives for participants.

© Springer Nature Switzerland AG 2020
H. Ayaz (Ed.): AHFE 2019, AISC 953, pp. 84–90, 2020.
https://doi.org/10.1007/978-3-030-20473-0_9

TOU pricing plans charge consumers based on their energy usage during busiest (peak) and not so busy (off-peak) times of daily consumption. Specifically, consumers are charged more during peak than non-peak hours. The total monthly charges are based on the duration of consumption during peak times, and the rates for peak and off-peak hours. Thus, we included peak duration, peak rate, and off-peak rate as attributes of each utility plan in our study. The assumption behind traditional conjoint models is that after evaluating the options, the choices that a respondent makes are a result of the trade-offs that reflect the value of attributes involved in the choices [3].

Response time is often used as a proxy to assess the conflict during choice [4]. However, response latencies alone might not be sufficient to reflect the conflict during choice. Thus, we also incorporate functional near infrared spectroscopy (fNIRS) in our follow-up study based on the research design used in this current study to obtain insights regarding the conflict in respondents' decision-making process [5–7].

An important factor that influences decision making and addresses choice conflict is the participant's level of engagement with the task, especially in a cognitively demanding pricing choice task. A typical scenario in a conjoint study usually asks respondents to select an option that they like more, which might not be enough to motivate participants to make realistic decisions. Hence, we also included additional role play scenarios that motivated consumer performance and engagement.

We tested three motivation role play scenarios for the conjoint study. Our results indicate that the valuation of the attributes in a conjoint model can be influenced by the types of incentives offered and the instructional message.

2 Method

2.1 Participants

Seven hundred and forty-eight participants were recruited from an online sample and were provided a token compensation for their participation. Participants who stated that they do not pay utility bills were excluded from the analysis. The final sample included 576 participants (321 females; M_{age} = 38 years).

2.2 Scenarios

In a conjoint study Ding et al. [8] required participants to make choices of menu items from a local Chinese restaurant menu. To pay for their choices the participants were able to use a larger monetary incentive for the study that was provided to participants to participate in this study. They found that such an incentive-aligned conjoint study was more robust in predicting the success of the restaurant menu than the one that did not use such an incentive. Following the procedures used in this study, we created three role play scenarios to assess the effects of a performance-based monetary incentive on the choice of utility plans. The three role play scenarios created different levels of engagement with the task that required participants to choose between utility plans (Appendix A). One scenario provided a monetary incentive linked to performance while the other two scenarios provide no monetary incentive and differed on diagnostics used to assess the utility plans (liking vs. savings).

2.3 Research Design

The online survey took approximately twenty minutes to complete. In addition to providing demographic information (age, gender, education level, and income level) and details about the residence (ownerships, type, and size), the participants indicated their knowledge and opinions toward their current utility provider. (Appendix B).

The participants were randomly assigned to one of the three role play scenarios. Each participant was then presented with two choices of utility plan. The choices varied on three attributes: peak rate, non-peak rate, and the duration of peak-hour. There were 28 choice pairs and to ensure statistical power and avoid fatigue, choice pairs were randomly divided into 4 groups. Each participant was randomly assigned to 1 of the 4 groups and saw 7 choice pairs (Appendix D).

3 Results

The analyses were run using SAS (version 9.4) and JMP (version 14) software. The default level of significance was set at $p < 0.05$.

3.1 Participants Profile

Over 50 utility providers were used by the participants. The average residence size was 1400 square feet and the average monthly utility bill was 103 dollars. More than half of the participants had heard of "Time-of-Use" (TOU) utility pricing plans and understood the plans.

3.2 Choice-Based Conjoint

The conjoint model in all three scenarios suggested that the participants valued all three attributes and preferred the lower level of all the attributes (Tables 1, 2, and 3). Specifically, non-peak rate appears to have the most weights, followed by peak rate and peak time duration, in participants' decision-making process across all three scenarios.

Table 1. Scenario 1 results.

Source	LogWorth	p	Estimate	Std Err
Non-peak rate	31.324	0.00000	−0.3734647	0.0962007
Peak rate	28.001	0.00000	−1.0642597	0.1026151
Peak time duration	4.069	0.00009	−1.1285338	0.1034626

However, the differences among the log worth of the attributes were not the same in each role play scenario. Additional analyses revealed that the difference between non-

Table 2. Scenario 2 results.

Source	LogWorth	p	Estimate	Std Err
Non-peak rate	35.158	0.00000	−0.5050288	0.1021649
Peak rate	25.201	0.00000	−1.0562915	0.1073825
Peak time duration	6.336	0.00000	−1.2516958	0.1097366

Table 3. Scenario 3 results.

Source	LogWorth	p	Estimate	Std Err
Non-peak rate	65.214	0.00000	−0.4687177	0.0962441
Peak rate	38.500	0.00000	−1.2621916	0.1055647
Peak time duration	6.129	0.00000	−1.6576451	0.1120546

Table 4. Scenario comparison results (parameter estimates).

Term	Estimate	Std Err
Non-peak rate [lower level: 5 cents/kWh]	0.675811663	0.0313722487
Peak rate [lower level: 15 cents/kWh]	0.566126212	0.0304181844
Peak time duration [lower level: 4 h]	0.225491902	0.0283890574
Scenario 1 * Non-peak rate [lower level]	0.675811663	0.0433454452
Scenario 1 * Peak rate [lower level]	−0.031793421	0.0424873503
Scenario 1 * Peak time duration [lower level]	−0.038008237	0.0397295137
Scenario 2 * Non-peak rate [lower level]	0.675811663	0.0446233421
Scenario 2 * Peak rate [lower level]	−0.035631250	0.0434643875
Scenario 2 * Peak time duration [lower level]	0.028257473	0.0409586158

Table 5. Scenario comparison results (likelihood ratio tests).

Source	L-R ChiSquare	DF	p
Peak time duration	64.724	1	<.0001***
Peak rate	401.078	1	<.0001***
Non-peak rate	568.484	1	<.0001***
Scenario * Peak time duration	0.963	2	0.6178
Scenario * Peak rate	2.471	2	0.2907
Scenario * Non-peak rate	13.254	2	0.0013**

*P < 0.05; **P < 0.01; ***P < 0.001

peak rate and peak rate was significant only in the scenario where participants were most engaged and were provided monetary compensation for their performance (Tables 4 and 5). This result suggests that the perceived value of the attributes was likely to be influenced by the monetary incentives provided to participants to enhance their engagement with the task.

4 Discussion

Our results indicate that offering an incentive influenced the evaluation of the attributes in our conjoint study. When a monetary incentive was tied to a performance on the task (Scenario 3), the attribute preferences were more pronounced and participants indicated that the non-peak rate was the most important attribute. Such a difference in attribute importance was relatively less salient in the other two role play scenarios. We did not observe differences in reaction times which has been traditionally used to assess decisional conflict in conjoint studies [4].

Based on these results, we modified the design for a follow up study using fNIRS. In the neuro-imaging study, we compensate each participant 50 dollars upon the completion of the study. Additionally, in the instructional message of the conjoint study, we emphasize the outcomes by stating that the results of the study will be weighed heavily by a local utility company in deciding new electricity plans to launch in the area in the next 6 months. By stressing the consequences of the choice and tying the outcomes with utility expenses, we expect to see higher levels of engagement from the participants. Additionally, we expect to see greater perceived value of the non-peak rate in the neuro-imaging study.

Appendix A: Role Play Scenarios

Scenario no.	Description
1	Which of the following two TOU plans do you like more?
2	Which of the following two TOU plans could help you save more on utility expense?
3	Which of the following two TOU plans could help you save more on utility expense? (top three participants who make decisions that will save the most will receive a $10 Amazon gift card)

Appendix B: Participant Demographics

Question no.	Description
1	Do you own or rent your residence?
2	What type of residence do you live in?
3	What is the size of the residence you live?
4	What types of utility services do you use?
5	What is your household composition like?
6	What is the average monthly electricity bill for your residence?
7	Describe your familiarity with "Time-of-Use" (TOU) pricing plan

Appendix C: Possible Choice Configurations (Pricing Plans)

Plan no.	Description
1	Non-peak rate: 5 cents/kWh; Peak rate: 12 cents/kWh; Peak time duration: 4 h
2	Non-peak rate: 5 cents/kWh; Peak rate: 12 cents/kWh; Peak time duration: 5 h
3	Non-peak rate: 7 cents/kWh; Peak rate: 12 cents/kWh; Peak time duration: 4 h
4	Non-peak rate: 7 cents/kWh; Peak rate: 12 cents/kWh; Peak time duration: 5 h
5	Non-peak rate: 5 cents/kWh; Peak rate: 15 cents/kWh; Peak time duration: 4 h
6	Non-peak rate: 5 cents/kWh; Peak rate: 15 cents/kWh; Peak time duration: 5 h
7	Non-peak rate: 7 cents/kWh; Peak rate: 15 cents/kWh; Peak time duration: 4 h
8	Non-peak rate: 7 cents/kWh; Peak rate: 15 cents/kWh; Peak time duration: 5 h

Appendix D: Example of Choices Presented (Plan 4 Vs. Plan 5)

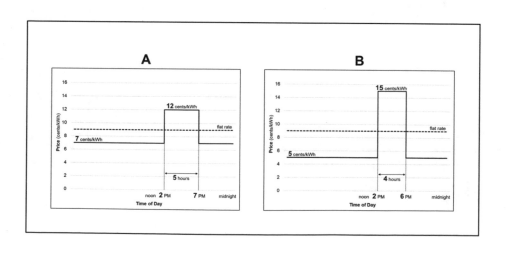

References

1. Camerer, C.F., Hogarth, R.M.: The effects of financial incentives in experiments: a review and capital-labor-production framework. J. Risk Uncertain. **19**(1–3), 7–42 (1999)
2. Eckel, C.C., Grossman, P.J.: Volunteers and pseudo-volunteers: the effect of recruitment method in dictator experiments. Exp. Econ. **3**(2), 107–120 (2000)
3. Green, P.E., Krieger, A.M., Wind, Y.: Thirty years of conjoint analysis: reflections and prospects. Interfaces **31**(3_Suppl), S56–S73 (2001)
4. Haaijer, R., Kamakura, W., Wedel, M.: Response latencies in the analysis of conjoint choice experiments. J. Mark. Res. **37**(3), 376–382 (2000)
5. Ayaz, H., Onaral, B., Izzetoglu, K., Shewokis, P.A., McKendrick, R., Parasuraman, R.: Continuous monitoring of brain dynamics with functional near infrared spectroscopy as a tool for neuroergonomic research: empirical examples and a technological development. Front. Hum. Neurosci. **7**, 871 (2013)
6. Di Domenico, S.I., Rodrigo, A.H., Ayaz, H., Fournier, M.A., Ruocco, A.C.: Decision-making conflict and the neural efficiency hypothesis of intelligence: a functional near-infrared spectroscopy investigation. Neuroimage **109**, 307–317 (2015)
7. de Winkel, K.N., Nesti, A., Ayaz, H., Bülthoff, H.H.: Neural correlates of decision making on whole body yaw rotation: an fNIRS study. Neurosci. Lett. **654**, 56–62 (2017)
8. Ding, M., Grewal, R., Liechty, J.: Incentive-aligned conjoint analysis. J. Mark. Res. **42**(1), 67–82 (2005)

Electrodermal Activity in Ambulatory Settings: A Narrative Review of Literature

Yigit Topoglu[1(✉)], Jan Watson[1], Rajneesh Suri[2,3], and Hasan Ayaz[1,3,4,5]

[1] School of Biomedical Engineering, Science and Health Systems, Drexel University, Philadelphia, USA
{yt422,ayaz}@drexel.edu
[2] LeBow College of Business, Drexel University, Philadelphia, PA, USA
[3] Drexel Business Solutions Institute, Drexel University, Philadelphia, PA, USA
[4] Department of Family and Community Health, University of Pennsylvania, Philadelphia, PA, USA
[5] Center for Injury Research and Prevention, Children's Hospital of Philadelphia, Philadelphia, PA 19104, USA

Abstract. Electrodermal activity (EDA) is a portable, non-invasive and wearable sensor that measures skin electrical properties to track correlates of autonomic nervous system activity. Although EDA utilization is sparse compared to some other biomedical signals in ambulatory settings, it can be a potentially helpful adjunct tool in neuroergonomics studies and mobile brain and body research. This paper summarizes EDA physiological principles and methodology including data acquisition, signal processing, and data analysis approaches. In addition, use of EDA in diverse neuroergonomic application areas, such as in psychiatry, neurology, operator and consumer assessment, virtual reality and gaming have been outlined.

Keywords: Electrodermal activity (EDA) · Neuroergonomics · Skin conductance · Sweat gland activity · Sympathetic activity

1 Introduction

Neuroergonomics is the area that focuses on the relationship between the human brain, body and performance in everyday and work environments, as it combines human factors, cognitive neuroscience, neuroengineering and psychology to give valuable information on brain mechanisms and behavior in natural settings [1, 2]. Electrodermal activity (EDA), also known as galvanic skin response (GSR), is a measurement technique which finds the changes of electrical properties of the skin caused from sweat gland activity. Previous studies show that EDA is generally considered as an indicator of physiological arousal [3, 4]. Even though the usage of EDA is simple, compared to other biomedical signals, the EDA utilization in ambulatory experimental setups is sparse.

© Springer Nature Switzerland AG 2020
H. Ayaz (Ed.): AHFE 2019, AISC 953, pp. 91–102, 2020.
https://doi.org/10.1007/978-3-030-20473-0_10

The purpose of the present paper is to provide a brief overview of EDA physiological principles, data acquisition, signal processing and analysis methods of EDA, along with a summary of research directions in neuroergonomics that utilize EDA measurement as a primary or adjunct tool for neuroergonomic applications spanning from consumer neuroscience to human factors and clinical domains. Ambulatory settings here refer to measurement of participants during unrestricted movement and postures in realistic and real-world environments. Most common environments such as indoor office-like places as well as outdoors, and including postures such as sitting as well as standing, walking, laying down and even sleeping. Such diverse ambulatory measurements will allow a more in-depth analysis of human experience consistent with the Neuroergonomic approach.

2 EDA Principles and Methodology

2.1 Physiological Principles of EDA

Electrodermal activity measures the changes in electrical properties of the skin. The electrical properties are altered by the electrolytes inside sweat when sweat secretion occurs by the eccrine sweat glands [3, 5]. Eccrine sweat glands play a key role in regulating thermoregulation and is activated by sympathetic activity of the autonomic nervous system (ANS) [6]. ANS controls vegetative autoregulatory processes such as body temperature, heart rate and blood pressure, which take a major part in homoeostasis [4]. ANS has two major parts: sympathetic and parasympathetic activity. Sympathetic activity is responsible from fight-or-flight response, which reflect bodily arousal and associated with emotional expressions and behaviors in humans [4]. Previous studies show that bursts of sympathetic ANS activity and EDA signal is highly correlated [7]. Hence, EDA is considered as an indicator of emotions and arousal [2, 4, 6].

2.2 Data Acquisition

Measurement. EDA is typically measured with two electrodes attached to the skin passively monitoring the skin electrical properties in between. Previous studies show that EDA is best measured at palmar sites (see Fig. 1) that are defined as medial (shown in blue) and distal (shown in red) phalanges of middle and index finger, or thenar and hypothenar eminence areas (shown in green) [6, 8]. However, there exist studies that also measure EDA from feet (shown in purple) [9, 10] or wrist (shown in orange) [11]. The rationale for selecting these regions is that they have higher eccrine sweat gland density compared to other body areas. EDA sensors typically are standard Ag/AgCl type electrodes in order to minimize of polarization and bias potentials [12]. Other types of electrodes such as dry carbon/salt adhesive electrodes [13] or textile Ag/AgCl type electrodes [14] also have been proposed. An electrode gel with a specific chloride salt (NaCl or KCl) concentration (0.05–0.075 molar) is commonly applied to electrodes to reduce the impedance between the skin and the electrode [12]. The rationale behind having a specific chloride salt concentration is that the concentration ranges represent the NaCl concentration estimate of the sweat on skin surface [15].

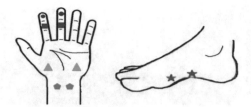

Fig. 1. Typical measurement sites for EDA.

For EDA measurement of the skin, two different methods are used: endosomatic and exosomatic measurement. In endosomatic method, the EDA is measured by using only skin potentials (SP) inside the skin, while exosomatic method applies an external direct current (DC) or alternating current (AC) through a circuit to measure changes in EDA, in either skin conductance (SC), skin resistance (SR), skin impedance (SZ) or skin admittance (SY) form [16]. In DC exosomatic measurements, SC (micro-Siemens (μS)) is recorded when the voltage is kept constant and SR is used when current kept constant. In AC exosomatic measurements, SZ is obtained when the voltage is kept constant, while SY is measured when current is constant [12].

Time Synchronization and Integration with Other Sensors. EDA can be easily used together concurrently with many other types of portable sensors, such as electrocardiogram (ECG) [17], electromyogram (EMG) [18], electroencephalogram (EEG) [19] and functional near-infrared spectroscopy (fNIRS) [20]. Hence, when acquiring all data, the synchronization of the sensors with each other is important, and there are varied approaches for this, depending on the EDA data acquisition configuration. One common solution is to collect data from different modalities on the same computer sharing the same high frequency counter/timer or collect the data on different computers and synchronize them via hardware cables such as RS232 serial ports, transistor-transistor logic (TTL), parallel ports, or ethernet local area network connection.

2.3 EDA Signal Processing and Data Analysis

EDA Signal. EDA is a time series signal with two components: Phasic and tonic activity. Phasic activity, also known as the skin conductance response (SCR), generally represents the fast fluctuations caused by the sympathetic arousal response to a stimuli [21]. SCR is divided into two parts: Event related SCR (ER-SCR) and non-specific SCR (NC-SCR). ER-SCR is the response with respect to some event, while NC-SCR represents the phasic changes are not related with any stimuli [21]. Generally, in most studies, event related SCRs are analyzed to capture stimulus response. The frequency of SCR varies between 0.05 to 1–2 Hz [22]. A typical SCR peak is shown in Fig. 2. In SCR, a latency occurs between the actual stimuli and the peak, which ranges between 1–3 s. The amplitude of SCR peaks ranges between 0.1 and 1 μS [6]. The latency is dependent on various features such as electrode placement and temperature [3]. The half recovery time of SCR, which is the time after the EDA value drops to 50%

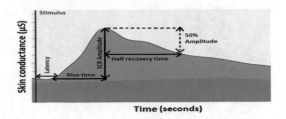

Fig. 2. An example of an SCR waveform (blue region) with tonic activity (grey region).

of the peak amplitude, ranges between 2–10 s, depending on the experimental circumstances [23].

The other component of EDA signal is tonic activity, also known as skin conductance level (SCL), which is the slowly changing component that can be affected by external and internal factors such as psychological state or skin properties [21]. An EDA signal with its SCL is shown in Fig. 3 as the gray foundation region. Amplitude of SCL ranges between 2 and 20 µS [6]. SCL indicates a constantly moving baseline for EDA, and the baseline values for SCL may not be the same in different individuals [24]. Due to interpersonal differences, SCL only analysis is not commonly used [3]. The frequency band of SCL varies between 0 to 0.05 Hz [22].

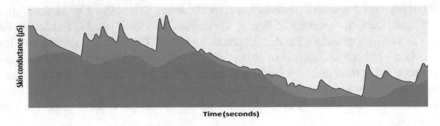

Fig. 3. An example of SCL waveform (gray region) with its EDA signal (black line) alongside SCR peaks (blue region).

Noise Removal and Filtering. For electrical noise removal, generally the raw EDA data is filtered with a low pass filter. The cut-off frequency of the low-pass filter is usually below 1 Hz, but it differs based on study parameters. In studies that focus on only SCR, a band-pass filter is used instead of the low-pass to get rid of not only noise, but also the low frequency component (0.05 Hz) SCL and very low frequency slow drifts. The low cut-off frequency of band-pass filter is usually around 0.05 Hz. It is shown that band-pass filtering with low cut-off frequency between 0.035 and 0.06 Hz improves event related SCR data analysis [25].

Motion Artifact Detection and Removal. One of the main issues for ambulatory measurement of EDA is the motion artifact detection and removal. Artifacts in EDA signals occur either due to the movement of electrodes with respect to skin, or muscle activity near recording sites [3]. The most basic approaches for removing artifacts are low pass filtering and manual inspection. Although low pass filtering can have limited effects, as it simply smears the spikes to a larger time period. It is also effective only on low amplitude artifacts and not effective on high amplitude artifacts, which are more frequent in ambulatory experiments [26]. Manual inspection is a very simple but effective method if done by expert reviewer. However, in the case of collecting days of EDA data, it may not be feasible [26].

Taylor et al. developed an approach that finds artifacts using support vector machines (SVM) and discrete Haar wavelet transform (DWT) to each 5-second long EDA epochs using features such as raw SC, filtered SC and wavelet coefficients [27]. This method is very good at detecting edges and sudden changes, while it is susceptible for other artifacts such as ringing effects due to the time-invariance of DWT [28] and it may struggle to find artifacts in a case of multiple responses in a certain section [29].

Chen et al. created an approach that uses stationary wavelet transform (SWT), Gaussian mixture models and adaptive thresholding methods [28]. SWT is used for transforming EDA to scaling and wavelet coefficients. After the transform, a Gaussian mixture model is created with respect to wavelet coefficients and thresholding is calculated for each EDA time window. Compared to DWT, SWT provides bigger artifact attenuation in EDA data [28], but artifacts are still needed to be eliminated separately in time windows.

Kelsey et al. implemented an approach that detects and removes artifacts automatically using optimal filtering and sparse recovery, specifically batch orthogonal matching pursuit (BOMP) method [26]. Later, they improved the approach by adding a dictionary to avoid overfitting BOMP using a test set and updating the dictionary after each analysis [29].

Decomposition of EDA to SCR and SCL. As explained above, in EDA data analysis, SCR is more commonly analyzed to assess response to rapid stimulus trains. Hence, finding new methods for the decomposition of SCR and SCL is critical. One of the earlier methods is the traditional analysis, where the significance of SCR is determined by amplitude of SCR peaks [30]. Despite its simplicity, this analysis has disadvantages as it uses only a few features such as latency and peak amplitude [31].

Different mathematical models have been developed for analyzing EDA using more features for more accurate analysis, called model-based analysis. Alexander et al. proposed a method that is based on deconvolving SC, isolating single peaks in the driver function and convolving peaks extracted with a biexponential function to get SCR [32]. This method has an advantage of separating overlapping SCR peaks [32], but it assumes a perfect impulse response function (IRF) for EDA, which causes the ignorance of tonic activity reflection on phasic activity [33].

Benedek and Kaernbach developed a model that utilizes nonnegative deconvolution to decompose EDA [34]. Later, they improved the model, naming continuous decomposition analysis (CDA), which separates SCR and SCL using deconvolution of SC data with its impulse response function (IRF) over time and estimating tonic and phasic activity from the result of deconvolution [33]. CDA gives a more straightforward scoring about phasic activity compared to nonnegative deconvolution while being less computationally intensive [33], but it is slow because of the optimization process and not robust to artifacts [29].

Bach et al. suggested a model that estimates SCR by applying general linear convolution model (GLM) to SCR time-series data to extract SCR and SCL [31]. Later, they improved the algorithm, using non-linear models for convolution and high pass filtering the raw SC data [25]. In terms of analysis, GLM method is more consistent and sensitive compared to Benedek's CDA [35]. However, this method relies on a priori location info of the stimuli that initiates SCR, making it unsuitable in real-time settings [29].

Greco et al. [36], produced a model that builds the IRF of the EDA signal not as a moving average model, but as an autoregressive moving average (ARMA) model and utilizes maximum a posteriori (MAP) estimation and convex approximation method, called cvxEDA. This model is robust to noise and provides a better discrimination than CDA, while not needing any additional processing steps [36]. The drawback of cvxEDA is that the noise is assumed to be white Gaussian noise, meaning it can be susceptible to motion artifacts [29].

Feature Extraction. In EDA analysis, feature extraction is done to summarize data by extracting its salient representative features (characteristics). For EDA feature extraction, generally time-domain data is used. The features extracted from phasic and tonic time domain EDA include number of significant SCR peaks, maximum, standard deviation, area under curve (AUC) and mean values of SCL and SCR, and frequency and mean of NS-SCR over a certain time window [30]. Recently, a study made by Posada-Quintero et al. proposed an EDA analysis method by extracting features from power spectral density (PSD) of SC data for sympathetic activity assessment of autonomic nervous system [37]. In the study, PSD of EDA shows that the sympathetic nervous activities can be defined within 0.045–0.25 Hz in the PSD and can be used as a reliable feature for EDA analysis, named as $EDASymp_n$.

2.4   Available Tools and Software for EDA

Most EDA data analysis is performed with custom-written scripts in MATLAB. However, there are studies that focused developing EDA signal processing and analysis tools so other researchers without expertise can benefit. Bach et al. presented a MATLAB software called SCRalyze [25, 31] which is based on general convolution modelling. SCRalyze software is migrated to PsPM, which is another MATLAB application Bach et al. created. Benedek and Kaernbach developed the MATLAB compatible software Ledalab [33, 34], which uses discrete and continuous decomposition analysis (DDA/CDA). Greco et al. also developed the software cvxEDA, using the method with the same name [36], which is compatible with both in Python and MATLAB. For EDA noise and artifact detection software, Taylor et al. developed an open source program named EDA Explorer using SVMs for data analysis, compatible with only Python [27].

3 EDA Applications

In this section, representative EDA use is outlined from diverse fields such as psychiatry, neurology, consumer assessment research (e.g. neuromarketing and consumer neuroscience), operator assessment (e.g. cognitive workload and training), virtual reality and gaming applications.

3.1 Operator Assessment

EDA has been used in human to machine and human to human interaction studies for operator assessment. In these studies, different operators were selected such as pilots, drivers, college students and cadets. Ghaderyan et al. focused on estimation of automatic workload [38] and cognitive load [39] of college students using EDA. Wobrock et al. designed a method for continuous mental effort evaluation during 3D manipulation tasks [19]. Dehzangi et al. examined driver distraction detection based on EDA [40]. Other EDA applications for operator assessment include training of military tactics of ROTC cadets [41], and analysis of mental workload of pilots using EDA, HR, EEG and eye blink data in flight conditions [42].

3.2 Virtual Reality and Gaming Applications

With the recent and gradual rise of gaming and Virtual Reality (VR) industry, EDA has been started to be integrated to studies that utilize games and VR technology. A VR game called SuperDreamCity that can respond to EDA changes have been created to analyze human physiology dynamics in immersive environments [43]. Another game called PhysiRogue is designed using EDA and EMG [44]. Drachen et al. uses EDA to analyze the correlation between HR, EDA and player experience in first-person shooter games [45]. Another VR study also utilizes EDA alongside heart rate for Quality of Experience evaluation method in VR environment [46].

3.3 Consumer Assessment

Since EDA can measure arousal, it is an emerging tool for the assessment of consumer decisions, such as neuromarketing and consumer neuroscience. The following is a brief list of representative studies: Leiner et al. proposed a method that measures audience arousal during media exposure using the positive changes in EDA [47]. Silveira et al. suggested an approach that predicts audience responses to movie content by looking at significant SCR peaks [48]. Wu et al. presented a study focused on prediction of online shopping satisfaction using EDA of 18 undergraduate students [49]. Holper et al. assessed objective versus subjective risk during risky financial decisions using both fNIRS and EDA signals [50]. Carbon and Leder used EDA with repeated evaluation technique for design innovativeness evaluation from the consumer perspective [51].

3.4   Psychiatric Applications

Since EDA is considered a reflection of bodily arousal which is strongly linked with affective state, it is used for many psychiatric studies related to mental health (see Table 1 for examples). In psychiatric studies, EDA is generally used to assess the relation between the EDA measures to clinical outcomes in mental disorders, as well as stress and monitoring and therapy of intrafamilial relationships.

Table 1. Examples of psychiatric disorder studies using EDA.

	Author	Participants	Scope of the study
Mental disorders	Greco et al. [52]	10 bipolar patients, 10 healthy subjects	EDA in bipolar patients in an ad-hoc affective elicitation experiment
	Myslobodsky and Horesh [53]	19 depression patients, 14 healthy subjects	Bilateral EDA in depressive patients during verbal and visual tasks
	Fung et al. [54]	503 participants	EDA in psychopathy-prone teenagers while responding to an audial burst
	Zahn et al. [55]	130 patients	Neuroleptics effects on autonomic activity in schizophrenia by analyzing EDA and heart rate
Stress	Liu and Du [9]	24 drivers	Stress level detection of drivers based on EDA
	Hernandez et al. [56]	9 call center employees	Stress recognition in call center using EDA with person-specific models
	Posada-Quintero et al. [57]	18 healthy subjects	EDA analysis under water due to cognitive stress
	Choi et al. [58]	10 subjects	Ambulatory stress monitor using EDA and heart rate variability
Other Psychiatric Applications	Paananen et al. [59]	4 cases, each with 2 clients and 2 therapists	EDA in Couple therapy for Intimate partner violence
	El-Sheikh et al. [60]	157 children	Child Maladjustment due to Marital Conflict by analyzing SCL reactivity

3.5   Neurological Applications

EDA is used in studies that focus on neurological and neurodevelopmental disorders. Prince et al. worked on EDA measurement of participants with autism to find the relationship between autism and bodily arousal [61]. Lane et al. made a study on EDA monitoring of the attention deficit hyperactivity disorder (ADHD) children with sensory over-responsivity (SOR) to find the correlation between SOR, ADHD, and anxiety [62]. Stevens and Gruzelier studied on the EDA difference on children that are normal, having autism spectrum disorder and having intellectual disability [63]. Other studies about neurological disorders are monitoring and detection of epileptic seizures by Poh et al. [64, 65], EDA evaluation of high and low alexithymia neurotic patients by Rabavilas [66], and improvement of the dementia detection due to anxiety using EDA by Melander et al. [67].

4 Conclusion

In this review, EDA principles and its utilization in diverse application areas are outlined. EDA methods including data acquisition, signal processing and data analysis were discussed. EDA is a practical biomedical sensor for ANS that has seen utilization throughout various research areas such as psychiatry, neuromarketing, consumer neuroscience, virtual reality, human factors, user experience and multimedia gaming. EDA can provide valuable information about the participant's affective state in ambulatory settings with wearable and miniaturized sensors, consistent with the Neuroergonomic approach.

References

1. Ayaz, H., Dehais, F.: Neuroergonomics: The Brain at Work and in Everyday Life. Elsevier, Academic Press, London (2019)
2. Curtin, A., Ayaz, H.: The age of neuroergonomics: towards ubiquitous and continuous measurement of brain function with fNIRS. Japan. Psychol. Res. 60(4), 374–386 (2018)
3. Boucsein, W.: Electrodermal Activity, 2nd edn. Springer Science + Business Media, New York (2012)
4. Critchley, H.D.: Electrodermal responses: what happens in the brain. Neurosci. Rev. J. Bringing Neurobiol. Neurol. Psychiatry 8, 132–142 (2002)
5. Carlson, N.R.: Physiology of Behavior. Allyn & Bacon, Boston (1994)
6. Dawson, M.E., Schell, A.M., Filion, D.L.: The electrodermal system. Handb. Psychophysiol. 2, 200–223 (2007)
7. Wallin, B.G.: Sympathetic nerve activity underlying electrodermal and cardiovascular reactions in man. Psychophysiology 18, 470–476 (1981)
8. Sharma, M., Kacker, S., Sharma, M.: A brief introduction and review on galvanic skin response. Int. J. Med. Res. Prof. 2, 13–17 (2016)
9. Liu, Y., Du, S.: Psychological Stress Level Detection Based on Electrodermal Activity (2017)
10. Handler, M., Nelson, R., Krapohl, D., Honts, C.: An EDA primer for polygraph examiners. Polygraph 39(2), 68–108 (2010)
11. Poh, M.-Z., Swenson, N.C., Picard, R.W.: A wearable sensor for unobtrusive, long-term assessment of electrodermal activity. IEEE Trans. Biomed. Eng. 57, 1243–1252 (2010)
12. Sharma, M., Kacker, S., Sharma, M.: A brief introduction and review on galvanic skin response. Int. J. Med. Res. Prof. 2, 13–17 (2016)
13. Posada-Quintero, H.F., Rood, R., Noh, Y., Burnham, K., Pennace, J., Chon, K.H.: Dry carbon/salt adhesive electrodes for recording electrodermal activity. Sens. Actuators, A 257, 84–91 (2017)
14. Lanata, A., Valenza, G., Scilingo, E.P.: A novel EDA glove based on textile-integrated electrodes for affective computing. Med. Biol. Eng. Compu. 50, 1163–1172 (2012)

15. Boucsein, W., Fowles, D.C., Grimnes, S., Ben-Shakhar, G., Roth, W.T., Dawson, M.E., Filion, D.L.: Publication recommendations for electrodermal measurements. Psychophysiology **49**, 1017–1034 (2012)
16. Cowley, B., Torniainen, J.: A short review and primer on electrodermal activity in human computer interaction applications (2016)
17. Bandara, D., Song, S., Hirshfield, L., Velipasalar, S.: A more complete picture of emotion using electrocardiogram and electrodermal activity to complement cognitive data. In: Foundations of Augmented Cognition: Neuroergonomics and Operational Neuroscience, pp. 287–298. Springer International Publishing (2016)
18. Dimberg, U.: Facial electromyographic reactions and autonomic activity to auditory stimuli. Biol. Psychol. **31**, 137–147 (1990)
19. Wobrock, D., Frey, J., Graeff, D., De La Rivière, J.-B., Castet, J., Lotte, F.: Continuous mental effort evaluation during 3d object manipulation tasks based on brain and physiological signals. In: Human-Computer Interaction, pp. 472–487. Springer (2015)
20. Holper, L., Scholkmann, F., Wolf, M.: The relationship between sympathetic nervous activity and cerebral hemodynamics and oxygenation: a study using skin conductance measurement and functional near-infrared spectroscopy. Behav. Brain Res. **270**, 95–107 (2014)
21. Aslanidis, T.: Electrodermal activity: applications in perioperative care (2014)
22. Ishchenko, A., Shev'ev, P.: Automated complex for multiparameter analysis of the galvanic skin response signal. Biomed. Eng. **23**, 113–117 (1989)
23. Cacioppo, J.T., Tassinary, L.G., Berntson, G.: Handbook of Psychophysiology. Cambridge University Press, Cambridge (2007)
24. Braithwaite, J.J., Watson, D.G., Jones, R., Rowe, M.: A guide for analysing electrodermal activity (EDA) & skin conductance responses (SCRs) for psychological experiments. Psychophysiology **49**, 1017–1034 (2013)
25. Bach, D.R., Friston, K.J., Dolan, R.J.: An improved algorithm for model-based analysis of evoked skin conductance responses. Biol. Psychol. **94**, 490–497 (2013)
26. Kelsey, M., Palumbo, R.V., Urbaneja, A., Akcakaya, M., Huang, J., Kleckner, I.R., Barrett, L.F., Quigley, K.S., Sejdic, E., Goodwin, M.S.: Artifact detection in electrodermal activity using sparse recovery. In: Compressive Sensing VI: From Diverse Modalities to Big Data Analytics, p. 102110D. International Society for Optics and Photonics (2017)
27. Taylor, S., Jaques, N., Chen, W., Fedor, S., Sano, A., Picard, R.: Automatic identification of artifacts in electrodermal activity data. In: 2015 37th Annual International Conference of the IEEE Engineering in Medicine and Biology Society (EMBC), pp. 1934–1937. IEEE (2015)
28. Chen, W., Jaques, N., Taylor, S., Sano, A., Fedor, S., Picard, R.W.: Wavelet-based motion artifact removal for electrodermal activity. In: 2015 37th Annual International Conference of the IEEE Engineering in Medicine and Biology Society (EMBC), pp. 6223–6226. IEEE (2015)
29. Kelsey, M., Akcakaya, M., Kleckner, I.R., Palumbo, R.V., Barrett, L.F., Quigley, K.S., Goodwin, M.S.: Applications of sparse recovery and dictionary learning to enhance analysis of ambulatory electrodermal activity data. Biomed. Signal Process Control **40**, 58–70 (2018)
30. Greco, A., Valenza, G., Scilingo, E.P.: Advances in Electrodermal Activity Processing with Applications for Mental Health: From Heuristic Methods to Convex Optimization. Springer Publishing Company, Incorporated (2016)
31. Bach, D.R., Flandin, G., Friston, K.J., Dolan, R.J.: Time-series analysis for rapid event-related skin conductance responses. J. Neurosci. Methods **184**, 224–234 (2009)
32. Alexander, D.M., Trengove, C., Johnston, P., Cooper, T., August, J., Gordon, E.: Separating individual skin conductance responses in a short interstimulus-interval paradigm. J. Neurosci. Methods **146**, 116–123 (2005)

33. Benedek, M., Kaernbach, C.: A continuous measure of phasic electrodermal activity. J. Neurosci. Methods **190**, 80–91 (2010)
34. Benedek, M., Kaernbach, C.: Decomposition of skin conductance data by means of nonnegative deconvolution. Psychophysiology **47**, 647–658 (2010)
35. Bach, D.R.: A head-to-head comparison of SCRalyze and Ledalab, two model-based methods for skin conductance analysis. Biol. Psychol. **103**, 63–68 (2014)
36. Greco, A., Valenza, G., Lanata, A., Scilingo, E.P., Citi, L.: cvxEDA: a convex optimization approach to electrodermal activity processing. IEEE Trans. Biomed. Eng. **63**, 797–804 (2016)
37. Posada-Quintero, H.F., Florian, J.P., Orjuela-Cañón, A.D., Aljama-Corrales, T., Charleston-Villalobos, S., Chon, K.H.: Power spectral density analysis of electrodermal activity for sympathetic function assessment. Ann. Biomed. Eng. **44**, 3124–3135 (2016)
38. Ghaderyan, P., Abbasi, A.: An efficient automatic workload estimation method based on electrodermal activity using pattern classifier combinations. Int. J. Psychophysiol. **110**, 91–101 (2016)
39. Ghaderyan, P., Abbasi, A., Ebrahimi, A.: Time-varying singular value decomposition analysis of electrodermal activity: a novel method of cognitive load estimation. Measurement **126**, 102–109 (2018)
40. Dehzangi, O., Rajendra, V., Taherisadr, M.: Wearable driver distraction identification on-the-road via continuous decomposition of galvanic skin responses. Sensors (Basel) **18**, 503 (2018)
41. Boyce, M.W., Goldberg, B., Moss, J.D.: Electrodermal activity analysis for training of military tactics. In: Proceedings of the Human Factors and Ergonomics Society Annual Meeting, pp. 1339–1343. Sage, Los Angeles, CA (2016)
42. Wilson, G.F.: An analysis of mental workload in pilots during flight using multiple psychophysiological measures. Int. J. Aviat. Psychol. **12**, 3–18 (2002)
43. Friedman, D., Suji, K., Slater, M.: SuperDreamCity: an immersive virtual reality experience that responds to electrodermal activity. In: International Conference on Affective Computing and Intelligent Interaction, pp. 570–581. Springer (2007)
44. Toups, Z.O., Graeber, R., Kerne, A., Tassinary, L., Berry, S., Overby, K., Johnson, M.: A design for using physiological signals to affect team game play (2006)
45. Drachen, A., Nacke, L.E., Yannakakis, G., Pedersen, A.L.: Correlation between heart rate, electrodermal activity and player experience in first-person shooter games. In: Proceedings of the 5th ACM SIGGRAPH Symposium on Video Games, pp. 49–54. ACM (2010)
46. Egan, D., Brennan, S., Barrett, J., Qiao, Y., Timmerer, C., Murray, N.: An evaluation of Heart Rate and ElectroDermal Activity as an objective QoE evaluation method for immersive virtual reality environments. In: 2016 Eighth International Conference on Quality of Multimedia Experience (QoMEX), pp. 1–6. IEEE (2016)
47. Leiner, D., Fahr, A., Früh, H.: EDA positive change: a simple algorithm for electrodermal activity to measure general audience arousal during media exposure. Commun. Methods Measures **6**, 237–250 (2012)
48. Silveira, F., Eriksson, B., Sheth, A., Sheppard, A.: Predicting audience responses to movie content from electro-dermal activity signals (2013)
49. Wu, Y., Liu, Y., Su, N., Ma, S., Ou, W.: Predicting online shopping search satisfaction and user behaviors with electrodermal activity. In: Proceedings of the 26th International Conference on World Wide Web Companion, pp. 855–856. International World Wide Web Conferences Steering Committee (2017)
50. Holper, L., Wolf, M., Tobler, P.N.: Comparison of functional near-infrared spectroscopy and electrodermal activity in assessing objective versus subjective risk during risky financial decisions. NeuroImage **84**, 833–842 (2014)

51. Carbon, C.-C., Michael, L., Leder, H.: Design evaluation by combination of repeated evaluation technique and measurement of electrodermal activity. Res. Eng. Design **19**, 143–149 (2008)
52. Greco, A., Valenza, G., Lanata, A., Rota, G., Scilingo, E.P.: Electrodermal activity in bipolar patients during affective elicitation. IEEE J. Biomed. Health Inf. **18**, 1865–1873 (2014)
53. Myslobodsky, M.S., Horesh, N.: Bilateral electrodermal activity depressive patients. Biol. Psychol. **6**, 111–120 (1978)
54. Fung, M.T., Raine, A., Loeber, R., Lynam, D.R., Steinhauer, S.R., Venables, P.H., Stouthamer-Loeber, M.: Reduced electrodermal activity in psychopathy-prone adolescents. J. Abnorm. Psychol. **114**, 187–196 (2005)
55. Zahn, T.P., Pickar, D., van Kammen, D.P.: Neuroleptic effects on autonomic activity in schizophrenia: between-group and within-subject paradigms and comparisons with controls. Schizophr. Bull. **27**(3), 503–515 (2001)
56. Hernandez, J., Morris, R.R., Picard, R.W.: Call center stress recognition with person-specific models. In: International Conference on Affective Computing and Intelligent Interaction, pp. 125–134. Springer (2011)
57. Posada-Quintero, H.F., Florian, J.P., Orjuela-Cañón, A.D., Chon, K.H.: Electrodermal activity is sensitive to cognitive stress under water. Front. Physiol. **8**, 1128 (2018)
58. Choi, J., Ahmed, B., Gutierrez-Osuna, R.: Development and evaluation of an ambulatory stress monitor based on wearable sensors. IEEE Trans. Inf. Technol. Biomed. **16**, 279–286 (2012)
59. Paananen, K., Vaununmaa, R., Holma, J., Karvonen, A., Kykyri, V.-L., Tsatsishvili, V., Kaartinen, J., Penttonen, M., Seikkula, J.: Electrodermal activity in couple therapy for intimate partner violence. Contemp. Fam. Ther. **40**, 138–152 (2018)
60. El-Sheikh, M., Keller, P.S., Erath, S.A.: Marital conflict and risk for child maladjustment over time: skin conductance level reactivity as a vulnerability factor. J. Abnorm. Child Psychol. **35**, 715–727 (2007)
61. Prince, E.B., Kim, E.S., Wall, C.A., Gisin, E., Goodwin, M.S., Simmons, E.S., Chawarska, K., Shic, F.: The relationship between autism symptoms and arousal level in toddlers with autism spectrum disorder, as measured by electrodermal activity. Autism **21**, 504–508 (2017)
62. Lane, S., Reynolds, S., Thacker, L.: Sensory over-responsivity and ADHD: differentiating using electrodermal responses, cortisol, and anxiety. Front. Integr. Neurosci. **4**, 8 (2010)
63. Stevens, S., Gruzelier, J.: Electrodermal activity to auditory stimuli in autistic, retarded, and normal children. J. Autism Dev. Disord. **14**, 245–260 (1984)
64. Poh, M.-Z., Loddenkemper, T., Swenson, N.C., Goyal, S., Madsen, J.R., Picard, R.W.: Continuous monitoring of electrodermal activity during epileptic seizures using a wearable sensor. In: 2010 Annual International Conference of the IEEE Engineering in Medicine and Biology Society (EMBC), pp. 4415–4418. IEEE (2010)
65. Poh, M.Z., Loddenkemper, T., Reinsberger, C., Swenson, N.C., Goyal, S., Sabtala, M.C., Madsen, J.R., Picard, R.W.: Convulsive seizure detection using a wrist-worn electrodermal activity and accelerometry biosensor. Epilepsia **53**, e93–e97 (2012)
66. Rabavilas, A.D.: Electrodermal activity in low and high alexithymia neurotic patients. Psychother. Psychosom. **47**, 101–104 (1987)
67. Melander, C., Martinsson, J., Gustafsson, S.: Measuring electrodermal activity to improve the identification of agitation in individuals with dementia. Dement. Geriatr. Cogn. Disord. Extra **7**, 430–439 (2017)

Reliability of Consumer Choices for Conflicting Price Promotions

Amanda Sargent[1](✉), Jan Watson[1], Yigit Topoglu[1], Hongjun Ye[2],
Wenting Zhong[2], Hasan Ayaz[1,3,4,5], and Rajneesh Suri[2,3]

[1] School of Biomedical Engineering, Science and Health Systems,
Drexel University, 3141 Chestnut St, Philadelphia, PA 19104, USA
{as3625,ljw437,yt422,ayaz}@drexel.edu
[2] LeBow College of Business, Drexel University, 3141 Chestnut St,
Philadelphia, PA 19104, USA
{hy368,wz326,surir}@drexel.edu
[3] Drexel Business Solutions Institute, Drexel University, Philadelphia,
PA 19104, USA
[4] Department of Family and Community Health, University of Pennsylvania,
3101 Walnut St, Philadelphia, PA 19104, USA
[5] Center for Injury Research and Prevention,
Children's Hospital of Philadelphia, 3401 Civic Center Blvd,
Philadelphia, PA 19104, USA

Abstract. Decisional conflict arises when all options of a multiple dimensional decisions task has equal or close to equal expected utility, causing additional mental effort. In this study, we investigated the potential of capturing decisional conflict related mental effort in a realistic complex decision-making task. In a binary choice paradigm, participants made decisions related to electricity supply plans. The study presented in each trial a choice with two different utility plans where variables related to fixed rate or time-of-use plans with peak-rate value and duration were compared against each other. We monitored the anterior prefrontal cortex of participants during binary decision-making, to assess the level of conflict using functional near infrared spectroscopy (fNIRS). Results indicate that fNIRS is able to measure the difference between conflict and no conflict decision making processes consistent with the neural efficiency hypothesis.

Keywords: Functional near infrared spectroscopy · Consumer behavior · Price promotions · Decisional conflict

1   Introduction

Many decisions made in the real world involve higher order executive functions [1]. Assigning a subjective utility value to a decision parameters is one of the fundamental forms of human behavior [2]. Individuals come up with different strategies when making a decision. Some choices may need to be simplified whereas others may require more analytical thought [2, 3]. Other decisions require a reward or incentive to provide an optimal decision making [4]. Across all cases, decision makers use the strategy that

© Springer Nature Switzerland AG 2020
H. Ayaz (Ed.): AHFE 2019, AISC 953, pp. 103–109, 2020.
https://doi.org/10.1007/978-3-030-20473-0_11

maximizes outcomes regardless of constraints. When a decision involves a risk, people tend to place more weight on unlikely events than likely events [5]. People also tend to weigh the losses more heavily than the potential rewards [6]. Indirect evidence has correlated fatigue with a greater willingness to take risks [7]. Making a decision involving risk is a complex process that involves evaluation of both the outcome and the risk associated [8].

In this study, we monitored the anterior prefrontal cortex (PFC), primary brain area related to higher executive functions, while participants performed binary decision-making using functional near infrared spectroscopy (fNIRS) [9]. fNIRS uses near infrared light to monitor oxygenated and deoxygenated hemoglobin changes at the outer cortex [10, 11]. fNIRS can measure outer cortex regions such as PFC or motor cortex relevant to the cognitive function. Although fNIRS cannot observe deep brain sub-cortical regions, the cortical areas that it monitors provide relevant and comparable information regarding the cognition and specific to the context of our study, preference and decisional conflict [9, 12]. The PFC plays a crucial role in purchase decisions and integrates several brain regions which are vital for decision making processes [4, 13–18]. The medial prefrontal cortex has been identified as one area in the brain that is correlated with risky decision making, however the neural processes associated with have not been identified [19]. The ventromedial prefrontal cortex has also been identified as critically involved in determining the value of different outcomes of a decision. During decision making processes, there is lower brain activation during low-conflict or no conflict decisions (easier condition) and increased activation during high conflict decisions (more difficult condition) that has been measured using fNIRS [9].

The overall objective of the present study is to identify factors affecting reliability of consumer decisions and specifically if mental effort related to decisional conflict can be captured with fNIRS during binary choice tasks. Participants were instructed at the beginning of the study that the decisions they make may affect their current utility bills. The study presented a binary choice between two competing utility plans. Participants were instructed to decide which plan was better for them. We hypothesize that participants will require more effort to decide between plans that have conflicting parameters to compare thus increasing the conflict level. To test this, we developed a study to investigate decision making where participants are presented with binary decisions while simultaneously measuring their brain activation using fNIRS.

2   Method

2.1   Participants

Twenty-seven participants (16 females, mean age = 29.4 ± 10.8 years) volunteered for the study. All confirmed that they met the eligibility requirements of being right-handed with vision correctable to 20/20, did not have a history of brain injury or psychological disorder, and were not on medication affecting brain activity. Prior to the study all participants signed consent forms approved by the Institutional Review Board of Drexel University.

2.2   Experimental Procedure

The experiment was performed over one sixty-minute session. After giving consent, the participants were fitted with sensor devices described in the next section. After the device set up participants answered a series of binary questions to determine their preference between two utility plans.

Participants completed a series of practice questions to familiarize them to the choice survey format. After the practice questions, participants were asked if they had any questions and then proceeded to begin the choice task. In this section, participants made a series of binary choices between time of use (TOU) electricity plans shown relative to a traditional fixed rate plan or a different TOU plan. The binary choices varied based on three different parameters, the peak rate, non-peak rate, and duration. In each choice either 1, 2 or 3 parameters were altered to increase the conflict level in the decision. After responding to each binary choice, participants were asked if they understood the choice, how attractive the choice was, if the choice was difficult, and if they needed more time to make the decision. A diagrammatic version of the protocol can be seen in Fig. 1 and an example of the binary choices can be seen in Fig. 2.

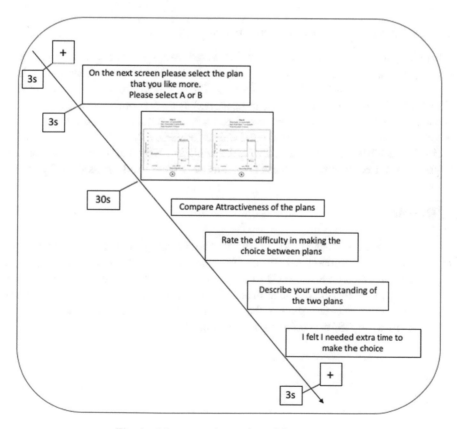

Fig. 1. Diagrammatic version of the protocol

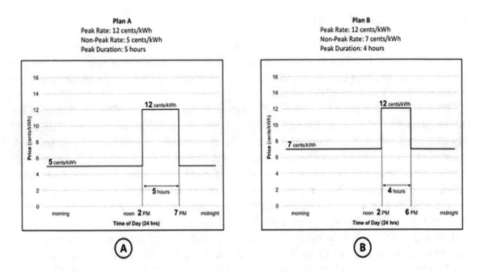

Fig. 2. Example of binary choice between utility plans

2.3   fNIRS Signal Acquisition and Analysis

Prefrontal cortex hemodynamics were measured using a continuous wave fNIRS system (fNIR Devices LLC, fnirdevices.com) that was described previously [20]. Light intensity at two near-infrared wavelengths of 730 and 850 nm was recorded at 2 Hz using COBI Studio software [21]. The headband system contains four LEDs and ten photodetectors for a total of sixteen optodes. Data was passed through a finite impulse response hamming filter of order 20 and cutoff frequency 0.1 Hz. Time synchronized blocks for each trial were processed with the Modified Beer-Lambert Law to calculate oxygenation for each optode. Linear mixed models were used for statistical analysis.

3   Results

First, we looked at the self-report surveys for the difficulty in deciding between plans. We looked at the comparison between two parameter and three parameter changes. Responses were rated on a scale of 1 to 7, 1 being extremely low and 7 being extremely high. There was a marginally significant difference in self-report measures on how difficult it was to decide between the two plans for both conflict and no conflict scenarios ($F_{1,214} = 2.3$, $p = 0.13$) as seen in Fig. 3.

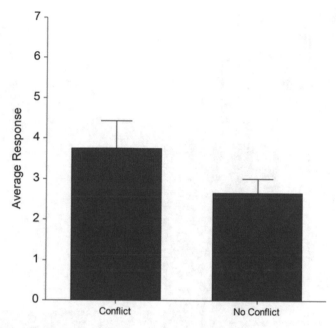

Fig. 3. Self-reported results for difficultly level deciding between the two plans Whiskers are SEM.

Next, we looked at pre-frontal cortex hemodynamics are summarized in Fig. 4. We looked at the comparison between two parameter and three parameter changes when deciding between each plan. There was a significant difference in between conflict and no conflict ($F_{1,78.4}$ = 4.907, p = 0.029) as seen in Fig. 4.

4 Discussion

In this study, we have investigated if fNIRS can capture decisional conflict related effort during a realistic binary decision making related to the choice of electric utility plans. Participants made decisions about which utility plan they preferred while their prefrontal cortex was monitored with fNIRS. Different factors (peak rate, non-peak rate and duration) were altered to increase the conflict level between the choices.

In self-report responses, as expected participants had more difficulty deciding between conflict choices when compared to no conflict decisions.

Pre-frontal cortex hemodynamics results further demonstrate that participants are requiring more effort to decide between choices that have conflict compared to no conflict decisions. These results confirm the literature that more energy is required during high conflict decisions [9].

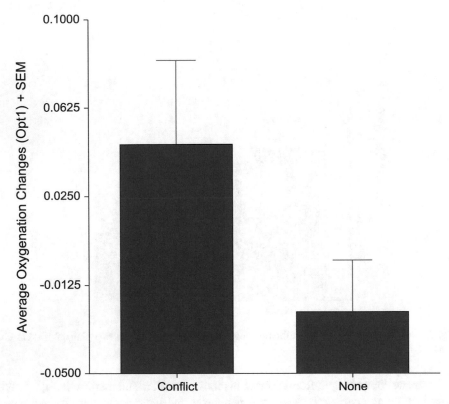

Fig. 4. Prefrontal cortex hemodynamic changes between conflict and no conflict choices. Whiskers are the standard error of the mean (SEM).

In conclusion, we have demonstrated that we are able to capture a difference between conflict and no conflict binary decision-making processes for utility plans using fNIRS consistent with previous fNIRS results [9]. Although preliminary, results here highlight the potential use of fNIRS in real-world decision-making assessment. As fNIRS can be miniaturized and ultra-portable sensors can be built with battery operated and wireless approaches, it is suitable for real-world measurement in natural settings [22].

References

1. Tremblay, S., Sharika, K.M., Platt, M.L.: Social decision-making and the brain: a comparative perspective. Trends Cogn. Sci. **21**(4), 265–276 (2017)
2. Yoon, C., et al.: Decision neuroscience and consumer decision making. Mark. Lett. **23**(2), 473–485 (2012)
3. Venkatraman, V., Payne, J.W., Bettman, J.R., Luce, M.F., Huettel, S.A.: Separate neural mechanisms underlie choices and strategic preferences in risky decision making. Neuron **62**(4), 593–602 (2009)
4. Rouault, M., Drugowitsch, J., Koechlin, E.: Prefrontal mechanisms combining rewards and beliefs in human decision-making. Nat. Commun. **10**(1), 301 (2019)

5. Tom, S.M., Fox, C.R., Trepel, C., Poldrack, R.A.: The neural basis of loss aversion in decision making under risk. Science **315**(5811), 515–518 (2007)
6. Barkley-Levenson, E.E., Van Leijenhorst, L., Galvan, A.: Behavioral and neural correlates of loss aversion and risk avoidance in adolescents and adults. Dev. Cogn. Neurosci. **3**, 72–83 (2013)
7. Harrison, Y., Horne, J.A.: The impact of sleep deprivation on decision making. A review. J. Exp. Psychol. Appl. **6**(3), 236–249 (2000)
8. Xue, G., Lu, Z., Levin, I.P., Weller, J.A., Li, X., Bechara, A.: Functional dissociations of risk and reward processing in the medial prefrontal cortex. Cereb. Cortex **19**(5), 1019–1027 (2009)
9. Di Domenico, S.I., Rodrigo, A.H., Ayaz, H., Fournier, M.A., Ruocco, A.C.: Decision-making conflict and the neural efficiency hypothesis of intelligence: a functional near-infrared spectroscopy investigation. Neuroimage **109**, 307–317 (2015)
10. Franceschini, M.A., Boas, D.A.: Noninvasive measurement of neuronal activity with near-infrared optical imaging. NeuroImage **21**(1), 372–386 (2004)
11. Izzetoglu, M., Izzetoglu, K., Bunce, S., Ayaz, H., Devaraj, A., Onaral, B., Pourezzaei, K.: Functional near-infrared neuroimaging. IEEE Trans. Neural Syst. Rehabil. Eng. **12**(2), 153–159 (2005)
12. Shimokawa, T., Misawa, T., Suzuki, K.: Neural representation of preference relationships. NeuroReport **19**(16), 1557–1561 (2008)
13. Knutson, B., Rick, S., Wimmer, G.E., Prelec, D., Loewenstein, G.: Neural predictors of purchases. Neuron **53**(1), 147–156 (2007)
14. Plassmann, H., Kenning, P., Deppe, M., Kugel, H., Schwindt, W.: How choice ambiguity modulates activity in brain areas representing brand preference: evidence from consumer neuroscience. J. Consum. Behav. **7**(4–5), 360–367 (2008)
15. Deppe, M., Schwindt, W., Kugel, H., Plassmann, H., Kenning, P.: Nonlinear responses within the medial prefrontal cortex reveal when specific implicit information influences economic decision making. J. Neuroimaging **15**(2), 171–182 (2005)
16. Gonzalez, C., Dana, J., Koshino, H., Just, M.: The framing effect and risky decisions: examining cognitive functions with fMRI. J. Econ. Psychol. **26**(1), 1–20 (2005)
17. Fleming, S.M., Huijgen, J., Dolan, R.J.: Prefrontal contributions to metacognition in perceptual decision making. J. Neurosci. **32**(18), 6117–6125 (2012)
18. Bang, D., Fleming, S.M.: Distinct encoding of decision confidence in human medial prefrontal cortex. Proc. Natl. Acad. Sci. **115**(23), 6082–6087 (2018)
19. Bechara, A., Damasio, A.R.: The somatic marker hypothesis: a neural theory of economic decision. Games Econ. Behav. **52**(2), 336–372 (2005)
20. Ayaz, H., Izzetoglu, M., Izzetoglu, K., Onaral, B.: The use of functional near-infrared spectroscopy in neuroergonomics. In: Neuroergonomics, pp. 17–25. Academic Press (2019)
21. Ayaz, H., Shewokis, P.A., Curtin, A., Izzetoglu, M., Izzetoglu, K., Onaral, B.: Using MazeSuite and functional near infrared spectroscopy to study learning in spatial navigation. J. Vis. Exp. **56**, 3443 (2011)
22. Curtin, A., Ayaz, H.: The age of neuroergonomics: towards ubiquitous and continuous measurement of brain function with fNIRS. Jpn. Psycholo. Res. **60**(4), 374–386 (2018)

Mental State Assessment

Towards Trustworthy Man-Machine Synchronization

Johan de Heer$^{(\boxtimes)}$ and Paul Porskamp

Thales Research and Technology, Gebouw N, Haaksbergerstraat 67,
7554 NB Hengelo, The Netherlands
{Johan.deHeer,Paul.Porskamp}@nl.thalesgroup.com

Abstract. Hyperscanning is a method by which multiple individuals can interact with one another while their brains are simultaneously scanned. Inter-personal brain activity synchronization has found to be indicative for levels of collaboration, trust, coordination and leadership among team members. Here it is suggested that systems that team with humans need to provoke specific responses as humans do to synchronize. Entities 'in sync' perform better. Therefore, understanding social interactions between humans in terms of synchronization may be the prerequisite for developing and enhancing true trustworthy man-machine teaming.

Keywords: Hyperscanning · Man-machine cooperation ·
Human behavior analytics

1 Introduction

We are at the beginning of a new technical (r)evolution towards systems with autonomous functions. Based on AI, machine learning and deep learning frameworks the technical and operational potential of these systems is considerable. In space, air, land, maritime, cyber domains autonomous functions are envisioned a key technology in operational theatres in the military domain [1]. To a large extent, the civil domain is actually driving the technology roadmap towards full autonomy and faces many challenges. A step-wise implementation expects to see changes in human interactions with systems that evolve with increasing autonomous functions. From human-in-the-loop to human-on-the-loop, and finally human-out-of-the-loop across a variety of tasks in several domains.

Man-machine harmonization and teaming is becoming a challenge in itself. From both a system and human factors perspective concepts such as mutual trust, cooperation, coordination and even leadership between man and machines will become more important than ever. Do human 'operators' trust systems that make decisions on their own, do we allocate specific tasks to systems because they may do a more efficient job, who leads, follows, coordinates subsequent activities and tasks? And vice versa, do systems trust humans regarding their (moral)judgment and decision-making capabilities, which may be characterized by various perceptual and cognitive biases [2]?

We suggest that psychological constructs such as trust, cooperation, coordination and leadership should be viewed from both the system and human perspective. For that

© Springer Nature Switzerland AG 2020
H. Ayaz (Ed.): AHFE 2019, AISC 953, pp. 113–118, 2020.
https://doi.org/10.1007/978-3-030-20473-0_12

we should define and implement a sense - think – act architecture that make the dynamics of these constructs transparent in real-time and in changing environments.

In our work in progress applied research, we first start by focusing on the human side. How do humans actually team up? Could we somehow measure the level of trust, cooperation, coordination and leadership between individuals that make up a team that is performing a task together. Could we show that if these construct levels increase then performance enhances as well? What happens if one or more team members take over some task aspects from other team members? A clear understanding how humans harmonize may be a prerequisite in finding pointers to implement trustworthy man-machine teaming that increase mutual performances. Also hypothesized by [3]; we expect as well that the performance of human-system teams is improved when a system teammate emulates the effective teamwork behaviors observed in human teams. The application of our research is in human-machine teaming and harmonization in high-intensity domains such as military tactical teams and aviation crews. Specifically, in these domains team dynamics have a significant impact on performance.

2   Technical Approach and Methods

One may examine and quantify concepts of trust, cooperation, coordination, and leadership in various ways. On the measurement side concepts like Quantified Self and Quantified Teams are emerging. QS and QT may include a variety wearable sensors that measure all kinds of physiological and behavioral data that is used to provide insight in the human 'state'. For example, data science and analytics in sports and health is gaining more and more influence. All relevant data around a sport is gathered and used for analyses to predict optimal performances and suggest optimal training programs. In cycling data science experts support to define optimal strategies for a race. Who leads who follows? How many times do we change runners? Similar examples can be found in skating, soccer, tennis, baseball, et cetera. Of course, the difficulty is in the modelling; relating multi-modal measurements to biological, psychological and behavioral constructs.

Likewise, in our domain of interest we could measure, examine and predict how human teams are teaming up and harmonizing their activities with technical systems. The approach we take is first to observe human teaming and during several empirical cycles add or replace humans by systems that may or may not increase in autonomous functions.

We have an interest to look at the potential of examining neural activities and relating those to performance metrics. Hyperscanning technology permits the study of the brain that underlie social interactions [4]. Hyperscanning is a method by which multiple individuals can interact with one another while their brains are simultaneously scanned. Several hyperscanning architectures have been employed using EEG, fMRI and fNIRS brain activity recordings. In this paper we will not focus on the brain recording technique that is actually used but instead focus on the temporal synchronization of the data series. To qualify and quantify inter-brain coupling for hyperscanning various analyzing techniques are used.

In Fig. 1 [taken from 5] a simplified schematic representation of the two main steps for – in this case - fNIRS-based hyperscanning applications is visualized. Data is recorded via a simultaneous measurement of optical imaging data from multiple participants in an experimental hyperscanning study. During the first step information on individual hemodynamic activities for each participant is obtained. Waveforms and activation maps are often used to represent brain data. In the second step of the analysis interpersonal synchronization and inter-brain coupling in hyperscanning is explored with e.g., correlation analysis, coherence analyses, Granger causality modelling.

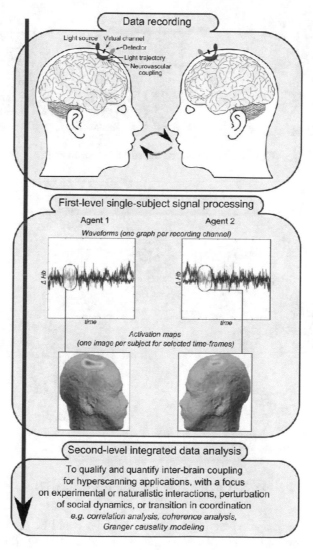

Fig. 1. Framework of the study design from [5]

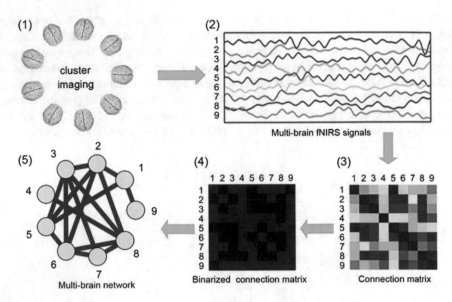

Fig. 2. Framework of the study design from [6]

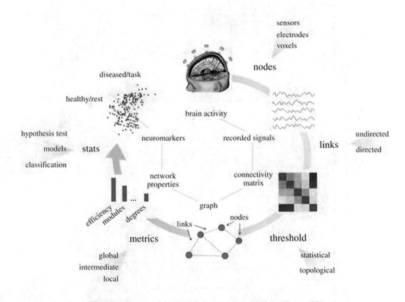

Fig. 3. Framework of the study design from [7]

In addition, as proposed by [6] a network analysis could be the next step in the pipeline (see flowchart in Fig. 2). Note that each step of the pipeline requires consideration to the selection of the best fitting method given theoretical and practical constraints, including the research hypothesis. In Fig. 2 brain activities from 9 participants were simultaneously collected using fNIRS (step 1). In step 2 the oxynated hemoglobin (HbO) signals from several region of interests (ROI) is averaged. In step 3 the network edge weights are obtained by calculating the Pearson correlation coefficients between the ROI-averaged time courses forming a 9×9 connection matrix. In step 4 the connection matrix is transformed into a binary network. Finally, the graph is visualized, and graph metrics can be used for further statistical analysis.

A similar more elaborated pipeline for modeling and analyzing functional brain networks was described by [7] and is depicted in Fig. 3. Specific brain areas resemble the network nodes. The links of the network are based on functional connections between the activity of brain nodes, and put in a connectivity matrix. Filtering is used to select the most relevant links. The topology of the resulting brain graph is quantified by different metrics. These graph metrics can be input to statistical analyses to look for significant differences between experimental conditions.

3   Some Research Findings

It has been found in hyperscanning studies that classification of human trust [8], cooperation, coordination and leadership is possible, in some cases even under high threat situations. For example, inter-brain activity coherence was found between participants that indicated better cooperation performance [9]. Looking at interpersonal neural synchronization (INS) of gamma band oscillations it was that found that INS was enhanced when people are under high threat; increased gamma interbrain synchrony is associated with higher coordination [10]. It has been shown that INS is significantly higher between leaders and followers than between followers and followers, suggesting that leaders emerge by synchronizing their brain activity with that of the followers [11]. And it has been found that entities 'in sync' perform better [12].

We intend to replicate these and other findings in more applied settings with appropriate tasks in our domain of interest.

4   Conclusion

Over the last decade the hyperscanning methodology showed to be a valuable paradigm to study social cognition. However, nearly all studies are performed in academic labs in controlled environments. Most of these studies is with dyads (2 persons mutually interacting). Notably, hyperscanning fNIRS allows for investigations to examine brain activity in natural settings thereby increasing its practical applicability. We have planned series of experimental studies to examine the applicability of the hyperscanning approach in the military domain in more natural settings. We concentrate on teams of 2, 4, 8 people in interacting with systems that are assumed to evolve in autonomous functions. In particular, we take command and control environments as use cases. This work in progress paper adopts the methodological framework depicted in Figs. 1, 2, and 3 to study team performance based on hyperscanning brain activity studies. We will focus on psychological constructs such as trust, cooperation, coordination and leadership. We will examine human to human interactions as well as team-system interactions. We will test our hypothesis that the performance of human-system teams is improved when a system teammate emulates the effective teamwork behaviors observed in human teams. Future research directs us towards adaptations to improve synchronization between the two entities; we expect this to go towards human and/or system enhancement technologies.

References

1. European Defence Agency: Remote Defence: Unmanned and autonomous systems take hold in military toolboxes. EDM European Defence Matters, Issue #16 (2018)
2. Kahneman, D.: Thinking, Fast and Slow. Macmillan, London (2011). ISBN 978-1-4299-6935-2
3. Shah, J., Kim, B., Nikolaidis, S.: Human-inspired techniques for human-machine team planning. In: AAAI Technical Report FS-12-04 Human Control of Bio-inspired Swarms (2012)
4. Hirsch, J.: The grand challenge to understand the brain: neuroimaging by functional near-infrared spectroscopy. Shimadzu J. 5(1) (2017). https://www.ssi.shimadzu.com/sites/ssi.shimadzu.com/files/Products/literature/Life_science/fnirs-article.pdf
5. Crivelli, D., Balconi, M.: Near-infrared spectroscopy applied to complex systems and human hyperscanning networking. Appl. Sci. 7, 922 (2017). https://doi.org/10.3390/app7090922
6. Duan, L., Dai, R., Xiao, X., Sun, P., Li, Z., Zhu, C.: Cluster imaging of multi-brain networks (CIMBN): a general framework for hyperscanning and modeling a group of interacting brains. Frontiers in Neuroscience, Technology Report. https://doi.org/10.3389/fnins.2015.00267. Accessed 28 July 2015
7. De Vico Fallani, F., Richiardi, J., Chavez, M., Achard, S.: Graph analysis of functional brain networks: practical issues in translational neuroscience. Phil. Trans. R. Soc. B369: 20130521. https://doi.org/10.1098/rstb.2013.0521. Accessed 1 Sept 2014
8. Akash, K., Hu, W., Jain, N., Reid, T.: A classification model for sensing human trust in machines using EEG and GSR. ACM Trans. Interact. Intell. Syst. 8(4), 27 (2018)
9. Cui, X., Bryant, D.M., Reiss, A.L.: NIRS-based hyperscanning reveals increased interpersonal coherence in superior frontal cortex during cooperation. Neuroimage 59(3), 2430–2437 (2012)
10. Mu, Y., Han, S., Gelfand, M.J.: The role of gamma interbrain synchrony in social coordination when humans face territorial threats. Soc. Cogn. Affect. Neurosci. 2017, 1614–1623 (2017)
11. Jiang, J., Chen, C., Dai, B., Shi, G., Ding, G., Liu, L., Lu, C.: Leader emergence through interpersonal neural synchronization. Proc. Natl. Acad. Sci. 112(14), 4274–4279 (2015)
12. Szymanski, C., Pesquita, A., Brennan, A., Perdikis, D., Enns, J., Brick, T., Muller, V., Lindenberger, U.: Teams on the same wavelength perform better: inter-brain phase synchronization constitutes a neural substrate for social facilitation. Neuroimage 152, 425–436 (2017)

Mental Health, Trust, and Robots: Towards Understanding How Mental Health Mediates Human-Automated System Trust and Reliance

Jordan R. Crawford$^{(\boxtimes)}$, Ella-Mae Hubbard, and Yee Mey Goh

Intelligent Automation, Loughborough University, Loughborough,
Leicestershire LE11 3GR, UK
{J.Crawford,E.Hubbard,Y.Goh}@Lboro.ac.uk

Abstract. This paper proposes a conceptual model derived through current ongoing research that incorporates the potential relationship that mental health may have on trust and reliance calibration in automated systems (AS). Understanding the variables involved in the human-AS interaction allows system designers to better achieve trust calibration and avoid AS misuse and disuse. However, most of the research area is saturated with understanding how external and internal (both to the human) short-term cognitive symptoms mediate this critical relationship. Therefore, the present paper extends human-AS trust literature with common mental disorders (CMDs) as outlined by the Adult Psychiatric Morbidity Survey (APMS) and incorporates them into existing models within the engineering and psychology subject areas to begin to understand what this relationship may look like. It is hoped that this paper will expand the scope of human factors (specifically human-AS trust) to include mental disorders.

Keywords: Mental health · Common mental disorders · Automated systems · Trust · Reliance

1 Introduction

As technology evolves, it becomes more a part of an individual's life. Whether they are needed to work alongside it at work, travel home in it or relax at home using it, people are increasingly required to co-exist with automated systems (AS) growing in their reach and capabilities. As humans assume an overseer role [1–3], the relationship between human and machine is becoming just as important as the interpersonal one always has been. This paper discusses mental health in this context in order to begin to understand its importance in ensuring the fruitfulness of the human-AS relationship.

In order for AS to be successful in its deployment, it is widely considered that it has to be trusted to be fully adopted by the end user(s), or it will not be used to its fullest potential [4]. Though in more detail, individuals must appropriately calibrate their trust to form the correct level of system reliance in order to avoid detrimental cost to the system. AS misuse and disuse, as introduced by Parasuraman and Riley [3] as excessive underreliance and overreliance respectively, have had prevalent examples in personal and non-personal navigation [5, 6] due to the increased access to satellite

© Springer Nature Switzerland AG 2020
H. Ayaz (Ed.): AHFE 2019, AISC 953, pp. 119–128, 2020.
https://doi.org/10.1007/978-3-030-20473-0_13

technology. Therefore, it is essential that researchers understand human-AS reliance decisions in as much detail that is possible in order to improve these reliance decisions. While past research has focused on cognitive decision-making factors in this relationship [7–9], it has begun to move towards more affective consideration [1, 10]. However, this paper proposes that a deeper look into human psychology is needed at the potential sources of undesirable cognitive and affective variation in this area. In this case, individual quality of mental health. This project incorporates mental health factors into conceptual human-AS trust and reliance models to better understand the varying and complex research in this area.

According to the Adult Psychiatric Morbidity Survey (APMS) [11] carried out in 2014 (conducted every seven years by the NHS), one in six (17%) of the working age population of Great Britain meet diagnostic criteria for a common mental disorder (CMD). The most common of these issues are anxiety, depression or a combination of the two [11], though there is an extensive spectrum of distinct diagnoses which result in different symptoms and require different approaches of care. While causing emotional distress and interference with day-to-day life, CMD's do not usually affect insight or cognition [11]. Due to its sizable portion of the working population (Fig. 1) resulting in the cumulative cost of CMD's to be larger than other psychiatric disorders, the present working paper presents interest in incorporating all CMD diagnoses deemed appropriate by the APMS [11].

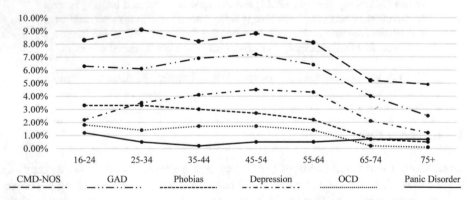

Fig. 1. APMS prevalence of CMD by age [12, Fig. 1b] shows greater pervasiveness in the working age (16–64). GAD, Phobias, Depressive Episode, OCD, and Panic Disorder are excluded from CMD-NOS.

APMS classifications, employed by the Mental Health Foundation [12], follow an amalgamated classification of the non-psychotic diagnoses and symptoms in the World Health Organization 10th International Classification of Disease (ICD-10) [13]. These include *generalised anxiety disorder* (GAD), *depression, phobias, obsessive-compulsive disorder* (OCD), *panic disorder*, and those *common mental disorders that are not otherwise specified* (CMD-NOS). It should be noted that 'CMD-NOS' is defined as individuals showing adequate symptom criteria to be considered a mental health disorder with 'mixed-anxiety and depression', but fell short of the criteria for any specific

CMD [11]. Therefore, it follows that individuals with this diagnosis therefore cannot be classed as having any other CMD, whereas the other five disorders (with the exception of panic disorder and phobias which have mutually exclusive diagnostic criteria [13]) can appear in more than one case.

2   The Model

The conceptual model proposed (Fig. 2) and the following explanations of the types of trust is developed from the model of trust and reliance in AS by Hoff and Bashir [7], in the review that considers all empirical research regarding human-AS trust and reliance. This model is similar to those proposed in other reviews and meta-analyses [8, 9] though is visually more effective than its counterparts due to its detailed inclusion of knowledge within a single intuitive model ideal for the purpose of this paper. The model considers trust in three parts, *dispositional trust*, *situational trust*, and *learned trust* [7].

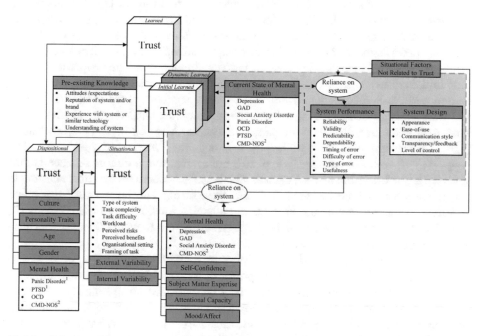

Fig. 2. Conceptual model developed from Hoff and Bashir [7] (While meeting criteria for its designated category, these factors do not hold exclusive symptoms to its designated category.). *Within* a single interaction (CMD-NOS is depicted in the proposed model in all types of trust since as defined as a catch-all mixed-anxiety and depression mental disorder [11], it is possible that it is both dispositional and situational whilst also influencing dynamic trust.), the shaded area depicts processes taking place and dashed arrows are elements that can change [7].

Dispositional trust is concerned with an individual's overall tendency to trust AS, independent of context or a specific system, as categorised by culture, age, gender, and personality traits. Dispositional trust factors are influenced by long-term tendencies arising from biology and the environment the individual is subjected to. As a result, these factors are less likely to change over time than those trust moderators of situational and learned trust. They can also alter trust formation in any situation, and therefore are not to be confused then with certain characteristics such as self-confidence, which is context dependent.

Situational trust concerns us with the context of the interaction between an individual and the AS. Figure 2 shows that situational trust involves both external environmental factors and the internal mental state of the individual. Therefore, different to personality traits in dispositional trust, internal variability fluctuates depending on their current mental state or their expertise.

Learned trust concerns us with attitudes that are gained from past experiences relevant to a specific AS. This factor of trust is directly influenced by the individuals pre-existing knowledge of the system and the AS's performance, perceptions of the latter influenced by system design. Note dynamic learned trust in Fig. 2, since during an individual's interaction with an AS, their trust and reliance strategy may dynamically fluctuate to reflect their reaction to the system's real-time performance.

This model suggests that dispositional trust, situational trust and initial learned trust form the initial reliance strategy that an individual enters a relationship with an AS. This initial reliance strategy, along with system design are the two sources that dictate system performance. System performance is then recognised by the individual in their real time dynamic learned trust, which, along with situational factors not related to trust, then develops reliance on the particular system. Also following Hoff and Bashir [7], Understanding that, as outlined by Lee and See *"trust guides—but does not completely determine—reliance"* [4, p. 51], moderators that influence reliance but do not have any relationship to trust, such as situational awareness [9] and time constraints [14] to name a couple are included. Such that these factors may have a mediating effect of reliance behavior, but do not have effect on human trust. For example, Endsley [9] considers situational awareness to be a moderator of the human-AS interaction performance but not of AS trust.

This working paper considers the influence of CMDs on human-AS trust and reliance under each of the different categories of human-AS trust, *dispositional, situational*, and *learned* as proposed in the model presented. More specifically, CMDs as outlined throughout this paper and as shown in Fig. 1 are considered to mediate trust and reliance within dispositional trust, situational trust (internal variability) and as a filter within the dynamic learned trust and reliance relationship. While initial learned trust may have an indirect influence on how mental health mediates human-AS trust and reliance behavior through dictating an individual's initial reliance strategy, it does not include specific mental health factors itself. This is because the initial learned trust node describes purely cognitive factors associated with a particular system [7]. This includes attitudes/expectations, knowledge of the system's reputation, experience with the system (or a similar technology), and understanding of the system, and CMDs do not usually affect insight or cognition [11].

2.1   Dispositional Trust

CMDs as identified by the APMS [11] can be organised into dispositional consideration consisting of panic disorder, PTSD, OCD and CMD-NOS each for their own unique reason. Note that the satisfaction criteria for dispositional factors as defined by Hoff and Bashir [7] is that they must both alter trust in every situation and be relatively stable over time. While these criteria when approaching factors such as age and gender are simple to decide upon, when applying them to mental conditions they become somewhat ambiguous. Therefore, these can be defined as possessing symptoms that remain consistent and are non-fluctuating from the point of the individual's conception of CMD of which may also alter trust in every situation. The ICD-10 [13] is utilised to define the following diagnosis criteria used.

Panic disorder diagnosis requires the inclusion of persistent worry in anticipation of an anxiety attack in which is present in all aspects of an individual's life. This may also include a consistent change in behaviour after conception of the disorder in comparison to their behaviour beforehand. However, while this qualifies panic disorder as a dispositional factor, a key symptom of panic disorder is panic attacks. Panic attacks occur irregularly and unexpectedly, sometimes through environmental triggers but often what appears to be random. The former criteria however are the distinction between an individual who suffers with panic disorder rather than one in which suffers only with panic attacks, regardless of the quantity or frequency. Recommendations for future human-AS trust and reliance research include testing the unique symptoms of panic disorder, namely the effect of consistent negative anticipation.

Post-traumatic stress disorder (PTSD), which may last up to an individual's entire life from its conception, has a diagnosis that requires the inclusion of persistent subjection to two or more of the following: difficulty falling asleep, emotional detachment, anhedonia, irritability or outbursts of anger, difficulty in concentrating, hyper-vigilance, or an exaggerated startle response. As a result, these symptoms satisfy the criteria to be considered dispositional. However, similar to panic disorder, PTSD is popularly categorised as a reaction to intrusive flashbacks, memories, or dreams regarding an experienced distressing event, which is inconsistent. Also, PTSD is sometimes categorised by its subject's avoidance of specific scenarios, which therefore does not satisfy dispositional criteria. Therefore, an analysis of the individual's personal experience with their PTSD is essential to understand its impact with AS on a case by case basis. For example, interesting future research may analyse the specific effect of PTSD as a result of a distressing experience with an AS, or otherwise test the dispositional symptoms of PTSD the subject possesses. Within general psychology literature, studies have found that those diagnosed with PTSD have an increased and persistent tendency to distrust, even towards those that are close to them [15]. Therefore, with research showing that human-AS interactions are highly comparable to those which are interpersonal [16], future directions in this area may identify whether this holds true with PTSD.

Obsessive-compulsive disorder (OCD) is characterised as a combination of recurrent and persistent obsessions (thoughts, ideas, or images) and compulsions (acts). It is an essential part of OCD that the individual understands the symptoms are excessive or unreasonable, they are unpleasant, and they are unsuccessfully resisted. As a result, carrying out their compulsions are not pleasurable but are followed with temporary relief of the tension or anxiety that drove the compulsion. OCD therefore causes a large amount of distress and interferes consistently with individual and social life, usually though distraction and/or wasting time. Therefore, future research may benefit from testing these symptoms as a part of OCD. This is because, for example, while distractions have been identified to have a mediating effect in the trust & reliance relationship in some cases [17, 18], they did not in an automated collision warning system task [19] of which the reason for this has not been definitively explained. However, OCD distractions are more extreme and unavoidable for those diagnosed with the disorder [13] to even what would appear to be minor problems to someone that does not suffer from it. Thus, outlining the potential for interesting comparative research in this area.

2.2 Situational Trust

CMDs as identified by the APMS [11] can be organised into situational consideration (internal to the individual) consisting of Depression, GAD, Social Anxiety Disorder and CMD-NOS. Internally variable traits can be seen as an extension of dispositional traits, in that while individuals have more enduring traits that arise as dispositional moderators, individuals also exhibit more impermanent traits that vary depending on any number of contextual, situational or time-related reasons. Identical to dispositional trust, The ICD-10 [13] is utilised to define the following diagnosis criteria used.

Depression is one of two core mental disorder sources, the other being anxiety [11], that may fluctuate in any one person from day-to-day. The other CMDs defined are on the spectrum of these two and as a result suffer from either depressive or anxiety symptoms. It is important to note that while individuals may have both depression and another CMD simultaneously, depression in this case is deliberately separated by the APMS to the other CMDs and considered its own disorder if not better explained by another CMD [13]. At this point, there has been research to suggest that depression is negatively correlated with interpersonal trust [20]. Therefore, if interpersonal and human-AS interactions are as similar as research suggests that they are [16], work is needed to identify if depression in this case holds consistent.

Generalised anxiety disorder (GAD) is separated from anxiety as a symptom and is diagnosed through prolonged feelings of tension, feelings of apprehension and worry about every-day scenarios. Common symptoms include a heightened startle response, difficulty concentrating, intense worry, persistent irritability, and difficulty sleeping. However, GAD is unstable and therefore is not hypothesised as a consistent moderator of human-AS trust, even for the same individual on different days. It does however, as its name suggests, permeate generally through all environmental circumstances. There has been some relevant research tackling some of the symptoms associated with GAD, such as lack of sleep [21]. However, considering GAD is the most prevalent CMD (particularly in the working age range as shown in Fig. 1), research is needed to assess the mental disorder in its entirety rather than its symptoms in isolation in order to understand the effects of them in combination. GAD symptoms such as feelings of apprehension, heightened startle response, worry, and irritability are all topics which may have an influential role in reactions and collaboration with imperfect AS.

Social anxiety disorder, or *social phobias* is predominantly the fear of scrutiny in comparatively smaller groups to large crowds (which distinguishes it from agoraphobia) and are usually associated with low self-esteem and a fear of criticism. This may result in the disorder being discrete, in that it involves the fear and avoidance of specific social situations unique to the individual subject or diffuse into almost all social situations outside of those which are already familiar. Aside from actively avoiding and not participating in certain social activities and interaction (including eye-contact in some cases), social anxiety disorder also results in symptoms such as hand tremors, nausea, or urgency of micturition. Human-AS interaction literature would therefore do well to assess social anxiety disorder and its common theme of avoidance of interaction and whether this applies to human-AS interaction, specifically if the symptoms influence trusting intentions and reliance behaviour. Human-AS collaboration (as opposed to human-human collaboration) may have many benefits for those that suffer from social anxiety due to the potential of creating AS with an absence of judgement and social pressure, such as eye contact and conversation. Also, an examination of discrete case by case examples versus individuals that have a diffused diagnosis would be interesting to understand the differences between the two in the context of human-AS trust and reliance.

2.3 Dynamic Learned

Learned trust is trust that is relative to characteristics of the AS drawn from an operator's evaluations of a system drawn from past experience or the current interaction [7]. Further, dynamic learned trust, being one of the interests of this paper, considers trust which changes over the course of an individual's interaction with an AS as a reaction to its real-time perceived performance [7]. It therefore identifies factors that influence how mental health may impact trust and reliance beliefs, attitudes, intentions, and behaviours [4, 22, 23]. The hypothesised dynamic learned trust relationship may be similar to Tomlinson and Mayer's model of trust repair [24], which incorporates Weiner's [25] attribution theory and Mayer et al.'s [26] feedback loop, where individuals evaluate whether their experience was positive or negative, influencing the way the individual feels, their expectations, and their behaviour going forward.

In order to avoid AS misuse and disuse, an equilibrium of reliance settling in the middle of these is the aim of any system design and implementation [4]. The conceptual model (Fig. 1) presented is effective at describing the importance of appropriate reliance calibration. This is because it can be considered that an individual's reliance strategy to be a function of their initial reliance strategy, generated from their initial trust, and their actual reliance on the AS, generated from their dynamic learned trust [7]. This identifies a key caveat to real time reliance in the form of misuse and disuse. In turn of system performance determining an individual's dynamic learned trust and therefore reliance strategy, this resulting reliance strategy may have a real time impact on perceived or even actual system performance caused by a resulting misuse or disuse [27]. This may logically result in the individual distrusting an AS more due to its poor performance, leading to the individual using the system less, resulting in the human-AS relationship being compromised and thus further poor performance due to the decreasing quality of collaboration. As a result, the interaction runs the risk of perpetuating and amplifying poor or excessive trust and reliance, highlighting the need to calibrate trust and reliance first time.

Further research will investigate the relationship between system performance and dynamic learned trust, and dynamic learned trust and reliance on the system respectively passing through a 'filter' of the individual's current state of mental health in the present interaction. This will follow the suggestion of models such as that proposed by Tomlinson and Mayer [24] in general trust psychology which identifies that positive and negative outcomes form affective emotional reactions, which in turn generate their response and future reliance intentions. Outcome based dynamic trust models have since appeared in human-AS trust literature [8, 9] but fail to include emotional cognitive processes. Thus Fig. 3 shown is a conceptual model of this relationship. This paper proposes that since research is starting to find that affective factors mediate trust and reliance throughout the human-AS interaction [1, 10], mental health factors that have more complex and extreme symptoms than solely affect (such as positive/negative mood) have an mediating effect in their own right. While CMDs do not usually affect insight and cognition [11], mental health is proposed to have a reactive or distractive mediation to AS performance.

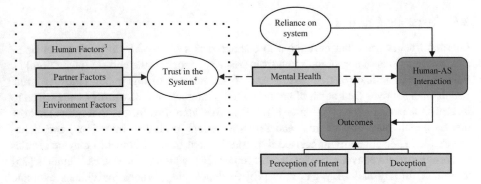

Fig. 3. A conceptual organization of trust influences highlighting trust development (adapted from [8, Fig. 1]). The dotted box represents the trust development part of the process and the dashed line represents trust calibration over time (Includes propensity to trust. Represents trust at a specific point in the human-AS interaction.).

3   Conclusion

If mental health mediates the human-AS trust relationship, current approaches to holistic system design are incomplete. Therefore, absence of consideration of such is likely to have a detrimental system cost due to the loss of calibration in trust and reliance, resulting in either misuse or disuse of AS [3]. During collaborative tasks of any nature, this relationship is then undermined. Further, with the large presence of CMDs within the working population (Fig. 1), the absence of consideration is also likely to have a high cumulative cost above all other mental health conditions [11], highlighting the need to prioritise the factors proposed within this paper. Following cognitive and affective literature, understanding mental health moderators brings us closer to understanding how humans are approaching the robotic future, and thus providing this future with priorities to ensure human flourishing.

The proposed model is considered conceptual, therefore research into these indicators is either rare or non-existent at the time of writing. As a result, some may have little effect. Contrastingly, others may have critical influential effects to individual trust and reliance (and therefore human-AS collaborative performance) that have otherwise been greatly neglected.

This conceptual model is a part of an ongoing research project of which is currently finalizing empirical evidence into GAD and Depression, though other factors described are largely unexplored.

References

1. Merritt, S.M.: Affective processes in human-automation interactions. Hum. Factors **53**, 356–370 (2011)
2. Lee, J.D., Moray, N.: Trust, self-confidence, and operators' adaptation to automation. Int. J. Hum. Comput. Stud. **40**, 153–184 (1994)
3. Parasuraman, R., Riley, V.: Humans and automation: use, misuse, disuse. Abuse. Hum. Factors **39**, 230–253 (1997)
4. Lee, J.D., See, K.A.: Trust in automation: designing for appropriate reliance. Hum. Factors **46**, 50–80 (2004)
5. Speake, J., Axon, S.: "I never use 'maps' anymore": engaging with sat nav technologies and the implications for cartographic literacy and spatial awareness. Cartogr. J. **49**, 326–335 (2012)
6. The Nautical Institute: Official Report No. 7042OR: Grounding of Cruise Ship (2002)
7. Hoff, K.A., Bashir, M.: Trust in automation: integrating empirical evidence on factors that influence trust. Hum. Factors **57**, 407–434 (2015)
8. Schaefer, K.E., Chen, J.Y.C., Szalma, J.L., Hancock, P.A.: A meta-analysis of factors influencing the development of trust in automation: implications for understanding autonomy in future systems. Hum. Factors **58**, 377–400 (2016)
9. Endsley, M.R.: From here to autonomy: lessons learned from human-automation research. Hum. Factors **59**, 5–27 (2017)
10. Stokes, C.K., Lyons, J.B., Littlejohn, K., Natarian, J., Case, E., Speranza, N.: Accounting for the human in cyberspace: effects of mood on trust in automation. In: Proceedings of the IEEE International Symposium on Collaborative Technologies and Systems, pp. 180—187. IEEE (2010)
11. McManus, S., Bebbington, P., Jenkins, R.: Mental health and wellbeing in England: adult psychiatric morbidity survey 2014. Technical report, NHS Digital (2014)
12. Mental Health Foundation: Fundamental facts about mental health. Technical report, Mental Health Foundation (2016)
13. World Health Organisation: The ICD-10 classification of mental and behavioural disorders: clinical descriptions and diagnostic guidelines. Technical report, World Health Organisation (1992)
14. Rice, S., Keller, D.: Automation reliance under time pressure. Cogn. Technol. **14**, 36–44 (2009)

15. Cias, C.M., Young, R., Barreira, P.: Loss of trust: correlates of the comorbidity of Ptsd and severe mental illness. J. Pers. Interpers. Loss. **5**, 103–123 (2009)
16. Madhavan, P., Wiegmann, D.A.: Similarities and differences between human-human and human-automation trust: an integrative review. J. Theor. Issues Ergon. Sci. **8**, 277–301 (2007)
17. Wickens, C.D.: Multiple resources and performance prediction. J. Theor. Issues Ergon. Sci. **3**, 159–177 (2002)
18. Phillips, R.R., Madhavan, P.: The effect of distractor modality and processing code on human-automation interaction. Cogn. Technol. Work **13**, 233–244 (2011)
19. Lees, M.N., Lee, J.D.: The influence of distraction and driving context on driver response to imperfect collision warning systems. Ergonomics **50**, 1264–1286 (2007)
20. Schneider, I.K., Konjin, E.A., Righetti, F., Rusbult, C.E.: A healthy dose of trust: the relationship between interpersonal trust and health. Pers. Relatsh. **18**, 668–676 (2011)
21. Reichenbach, J., Onnasch, L., Manzey, D.: Human performance consequences of automated decision aids in states of sleep loss. Hum. Factors **53**, 717–728 (2011)
22. Ajzen, I., Fishbein, M.: Understanding Attitudes and Predicting Social Behaviour. Prentice Hall, Upper Saddle River (1980)
23. Fishbein, M., Ajzen, I.: Belief, Attitude, Intention and Behavior. Addison Wesley, Reading (1975)
24. Tomlinson, E.C., Mayer, R.C.: The role of causal attribution dimensions in trust repair. Acad. Manag. Rev. **34**, 85–104 (2009)
25. Weiner, B.: An attributional theory of achievement motivation and emotion. Psychol. Rev. **92**, 548–573 (1985)
26. Mayer, R.C., Davis, J.H., Schoorman, D.F.: An integrative model of organizational trust. Acad. Manag. Rev. **20**, 709–734 (1995)
27. Manzey, D., Bahner, J.E., Hueper, A.: Misuse of automated aids in process control: complacency, automation bias and possible training interventions. In: Proceedings of the Human Factors and Ergonomics Society 50th Annual Meeting, pp. 220—224 (2006)

Eye Tracking-Based Workload and Performance Assessment for Skill Acquisition

Jesse Mark[1](✉), Adrian Curtin[1], Amanda Kraft[2], Trevor Sands[2],
William D. Casebeer[2], Matthias Ziegler[2], and Hasan Ayaz[1,3,4,5]

[1] School of Biomedical Engineering, Science, and Health Systems,
Drexel University, Philadelphia, PA, USA
{jam673,hasan.ayaz}@drexel.edu
[2] Advanced Technology Laboratories, Lockheed Martin, Cherry Hill, NJ, USA
[3] Department of Family and Community Health, University of Pennsylvania,
Philadelphia, PA, USA
[4] Center for Injury Research and Prevention,
Children's Hospital of Philadelphia, Philadelphia, PA, USA
[5] Drexel Business Solutions Institute,
Drexel University, Philadelphia, PA, USA

Abstract. The result of training to improve in a given skill is most often demonstrated by an increase in the relevant performance measures. However, a complementary and at times more informative measure is the mental workload imposed on the performer when doing the task. While a number of varied methods exist for measuring workload, we have chosen to explore physiological and neurological correlates for their low amount of impact and interference on subjects during an experiment. In this study, participants trained on a six-task cognitive battery over four weeks while being simultaneously recorded with remote eye tracking and a host of other neurophysiological instruments. In this preliminary analysis, we found that measures of saccades, fixations, and pupil diameters significantly correlated with task performance over time and at different difficulties, indicating the validity of our task battery as well as the specificity of workload-related eye tracking measures.

Keywords: Cognitive workload · Eye tracking · Multimodal · Task battery

1 Introduction

In all areas of life, but most importantly in high stress, high risk occupations such as performing surgery, piloting aircraft, controlling air traffic, or operating heavy machinery, it is critical to have the skills and mastery necessary to succeed and avoid human-derived error. This is the reason why such professions require years of specialized training, including tests and checks to confirm competency along the way. However, what these assessments often miss is the level of mental effort being invested in achieving passing marks [1, 2]. Moreover, at each stage of learning, as well as during actual performance in the field, cognitive workload should be neither too high nor too

© Springer Nature Switzerland AG 2020
H. Ayaz (Ed.): AHFE 2019, AISC 953, pp. 129–141, 2020.
https://doi.org/10.1007/978-3-030-20473-0_14

low in order to minimize potential human error [3]. When a person is not challenged enough, they are more susceptible to distraction or boredom, which can lead to careless mistakes [4, 5]. Likewise, when the task requirements are too difficult, performers are overloaded and cannot keep up with the necessary actions [6]. Therefore, in addition to judging a person's abilities solely by external output, it is more valuable to include measures of internal workload.

Cognitive workload as it is commonly known is how much effort a person must put into completing a task to satisfaction. More specifically, it is the interaction between a person's internal capabilities and a task's external difficulty [7]. This may be measured via surveys such as the NASA-TLX that inquire about different factors related to workload (such as frustration and time pressure) [8], but rely on an individual's subjectivity, which is both a benefit and a demerit. There are also indirect measures, which are based on cognitive load theory [9] and the limited mental resources available at any given time [10, 11]. By presenting a secondary task unrelated to the main goal that is simple in nature but effortful to achieve, the amount of cognitive leeway with respect to how much is already engaged can be calculated [12, 13]. But again, injecting an unrelated requirement on an already burdened performer interferes with main task performance. This leads to our use of physiological and neuroimaging modalities of workload measurement.

Eye tracking is a useful and highly neuroergonomic modality for gathering both mental workload and relevant practical attentional information. It is non-invasive and easy to set up below a computer screen, allowing for consistent data collection. It does not impose any physical burdens on users, and calibration is fast and simple. A variety of measures have been described in the literature that elucidate the efficiency of visual search and relate to mental workload, including fixation count, fixation duration, fixation rate, the fixation to saccades ratio, average saccade distance and velocity, peak saccade velocity, number of long fixations, and average pupil diameter [14–16]. Depending on task and screen layout, the meaning and relevance of each measure may change, so it is important to consider the implications of each when used for analysis purposes.

The neural correlates of mental workload have been well studied, including in neuroergonomic contexts allowing for more realistic experiments that are applicable outside of a laboratory setting [1, 17, 34]. Out of the available modalities, electroencephalography (EEG) and functional near-infrared spectroscopy (fNIRS) paint a complementary and comprehensive picture of cortical brain activity for multimodal studies [18, 19]. EEG records fast brain signals of the electrophysiological activity of neurons, but lacks high spatial resolution due to the dispersion of those signals. In contrast, fNIRS measures hemodynamics by the difference in hemoglobin absorption spectrums, which can be applied to workload with the neural efficiency hypothesis that states higher cognitive activity requires more energy and oxygen [20]. Together, EEG and fNIRS provide the backbone for a multimodal assessment of task-induced workload.

A plethora of other physiological measures add to our available information to draw on. In this study, we included electrocardiography (ECG) and photoplethysmography (PPG) to record the heart and pulse, as well as electrooculography (EOG). Heart rate variability has been shown to correlate with differing amounts of workload [21], and a combination of eye tracking factors have also proven useful in classifying mental workload [14]. By incorporating all of the above modalities of brain and body measurements, we can open the door to a more holistic understanding of the brain and body at work [35, 36].

Using these techniques, we sought to examine six distinct cognitive domains with a task battery based on classical experimental paradigms. Each of these was designed with two levels of difficulty intended to elicit high and low workload states, with one exception. The chosen tasks included modifications of the following: A variant of the go-no go task, where inhibited responses to stimuli elicit higher workload [22, 23]; The conjunctive continuous performance task (CCPT), where sustained attention over time induces mental load [24]; A spatial N-back, where higher numbers of objects to remember is more difficult [25, 26]; A situation awareness task where subject must notice discrepancies in prerendered videos with either a low or high amount of objects of interest [27, 28]; The trail making test, where either a simple numeric or mixed alphanumeric series of targets must be located in order [29, 30]; And a modified synthesis of the balloon analogue risk task and the "just one more" test of optimizing risk and reward under differing conditions [31, 32]. All of the above are well studied in the literature, but have not been analyzed under such thorough conditions as we have used.

We have collected a rich database of over five different mental workload-correlated measures of neurophysiological activity for a six-task battery. In this paper, we present our initial findings from the performance results as well as eye tracking-based measures.

2 Methods

2.1 Participants

Twenty-three participants between the ages of 18 to 48 (7 males, mean age 23 years) volunteered for the study. All subjects confirmed via survey given in person that they met the eligibility requirements of being right-handed with vision correctable to 20/20, did not have a history of brain injury or psychological disorder, were not on medication affecting brain activity, and were United States citizens or permanent residents. Prior to the study all participants signed consent forms approved by the Institutional Review Board of Drexel University.

2.2 Experimental Procedure

The experiment was performed over four sessions, once a week for four weeks, each lasting between 60–90 min. Subjects were seated upright in front of a computer with a standard mouse and keyboard one meter away from the monitor. They were fitted with an fNIR Devices Model 1200 headband over the forehead, a Cognionics HD-72 dry electrode cap, and a Cognionics extension providing sticky electrodes for the ECG (3 electrodes), EOG (4 electrodes), and PPG ear clip. Eye tracking was calibrated using the Smart Eye Aurora system recording gaze location and pupil diameter. The six tasks described below were coded using the Python extension PsychoPy. Tasks were preceded by instructions and practice trials for each difficulty condition where subjects could familiarize themselves with the procedure and ask clarifying questions. Each task was designed to take 5–8 min to complete, and over the course of the full experiment was performed three times total in pseudorandom balanced order (Fig. 1).

Day 1	TMT	SA	WM	IC		
Day 2	WM	IC	VIG	RK	N/A	
Day 3	RK	VIG	TMT	SA		
Day 4	IC	RK	SA	WM	TMT	VIG

Fig. 1. Sample pseudorandom task order balanced within and between sessions. Task order was also balanced over all subjects to ensure the most even distribution possible.

Working Memory. A modification of the spatial N-back. A blank green circular radar was displayed on screen for 1 s, which was then followed by 1 s of triangular targets. A static noise screen covered the radar for 3 s as a retention period, and then a blank radar was displayed for up to 5 s, during which subjects had to click as close as possible to the locations of the targets previously displayed. The low and high work-load conditions were differentiated by either 5 or 7 targets total. Four consecutive trials of one condition comprised a block, and three blocks of each condition was presented in random order. Performance measures include distance offset error and response time.

Vigilance. A modification of the conjunctive continuous performance task. During a single block lasting five minutes, images of various shapes (flag, circle, square, triangle) and patterns (stripes, dots, crosses, grids, or solid) flashed for 100 ms one at a time with an inter-stimuli interval between 500–1000 ms. About 30% of the total stimuli were the target image (striped flag), for which participants had to click the enter key as quickly as possible. All other images did not require an active response. Performance measures included accuracy, error rates, and response time.

Risk Assessment. The goal of this task was to collect as many resources as possible without "crashing" a virtual search vehicle. Subjects were given three trials of each condition in a row per block, three blocks each condition for a total of six. In each trial, a screen displaying a map with six highlighted areas of interest was shown. Subjects had five seconds to click each area to search it, revealing whether it contained a resource or caused a crash, losing the current resources and ending that trial. At any point the subjects could click the space key to end searching for the current trial, keeping all collected resources. In the low risk condition, each area had a low chance of crashing and provided one resource. In the high risk condition, each area clicked returned double resources (1, 2, 4, 8, 16, 32) but had a higher chance of crashing. In addition, taking more than 5 s to decide caused a crash. Collected resources carried over between trials and blocks. Performance measures included total clicks, clicks without a crash, crashes, earnings, and response time.

Shifting Attention. A digitized trail making test with three blocks each of two difficulty conditions for a total of six. Subjects had 30 s per block to click 24 circles in either numeric ascending order (easy) or alphanumeric alternating order i.e. 1-A-2-B (hard). Targets were randomly located around the screen; the first and last targets in order were colored orange, and all others were colored white. Correct clicks turned the target gray and incorrect clicks turned it red until the subject went back and rectified the error. Performance measures included number correct total, number of errors, reaction time, and velocity of mouse movement.

Situation Awareness. Subjects viewed 30 s prerecorded videos of a top-down aircraft mission simulator of either one plane (easy condition) or three planes (hard condition) flying various paths. On the right side of the screen were vehicle dashboards for each plane indicating their airspeed, fuel levels, and heading. The goal of this task was to determine which, if any, of these dashboard levels did not accurately match the actual condition of the planes. Each plane could either have zero or one discrepancies. After each video, subjects used the keyboard to indicate which, if any, planes had discrepancies, and which measure (speed, fuel, heading) was wrong. Performance measures were accuracy and response time.

Inhibitory Control. Modified go-no go task with "go" condition called "ignore" and "stop" condition called "inhibit". Each condition was repeated for three blocks for a total of six per session. In both conditions, a series of stimuli were presented requiring subjects to either click enter or to refrain from clicking. Each stimulus began with a plus symbol lasting for 500 ms, indicating the subject should try to click. In the easy ignore condition, half of these lasted for only 100 ms, immediately changing to 400 ms of a flag symbol, which the subjects should ignore and press enter as if it did not appear. In the hard inhibit condition, instead of flags, a skull appeared in half of the stimuli, indicating the subject should stop their clicking. Subjects were told to always try to click enter upon confirmation of the plus symbol and inhibit their response for skulls, not to "wait and see" if it changed. The inter-stimuli interval varied between 750 ms–1.25 s. Performance measures included accuracy, false positive rate, false negative rate, and response time.

2.3 Metrics

Performance Measures. Each task provided some measure of accuracy, whether in the form of correct answers, amount of crashes or incorrect clicks, or total amount finished within the allotted time. Other measures such as mouse movement velocity were included for relevant tasks. Certain task responses were then converted into more meaningful metrics such as sensitivity index (d').

Eye Tracking. The Smart Eye Aurora remote eye tracking system recorded eye gaze and pupil diameter at 60 Hz using OGAMA software (open gaze and mouse analyzer©). Raw data was then post-processed into blinks, gaze path, saccades, and fixations based on the methods described in [14] and converted into millimeters based on screen size and pixel density. The following three processing methods were used [14]:

The *pupil diameter* of both eyes over the time span of a task block was averaged to create a single data point per block. Higher pupil diameter correlates with higher workload, and is measured in millimeters. The *peak saccade velocity*, calculated based on screen size, resolution, and distance from subject was measured in millimeters per second, and was limited to speeds between 524–6,108 mm/s (higher than the maximum speed counts as a blink). Peak velocity decreases as workload increases. Finally, the *frequency of long fixations* was calculated as the total number of sequences of gaze positions separated by saccades or blinks where the distance between subsequent points was less than 8.73 mm for at least 500 ms (for data recorded at 60 Hz). The number of long fixations increases with workload. Also considered but found to be less significant was mean and median fixation duration, rate of fixations, average saccade distance, and the fixation to saccade ratio. Due to some signal interference from other recording modalities, not every block for every subject could be used in data analysis. Therefore, eye tracking measures were statistically analyzed using repeated measures linear mixed models using Bonferroni correction with alpha 0.05. AIC was minimized to find the best model of fixed effects.

3   Results

Behavioral performance data was analyzed in conjunction with related eye tracking-based cognitive workload measures. All data was statistically analyzed using repeated measures linear mixed models using as described above.

Working Memory. Position error, or average distance between response and actual locations of targets, was higher in the hard difficulty than easy ($F_{1,375}$ = 125.6, p < 0.001), but did not change between sessions. There is also a consistent but non-significant decrease in pupil diameter-indicated workload over sessions for both conditions ($F_{2,6558}$ = 1.1, p > 0.1) (Fig. 2).

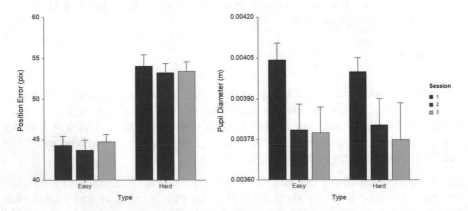

Fig. 2. Spatial N-back working memory performance (left) and eye tracking (right) indicate larger error in the more difficult condition and a consistent decrease in average pupil diameter over time. Error bars are standard error (SE).

Vigilance. The mean accuracy, the combined ratio of true positive clicks for targets and true negative non-clicks for non-targets, increases slightly before decreasing in session 1, and overall decreases over time for sessions 2 and 3. There is a significant difference in accuracy per block overall ($F_{7,372}$ = 2.1, P < 0.05). Pupil diameter overall decreases per session. In each session it slowly rises over time, but the starting diameter begins smaller each session (Fig. 3).

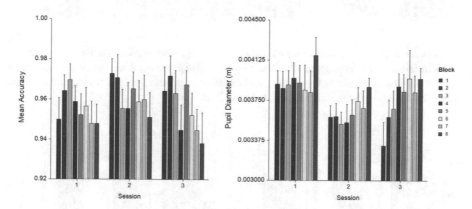

Fig. 3. Conjunctive continuous performance task for vigilance results in a decrease of accuracy (left, proportion of correct responses) and an increase in pupil diameter (right) over time for each session. Error bars are standard error (SE).

Risk Assessment. The average earnings per block for the easy condition raised over time while it fluctuated for the hard condition, as well as having more variability in the results due to the doubling nature of the earnings in the hard condition (by type, $F_{1,388}$ = 1.7, p > 0.1). The number of crashed trials steadily decreased for the easy condition and remained higher and more consistent for hard, and there was a significant difference between conditions ($F_{1,402}$ = 60, p < 0.001). There was also a slightly significant difference in sessions ($F_{2,402}$ = 2.4, P < 0.1). Peak saccade velocity, which inversely correlates with workload, increased over time for easy and decreased for hard, with a slightly significant difference in session by task interaction ($F_{2,185}$ = 2.6, p = 0.07) (Fig. 4).

Shifting Attention. The maximum correct clicks for the easy condition had a ceiling effect at 24, but steadily increased over time for the hard condition. Significant differences were found between session ($F_{2,386}$ = 6.3, P < 0.01), condition type ($F_{1,386}$ = 57.5, p < 0.001), and session by type interaction ($F_{2,386}$ = 3.6, p = 0.03). Mouse velocity increased per session for both conditions, but remained higher in the easy condition. Differences were found between session ($F_{2,386}$ = 44, p < 0.001), condition ($F_{1,386}$ = 364, p < 0.001), and interaction ($F_{2,386}$ = 4.4, p = 0.01). The frequency of long fixations dropped on session 3 for easy but was equal and consistent for hard, and low but non-significant differences were found between types ($F_{1,142}$ = 2.1, p = 0.15) (Fig. 5).

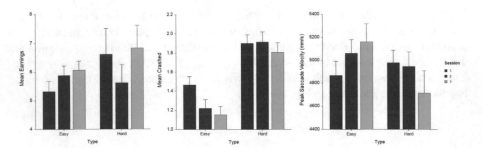

Fig. 4. Risk assessment task shows an increase of earnings over sessions for the easy condition, but more inconsistent earnings for the hard condition (left), a decreasing number of crashes for easy but stable amount for hard (middle), and an increase in peak saccade velocity for easy and decreasing for hard (right). Error bars are standard error (SE).

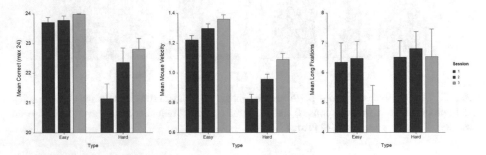

Fig. 5. The trail making test for shifting attention shows a ceiling effect for performance in the easy condition but improvements for hard (left), a steady increase in mouse velocity for both conditions, but overall higher velocity for easy (middle), and slightly decreasing long fixation frequency for easy and consistent for hard (right). Error bars are standard error (SE).

Situation Awareness. The easy condition shows fluctuating but overall higher accuracy of answering the questions related to each video, whereas the hard condition shows lower but steadily increasing accuracy, with a significant difference found between condition ($F_{1,208} = 5.1$, $p = 0.02$). Peak saccade velocity increases steadily over sessions for both conditions indicating decreasing workload, and a significant difference was found between sessions ($F_{2,141} = 2.8$, $p = 0.06$) (Fig. 6).

Inhibitory Control. The sensitivity index, or signal to noise ratio, increases over sessions for the easy condition but remains low and constant for hard, with a significant difference found between difficulties ($F_{1,386} = 10.7$, $p < 0.01$). Mean response time decreases slightly over time for both conditions, but is overall lower in hard (by sessions, $F_{2,383} = 4.3$, $p = 0.01$; by condition, $F_{1,383} = 160$, $p < 0.001$). There is a subjective but non-significant change in the condition by session interaction suggesting that the ignore condition is more difficult at the start due to unfamiliarity, but is quickly outpaced by the difficulty of the inhibit condition as skill increases (Fig. 7).

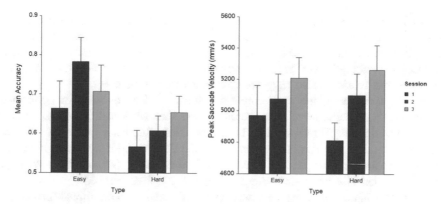

Fig. 6. Video monitoring for situation awareness shows varying question and answer accuracy for easy and improving but overall lower scores for hard (left) and a steeper slope of increasing peak saccade velocity over time for hard (right). Error bars are standard error (SE).

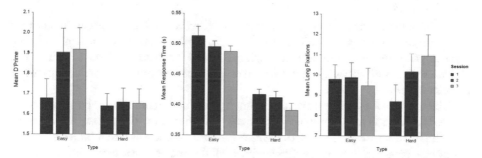

Fig. 7. Modified go/stop task for inhibitory control shows an increasing sensitivity index (d') for the easy condition but low and consistent for hard (left), slightly decreasing response times for both but overall lower for hard (middle), and an increasing ratio of hard to easy long fixation frequencies over time (right). Error bars are standard error (SE).

4 Discussion

In this study, we used a comprehensive multimodal neurophysiological recording setup including fNIRS, EEG, ECG, EOG, PPG, and eye tracking while subjects performed a task battery of six tasks targeting distinct cognitive domains during four sessions over the course of one month. For this paper, we have analyzed the performance data for all tasks in conjunction with three eye tracking-related measures of mental workload: pupil diameter, peak saccade velocity, and frequency of long fixations.

In the working memory task, there was a clear difference in performance between the easy and hard conditions, but it did not change over time. Both difficulties displayed a decrease in workload by pupil diameter measures, showing that subjects required less mental resources to attain the same results. This is strong evidence of the benefits of physiological workload measurements as a correlate of skill acquisition.

For the vigilance task, which did not have separate conditions but increased in effective difficulty as focus and concentration dropped, performance went down as pupil diameter increased for every session. Thus, even as skill increases, we can see an increase in workload over time. Interestingly, we also see a decrease in starting workload over sessions, also suggesting an increase in expertise.

In risk assessment, the easy condition shows a clear increase in performance measures as peak saccade velocity increases, and thus workload decreases, implying skill improvements. However, the hard condition fluctuates more as workload increases. Based on the learning plateau theory [33], this could indicate that not enough time was yet invested to reach the next level of expertise for this condition. We expect workload to decrease again once this new plateau has been reached.

Shifting attention with the trail making test showed marked improvements in all performance measures for both conditions, but better results for the easy levels, as expected. At the same time, workload based on long fixation measures remained constant for the hard condition and only decreased on the third session for easy. This could indicate that this third session was the turning point for expertise acquisition for the easy condition.

With situation awareness, performance was higher but inconsistent in the easy condition and lower but steadily rising for hard. Both of these were accompanied by what appears to be a decrease in workload based on peak saccade velocity. This indicates that skill steadily increased over time as performance also increased, a promising result.

The modified go-stop task comparing ignore and inhibitory control responses showed an increase in performance for the easy condition in both accuracy and response time, but a lesser response in performance for hard. Workload based on long fixation frequency remained mostly constant with a slight decrease for easy, but a constant increase over time for hard. Similarly to risk assessment, this suggests that the learning plateau for increased skill level has yet to be reached for the hard condition. In time, there should be a performance increase with an accompanying workload decrease.

Overall, performance results for each of the six tasks correlated well with our expectations based on the literature these tasks were taken from. Eye tracking measures of workload also showed significant differences based on difficulty level and time, explained by the acquisition of skill through training. As a next step, we will analyze the comprehensive suite of neuro and physiological measures of cognitive workload correlates.

Acknowledgments. This research was supported by the Air Force Research Laboratory's Human Performance Sensing BAA call 002 under contract number FA8650-16-c-6764. The content of the information herein does not necessarily reflect the position or the policy of the sponsor and no official endorsement should be inferred.

References

1. Parasuraman, R.: Neuroergonomics: research and practice. Theor. Issues Ergon. Sci. **4**(1–2), 5–20 (2003)
2. Parasuraman, R., Wilson, G.F.: Putting the brain to work: neuroergonomics past, present, and future. Hum. Factors **50**(3), 468–474 (2008)
3. Fedota, J., Parasuraman, R.: Neuroergonomics and human error. AU Theor. Issues Ergon. Sci. **11**(5), 402–421 (2010)
4. Ayaz, H., Pakir, M., Izzetoglu, K., Curtin, A., Shewokis, P.A., Bunce, S.C., Onaral, B.: Monitoring expertise development during simulated UAV piloting tasks using optical brain imaging. In: 2012 IEEE Aerospace Conference (2012)
5. Ayaz, H., Onaral, B., Izzetoglu, K., Shewokis, P.A., McKendrick, R., Parasuraman, R.: Continuous monitoring of brain dynamics with functional near infrared spectroscopy as a tool for neuroergonomic research: empirical examples and a technological development. Front. Hum. Neurosci. **7**, 871 (2013)
6. Afergan, D., Peck, E.M., Solovey, E.T., Jenkins, A., Hincks, S.W., Brown, E.T., Chang, R., Jacob, R.J.K.: Dynamic difficulty using brain metrics of workload. ACM, Toronto, Ontario, Canada (2014). https://doi.org/10.1145/2556288.2557230
7. Hancock, P., Chignell, M.H.: Toward a theory of mental work load: stress and adaptability in human-machine systems. In: Proceedings of the International IEEE Conference on Systems, Man and Cybernetics, 378–383 (1986)
8. Hart, S.G.: NASA-task load index (NASA-TLX); 20 years later. In: Proceedings of the Human Factors and Ergonomics Society Annual Meeting. Sage (2006)
9. Debue, N., van de Leemput, C.: What does germane load mean? An empirical contribution to the cognitive load theory. Front. Psychol. **5**, 1099 (2014)
10. John, M.S., Kobus, D.A., Morrison, J.G.: A multi tasking environment for manipulating and measuring neural correlates of cognitive workload. In: Proceedings of the IEEE 7th Conference on Human Factors and Power Plants (2002)
11. Miller, G.A.: The magical number seven, plus or minus two: some limits on our capacity for processing information. Psychol. Rev. **63**(2), 81 (1956)
12. Solovey, E.T., Zec, M., Garcia Perez, E.A., Reimer, B., Mehler, B.: Classifying driver workload using physiological and driving performance data: two field studies. In: Proceedings of the 32nd Annual ACM Conference on Human Factors in Computing Systems. ACM (2014)
13. Mehler, B., Reimer, B., Coughlin, J.F.: Sensitivity of physiological measures for detecting systematic variations in cognitive demand from a working memory task: an on-road study across three age groups. Hum. Factors **54**(3), 396–412 (2012)
14. Ahlstrom, U., Friedman-Berg, F.J.: Using eye movement activity as a correlate of cognitive workload. Int. J. Ind. Ergon. **36**(7), 623–636 (2006)
15. Jacob, R.J.K., Karn, K.S.: Eye tracking in human-computer interaction and usability research: ready to deliver the promises. In: Hyönä, J., Radach, R., Deubel, H. (eds.) The Mind's Eye, pp. 573–605. North-Holland, Amsterdam (2003)
16. Goldberg, J.H., Kotval, X.P.: Computer interface evaluation using eye movements: methods and constructs. Int. J. Ind. Ergon. **24**(6), 631–645 (1999)
17. Parasuraman, R., Christensen, J., Grafton, S.: Neuroergonomics: the brain in action and at work. Neuroimage **59**(1), 1–3 (2012)

18. Ayaz, H., Willems, B., Bunce, S., Shewokis, P.A., Izzetoglu, K., Hah, S., Deshmukh, A., Onaral, B.: Estimation of cognitive workload during simulated air traffic control using optical brain imaging sensors. In: Schmorrow, D.D., Fidopiastis, C.M. (eds.) Foundations of Augmented Cognition. Directing the Future of Adaptive Systems: 6th International Conference, FAC 2011, Held as Part of HCI International 2011, Proceedings, pp. 549–558, Springer, Berlin, Heidelberg, Orlando, FL, USA, 9–14 July 2011

19. Liu, Y., Ayaz, H., Shewokis, P.A.: Multisubject "learning" for mental workload classification using concurrent EEG, fNIRS, and physiological measures. Front. Hum. Neurosci. **11**, 389 (2017)

20. Di Domenico, S.I., Rodrigo, A.H., Ayaz, H., Fournier, M.A., Ruocco, A.C.: Decision-making conflict and the neural efficiency hypothesis of intelligence: a functional near-infrared spectroscopy investigation. Neuroimage **109**, 307–317 (2015)

21. Durantin, G., Gagnon, J.F., Tremblay, S., Dehais, F.: Using near infrared spectroscopy and heart rate variability to detect mental overload. Behav. Brain Res. **259**, 16–23 (2014)

22. Rodrigo, A.H., Domenico, S.I.D., Ayaz, H., Gulrajani, S., Lam, J., Ruocco, A.C.: Differentiating functions of the lateral and medial prefrontal cortex in motor response inhibition. NeuroImage **85**, Part 1(0), 423–431 (2014)

23. Logan, G.D., Van Zandt, T., Verbruggen, F., Wagenmakers, E.J.: On the ability to inhibit thought and action: general and special theories of an act of control. Psychol. Rev. **121**(1), 66–95 (2014)

24. Shalev, L., Ben-Simon, A., Mevorach, C., Cohen, Y., Tsal, Y.: Conjunctive continuous performance task (CCPT)–a pure measure of sustained attention. Neuropsychologia **49**(9), 2584–2591 (2011)

25. McKendrick, R., Ayaz, H., Olmstead, R., Parasuraman, R.: Enhancing dual-task performance with verbal and spatial working memory training: continuous monitoring of cerebral hemodynamics with NIRS. NeuroImage **85**, Part 3(0), 1014–1026 (2014)

26. Owen, A.M., McMillan, K.M., Laird, A.R., Bullmore, E.: N-back working memory paradigm: a meta-analysis of normative functional neuroimaging studies. Hum. Brain Mapp. **25**(1), 46–59 (2005)

27. Wickens, C.D.: Situation awareness and workload in aviation. Curr. Dir. Psychol. Sci. **11**(4), 128–133 (2002)

28. Endsley, M.R.: Design and evaluation for situation awareness enhancement. Proc. Hum. Factors Soc. Annu. Meet. **32**(2), 97–101 (1988)

29. Hagen, K., Ehlis, A.-C., Haeussinger, F.B., Heinzel, S., Dresler, T., Mueller, L.D., Herrmann, M.J., Fallgatter, A.J., Metzger, F.G.: Activation during the trail making test measured with functional near-infrared spectroscopy in healthy elderly subjects. NeuroImage **85**, Part 1, 583–591 (2014)

30. Müller, L.D., Guhn, A., Zeller, J.B.M., Biehl, S.C., Dresler, T., Hahn, T., Fallgatter, A.J., Polak, T., Deckert, J., Herrmann, M.J.: Neural correlates of a standardized version of the trail making test in young and elderly adults: a functional near-infrared spectroscopy study. Neuropsychologia **56**, 271–279 (2014)

31. Aklin, W.M., Lejuez, C.W., Zvolensky, M.J., Kahler, C.W., Gwadz, M.: Evaluation of behavioral measures of risk taking propensity with inner city adolescents. Behav. Res. Ther. **43**(2), 215–228 (2005)

32. Crowley, T.J., Raymond, K.M., Mikulich-Gilbertson, S.K., Thompson, L.L., Lejuez, C.W.: A risk-taking "set" in a novel task among adolescents with serious conduct and substance problems. J. Am. Acad. Child Adolesc. Psychiatry **45**(2), 175–183 (2006)

33. Ebbinghaus, H.: Memory: a contribution to experimental psychology. Ann. Neurosci. **20**(4), 155–156 (2013)

34. Ayaz, H., Dehais, F.: Neuroergonomics: The Brain at Work and Everyday Life, 1st edn. Elsevier, Academic Press, London (2019)
35. Gramann, K., Ferris, D.P., Gwin, J., Makeig, S.: Imaging natural cognition in action. Int. J. Psychophys. **91**(1), 22–29 (2014)
36. Gramann, K., Fairclough, S.H., Zander, T.O., Ayaz, H.: Editorial: trends in neuroergonomics. Front. Hum. Neurosci. **11**(165) (2017). https://doi.org/10.3389/fnhum.2017.00165

Promoting Soldier Cognitive Readiness for Battle Tank Operations Through Bio-signal Measurements

Jari Laarni[1(✉)], Satu Pakarinen[2], Mika Bordi[3], Kari Kallinen[4],
Johanna Närväinen[1], Helena Kortelainen[1], Kristian Lukander[2],
Kati Pettersson[2], Jaakko Havola[3], and Kai Pihlainen[5]

[1] VTT Technical Research Centre of Finland Ltd., Espoo, Finland
jari.laarni@vtt.fi
[2] Finnish Institute of Occupational Health, Työterveyslaitos, Helsinki, Finland
[3] Savox Communications Ltd., Espoo, Finland
[4] Finnish Defence Forces Technical Research Centre, Tuusula, Finland
[5] Training Division of the Defence Command Finland, Tuusula, Finland

Abstract. This paper will present the progress in developing a concept and a demonstrator system for the assessment of fatigue, acute stress and combat/cognitive readiness in military domain. A battle-tank crew's acute stress is measured with electrocardiography recordings of heart rate/heart rate variability. Cognitive performance is measured with a battery of cognitive tests, and task performance is estimated by soldiers' self-ratings, trainers' evaluations and objective measures from simulator data. The project consists of several test sessions in which cognitive and physiological indices of stress are measured while military conscripts perform battle-tank exercises both in simulator and field settings. The effect of task difficulty, sleep deprivation and operator role on performance are investigated. Different versions of the demonstrator system are also evaluated. The project results will be primarily used for the development of a bio-signal monitoring system, evaluation of transfer of simulator training to real-life exercises and improvement of military aptitude testing.

Keywords: Cognitive readiness · Psychophysiology · Military ·
Wearable system

1 Introduction

Soldier resilience can be defined as a soldier's ability to adapt to adverse battlefield conditions. Cognitive readiness is an essential element of soldier resilience. Cognitive readiness is cognitive preparation a practitioner or a team has to possess in order to be able to maintain competent performance in challenging environments. Self-knowledge and -awareness both at the individual and team-level are critical elements in establishing cognitive readiness. Since sensor technology provides new opportunities for monitoring of human physiology, there is a trend for acquiring self-knowledge through self-monitoring technology. Monitoring technology has also been applied in the military domain with the ultimate aim of improving soldier readiness. On the other hand, it

© Springer Nature Switzerland AG 2020
H. Ayaz (Ed.): AHFE 2019, AISC 953, pp. 142–154, 2020.
https://doi.org/10.1007/978-3-030-20473-0_15

is well known that it is very difficult to mitigate acute stress or augment cognition with the help of technology. There are thus huge challenges in developing a simple and robust bio-signal monitoring and feedback system that would also be easy to use for cognitive readiness enhancement purposes in the military domain.

A majority of research on soldier physical and mental workload is conducted either with heavily-trained special-force soldiers or with infantry soldiers who are easily available. Less is known about the fatigue, workload and other complications of other groups of soldiers with a key role in assisting infantry, such as armored battle tank crews. One of the few exceptions is the dissertation of Scriven [1] in which she studied a main battle tank crew's activities and workload. For example, a comparison of three-man-crew and four-man-crew configurations showed that the former configuration caused increased stress and fatigue, partly because the three-man crew tried to do too many things simultaneously.

In the current study we focus on battle-tank crews' acute stress and its impact on cognitive performance. More specifically, our first aim is to investigate the impact of physical and mental workload and acute stress on crew-members' cognitive performance using sensor technology. The second aim is to investigate the usability and functionality of current sensor technology in monitoring soldier performance in battle-tank operations, and how sensor information can be processed and transferred to different echelons of the military hierarchy. The third main objective is to develop cognitive readiness training for military operations and develop tools for predicting the completion of military service.

This paper will present the progress of the project and development of a concept and a demonstrator system for supporting assessment of workload, fatigue, acute stress and combat/cognitive readiness in the military domain. A battle-tank crew's acute stress is measured with electrocardiography recordings of heart rate and heart rate variability. Sleep quantity and sleep structure are measured with electrocardiogram (ECG), electro-oculography (EOG), and subjective quality of sleep with a diary. Patterns of physical activity and inactivity are measured with both wrist and chest worn 3d-accelerometers. Cognitive performance is measured with a battery of cognitive tests, and task performance is estimated by soldiers' self-ratings, evaluations by trainers and objective measures from combat simulator data. Also, battle tank crews' communication is recorded and analyzed in order to evaluate team behavior and dynamics.

1.1   Stress and Cognitive Performance in Military Operations

Stress can be defined as a relationship between an individual and his/her environment that is self-evaluated as demanding, and possibly threatening his/her well-being [2]. Combat stress is a blend of several kinds of stressors encountered in combat situations such as physical strain, fear of death, noise, fatigue, sleep deprivation, mental workload and time pressure [3]. The effects of stressors have been divided into five main categories, physiological, emotional, social, cognitive and performance effects. We are mainly interested in cognitive effects, including factors such as narrowing of attention, tunnel vision, disruption of working memory and response rigidity [3].

According to the theory of cognitive appraisal, the evaluation of the meaning and significance of an environmental stimulus and one's capacity to meet the demands are essential steps in coping with stressful events. Cognitive readiness can be considered as an individual's ability to anticipate the occurrence of environmental stressors, make accurate evaluations of threats and realistic expectations of one's ability to manage them. The four-stage model of successful coping with combat stress shown in Fig. 1 is based on Lazarus and Folkman's theory, and it is adapted and modified from [4].

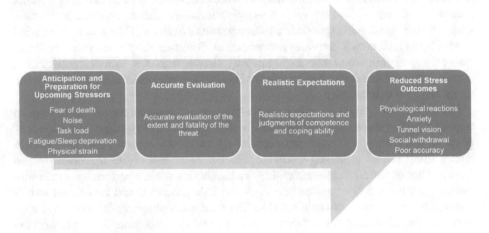

Fig. 1. Four-stage model of successful coping with combat stress (modified from [4])

1.2 Basic Features of a Wearable Bio-signal Monitoring System

Monitoring an individual's real-time physiological status makes it possible to estimate his/her current state and predict health and performance [5]. There are several commercially available wearable bio-signal monitoring technologies, such as heart rate monitors that come with different kinds of route, altitude and movement trackers, delivered by Suunto, Polar, and Garmin, and multi-parametric measurements systems, such as EQ 02 Life Monitor (Equivital, UK). Some of these latter systems are specifically dedicated to professional use, e.g., in military context, such as Black Ghost (Equivital, UK; http://www.equivital.com/products/military) and PADIS (Axiamo, SZ; http://www.axiamo.com/padis/). Both Black Ghost and PADIS consist of several wearable sensors measuring, e.g., heart rate and motion, and an application to performance evaluation.

Bio-signal Monitoring in Soldier Systems. In addition to the research using commercial wearable products and civilian subjects, several research projects in the military domain have been launched in recent years in which a large set of sensors have been tested (e.g., [6]). A typical list includes sensors for measuring heart rate, electrocardiogram, respiratory rate, blood pressure, electromyography, skin and core temperature, and motion tracking; used alone or in different combinations.

Collected data is typically first analyzed by using algorithms and models enabling comparisons to earlier data collected by/from the individual as well as norm values and performance metrics, after which high-level feedback is provided to the user or to the commanders [5]. These systems can thus be used, e.g., in diagnosing the physiological state of soldiers during training or military missions and making predictions about their ability to continue their missions. The bio-signal monitoring system can be further integrated into a wearable soldier system consisting of several other subsystems such as battle gear, helmet and weapon subsystem.

Requirements for a Wearable Bio-signal Monitoring System. Some key requirements for a wearable monitoring system are measurement accuracy and reliability, ability to present (high level) real-time readiness information, data transfer reliability, durability and ruggedness, small size and weight, long battery life, and easiness to wear, use and maintain. For online streaming, the device should provide fault-tolerant, reliable and secure data transfer. It is also important that the system does not cause harm and compromise soldier combat performance, but, quite the opposite, provides some added value to the soldiers themselves or to their leaders by enabling accurate predictions that cannot be made by more conventional means (e.g., by visual inspection).

2 Objectives

A starting point for our study is that with a shift from a platform-centric to a more network-centric warfare characterized, e.g., by information overload and longer mission duration, there is a need for more accurate information about the performance of an individual soldier. The main research question is then how real-time physiological status monitoring would benefit soldier performance and readiness prediction ability.

The first phase of the project will focus on studying soldier performance of one specific military branch, the armored forces. Battle tank crews were selected for the test for two reasons. On the one hand, they have a key role in supporting infantry, and they are responsible for costly military equipment, and on the other hand, relatively little is known about the effect of workload, stress, and fatigue on battle tank soldiers' performance during tank operations.

The aim of the project is to:

1. Identify the most demanding tasks for battle tank crew members by investigating how workload varies in the different phases of a military operation.
2. Investigate the effect of physical and mental workload on task performance in general and cognitive performance particularly during a battle tank combat exercise.
3. Evaluate the effectiveness of simulator training on soldier performance (i.e., transfer of training) and investigate how training effectiveness can be further increased.
4. Develop a cost-effective demonstrator system for combat/cognitive readiness assessment based on bio-signal measures.

3 Project Structure

In addition to conducting psychological experiments both in simulator and field environments, the aim is to develop a demonstrator system for real-time monitoring of soldier physiology. Several phases can be identified in the development process. The basic structure of the project is illustrated in Fig. 2.

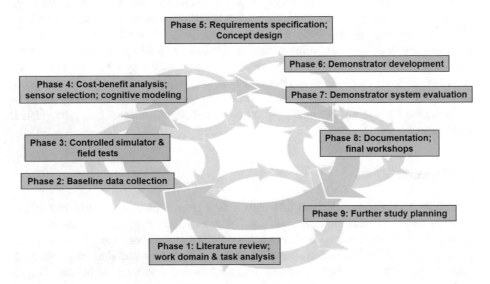

Fig. 2. Basic structure of the project

The basic idea is that wearable sensors are selected for the demonstrator system on the basis of controlled simulator and field studies led by the Finnish Defence Forces. In the first phase of the project, test activities are spread along a period of ten months, making it possible to follow up the development of the conscripts' skills and competencies with regards to battle tank operations.

Test participants were recruited among the conscripts of battle tank units from the Karelia Brigade. About fifty conscripts were originally recruited for the study, representing three different roles: drivers, gunners and tank commanders. A driver's task is, e.g., to place the tank in motion, drive it in different formations, cross obstacles and prepare the tank for operations. A gunner's task is, e.g., to prepare the gunner's station to operation, observe target areas, identify and prioritize targets and fire the guns. A tank commander has, e.g., to perform inspections and checks, execute navigation, create fire plans and receive and give orders. The gunner and driver are normally instructed to act according to the commander's orders.

The applied controlled studies include three kinds of test activities: baseline measurements, comparison of simulator and field trials, and a study concentrating on the effects of sleep deprivation on cognitive performance. Controlled simulator studies are conducted either in a classroom environment using a simulation software or during battle tank shooting training. Applied field studies are conducted to test wearable sensors in the field setting during a combative exercise. Based on the results of these studies, sensors and signals are selected and integrated into the demonstrator system.

At the next phase, the streams of sensor data are integrated into higher-level information about soldier cognitive readiness status, which will be displayed in intuitive and easily comprehensible way. Test results will also be used in cognitive modeling of fatigue from sleep loss by using the ACT-R cognitive architecture. The modeling work is carried out by adjusting parameters of procedural and declarative memory and the Utility/Utility Threshold in ACT-R.

The project will also include a cost-benefit analysis in order to determine the strengths and weaknesses of alternative sensor solutions and find the best option for further development.

At the final phase of the project, the demonstrator system will be field tested on a military field exercise.

4   Methods and Measures

The basic structure of the measurement process for all subtests is shown in Fig. 3. Before the mission, basic psychological processes are measured by questionnaires and computer-based behavioral tests. During the mission, bio-signals, team communication and response time, error rate, and performance accuracy data will be collected. After the mission, basic psychological processes are again measured.

Fig. 3. Measurement process

4.1   Psychophysiological Measures

Based on our literature review, on the earlier studies of the consortium and on the preliminary cost-benefit analysis, we concluded to record the cardiac (electrocardiography, ECG) and physical activity (actigraphy) and eye movements (electro-oculography EOG) during sleep. Cardiac activity is recorded with a sampling rate of 500 Hz using two disposable chest electrodes. The movement activity is recorded with three-dimensional accelerometers worn both at chest and in the wrist of the non-dominant hand with a sampling rate of 50 Hz. The EOG is recorded with 500 Hz sampling rate using two electrodes, attached diagonally: above and to the left of the right eye, and below and to the right of the left eye. The ECG, EOG, and chest actigraphy are measured with a Faros 360 device (Bittium, FI), and the wrist actigraphy with an ActiGraph GT9X Link (ActiGraph, US) device.

From the ECG, measures of heart rate (HR) and heart rate variation (HRV) are extracted. EOG is used in classifying different sleep stages [7]. Actigraph data is used for generating time series for movement vectors: From the EOG, measures of different sleep stages (N1, N2, and N3) and the amount of sleep in minutes are extracted. From the actigraphs, we calculate the time series for the total movement vectors. The above-mentioned signals are analyzed, e.g., by a mental workload classification method developed by [8].

4.2 Psychological and Cognitive Performance Measures

Soldiers' cognitive capabilities are measured using tests of working memory, vigilance, attention and problem solving before they participated in controlled simulator and field tests. Cognitive performance is also measured during simulator and field test trials using tests of working memory and attention control. Some psychological tests are also administered before performance tests measuring such things as personality, coping mechanisms, dispositional resilience, temperament, depression, and mindful attention awareness. In addition to the above-mentioned measures, workload, situation awareness and teamwork are measured by using questionnaire-type scales after the test trials.

5 Preliminary Results of the Simulator and Field Tests

The project consists of several test sessions in which cognitive and physiological indices of stress are measured while conscripts perform battle-tank exercises both in field and simulator settings. The effect of task difficulty, sleep deprivation and operator role etc. on the performance are investigated. Different versions of the bio-signal monitoring system prototype are evaluated as well.

Since the study is still in progress, only some preliminary results are available. According to military trainer evaluation, overall, the battle tank crews' performance in the military field exercise was not at a very good level, and there were quite large differences in performance between the crews. This was somewhat expected, as the conscripts were in the early stages of their training.

On the positive side, registration of bio-signals, and data collection in general, proceeded without problems. On the negative side, battle tank operations and soldier activities during them were not considered very stressful in neither simulator nor field conditions. For example, it is highly questionable whether the conscripts felt any combat stress during the military exercise. It is possible that since training was still ongoing, and the crewmen were not yet very proficient, they did not know what was required from them, and thus they were not able to be nervous about their performance. In addition, there was a lot of waiting time compared to the time spent in combat during the military exercise.

Seemingly, sleep deprivation was not as big a challenge in this particular military exercise as expected. Some signs of fatigue and sleep deprivation were recognized, but it is still unclear whether these factors had any impact on soldier performance. According to interviews performed by the researchers, apathy, boredom and feeling of hunger seemed to be more prevalent feelings, which may also have some impact on cognitive performance.

6 Demonstrator System Design

It has been emphasized that there are multiple factors that have to be considered in soldier system development. For example, [9] have recently presented a framework for soldier-centered design of systems identifying such elements. They have categorized these factors into three groups: cognitive processes, measurement systems and computation/modeling. Next, we will focus on the design and development of the measurement systems – less will be said about the other two groups of factors.

6.1 Task Analysis and Requirements Specification

The design methodology used in the development of the demonstrator system is divided into nine key phases (see Fig. 2). In order to be able to develop the proposed system for military purposes we have to define soldiers' physical, perceptual and cognitive activities, analyze task performance and describe the environments in which they perform military operations.

The key task is to identify and analyze the work, information and design requirements that the new system should support. To that aim, we have worked with military subject matter experts to help them to describe the critical aspects of battle tank operations and required physical and cognitive performance.

Based on these data we conducted a preliminary task analysis and modeled the critical tasks and provided some recommendations that will guide the design work.

6.2 Concept and Demonstrator Design

The requirements specification work resulted in a basic concept of a wearable bio-monitoring system. According to our present view, the demonstrator system will consist of the following main elements:

1. Wearable sensors for monitoring bio-signals;
2. Transmission path for data transfer;
3. Power solution for the wearable system;
4. Wearable solution for temporary data storage;
5. Data transmission channel from the wearable system to the platoon/company level;
6. Solution for a permanent data storage;
7. Data handling and analysis system;
8. Interface for information visualization.

Next, we will briefly present some of the main steps in demonstrator development.

Sensor Technology Development. Even though the measurement system is based on commercial sensors, some development work has been done, e.g., in configuring the electrodes and the measurement device (e.g., see Fig. 4).

Fig. 4. An adapter connecting the electrode configuration for EOG monitoring with the Faros (Bittium, Oulu, FI) measurement device

Local Data Transfer. Software will be developed to enable a Bluetooth interface between a Faros sensor and a soldier computer unit to transfer and store data from one unit to another.

Local Data Management. Local preprocessing of data will be executed at the soldier computer unit level in order to make it easier to transfer it to the following levels of the military hierarchy. At the same time, it is possible to combine data coming from different sources into a single totality of measurement data. Savox Warrior Core (Savox Communications, FI) is one possible mobile computer solution for local data management providing powerful application processing and several interfaces for devices, systems and services.

Integrated System. The aim is to integrate all the functions into a single chain of functions and validate the functionality and performance of the transmission paths. Several integration phases are included into the task in which a connection is built between the soldier computer unit (e.g., Savox Warrior Core) and a temporary storage (e.g., vehicle) or a command center.

The aim is that the sensor data is progressively accumulated, structured, organized and interpreted in order to provide a general overview picture of the troops' combat/cognitive readiness for higher echelons. For example, a special algorithm were developed by FIOH for the automated classification of sleep phases [7].

One of the key tasks in this phase is to design a clear and user-friendly interface to present combat/cognitive readiness information effectively to the commanders. It would also be desirable if the readiness information can be easily accessed through a mobile phone.

The details of the information presented on a computer screen will be determined together with the Defence Forces, but, for example, the following information can be displayed:

- The unit's physiological status and combat readiness with a stoplight format;
- Estimates of the progress of the unit's combat readiness;
- Estimates of data uncertainty;
- Map displaying the location of each unit.

Field Testing. After the development work is done, the demonstrator system is ready for user testing. At this phase, the demonstrator will be tested in field conditions together with the Finnish Defence Forces. Based on the results of the field test, the demonstrator system will be evaluated in terms of usability and functionality, and some further refinements are made. Results will be documented as a part of the project's final report.

7 Cost-Benefit Analysis

The aim of the cost-benefit analysis is to determine the strengths and weaknesses of alternative sensor solutions and find the best option for further development. The cost-benefit assessment (CBA) aims to figure out if the benefits created by a project or an investment are greater than the incurred costs [10]. Cost-benefit analysis has also been applied to evaluate the feasibility of security investments and technologies [11]. The analytic methods to assess investment appraisals include also life cycle costing that considers not only the original investment but also the costs incurring during the whole life cycle of a product [12].

The concept of Life Cycle Costing (LCC) was developed by U.S. Defense Forces in early 1960s in order to improve the effectiveness of the defense material acquisitions. LCC models for defense sector (e.g., [13]) help to implement longer planning period that covers acquisition, development, operation, support and maintenance costs. A crucial part of an LCC model is the cost breakdown structure, which is a hierarchical representation of costs related to the case in question. It divides a rather abstract life cycle cost value in to more concrete and thus more easily estimated cost elements. Table 1 presents an example of the cost breakdown structure.

Table 1. Example of main cost and benefit categories for security technology investments [11]

COSTS		
Acquisition cost / investment cost / capital cost	**Direct sustaining costs**	**Indirect sustaining costs**
Concept and definition phase	Operations costs	Indirect operations, maintenance and disposal costs
Design and development phase	Maintenance costs	
Non-recurring investment costs	Disposal costs	Societal costs
Recurring investment costs		Legal and regulatory costs
		Political costs

BENEFITS				
Direct economic benefits	Benefits of reduction of threats	Societal benefits	Legal and regulatory benefits	Political benefits

Life cycle costing is a data driven process, as the amount, quality and other characteristics of the available data often define what methods and models can be applied, and what analyses can be performed [13]. The evaluation of the cost-effectiveness of the demonstrator systems will concentrate on assessing the LCC of two or more alternative solutions. The cost breakdown structure will be developed along with the development of the demonstrator concept, the architecture and use scenarios.

Benefits may be economic or operational, but they may also include intangible elements like improved efficiency of an individual solder or a troop, or other immaterial issues. In the defense sector, the benefits are typically difficult to assess in monetary terms. For this reason, benefits of the soldier monitoring system will be identified on a qualitative level and the optional monitoring systems will be assessed by scoring the benefits.

8 Conclusions

We have presented the progress of the project, and, in particular, the content and progress of the field and simulator tests and the development of a concept and a prototype of a demonstrator system for the assessment of workload, fatigue, acute stress and combat/cognitive readiness in the military domain. The results of the project will be primarily used for the development of bio-signal monitoring system, further development of predictive cognitive modeling techniques, evaluation of transfer of simulator training to the real-life combat environment and improvement of military aptitude testing.

The project is still ongoing, but some lessons have already been learned. both about the measurement of acute stress, fatigue and workload in military exercises and about the design and development of the wearable bio-signal monitoring system. Regarding the measurement of stress and fatigue, the following lessons have been identified:

- It is very difficult to elicit combat stress in peacetime conditions even in combat training.
- Conscripts are not necessarily the optimal participants if the aim is to study the effect of combat stress, for example, because they may not have a clear view of their role or function in terms of success in combat, or they are not even able to differentiate good and poor performance.
- To achieve a reliable interpretation of the field data, the analysis needs prior individual data from a controlled quasi-experimental design comprising of a careful, systematic manipulation of acute stress and fatigue and necessary tests, conducted in simulator-type training facilities.

Some lessons have also been learned in developing the measurement system:

- By conducting user tests at different phases of the project makes the design process soldier-centered and participatory, which is very valuable in developing wearable soldier systems.

- More research is needed to determine what the effects of the introduction of the bio-signal soldier system on warfare are in general, and specifically on an individual soldier's/crew's way of combating.
- The big question is what is the real value of online monitoring of bio-signals and where the biggest value lies. And to what degree we are creating new problems while trying to solve old ones.
- In general, the wearable bio-signal monitoring system should be seen as one element of a complex system of systems and a larger military ecosystem, and its development should also be considered from this larger perspective.

The project will continue by studying soldier cognitive readiness with participants coming from other branches of the military and tailoring the wearable monitoring system for their special needs.

Acknowledgments. This work is funded by the Technology Program of the Finnish Defence Forces. We thank the Karelia Brigade, the military trainees and conscripts of the Brigade and members of the reference group for making this research possible.

References

1. Scriven, J.G.: Main battle tank crew in-tank activities and workload. Doctoral thesis (1993). https://dspace.lboro.ac.uk/2134/32084
2. Lazarus, R.S., Folkman, S.: Stress, Appraisal, and Coping. Springer, New York (1984)
3. Orasanu, J.M., Backer, P.: Stress and military performance. In: Driskell, J.E., Salas, E. (eds.) Stress and Human Performance, pp. 89–125. Erlbaum, Mahwah (1996)
4. Salas, E., Driskell, J.E., Huges, S.: Introduction: the study of stress and human performance. In: Driskell, J.E., Salas, E. (eds.) Stress and Human Performance, pp. 1–45. Erlbaum, Mahwah (1996)
5. Friedl, K.E.: Military applications of soldier physiological monitoring. J. Sci. Med. Sport **21**, 1147–1153 (2018)
6. Kutilek, P., Vlf, P., Viteckova, S., Smrcka, P., Krivanek, V., Lhotska, L., Hana, K., Doskocil, R., Navratil, L., Hon, Z., Stefek, A.: Wearable systems for monitoring the health condition of soldiers. Review and application. In: Proceedings of 2017 International Conference of Military Technologies (ICMT), pp. 748–752, May 31–June 2, 2017. IEEE, Brno, Czech Republic (2017)
7. Virkkala, J., Hasan, J., Värri, A., Himanen, S.L., Müller, K.: Automatic sleep stage classification using two-channel electro-oculography. J. Neurosci. Methods **166**, 109–115 (2007)
8. Henelius, A., Hirvonen, K., Holm, A., Korpela, J., Muller, K.: Mental workload classification using heart rate metrics. In: 2009 Annual International Conference of the IEEE Engineering in Medicine and Biology Society, pp. 1836–1839. IEEE (2009)
9. Oie, K.S., Gordon, K., McDowell, K.: The multi-aspect measurement approach: rationale, technologies, tools, and challenges for systems design. In: Savage-Knepshield, P., Martin, J., Lockett III, J., Allender, L. (eds.) Designing Soldier Systems. Current Issues in Human Factors. Ashgate, Farnham (2012)
10. Boardman, A.E., Greenberg, D.H., Vining, A.R., Weimer, D.L.: Cost-Benefit Analysis: Concepts and Practice, 3rd edn. Prentice Hall, Upper Saddle River (2006)

11. Räikkönen, M., Kunttu, S., Poussa, L., Blobner, C.: Benefits, risks and costs of security measures. A portfolio-oriented decision-support approach for political decision-making. In: Lauster, M. (ed.) Proceedings of the 8th Future Security Research Conference, pp. 442–444. Fraunhofer Verlag, Berlin, Stuttgart, 17–19 September 2013
12. IEC 60300-3-3 Ed.3.0: Dependability management – Part 3-3. Application guide – Life Cycle Costing (2017)
13. RTO: Methods and Models for Life Cycle Costing. RTO Technical Report. Final Report of Task Group SAS-054. Research and Technology Organisation North Atlantic Treaty Organisation. TR-SAS-054 (2007)

A Change in the Dark Room: The Effects of Human Factors and Cognitive Loading Issues for NextGen TRACON Air Traffic Controllers

Mark Miller, Sam Holley[✉], Bettina Mrusek, and Linda Weiland

Embry-Riddle Aeronautical University Worldwide College of Aeronautics,
Daytona Beach, FL, USA
holle710@erau.edu

Abstract. By 2020 all aircraft in United States airspace must use ADS-B (Automatic Dependent Surveillance-Broadcast) Out. This is a key component of the Next Generation (NextGen) Air Transportation System, which marks the first time all aircraft will be tracked continuously using satellites instead of ground-based radar. Standard Terminal Automation Replacement System (STARS) in the Terminal Radar Approach Control (TRACON) is a primary NextGen upgrade where digitized automation/information surrounds STARS controllers while controlling aircraft. Applying the SHELL model, the authors analyze human factors changes affecting TRACON controllers from pre-STARS technology through NextGen technologies on performance. Results of an informal survey of STARS controllers assessed cognitive processing issues and indicates the greatest concern is with movements to view other displays and added time to re-engage STARS.

Keywords: SHELL · Human factors · NextGen · STARS · TRACON · Cognitive loading · Distraction

1 Introduction

The Federal Aviation Administration (FAA) has mandated that by 2020 all aircraft flying in Controlled Airspace of the United States (U.S.) use an Automatic Dependent Surveillance Broadcast-Out (ADSB-Out) device to designate aircraft location in controlled airspace. This does not seem like a big change in the air traffic control (ATC) system, however the reality is that it marks a drastic shift in ATC operations moving from a ground-based radar system with navigational aids that is nearing capacity to a satellite-based system that can absorb the predicted future growth of the aviation industry. NextGen has been a work in progress for over 25 years. Most of the attention to NextGen has focused on how it will make flight operations safer and more efficient. One area of utmost importance and central to making NextGen ATC a success is the TRACON (Terminal Radar Approach Control) with STARS (Standard Terminal Replacement System) display. Currently the FAA requires STARS as a mandatory cornerstone to NextGen terminal area operations. The TRACON controllers must direct

© Springer Nature Switzerland AG 2020
H. Ayaz (Ed.): AHFE 2019, AISC 953, pp. 155–166, 2020.
https://doi.org/10.1007/978-3-030-20473-0_16

aircraft safely and efficiently, but many of the terminal airspaces in the U.S. have become crowded due to increased popularity of air travel along with the growth of air cargo. Consequently, ATC needs a viable solution.

2   Intent of Study on STARS TRACON Controller Cognition

The intent of this study was to analyze the dark room environment of controllers to gain an understanding of how human factors affects their work in relation to NextGen equipment upgrades like the STARS display. As the ADSB-IN/OUT is important to the NextGen cockpit, so too is the STARS display equipment to the controller. Of particular interest is how STARS controllers' job performance is affected cognitively by this technology. Although the combined STARS equipment and NextGen satellite ATC system should be a substantial gain in safety and efficiency for the controller, these potential gains could be negated by human error caused by a myriad of cognition issues. In terms of cognitive workload, the work of the TRACON controller is one of the most challenging in the world. Unlike their tower controller counterparts who can see aircraft and clear them to take off and land, the TRACON controller creates a three-dimensional cognitive map from a two- dimensional screen to simultaneously track aircraft. A cognitive slip could cause the controller to lose situational awareness of an aircraft under their control. This could lead to unfortunate accidents.

3   The Serious Threat to Aviation Safety

Although accidents involving TRACON controllers have become rare in the U.S., serious incidents still occur. With the pending future growth in the industry and ever shrinking skies looming ahead, accidents with TRACON involvement could pose a threat to the industry. An example of TRACON controllers heavily involved in an accident was the Avianca Flight 52 crash that occurred in 1990 on Long Island. The crash caused 73 fatalities when the aircrew failed to declare a fuel emergency to the New York TRACON controllers. Although the probable cause was rooted in failures to communicate from the cockpit, the National Transportation Safety Board [1] also determined that there were serious flaws in the ATC handling of the aircraft. Accidents like the Avianca crash illustrate how the controllers can err and contribute to an accident. Human factors is critical to TRACON operations and to gain a deeper understanding of the potential human error in the TRACON, it is important to evaluate how cognition has been affected by the upgrades in ATC technologies. The authors analyze how the NextGen system will influence controllers' cognitive performance.

4   The SHELL Model Revisited, Pre-STARS Analysis

To analyze TRACON controller human factors and their influences on cognition, the study utilized the basic SHELL model of human factors introduced in 1987 [2] to assess the TRACON controller's position before STARS, followed by an analysis with

STARS, and finally an analysis of STARS fully integrated with the NextGen satellite ATC system. The SHELL model of 1987 employed a block layout placing the human (controller) represented as Liveware (L) in the center, then surrounding it by four interfaces: Software (S), Hardware (H), Environment (E) and other Liveware (L).

Referencing the SHELL diagram in Fig. 1., the TRACON controller in the pre-STARS environment had a different set of hardware to work with ergonomically to control aircraft. Many controllers who started in the last century were exposed to older analog/CRT (Cathode Ray Tube) radar screens. When analyzing the L-H linkage of this older SHELL, the controllers had screens that could depict the target, but did not have the capability to filter weather. Their map overlays were not clear and accurate. Color was also limited on these older radar screens. In tracking aircraft, separation was not reliable because controllers had to predict aircraft vectors for where they thought the aircraft was going. The screen had very little adjustment for light, color or layout. This older radar equipment was manufactured ergonomically as one size for all controllers. There were no altitude or separation alerts to increase margins of safety. Communications via radio were critical to confirm aircraft location.

Human Factors Analysis and the SHELL Diagram

Fig. 1. SHELL model by Hawkins in 1987 featuring the Liveware-Liveware interface [2].

The L-E interface in pre-STARS was two-fold: one representing the physical environment and the other representing the artificial environment. The physical environment called for a dark room to see the radar screen and a cool room to maintain temperature at the correct level for the CRTs. The artificial environment kept distraction and noise levels to a minimum. Work rules included breaks and work shifts for rest. All FAA ATC policies were adhered to, including the team concept. Other variables affecting the controller's environment were volume of traffic handled, VFR or IFR conditions, weather severity and types of aircraft controlled for wake turbulence.

The L-S interface of SHELL in the pre-STARS era was challenging as the Software represented such things as FAA ATC procedures, ATC regulations, approach plates, weather, winds, ATIS and flight strip information. This information was gathered from many places and funneled to the controller. FAA ATC procedures and regulations were usually stored in several bulky volumes of FAA paper publications located in the TRACON radar room. The approach plate to back up what approach an aircraft was flying was a paperbound booklet. Weather was usually updated via the telephone and noted on paper nearby with wind conditions. The ATIS information for the aerodrome was broadcast via radio. Perhaps the most widely used piece of information of critical importance was that of the flight strips. These paper/plastic flight strips held aircraft information such as call sign, altitude, destination and aircraft type. Ergonomically these were challenging as they were hand created and used manually.

The most compelling part of the pre-STARS SHELL analysis is the relationship of the controller and how they interact with other Liveware in the L-L interface. The TRACON L-L interface was quite strong in the analog/CRT radar period. The perspective gained from analysis of the three previous interfaces (L-H, L-E, and L-S) is that the radar controller was not a one-person job. It required a team of controllers to work together safely and effectively to ensure that all the information was updated accurately and then disseminated to the radar controller to control the aircraft. At a minimum, there would be a radar controller, flight strip controller and manager. Teamwork in this older TRACON environment also required exceptional communication. While internal TRACON communications were based on teamwork and standard operating procedures, the external communications were accomplished through a complicated radio communications panel with multiple switches for the controller to manually switch. Training and qualification of new controllers was accomplished by studying procedures, radar training and on the job training with a qualified controller.

5   From Analog/CRT Radars to STARS Digital Equipment

Although challenged and susceptible to a multiple number of human factors issues in each interface of the SHELL diagram, the training, teamwork and professionalism seemed to work for pre-STARS radar controllers. Being that this model was already very work intensive as more aircraft took to the skies, future growth of the U.S. industry in the new deregulated environment of the 1980's and 1990's would force the TRACON to upgrade to safer and more efficient technologies. TRACONs would need to maintain the high level of professionalism and teamwork for future success. The shift to modern TRACON technologies was accomplished by phasing out the older analog/CRT displays with digital replacements like STARS. To help support the FAA's choice of the digital STARS display for future use with the NextGen satellite system, other equipment has also become digital to help the controller. With these multiple additions of computer automation/information becoming commonplace in support of the STARS controller, the original SHELL model of direct linkages now must be adjusted to account for computerized indirect linkages that now exist between the STARS controller (L at the center of SHELL) and the four interfaces caused by the computer automation/information. To accomplish a human factors analysis of the current STARS controller and account for the computer automation/information, the updated SHELL Model 2017 used by Miller [3] and shown in Fig. 2 was used.

The benefits of using SHELL 2017 are seen in the L-H analysis of the digital STARS controller, because the STARS display is not only digital, it is also highly computer-automated with extensive information. As a digital and optimally designed air traffic display, the STARS controller now customizes the display so they can

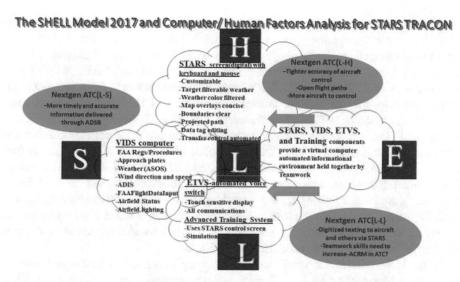

Fig. 2. The SHELL model 2017 adopted for the computer-automation/information of the STARS TRACON controller with predicted Nextgen ATC effects by Mark Miller in 2019.

interface ergonomically. The controller sets the brightness, the size of the screen and adjusts colors to their preference. They can also save the setting to reinstate it whenever they need to. To assist in the automation adjustments, the STARS display has a fully functioning computer keyboard and built in mouse. The target aircraft on the display screen can be filtered from the weather to become highly visible. Meanwhile, the software on the display screen can give the controller an accurate representation of weather intensity around aircraft by depicting it with six different colors. Airspaces and their boundaries are seen more accurately and can be enhanced. Map overlays are accurate. These efficiencies also come with gains in safety since operators can control aircraft more accurately with far less stress. Furthermore, the STARS controller has the aid of projected path software enhancements that allow them to see aircraft vectors in relation to other aircraft they are controlling. Software filters enhance margins of safety when activating low altitude warning and separation alerts. Wake turbulence distances have been greatly reduced. Perhaps one of the biggest improvements of the STARS system is the accuracy provided by an electronic Data Tag that appears with the aircraft as aircraft flight information. What was once a mounted paper flight strip identifying the flight and aircraft is replaced by information entered in electronic format directly into the computer scratch pad that is then digitally transformed next to the aircraft on the STARS screen. This digitized Data Tag of aircraft identification includes call sign, aircraft type, assigned altitude, airspeed, destination, service requested, airport and runway. The controller can update information on that Data Tag while controlling the aircraft through the STARS keyboard and mouse.

As the computer automation/information in the STARS digital display greatly enhances the interface between L-H, it accomplishes this through shifting automation and information from other interfaces. In the L-E interface the physical dark room remains dark to optimally use the STARS display. However, many aspects, like weather, have migrated to the STARS display or other computer automated/informational technologies in other SHELL interfaces. Managing the controllers' work and teamwork still remain strongly at play in the TRACON. Where the technological shift has affected TRACON operations the most is in the volume of traffic handled, the severity of weather around that traffic, and the types of aircraft being controlled for wake turbulence. The accuracy of the STARS display in visibly tracking aircraft and Data Tagging each aircraft along with the projected path automation and alerts means that a controller can efficiently increase the number of aircraft under their control. The same automation/information can add VFR aircraft to controller tasking along with IFR aircraft. The accurate weather environment depicted on the STARS controller's screen helps them see the most dangerous weather, differentiate the aircraft from the severe weather on the screen, and accurately vector the aircraft safely around the weather. This same accuracy also enabled reducing wake turbulence separation between aircraft as recently mandated by the FAA.

In the L-S environment, what was once a myriad of separate information resources of Software representing FAA ATC procedures, regulations, approach plates, weather, wind shear, ATIS and flight strip information is now placed in one computer source to the side of the controller via another computer screen called VIDS (Visual Information Display System). The VIDS displays many different icons to represent important areas of information now found in one computer information source. There are icons for FAA ATC procedures, regulations, approach plates, Automated Surface Observing System for weather, Wind Speed and Direction Indicator, ATIS, Airport Status Display, and the FAA Flight Data Input/Output (FDIO). Special attention needs to be given to the FAA FDIO icon as this is where the controller inputs the Flight Progress Strip data for the Data Tag on the STARS screen. VIDS brings many different sources of information to the controller in one automated location.

The L-L interface is now enhanced with automation in three major ways. First, through enhanced communications among all L-L participants by means of the automated communications suite called the Enhanced Terminal Voice Switch (ETVS). ETVS provides control of all frequencies, interphones, and landlines with touch-sensitive controls and displays instead of antiquated switches. Secondly, through what is the most significant feature of STARS computer automation enhancement, is its ability to complete the transfer of a tracked aircraft from one controller to another through an automated exchange. In receiving control of the transferred aircraft, the new controller sees the aircraft target flashing on their STARS control screen and clicks on it to accept control. The colors then change for the aircraft on both screens as both controllers receive indications of transfer of that aircraft. The third major way is through a training enhancement. Instead of classroom hours and on the job training with controllers, training controllers in the STARS systems is all about real world controlling via simulation. A STARS display is designated in the TRACON while other controllers are actually controlling real aircraft. Meanwhile, the trainee is in the same room doing simulated controls on an actual STARS display. This training is as close as a new controller can get to live operations without controlling real aircraft.

The shift in SHELL analysis from an era of controllers using analog/CRT radar displays with direct linkages shown in Fig. 1., as compared to that of the digital automated/informational displays used by the current controllers in Fig. 2., shows a clear trend toward adding significant levels of visual information to the controller enabled by computers. STARS, supported by VIDS and ETVS, clearly shows that the L-E direct linkage of the old TRACON is being outsourced and replaced by multiple computer automated/informational devices. The technology is creating a detailed virtual L-E for the STARS controller. However, the now indirect linkages of the L-H, L-S, and L-L replacing the L-E pose challenges for STARS TRACON controllers that invite human error. The depiction of the L-H, L-S, and L-L replacement of the L-E in Fig. 2. shows all these new connections as cognitive clouds overlapping each other. The new computer technologies are causing the indirect linkages of the STARS TRACON controller to become concatenated. This means that cognitive tasks are overlapping. To counterbalance the technology from overwhelming the STARS controller, training and teamwork are of utmost importance. Just as the modern commercial cockpit has turned to an Advanced Crew Resource Management culture that emphasizes teamwork skills integrated with new multiple technologies, the STARS TRACON should adopt more formal aspects of teamwork through the FAA to meet the challenges of the future. The STARS controllers seem to enjoy this technological configuration as a leap forward in efficiency, safety gains and confidence in operations. In terms of technologically integrating the STARS systems with NextGen, there is not a part of the ATC system better prepared to transition to the satellite-driven system than a STARS TRACON integrated with VIDS and ETVS. Yet as the new technologies seem to enhance their abilities, what will happen to STARS controller cognition when the current STARS model merges fully with NextGen?

6   NextGen Factors to Influence SHELL 2017

The SHELL 2017 STARS controller analysis, as depicted in Fig. 2. also has added to it in red NextGen effects for L-H, L-S, and L-L. With NextGen, the L-H interface of STARS gets an immediate boost from the constant satellite signal without interruptions. This makes controlling aircraft through STARS more accurate and the STARS controller gains more open flight paths for free flight. However, the biggest gain from this NextGen satellite accuracy is that the STARS controller will be able to efficiently, effectively and safely handle more aircraft. In the L-S interface the STARS controller quickly gains more accurate and up to date information from the VIDS that is now connected to an ADS-B system that is able to share that information with aircraft (ADS-B In screen). Perhaps the biggest potential human factors change affecting the STARS controller using the NextGen system will come in the form of digitized (texting) communications in the L-L interface. At some point STARS controllers will have the option to text other aircraft and the aircraft will be able to text reply back. This is currently being demonstrated through Datalink. In the future this might possibly be accomplished better ergonomically through modifying the Data Tags that currently allow for editing and also enable two-way digitized texting. The NextGen enhanced system will give STARS controllers an upgraded virtual environment from the L-E interface to the STARS, VIDS, and ETVS. Acknowledging the critical importance of approach and departure control in the ATC system, and that most aviation accidents occur on or near an airport, the NextGen upgrade added to the STARS controllers' arsenal will give promise to a safer, more efficient, path to future flight operations. In referencing the Avianca Flight 53 accident mentioned previously, it is inconceivable that this accident could occur in the STARS controlled NextGen satellite system. The aircraft would be controlled more effectively, but most importantly the inability of the Avianca crew to communicate a low fuel emergency would be negated with texting. Free flight accuracy would then line up the low fuel aircraft for an immediate landing. The fact that this new TRACON system could prevent accidents is noteworthy. However, with so many technologies converging to make the virtual environment more accurate, controller cognitive loading could be challenged.

7 Cognitive Loading in the STARS Environment

The authors have noted SHELL originally did not envision simultaneous, multi-dimensional interfaces with increased optical and cognitive loads. Consequently, the SHELL 2017 model was proposed to demonstrate cognitive load effects from overlapping interactions [3]. Cognitive load, for this study, refers to attentional or working memory resources dedicated to information processing. Evident in SHELL 2017 is clear evidence of concatenated cognitive tasks and neural loading that presents opportunity for capacity problems and competing resources. Figure 2 illustrates cognitive processing required for dynamic visual cues, icons, and text using multi-dimensional interfaces (screen, tablet, keyboard, and mouse). Cognitive loading effects with STARS require head shifts by operators, viewing separate screens displaying with disparate information, and processing that precipitates dynamic shifts in perceptual load, processing resources, and interface distractions.

Task load and workload are related, but not the same. Controllers may perceive changes in task load as a workload increase. In ATC, increasing the number of aircraft under control has affected cognitive workload negatively [4]. Likewise, the NextGen transition from audio communication to texting increases cognitive load. Controllers' responses to visual cues were found to be more accurate but slower than performance with auditory cues [5], suggesting controllers are more susceptible to overload with visual cues attributed to depletion of neural protein and working memory deficit [6].

With multiple STARS interfaces, image-processing considerations must be considered. Display density, target-background, and layout perspectives vary among interfaces used. Consequently, ergonomics issues for search time expended and compromised accuracy presume increased error rates [7]. As NextGen progresses, development may follow the progress of ATM in Europe which employs four-dimensional trajectories, extending the cognitive processing load. Among the tools used are automated systems electronic coordination and conflict detection for enroute traffic. Corver [8] found that in control centers using newer technology, the nature of cognitive error had changed. While effectively reducing cognitive errors related to detection, memory, and decisions, new tasks invited error in timely detection of relevant data. A shift from individual controller responses to a larger organizational framework may hold promise in addressing cognitive loading concerns. Examining SHELL interactions, Chang [9] found controllers were more influenced by organizational factors than individual differences which invites further study of team effects when changes occur in the STARS environment.

Transition has posed numerous challenges for the FAA, including the need to validate new requirements and provide automated tools for controllers, e.g., performance-based navigation, to merge and sequence aircraft [10]. Tasks for controllers have emphasized effective visual radar scanning, and a study of controllers performing monitoring tasks determined that effective visual scanning is their principal concern [11]. Results showed significant variation in visual scan patterns tied to particular tasks and which type of interface was used. The study identified that perceptual load effect required attentional control that restricted neural resource allocation for distractors.

Perceptual load (or, load theory) is based on models of dual-task information processing and indicates the extent of behavioral interference imposed by high rates of attentional demands. Perceptual resources are allocated first to task-relevant information and, if capacity remains, to less relevant information [12]. Controllers in the STARS environment, working with several interfaces of varying symbology, are subject to saturation from data and the need to interpret significance of display information, which invites the possibility of delayed controller comprehension. Selective attention assigns limited resources to significant information while filtering task-irrelevant ones and load theory suggests irrelevant stimuli are not processed under high perceptual load [13]. The problem for STARS controllers is that when irrelevant stimuli, e.g., a road crossing a runway, become relevant the attention may not be perceived cognitively since no neural resources are available for processing.

Distraction can impede our ability to detect and effectively process task-relevant stimuli in our environment. Cognitive load influences situation awareness (SA) and is affected adversely as demands increase [4]. Discussions about distraction involve a bottom-up response to unexpected stimuli and a top down capture of attention using working memory to filter out what is not relevant. These two routes recruit different neural resources like when bottom-up stimuli activate the ventral frontoparietal network that computes relevance and suppresses response to items not relevant [14]. To accommodate overlapping cognitive load shown in Fig. 2., synchronous neural processing must flow freely. However, when disrupted by distractions or loss of capacity from rapid updating of working memory, there are notable losses of sustained attention, mental sequencing, and integrity of an associated cognitive map [15].

Attention is related to working memory in two ways. One affects memory load, the other influences content, and both relate to perceptual load and distractor processing. Evidence suggests that perceptual load reduces distractor interference and working memory load increases distractor processing. However, working memory load restricts resources to resolve distractor interference and largely depends on the mode of information. When a distractor is being held in working memory it will interfere with processing targets under high perceptual load [16]. Display clutter is closely tied to effective performance and has been linked to impaired performance. Both the number and density of display entities has been recognized as a concern and contributes to the overall problem of clutter. For STARS, the task relevance of added icons and identifying data in supplemental interfaces presents a potential for mode confusion. While an experienced controller may work efficiently with multiple screen representations, when cognitive workload approaches maximum capacity, novel situations or heightened task difficulty the opportunity for error and unintended actions (or inactions) is likely to increase [17]. Eye tracking and gaze duration are concerns, particularly when multiple interfaces are used and controllers must shift attention to different layouts and areas of interest (AOI) (see Fig. 2). While VIDS aides the controller by aggregating information, the shift to distractor AOIs for verifying information challenges integrity of the STARS cognitive map held by the controller. Issues related to the number of fixations, scan path ratio, and duration can be problematic when upper limits are exceeded [18]. STARS does not include an embedded function for digitized communication which currently must be provided on a separate display. This also invites added cognitive issues for the controller.

8   Controller Assessment of STARS Cognitive Loading

The authors conducted a program review with TRACON STARS controllers at a facility using the current technology. Controllers responded to questions (available upon request) about attention to displays, operating conditions, and potential distractions. Incidence of degraded cognitive performance was assessed with respect to selective and divided attention between displays having disparate three-dimensional cognitive maps, with attendant problem resolution and decision actions during operations. Available controllers choosing to participate were asked to respond to an electronic survey. Responses were categorized and evaluated to identify potential impaired action, error, or other influences. The results enabled a glimpse of human factors and cognitive performance challenges that may confront NextGen controllers operating in the near future. The results are shown in Table 1.

Table 1. Responses of STARS controllers in a TRACON environment.

Item	N	Never	Sometimes	Often
Head turn required for other displays	11	0	6	5
Displays viewed peripherally	11	1	1	9
Added time/effort to re-engage STARS	11	4	1	6
Supervisor called for STARS assist	11	7	4	0
Physical actions other than for STARS	11	2	1	8
Missed item of importance	11	6	5	0
Annoyed by intrusion	11	5	6	0
Uncertainty after distraction	11	7	4	0
Distracted by non-flight activity	11	7	4	0
Read status message more than once	11	6	5	0

Results showed a mean of 4.2 years as controller, 2.9 years at TRACON, and 2.0 years at the STARS location. Findings indicated that for more than half the controllers head turns or body movements were required to view other interface displays, inviting disorientation and vestibular interference. Nearly all the controllers acknowledged scanning displays peripherally which added time. For distractions, nearly half the controllers reported they missed important items, were annoyed, or were uncertain about aircraft status after viewing other screens, requiring reading messages twice.

9 Conclusion

Precautions and recommendations to address cognitive workload related to added digitized communication messaging and visual tasks must be integrated into the NextGen controllers work environment. This study provides evidence that increased cognitive load can lead to distraction and delay in responding. Further investigation for understanding cognitive limits, team interactions, and NextGen changes is needed.

References

1. National Transportation Safety Board: 707 Fuel Exhaustion Avianca Flight 52 NTSB/AAR-04/91 (1991)
2. Hawkins, F.H.: Human Factors in Flight, 2nd edn. Ashgate, Aldershot (1987)
3. Miller, M.D.: Human factors computer information/automation beyond 2020 NextGen compliance: risk assessment matrix of situational awareness (cockpit computer use versus aviate, navigate, communicate). In: Presentation to FAA Aviation Safety Conference, Honolulu (2017)
4. Friedrich, M., Biermann, M., Gontar, P., Biella, M., Bengler, K.: The influence of task load on situation awareness and control strategy in the ATC tower environment. Cog. Technol. Work 20(2), 205–217 (2018)
5. Pant, R., Taukari, A., Sharma, K.: Cognitive workload of air traffic controllers in area control center of Mumbai enroute airspace. J. Psychosocial Res. 7(2), 279 (2012)
6. Ravassard, P., Kees, A., Willers, B., Ho, D., Aharoni, D., Cushman, J., Aghajan, Z., Mehta, M.: Multisensory control of hippocampal spatiotemporal selectivity. Science 340(6138), 1342–1346 (2013)
7. Neider, M., Zelinsky, G.: Cutting through the clutter: searching for targets in evolving complex scenes. J. Vision 11(14), 7 (2011)
8. Corver, S.C., Aneziris, O.N.: The impact of controller support tools in enroute air traffic control on Cognitive error modes: a comparative analysis in two operational environments. Saf. Sci. 71, 2–15 (2015)
9. Chang, Y., Yeh, C.: Human performance interfaces in air traffic control. Appl. Ergon. 41(1), 123–129 (2010)
10. Federal Aviation Administration: Fact Sheet-Standard Terminal Automation Re-placement system. U.S. Dept. Transport, Washington (2016)
11. Li, W., Kearney, P., Braithwaite, G., Lin, J.: How much is too much on monitoring tasks? visual scan patterns of single air traffic controller performing multiple remote tower operations. Int.. J. Indust. Ergon. 67, 135–144 (2018)
12. Giesbrecht, B., Sy, J., Bundesen, C., Kyllingsbæk, S.: A new perspective on the perceptual selectivity of attention under load. Ann. New York Acad. Sci. 1316(1), 71–86 (2014)

13. Yin, S., Liu, L., Tan, J., Ding, C., Yao, D., Chen, A.: Attentional control underlies the perceptual load effect: evidence from voxel-wise degree centrality and resting-state functional connectivity. Neuroscience **362**, 257–264 (2017)
14. Greene, C.M., Soto, D.: Functional connectivity between ventral and dorsal frontoparietal networks underlies stimulus-driven and working memory-driven sources of visual distraction. Neuroimage **84**, 290–298 (2014)
15. Lind-Kyle, P.: Heal Your Mind: Rewire Your Brain. Energy Psychology Press, Santa Rosa (2010)
16. Koshino, H.: Effects of working memory contents and perceptual load on dis-tractor processing: when a response-related distractor is held in working memory. Acta Physiol. **172**, 19–25 (2017)
17. Moacdieh, N., Sarter, N.: Display clutter: a review of definitions and measure-ment techniques. Hum. Factors **57**(1), 61–100 (2015)
18. Lohrenz, M., Trafton, J., Beck, M., Gendron, M.: A model of clutter for complex, multivariate geospatial displays. Hum. Factors **51**, 90–101 (2009)

The Neuromodulative Effects of Tiredness and Mental Fatigue on Cognition and the Use of Medication

José León-Carrión[1]([⊠]), Umberto León-Domínguez[2],
and Maria del Rosario Dominguez-Morales[3]

[1] Department of Ëxperimental Psychology, University of Seville, Seville, Spain
leoncarrion@us.es
[2] Department of Health Services, School of Psychology,
University of Monterrey, Monterrey, Mexico
umbertoleon@gmail.com
[3] Center for Brain Injury Rehabilitation (CRECER), Seville, Spain
rdominguez@neurocrecer.es

Abstract. Fatigue is defined by means of subjective and neurophysiological deficits that vary as a function of time and workload. Cognitive load and task difficulty mediate fatigue. Performance deficits normally demonstrate evidence of fatigue and awareness of mental effort, although overt deficits may not be a consequence of fatigue, as increased efforts normally compensate the decline of mental resources. Subjective mental fatigue leads to increased mental effort, impaired learning and stress. Neurophysiological mechanisms affecting mental fatigue include homeostatic and circadian components of fatigue. One example is the dopamine system. Mental fatigue provokes significant changes in the brain, altering the glutamate network, particularly in challenging situations, including chronic pain and neurodegenerative disease. Coping with fatigue often requires pharmaceutical measures to maintain, or even restore, operative cognitive functions. We review fatigue and tiredness management strategies, the use of hypnotic and other drugs, short and long-term efficacy of drugs and recommendations.

Keywords: Mental fatigue · Tiredness · Neuropsychology · Dopamine ·
Hypnotic drugs

1 Introduction

Fatigue is considered a decline in performance in any prolonged or repeated task that is not always observable [1, 2]. However, the individual experiences a subjective sensation that affects task performance. The effects of fatigue are also mediated by cognitive load, making task difficulty a factor in its onset. Cognitive load is defined as the ratio between the time needed to process specific information to complete a task and the time available to do so. Performance deficits normally demonstrate evidence of fatigue and awareness of mental effort. Overt performance deficits are not necessarily a

© Springer Nature Switzerland AG 2020
H. Ayaz (Ed.): AHFE 2019, AISC 953, pp. 167–172, 2020.
https://doi.org/10.1007/978-3-030-20473-0_17

consequence of fatigue, as increased efforts normally compensate the decline of mental resources [3].

Homeostatic and circadian components of fatigue are different neurophysiological mechanisms affecting mental fatigue [4]. One of them is the dopamine system, an inhibitory neurotransmitter that modulates brain activity and cortico-cortical circuits that are essential for cognitive control. Low levels of dopamine produce attentional and executive deficits. Dopaminergic neurons play a pertinent role in the regulation of homeostatic balance and inform the brain on changes in the internal environment. Dopamine depletion may not suppress the ability to perform tasks, but it restricts behavioral and mental flexibility.

Mental fatigue provokes significant changes in the brain, altering the glutamate network, particularly in challenging situations, including chronic pain and neurodegenerative disease. Mental work is compromised by mental fatigue, deteriorating a patient's capacity to prepare and guide action. Confronting fatigue requires coping mechanisms and motivation that arises from outside (extrinsic) or inside (intrinsic) the individual [5]. Legal and illegal drugs also may improve cognitive task performance over an extended period [6].

Given this background, fatigue should be defined by means of subjective and neurophysiological deficits that vary as a function of time and workload. Subjective mental fatigue leads to increased mental effort, impaired learning and stress [7–9]. Mental fatigue is reversed simply by resting or taking time away from the task. It is also necessary to differentiate between fatigue and sleepiness. Fatigue is manifested by subjective and physiological deficits and corresponding decrements in overt performance. Sleepiness does not cause fatigue nor does fatigue cause sleepiness. However, both are triggered, direct or indirectly, by work requiring continuous task completion without adequate rest, sleep or time off. Sleep deprivation usually causes time-on-task performance decrement even if the task is not cognitively challenging. Tasks that involve the processing of novel or inconsistent information require cognitive effort. Several variables significantly influence cognitive fatigue, including cognitive load, mental flexibility and lack of sleep (< 6 h/day). Sleep is normally restorative. A short nap after hard work reduces the effects of fatigue. Greater resistance to fatigue is usual in the mornings and the opposite after food intake or a recent meal.

2   Pharmaceutical Measures

People coping with the effects of fatigue must rely on pharmaceutical measures if they need to maintain, or even restore, operative cognitive functions. The most recommended measures to combat fatigue are:

2.1   Provigil (Modafinil)

Provigil is a stimulant medication used to combat fatigue and symptoms of sleep disorders, without the risk of serious side effects, including tolerance, dependence and addiction. Also known as Modafinil, this neurostimulant contains neuroprotective properties that promote alertness. Approved by the U.S. Food and Drug Administration

(FDA) to treat narcolepsy, it is also prescribed to treat excessive daytime sleepiness associated with narcolepsy and obstructive sleep apnea, as well as shift work sleep disorder. Although it may decrease sleepiness, it is not a cure. Provigil has been compared to Adderall and Ritalin, FDA approved amphetamines to treat attention deficit/hyperactivity disorder (ADHD) and certain sleep conditions. All three drugs are popular among healthy individuals without these disorders, consuming them for study or work. Unlike Adderall and Ritalin, Provigil does not produce euphoria. According to the Federal Drug Administration (FDA), this medication does not have the same potential to cause addiction as Adderall and Ritalin, classifying it as a Class IV controlled substance, whereas the other two drug are Class II, a more restricted category [10–12].

2.2 Hypnotics

The use of hypnotics in the evening is common to prevent insomnia and associated fatigue. Studies indicate that benzodiazepine receptor agonists are safe and effective in the initiation and maintenance of sleep. The newer version of benzodiazepine is recommended for reducing hangover effects. Hypnotics, such as Zolpidem, improve daytime sleep. Alertness appears to be higher after a Zolpidem-induced prophylactic nap than after no nap without post-nap problems. While the long-term efficacy of sleep medication is still under debate, some authors recommend daytime napping procedures to supplement inadequate nightly sleep [13].

2.3 Tryptophan

Tryptophan, an essential amino acid found in many foods and supplements, is known to boost sleep quality and mood. It is a precursor of Niacin/Vitamin B3, melatonin and serotonin, which helps one recover from the stress associated with our way of life. Because the body cannot make its own tryptophan, it must be taken in as part of the diet. However, it is not easy to obtain Tryptophan strictly from foods that are adequate in quantity and dosage. Low serotonin levels may lead to anxiety and mild depression, as well as snacking, uncontrolled eating at meals, etc. Taking tryptophan supplements can help those suffering from stress or insomnia. Pharmacologists have noted that tryptophan is extremely beneficial for our organism and that it can be found in the Casein protein that is digested slowly, working in synergy with L-theanine.

2.4 Zinc and Magnesium (ZMA) and Vitamin B

Zinc and Magnesium are very popular dietary supplements for improving sleep quality. Zinc is responsible for testosterone production, while magnesium is known to increase sleep quality. Vitamin B6 facilitates the absorption of ZMA in our body. Regardless of your workout or diet, ZMA helps increase hormone levels, promotes better sleep and contributes to brain function and lucidity. Magnesium bisglycinate is a form of the mineral that is naturally linked to the glycine amino acid, thereby facilitating its absorption.

2.5 Vitamin B

Research has shown that B vitamins are essential for myriad physical and mental functions [14]. Vitamin B1 (thiamine) helps improve overall welfare and mood. Vitamin B2 (riboflavin) helps reduce oxidative stress and fatigue. Vitamin B3 (niacin) supports brain functions and ensures optimum health. Vitamin B5 (pantothenic acid) helps synthesize important neurotransmitters in the brain. Vitamin B6 (pyridoxine hydrochloride) plays a key role in reducing fatigue and strengthening the immune system. Vitamin B8 (inositol) facilitates communication between nerve signals. The Vitamin B complex offers much more, including Choline, a dietary nutrient that is required for various physiological processes—from maintaining proper cell structure to supporting nervous system and brain function [15]. Iron supplements are often recommended as a solution to combat a low energy, especially among women. While it is true that women often have iron deficiencies, men are not immune to this deficit and could also benefit from these supplements.

2.6 Complete Multivitamin Complex

A Complete Multivitamin Complex is recommended for stressful and hectic lifestyles. This not only can enhance general health, but it can act as a "safety net" for nutrients in moments of stress. All-natural supplements also boost energy, including caffeine or Taurine tablets.

3 Recovery from Fatigue

Recovery from fatigue is one of the goals traditionally pursued by individuals whose work and emotional activity are under high-stress conditions. However, not all succeed, and many are not even aware of being fatigued or tired. Fatigue or tiredness is the logical consequence of carrying out a mental and/or physically task requiring perseverance, consistency and efficacy during a specific period. Recovery from fatigue is linked to the control of these three factors. A person will be more fatigued when their work develops with regularity and continuity without sacrificing efficiency. This exhaustion is often associated with intense, on-going tasks. In these circumstances, it is best to seek medical advice, as it can be a sign of the need for rest, something that will help boost motivation and improve our quality of life.

The following exercises and activities are recommended for combating fatigue:

1. Drink plenty of water, and avoid dehydration, at least two liters a day. Balancing lost electrolytes is also recommended before or after strenuous physical activity. Electrolytes are salts and minerals in the body that maintain fluid balance and blood pressure. Apart from sport drinks, foods that replenish electrolytes quickly include milk or yogurt, bananas, avocados, watermelon and coconut water.
2. Beverages containing amino acids help increase energy efficiency, for example, 500 mg to 3 g of L-carnitine and 10–30 g of Creatine. Five grams of l-glutamine, along with 100–200 mg of L-tyrosine can also aid recovery. L-lysine and L-arginine may also speed up recovery and increase muscle growth.

3. Get regular exercise. Daily exercise can boost your energy and reduce the risk of factors that contribute to fatigue, including obesity, heart disease, and diabetes. Exercise also increases the level of endorphins in the blood, which produces a greater sense of well-being.
4. Drink less alcohol. It may help you drop off to sleep more quickly, but drinking alcohol before bed is linked with less Rapid Eye Movement (REM), the more restful stage of sleep, and you're likely to wake up feeling groggy and unfocused.
5. Underlying medical conditions, including hormonal imbalances or anemia, may give rise to the sense of fatigue. Consult your doctor or other health care specialists for an accurate diagnosis. Proper nutrition in these cases is a priority.
6. Get sufficient sleep. Getting enough undisturbed sleep is key to reducing fatigue on a daily basis.
7. Our bodies need time to recover from activities, diseases or psychological trauma. Try to get 7–8 h of sleep at night. Go to bed at the same time to develop a sleep pattern, de-stress with relaxation techniques before going to sleep and avoid caffeine as much as possible, especially in the afternoon.
8. Psychological detachment or "switching off" from work is essential. Avoid job-related tasks during non-work hours, and evade thoughts on work-related tasks.

References

1. Lorist, M.M., Faber, L.G.: Considerations of the influence of mental fatigue on controlled and automatic cognitive processing and related neuromodulatory effects. In: Ackerman, P.L. (ed.) Cognitive Fatigue. Multidisciplinary Perspectives on Current Research and Future Applications, pp. 105–126. American Psychological Association, Washington, DC (2011)
2. Wesensten, N.J., Killgore, W.D., Balkin, T.J.: Performance and alertness effects of caffeine, dextroamphetamine, and modafinil during sleep deprivation. J. Sleep Res. 14(3), 255–266 (2005)
3. Saper, C.B., Scammell, T.E., Lu, J.: Hypothalamic regulation of sleep and circadian rhythms. Nature 27, 1257–1263 (2005)
4. Massar, S.A.A., Csathó, A., Van der Linden, D.: Quantifying the motivational effects of cognitive fatigue through effort-based decision making. Front. Psychol. 9, 843 (2018)
5. Myers, D.G.: Psychology: eighth edition in modules. In: Plotnik, R., Kouyoumjian, H. (eds.) Introduction to Psychology. Worth Publishers, New York (2010)
6. Van Dongen, H.P., Belenky, G., Krueger, J.M.: A local, bottom-up perspective on sleep deprivation and neurobehavioral performance. Top. Med. Chem. 11, 2414–2422 (2011)
7. Hockey, G.R.: The Psychology of Fatigue: Work Effort and Control. Cambridge University Press, New York (2013)
8. Ackerman, P.L. (ed.): Cognitive fatigue: multidisciplinary perspectives on current research and future applications. American Psychological Association, Washington, DC (2011)
9. Baranski, J.V., Pigeau, R., Dinich, P., Jacobs, I.J.: Effects of modafinil on cognitive and meta-cognitive performance. Hum. Psychopharmacol. 19, 323–332 (2004)
10. Baranski, J.V., Gil, V., McLellan, T.M.: Effects of modafinil on cognitive performance during 40 hr of sleep deprivation in a warm environment. Military Psychol. 14, 23–47 (2002)

11. McLellan, T.M., Ducharme, M.B., Canini, F., Moroz, D., Bell, D.G., Baranski, J.V., Gil, V., Buguet, A., Radomski, M.W.: Effect of modafinil on core temperature during sustained wakefulness and exercise in a warm environment. Aviat. Space Environ. Med. **73**, 1079–1088 (2002)
12. Gerrard, P.: Mechanisms of modafinil: a review of current research. Neuropsychiatr Dis Treat **3**, 349–364 (2007)
13. Takahashi, M.: The role of prescribed napping in sleep medicine. Sleep Med. Rev. **7**, 227–235 (2003)
14. Kennedy, D.O.: B Vitamins and the brain: mechanisms, dose and efficacy—a review. Nutrients **8**, 68 (2016)
15. Leon-Carrion, J., Dominguez-Roldan, J.M., Murillo-Cabezas, F., Dominguez-Morales, M.R., Muñoz-Sanchez, M.A.: The role of citicholine in neuropsychological training after traumatic brain injury. NeuroRehab **14**, 33–40 (2000)

Study on Mental Workload Assessment of Tank Vehicle Operators

Zhongtian Wang[✉], Qanxiang Zhou, and Yu Wang

Beijing University of Aeronautics and Astronautics, Beijing 100191, China
13161395752@163.com

Abstract. In order to monitor and evaluate the mental load of the operators in the operation of tank vehicle more effectively, and ensure the safe and efficient operation of tank man-machine-environment system. Using the tank vehicle operation simulation experiment platform, the tank firing experiment task was simulated according to three links of perception, judgment and decision-making, and motion execution. The changing trend of the tank vehicle operator's mental load was analyzed from two aspects of subjective performance and EEG signal characteristics. After the completion of the simulated tank shooting experiment task, the subjects' thinking clarity, sleepiness, fatigue comprehensive assessment, emotional state and motivation level were all significantly improved ($P < 0.01$); with the increase of task difficulty, the scores of NASA-TLX load scale in different task loads were improved, and after weight comparison, the total score and mental demand both had very significant difference between low load and medium load, and between low load and high load ($P < 0.01$), and had significant difference between medium load and high load ($P < 0.05$) There was no significant difference in performance level between high, medium and low task loads ($P > 0.05$). There was significant difference in time demand, effort and frustration between low and high load ($P < 0.05$), and no significant difference between low and medium load, medium and high load ($P > 0.05$). The number of firing times per unit time with different task difficulty was significantly different between low and medium load, low and high load ($P < 0.01$), but no significant difference between medium and high load ($P > 0.05$). For EEG power spectrum characteristics, α, β, α/θ, β/θ, $(\alpha + \beta)/\theta$ index were all sensitive to mental load level, but only α and β were significantly. Subjective questionnaires and performance data can objectively reflect the level of mental load in different difficulty tasks. For different time periods of tasks, the times of pressing shoot keys can reflect the changing trend of mental load in the process of tasks. Among EEG indicators, α and β can effectively evaluate the level of mental workload.

Keywords: Mental workload · Subjective performance · EEG · Tank operator

1   Introduction

Mental workload is used to describe the ability of an operator to process information while performing a task, or to describe the mental state of the operator when performing a task. Mental load is often difficult to define unilaterally and simply.

© Springer Nature Switzerland AG 2020
H. Ayaz (Ed.): AHFE 2019, AISC 953, pp. 173–184, 2020.
https://doi.org/10.1007/978-3-030-20473-0_18

Generally speaking, the mental workload of operators is related to task requirements, personnel capacity and external environment. Many scholars at home and abroad have also proposed different definitions of the concept of mental workload, Rouse [1] emphasized that there was a great relationship between mental workload and individual differences. For the same flight task, the mental workload of different personnel is different. O'Domrell [2] defined information processing capabilities for specific tasks as mental workload. Jianqiao [3] believes that mental load is related to the use of information resources in the process of task execution. The lower the capacity of dealing with idle information resources, the higher the mental load was. In complex man-machine-environment system of tank vehicle, operators need to perform complex and flexible combat tasks for a long period. They are usually in a state of high mental workload. It is easy to induce mental fatigue and make mistakes under such working conditions. In extreme cases, it may lead to aircraft damage and death. Therefore, to carry out evaluation of mental workload on the tank vehicle operator, is of great significance to monitor and evaluate the operator's mental load state more effectively and to ensure the safe and efficient operation of the man-machine-environment system of tank vehicles.

2 Objects and Methods

2.1 Experimental Object

18 volunteers were randomly recruited as subjects, all the subjects were male, aged between 21 and 30, with bachelor's degree or above. Of all the 18 subjects recruited, 2 participated in the preliminary experiment and 16 participated in the formal experiment. All subjects were right-handed and had no disease history of psychological, mental, neurological and color blindness. They also had no history of drug abuse. All subjects were in good physical and mental state. Their visual acuity was normal or their corrected visual acuity was above 1.0.

2.2 Experimental Method

General Design of Experiment. The tank firing simulation mission system is used to complete the experiment. The system mainly includes EEG acquisition system, task presentation and data acquisition computer, control handle and so on. The operation system of task presentation and EEG acquisition computers are Windows 7 Pro system. Mission presentation computer equipped with simulated tank firing platform. During the experiment, the computer automatically saves the performance data of the experiment task. The EEG acquisition computer is equipped with Recorder 2.1 EEG recording software for recording EEG data.

The Lestat PXN-8103 handle was used to manipulate the handle during the experiment, as shown in Fig. 1. In the process of manipulation, the subjects were required to hold the handle with both hands and use the direction key of the handle to manipulate the movement of the sight bead. The Y key of the handle is used to control tank firing.

Fig. 1. PXN-8103 operating handle

Design of experiment using the scheme of single factor repeated measurement. The experiment task of the same difficulty level needs to be repeated three times. Task difficulty level is divided into three levels: low load, medium load and high load. For tank firing tasks of the same difficulty level, the subjects need to complete three times, and rest at least 3 min after each task is completed. To avoid the effect of the same order of experiments on the results of experiments, the difficulty sequence of three tank firing tasks is disrupted by Latin square design method (For example: Subject I operated the experimental tasks in the order of high load, medium load and low load, Subject II operated in the order of high load, low load and medium load). In the whole process of experiment, the subjects need to complete 3*3 times of experiment tasks.

Tanks that simulate tank firing missions are divided into enemy tanks and friendly tanks. The friendly tanks include operator and other tanks. The enemy tanks can be divided into two types: the enemy type I tanks and the enemy type II tanks. Operators can only shoot enemy tanks in the middle sector area (field of view), and only enemy tanks in the sector area have artillery attack capability against operator tanks. There are 15 enemy type I tanks and 15 friendly tanks under low load condition. The experimental time is limited to 5 min. There are 15 enemy type I tanks and 15 enemy type II tanks under medium load condition, and the experimental time is limited to 7 min. The number of enemy tanks under high load condition is the same as that under medium load condition, but the number of friendly tanks is 15 more than that under medium load condition, and the experimental time is limited to 8 min. The experimental task operation interface is shown in Fig. 2.

Fig. 2. Experimental interface of simulated tank firing mission

Mental Workload Measurement Method. There are three main methods to measure mental workload: subjective measurement, task measurement and physiological measurement.

Subjective measurement refers to the method by which operators state their feelings in the process of tasks and score them subjectively according to established rules after completing specific tasks. The main scales used in this method are NASA-TLX, SWAT, Mental State Inventory (MSI), Attention Allocation Scale and so on [4]. The mental state scale includes six dimensions: clarity of thinking, concentration of energy, sleepiness, comprehensive fatigue assessment, emotional state and motivation level. Each dimension is divided into 10 grades. The higher the grade, the higher the score of the corresponding scale, the worse the state of the subjects in the corresponding dimensions. Scores range from 1 to 10. NASA-TLX mental workload scale consists of six dimensions: mental demand, physical demand, time demand, performance level, effort level and frustration level. Each dimension is represented by a scale line of 20 equal parts, and the two ends of the scale line are marked with low and high characters respectively. Each dimension is described in detail [5]. This experiment mainly used NASA-TLX Mental Load Scale and MSI Mental State Scale to evaluate subjective performance.

Task measurement method mainly includes two methods: chief task measurement method and sub-task measurement method. The chief task measurement method is a method to measure the performance level of the operators in performing a shief task, while the sub-task measurement is in performing a sub-task. Operators need to do their best to complete the sub-task under the condition of ensuring the smooth completion of the chief task [6].

Physiological measurement is a method to monitor the state of mental load by measuring physiological indicators. The main physiological parameters used in this method are heart rate and its variability, eye movement, spontaneous electroencephalography (EEG) and event-related potentials (ERPs) [7].

Before and after each experiment, subjects were asked to fill in MSI and NASA-TLX mental load scale respectively. During the process of task, EEG signals were real-time collected by EEG acquisition system.

Processing Method of Experimental Results. One-way ANOVA was used to analyze the scores of each dimension of MSI scale before and after the experiment. Mean and standard deviation of NASA-TLX scale scores of six dimensions under three mental loads were calculated for all subjects, and the variance analysis of repeated measurements of scores of six dimensions under three mental workloads was carried out. For further comparison, NASA TLX scores of six dimensions (low-medium, low-high, medium-high) were tested by paired t test.

For attention allocation, after each task was completed, the subjects were asked to score the attention distribution between the tank's entry and exit boundary and the firing tank according to the three experiments. There are 10 grades in the score between the tank's entry and exit boundary and the tank's firing tank. Each grade has one point. The higher the score, the more attention the subjects pay to the tank's firing.

For the analysis of EEG indicators, the typical electrodes on the corresponding brain regions are selected and their eigenvalues are averaged. For each EEG eigenvalue, the results include two factors: factor A: is task difficulty of mental load (low, medium and high levels), and factor B: is time factor (5 time points). Each EEG eigenvalue was processed by two-factor repeated measurement variance analysis. The electrodes in frontal (F), central (C), parietal (P) and occipital (O) regions were taken for analysis. When comparing the difficulties of the same task, the collected EEG signals were divided into 5 segments according to the time length. Average value of eigenvalues on corresponding electrodes were extracted. Then the correlation of EEG parameters at five time points was analyzed. SPSS 22.0 was used to analyze the experimental data (Fig. 3).

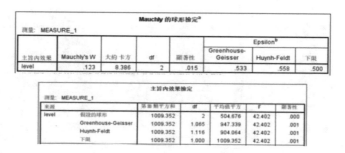

Fig. 3. Analysis of variance of repeated measurements of NASA-TLX score

3 Experimental Result

3.1 ANOVA Results of Subjective Performance Data Scores

ANOVA Results of the Effect of Experimental Task Difficulty on Mental Load of Subjects. For NASA-TLX mental workload scale, the result of Mauchly test was $P = 0.015 < 0.05$, indicating that the data did not obey the spherical hypothesis and the degree of freedom needed to be corrected. The value of Wilks' Lambda is between 0 and 1, and the closer it approaches 0, the more it rejects the original hypothesis. SPSS statistical analysis showed that the value of Wilks' Lambda was 0.035, which effectively rejected the hypothesis that the difficulty of the experiment task had no effect on the mental load level of the subjects. The results of variance analysis after correction showed that $F = 42.402$, $P = 0.001 < 0.05$, indicating that the difficulty of experiment task had a significant impact on the scores of NASA-TLX mental load scale. From Fig. 4, it can be seen that there are significant differences in psychological load levels between low and medium loads, low and high loads, and between medium and high loads of tank firing tasks. From the comparison of the statistical results in Fig. 5, it can be seen that with the increase of task difficulty, the scores of NASA-TLX load scale in different task loads have increased.

ANOVA Results of Mental State Before and After the Experiment. One-way ANOVA was used to analyze the scores of MSI. The results of variance analysis (Fig. 6) showed that $F = 57.040$, $P = 0.001 < 0.05$, indicating that there was

(I) 任务难度水平	(J) 任务难度水平	平均差異 (I-J)	標準誤談	顯著性b	95% 差異的信賴區間b	
					下限	上限
1.0	2.0	-7.402*	1.992	.004	-11.840	-2.964
	3.0	-18.235*	1.992	.000	-22.673	-13.797
2.0	1.0	7.402*	1.992	.004	2.964	11.840
	3.0	-10.833*	1.992	.000	-15.271	-6.395
3.0	1.0	18.235*	1.992	.000	13.797	22.673
	2.0	10.833*	1.992	.000	6.395	15.271

Fig. 4. The results of analysis of variance of NASA-TLX score

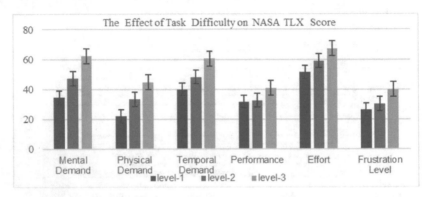

Fig. 5. Statistical results of the impact of task difficulty level on NASA score

significant difference in the scores of MSI mental state scale before and after the completion of the simulated tank firing experiment task. The main effect of experimental situation on mental state of subjects is significant (Fig. 7). From the statistical comparison results in Fig. 8, it can be seen that the level of thinking clarity, sleepiness, fatigue comprehensive assessment, emotional state and motivation of the subjects have significantly improved after the tank firing task.

單變量測試

因變數: 得分

	平方和	df	平均值平方	F	顯著性
比對	6.866	1	6.866	57.040	.001
誤談	.602	5	.120		

F 測試 狀態 的效果。此測試是以已估計邊際平均值中的線性獨立成對比較為基礎。

成對比較

因變數: 得分

(I) 狀態	(J) 狀態	平均差異 (I-J)	標準誤談	顯著性b	95% 差異的信賴區間b	
					下限	上限
1.0	2.0	-1.513*	.200	.001	-2.028	-.998
2.0	1.0	1.513*	.200	.001	.998	2.028

根據估計的邊際平均值

*. 平均值差異在 .05 層級顯著。

b. 調整多重比較: 最小顯著差異 (等同於未調整)。

Fig. 6. The results of variance analysis of MSI

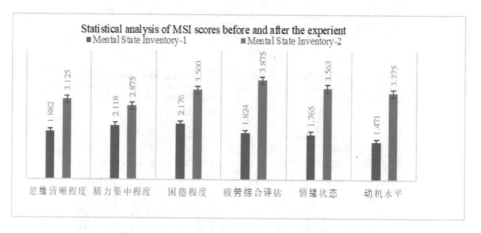

主旨間效果檢定

因變數: 得分

来源	第 III 類平方和	df	平均值平方	F	顯著性
修正的模型	7.205ª	6	1.201	9.975	.012
截距	82.945	1	82.945	689.079	.000
指标	.339	5	.068	.562	.728
状态	6.866	1	6.866	57.040	.001
錯誤	.602	5	.120		
總計	90.751	12			
校正後總數	7.806	11			

a. R 平方 = .923 (調整的 R 平方 = .830)

Fig. 7. Main effect analysis results

Fig. 8. The statistical comparison results of MSI

Compared Results of Mental Workload Status under Three Level Mental Workloads. The NASA TLX scores of six dimensions under three mental loads were compared in pairs. The comparison results are shown in Tables 1–3. From the results of pairwise comparison, we can see that there are significant differences in mental and physical demands between high and medium, medium and low, and high and low task loads ($P < 0.05$). However, there is no significant difference in performance level among high, medium and low task loads ($P > 0.05$). The time requirement ($P = 0.026 < 0.05$), effort ($P = 0.035 < 0.05$) and frustration ($P = 0.024 < 0.05$) were significantly different between low load and high load, but there was no significant difference between low load and medium load, middle load and high load.

3.2 Comparisons of Unit Shooting Times Under Different Task Difficulty

In order to detect the dynamic change of task difficulty in the course of tank firing task, the number of shooting times in unit time was extracted. The time of each task was divided into five sections, and the number of shooting times in each period was calculated. Because the time required for each task is different, and the completion time of each task is different, it is necessary to normalize the number of shooting times to get the number of shooting times per unit time of each task difficulty. Matched t-test was used to test the number of shooting times per unit time under each task difficulty (Table 4).

Table 1. Comparison results of NASA-TLX between low load and medium load

Dimensionality	T	df	95% confidence interval of mean value		Sig.
			Upper limit	Lower limit	
Mental demand	−3.743	16	−19.8105	−5.4836	0.002*
Physical demand	−2.514	16	−20.6015	−1.7515	0.023*
Temporal demand	−1.487	16	−19.9788	3.5082	0.157
Performance level	−0.209	16	−9.8448	8.0801	0.837
Effort level	−1.552	16	−18.0942	2.8001	0.140
Frustration level	−0.550	16	−18.5502	10.9032	0.590

Table 2. Comparison results of NASA TLX between low load and high load

Dimensionality	T	df	95% confidence interval of mean value		Sig.
			Upper limit	Lower limit	
Mental demand	−3.786	16	−43.1282	−12.1659	0.002*
Physical demand	−3.521	16	−36.2842	−9.0100	0.003*
Temporal demand	−2.451	16	−38.3963	−2.7802	0.026*
Performance level	−1.294	16	−24.0589	5.8236	0.214
Effort level	−2.305	16	−18.0942	2.8001	0.035*
Frustration level	−2.497	16	−25.0178	−2.0411	0.024*

Table 3. Comparison results of NASA TLX between medium load and high load

Dimensionality	T	df	95% confidence interval of mean value		Sig.
			Upper limit	Lower limit	
Mental demand	−2.268	16	−29.0219	−0.9781	0.038*
Physical demand	−2.271	16	−22.1773	−0.7639	0.037*
Temporal demand	−2.037	16	−25.2105	−0.5047	0.059
Performance level	−1.401	16	−20.6956	4.2250	0.180
Effort level	−1.213	16	−22.6333	6.1627	0.243
Frustration level	−1.749	16	−21.4680	2.0563	0.099

The results showed that there were significant differences in shooting times per unit time between low load and medium load, low load and high load tasks ($P = 0.000 < 0.05$), and there was no significant difference in shooting times per unit time between medium load and high load ($P = 0.982 > 0.05$). From the statistical results shown in Fig. 9, it can be seen that with the increase of task difficulty, the average number of shootings per unit time showed an upward trend.

Table 4. Comparison of shooting times in unit time under different task difficulties

Group	T	df	95% confidence interval of mean value		Sig.
			Upper limit	Lower limit	
Low load-medium load	−5.851	15	−77.169	−35.956	0.000*
Medium load-high load	−5.621	15	−78.184	−35.191	0.000*
Medium load-high load	−0.023	15	−11.861	11.611	0.982

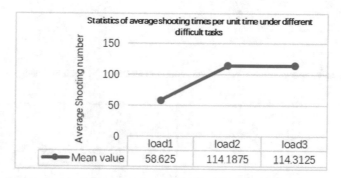

Fig. 9. Comparison of shooting times per unit time under different task difficulty

3.3 EEG Signal Processing Results of Subjects in Tank Shooting Mission

Each EEG eigenvalue was analyzed. The results are shown in Fig. 10.

As can be seen from Fig. 10, the trend of the relative energy value of α increased first and then reduced, and further increased with the task under low load state, and increased first and then reduced with the task under medium load and high load state. The trend of the relative energy value of β shows the tendency of increasing first and decreasing then with the task as a whole. The relative energy value of α/θ is increasing with the task when it is in low load state, and first increasing and then decreasing with the task in the middle load state, and decreasing first and increasing then in the high load state. The trend of the relative energy value of β/θ increased first and then reduced, and further increased with the task under medium load state, while reduced first and then increased, and further reduced under high load state, and increased first and then reduced under low load state. The relative energy value of $(\alpha + \beta)/\theta$ increased first and then reduced under high load state, and reduced first and then increased, and further reduced under both medium and low load state.

The comparison results of the energy values of each parameter in the pillow area and the frontal area under different task difficulties are shown in Figs. 11, 12 and Table 5.

As can be seen from Figs. 12 and 13, the energy values of α, β, α/θ, β/θ, $(\alpha + \beta)/\theta$ parameters all increase with the difficulty of the task. For the relative energy value of θ, two brain regions showed the highest energy value in the middle load. In pairs, it is found that the significance of θ energy value with different task difficulty is not

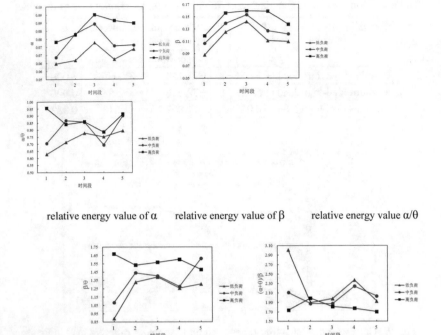

relative energy value of α   relative energy value of β   relative energy value α/θ

relative energy value of β/θ   relative energy value of(α+β)/ θ

Fig. 10. Processing results of EEG eigenvalues

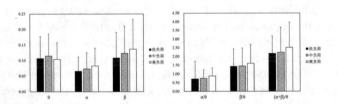

Fig. 11. Comparison results of each energy values in the frontal area with different task difficulty

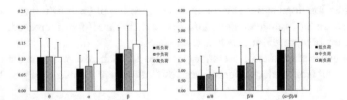

Fig. 12. Comparison results of each energy values in the pillow area with different task difficulty

Table 5. Comparison results of the energy values in different brain areas in different task difficulties

			θ	α	β	α/θ	β/θ	(α + β)/θ
F	Low	Medium	/	<0.05	<0.05	/	/	/
	Low	High	/	<0.05	<0.05	<0.05	/	<0.05
	Medium	High	/	<0.05	<0.05	<0.05	/	<0.05
O	Low	Medium	/	<0.05	<0.05	/	/	/
	Low	High	/	<0.05	<0.05	<0.05	<0.05	<0.05
	Medium	High	/	<0.05	<0.05	/	<0.05	<0.05

obvious, and the energy values of α and β are significantly different (p < 0.05) between low load and medium load, low load and high load, and medium load and high load, which means that these two parameters have good sensitivity to different task difficulty. For α/θ, β/θ and (α + β)/θ, the significance of middle load and low load in frontal area and pillow area is not obvious, especially the difference of β/θ value of frontal area in all task difficulty comparison is not significant. This indicate that in EEG index, the energy value of α and β can effectively evaluate the level of mental load.

4 Conclusion

Statistical analysis of subjective performance shows that subjective questionnaires and performance data can objectively reflect the level of mental load of different task difficulties, and the three task difficulties can distinguish the different mental load needs of the subjects. Subjective and performance data can be used as a basis for assessing the level of mental load. The number of shooting per unit time corresponding to the difficulty of different tasks can dynamically monitor the changes of mental load in the process of tasks. Specifically, from the analysis results the of the average shooting times under different task difficulty, it is not difficult to see that with the increase of task difficulty, the mental load level of the subjects increased correspondingly, and the average shooting times per unit time increased also. This is because from the low load tank firing mission to the medium load tank firing mission, friendly tanks disappeared, and 15 enemy type II tanks increased, the task difficulty increased significantly. Without the help of friendly tanks, and the number of enemy tanks doubled, the subjects themselves need to identify and hit two kinds of enemy tanks independently. The amount of information processed per unit time increased, and the decision-making mode was adjusted greatly. Therefore, the average number of unit shootings from low to medium load tasks had significant difference, and the upward trend was obvious. From medium load to high load, the number of friendly tanks increased, which helped the subjects to share part of the task of hitting enemy tanks, but the number of enemy tanks did not change. S compared with medium load tasks, the only task that the subjects increased was to protect friendly tanks from being hit. With the help of friendly tanks, the difficult degree of deal with enemy tanks was relatively reduced. So the average number of shootings per unit time of subjects from middle to high load tasks increased only slightly, not significant. There was no significant difference in the average number of shooting times per unit time between middle and high load tasks. It shows that the average number of shooting times per unit time can reflect the changing trend of mental load. Therefore, with the increase of task difficulty, the average number of shootings per unit time increases, and the mental load of the subjects show an upward trend.

Studies on EEG changes caused by load show that when the subjects are in a higher mental load state, slow wave increases (δ, θ) and fast wave decreases (α, β). At present, many studies regard the ratio of rhythmic waves, such as $(\alpha+\theta)/\beta$, α/β, $(\alpha+\theta)/(\alpha+\beta)$ and (θ/β), as the detection indicators of mental load [8]. In this paper, the energy values of five parameters of α, β, β/θ, α/θ and $(\alpha + \beta)/\theta$ in frontal area, central area, top area and pillow area under different task difficulty are measured and analyzed. In the experiment, when the subjects carried out the simulated tank shooting task, they mainly involved visual processing, decision judgment and motion control, while the pillow area was related to the visual cognitive function, and the frontal area was related to motion control and reasoning. The results of statistical analysis show that the five parameters of α, β, β/θ, α/θ and $(\alpha + \beta)/\theta$ in frontal area and pillow area have good sensitivity to the level of mental load, which is consistent with the situation of this experiment. Further comparative analysis between different loads shows that α, β two indexes can effectively evaluate the level of mental load.

References

1. Rouse, W.B., Edwards, S.L., Hammer, J.M.: Modeling the dynamics of mental workload and human performance in complex systems. IEEE Trans. Syst. Man Cybern. **23**, 1662–1671 (1993)
2. O'Donnell, R.D., Eggemeier, F.T.: Workload assessment methodology. Handb. Percept. Hum. Perfor. **2**, 42–49 (1986)
3. Jianqiao, L.: Mental workload and its measurement. J. Syst. Eng. **10**(3), 119–123 (1995)
4. Gao, Q., Wang, Y., Song, F., et al.: Mental workload measurement for emergency operating procedures in digital nuclear power plants. Ergonomics **56**(7), 1070–1085 (2013)
5. Hart, S.G., Staveland, L.E.: Development of NASA-TLX (task load index): results of empirical and theoretical research. Adv. Psychol. **52**(6), 139–183 (1988)
6. Zhongqi, L., Xiugan, Y., Liu, T., et al.: Mental workload measurement technology in aeronautical ergonomics. Ergonomics **9**(2), 19–22 (2003)
7. Jap, B.T., Lal, S., Fischer, P., et al.: Using EEG spectral components to assess algorithms for detecting fatigue. Expert Syst. Appl. Int. J. **36**(2), 2352–2359 (2009)
8. Lal, S.K., Craig, A.: Driver fatigue: electroencephalography and psychological assessment. Psychophysiology **39**(3), 313–321 (2002)

Comparative Analysis of Machine Learning Techniques in Assessing Cognitive Workload

Colin Elkin[✉] and Vijay Devabhaktuni

ECE Department, Purdue University Northwest, Hammond, IN, USA
{celkin,vjdev}@pnw.edu

Abstract. Cognitive workload refers to the amount of mental capacity that is stored in working memory and remains a vital yet challenging aspect of many human factors applications. This paper presents a comprehensive analysis of alternatives for different machine learning techniques used for assessing mental workload. This methodology consists of five techniques and four datasets. Each dataset consists of physiological factors such as electrodermal activity (EDA) or heart rate data as well as subjective factors. When evaluating our techniques, we compare accuracy, runtime, and F1 score across multiple different method-specific experimental parameters. Ultimately, the results indicate that ANNs perform best on datasets with large numbers of inputs and classification options, while decision trees are most ideal under a lower number of output possibilities. By understanding these individual strengths and weaknesses, we can ultimately improve the balance of ideal human performance through early detection of cognitive overload or underload.

Keywords: Cognitive workload · Machine learning ·
Artificial neural networks · Support vector machines · K-nearest neighbors ·
Decision trees · Random forests

1 Introduction

The ability to analyze and predict cognitive workload remains an interesting and open-ended research challenge in many applications. In addition, machine learning is a rapidly growing interdisciplinary field that has greatly aided in the prediction and classification of mental load. Many machine learning (ML) techniques that have contributed to this field include artificial neural networks (ANNs) [1, 2], support vector machines (SVMs) [3–6], k-nearest neighbors (KNN) [7], decision trees (DTs) [7], and random forests (RFs).

In general, cognitive load can be perceived is the amount of mental effort that is stored in working memory. It can be perceived as a balancing act between overload (e.g. driving in downtown Washington traffic while talking to passengers, using a smartphone, and listening to the radio) and underload (e.g. driving on a rural highway with other no other vehicles on the road, no passengers, and the radio turned off). Thus the overall goal is to develop mechanisms to reduce potential overload while alerting individuals of potential underload. Based on prior research efforts, detection and prevention of mental overload appear to be the more established objectives, particularly in

© Springer Nature Switzerland AG 2020
H. Ayaz (Ed.): AHFE 2019, AISC 953, pp. 185–195, 2020.
https://doi.org/10.1007/978-3-030-20473-0_19

applications such as general cognitive tests [8], science education [9], and even ordinary driving [10–12].

Many works have succeeded in comparing ML techniques to some capacity, including ANNs with SVMs [13] and different training techniques for supervised and unsupervised ANNs [14]. This paper extends these milestones by comparing five different techniques across four different physiological datasets.

2   Methods and Materials

A generalized flowchart of the analysis of alternatives process is presented in Fig. 1.

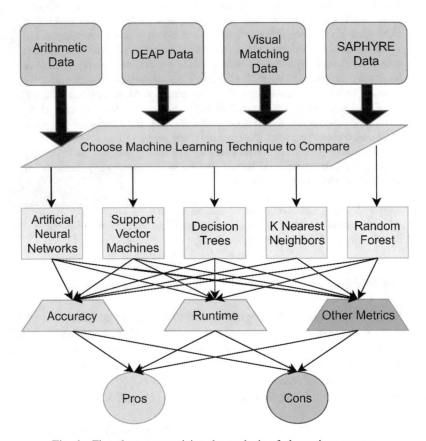

Fig. 1. Flowchart summarizing the analysis of alternatives process.

The initial step is to select the dataset to be tested. A description of the datasets used is provided in the next subsection. Next is to select a machine learning technique to be evaluated. In doing so, three method-specific parameters are compared: two of which are numerical and iterative, and the third of which is more subjective and typically consists of two options. The following step is to evaluate the method based on fundamental metrics such as accuracy and computational runtime. Other metrics for consideration include F1 score, precision, and recall. In viewing and analyzing the results of these metrics, the final step is to determine the pros and cons of the given machine learning method.

We wrote the overall process in Python and incorporated the scikit-learn package [15] for access to built-in machine learning functions. The experiment was performed using the Anaconda Python environment on a Windows 10 desktop with an Intel Core i5-3570 3.40 GHz quad core CPU and 16 GB of RAM. In addition, the results were presented as the averages of 10 independent trials for redundancy and consistency.

2.1 Data Acquisition

To thoroughly compare our candidate ML techniques, we used four different physiological datasets. Each dataset involves a classification problem of some set of cognitive activities. For instance, adding two repeatedly or subtracting seven repeatedly requires increasingly high amounts of mental load, and thus the goal here is to determine which activity took place at a given time step given the physiological data values at that time and accurately classify which activity (including which specific iteration of add two or subtract seven) took place. By doing so and by interpreting the amount of cognitive load anticipated in each activity, meaningful workload assessment can take place.

One dataset utilized in this research is based on an experiment by [16] in 2016. Natarajan et al. collected the data for the purposes of evaluating fear conditioning and cognitive load. In turn, the data has been formatted so that it focuses entirely on the latter. To address cognitive workload, the data measures electrodermal activity (EDA) during a series of tests in which a human subject is required to either add 2, subject 7, or rest for 30-s periods. These mental tasks were chosen for the specific purpose of assessing cognitive workload, based on an individual's ability to remember a number and repeatedly, as well as interchangeably, add to or subtract from it. For this research, we repurposed the data to ask the question: *Given a subject number, his/her EDA, and a time step, what task was he/she performing?*

We obtained another set named Database for Emotion Analysis using Physiological Signals (DEAP) from [17], which examines emotional response to different music videos. Given the sheer amount of visual creativity involved in many music videos, it can be established that the amount of cognitive load required to absorb such videos can vary rapidly. The inputs consist of EEG in 40 channels as well as time step and subject number, while the output is simply the video being watched on the condition that each video has a varying level of cognitive load required to adequately absorb what is being watched. Therefore, this data can be modeled based on the question: *Given a time step, a subject number, and the subject's EEG, what task was he or she performing?*

A third dataset was derived from [18], which examines genetic predisposition among alcoholic and control subjects based on EEG correlation. Over the course of 122 subjects and 120 trials, the cognitive activity consisted of being shown either a single image, two matching images, or two non-matching images. Each of these three activities was considered as separate classifications between alcoholic subjects and control subjects, hence six possible output choices. Thus, this classification problem is modeled from the modified question: *Given a time step, a subject number, a trial number, and the subject's EEG across 64 channels, what task was he or she per-forming?* Figure 2 depicts an overview of this data's machine learning process as well as the inputs and outputs.

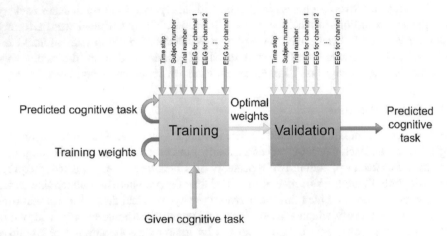

Fig. 2. Flowchart of the general machine learning process for the visual matching dataset.

Finally, one last dataset comes from the project Sliding Scale Autonomy through Physiological Rhythm Evaluations (SAPHYRE) as detailed in [19] and consists of both physiological (in this case, heart rate) and subjective (e.g. altitude, throttle, rudder) inputs of pilots. From here, the output classification is a combination of flight route, flight path, and pilot skill level, which is given by

$$C = 5\mathrm{sp} + \mathrm{r}. \tag{1}$$

where C is the output classification, s denotes a pilot's skill level (1 for expert, 0 for novice), p is the path number, and r represents the route number. This is on the basis that cognitive workload levels could vary with skill level. In other words, a more experienced pilot may need less mental load to complete the same task undertaken by a novice pilot. This is also based on the fact that the complexity of different routes or paths could also contain varying levels of cognitive load. Thus, the overall problem given by this data models the question: *Given the above inputs, what are the route, path, and pilot skill level?*

2.2   Evaluation Metrics

Three metrics were selected for evaluation: accuracy, computational runtime, and F1 score. Accuracy is divided between training and validation data. In either case, it is the ratio of correct classifications to total number of classifications possible. Here, we focus primarily on the former. The F1 score is determined by the formula

$$F1 = 2 * (\text{precision} * \text{recall})/(\text{precision} + \text{recall}). \qquad (2)$$

With exception of runtime, all of the evaluation scores are normalized values, in which 1 is a perfect score and 0 is the lowest possible.

3 Experimental Results

This section provides the objective, numerical results for all machine learning methods and datasets. Each dataset is separated by subsection.

3.1 SAPHYRE Data

The average results for validation accuracy (denoted as "Val"), computational runtime, and F1 score for each ML technique and for each given subjective parameter are provided below in Table 1.

Table 1. Summary of results for SAPHYRE dataset.

Technique	Mean val	Mean runtime (s)	Mean F1 score
ANN SGD	.565	28.0	.429
ANN L-BGFS	.511	69.3	.376
SVM linear	.960	9.47	.940
SVM RBF	.981	34.7	.977
KNN uniform	.966	10.2	.959
KNN distance	.977	9.09	.973
DT Gini	.989	8.34	.982
DT entropy	.989	7.59	.984
RF Gini	.992	9.03	.988
RF entropy	.993	8.42	.989

For validation accuracy, the overall values for each method held up quite well, with exception of the ANN, which spanned a wide variety of results. Overall, RF produces the highest possible accuracy values as well as the highest minimum and average values. DT is a close second, followed by a near tie between KNN and SVM. From here, it is also beneficial to observe the results from a more graphical context, as indicated in the next few figures.

Figure 3 presents the three evaluation metrics for ANN on this dataset with the SGD parameter. As can be observed, the accuracy and F1 score vary by a wide amount. The SVM results under the linear kernel are given below in Fig. 4.

As observed here, validation accuracy and F1 score vary proportionally by the penalty term and have little change with respect to the gamma value. The runtime follows a similar approach for the latter axis but carries an inverse proportionality on the former.

(a)   (b)   (c)

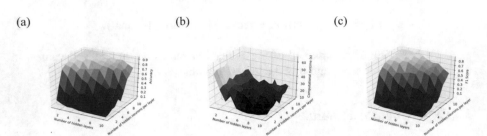

Fig. 3. Surface plots of (a) validation accuracy, (b) computational runtime, and (c) F1 score for ANN on SAPHYRE data in terms of number of hidden layers and number of hidden neurons per layer.

(a)   (b)   (c)

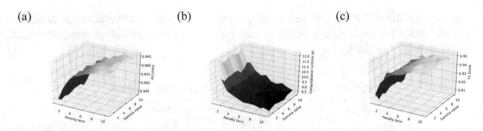

Fig. 4. Surface plots of (a) validation accuracy, (b) computational runtime, and (c) F1 score for SVM on SAPHYRE data in terms of penalty term and gamma value.

KNN results under the distance parameter are then provided in Fig. 5. Here, the computational runtime increases with the number of neighbors and decreases with leaf size. The other two metrics, however, bear similarity to the SVM runtime. Figure 6 shows the decision tree results with the gini criterion, in which there are no noticeable patterns in runtime. The other two metrics bear inverse proportionality to the minimum samples required for leaf and are virtually unaffected by the minimum samples for split. We also observe, however, there is less than one percent variation in these two metrics.

(a)   (b)   (c)

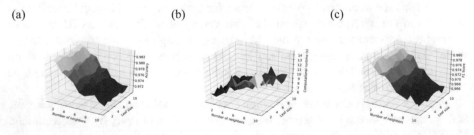

Fig. 5. Surface plots of (a) validation accuracy, (b) computational runtime, and (c) F1 score for KNN on SAPHYRE data in terms of number of neighbors and leaf size.

(a)   (b)   (c)

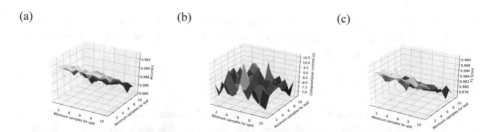

Fig. 6. Surface plots of (a) validation accuracy, (b) computational runtime, and (c) F1 score for DT on SAPHYRE data in terms of minimum number of samples for split and for leaf.

These same trends are also visible in Fig. 7.

(a)   (b)   (c)

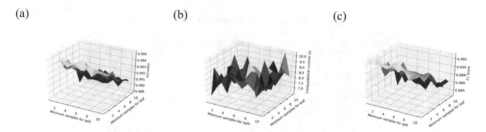

Fig. 7. Surface plots of (a) validation accuracy, (b) computational runtime, and (c) F1 score for RF on SAPHYRE data in terms of minimum number of samples for split and for leaf.

3.2   Arithmetic Data

We then present the results for the arithmetic data in Table 2 in a manner similar to that of the previous table.

Table 2. Summary of results for arithmetic dataset.

Technique	Mean val	Mean runtime (s)	Mean F1 score
ANN SGD	.706	8.68	.567
ANN L-BGFS	.703	10.8	.570
SVM Linear	.436	7.21	.121
SVM RBF	.983	4.38	.979
KNN uniform	.994	.348	.986
KNN distance	1.00	.353	.990
DT Gini	.999	.210	.999
DT entropy	.999	.216	.999
RF Gini	.999	.318	.998
RF entropy	.999	.368	.997

In terms of accuracy alone, seven of the ten given technique combinations show highly promising results for this dataset by providing at least 99% average accuracy. As with the previous dataset, decision trees provide the best computational runtime. Combined with their high accuracy, this method serves as the best balance of accuracy and runtime for this dataset.

3.3   DEAP Data

From here, Table 3 presents the results for the DEAP dataset. The key observation here is that none of the given techniques are able to model this data with stellar accuracy. ANN comes close by producing over 70% accuracy, but all other methods remain below 20%. In addition, the F1 score does not hold well in any of the cases. ANN also provides the best success in runtime, thereby serving as the best candidate technique for this data.

Table 3. Summary of results for DEAP dataset.

Technique	Mean val	Mean runtime (s)	Mean F1 score
ANN SGD	.724	6.45	.281
ANN L-BGFS	.722	18.1	.305
SVM linear	.098	54.1	.090
SVM RBF	.165	84.9	.163
KNN uniform	.151	64.5	.145
KNN distance	.160	50.2	.159
DT Gini	.175	16.6	.173
DT entropy	.178	21.6	.177
RF Gini	.174	18.3	.173
RF entropy	.173	21.4	.171

3.4   Visual Matching Data

Finally, Table 4 provides the results for the EEG visual matching dataset.

Table 4. Summary of results for visual matching dataset.

Technique	Mean val	Mean runtime (s)	Mean F1 score
ANN SGD	.724	6.20	.280
ANN L-BGFS	.723	14.6	.304
SVM linear	.724	36.7	.280
SVM RBF	.723	138	.376
KNN uniform	.702	77.7	.359
KNN distance	.693	74.5	.379
DT Gini	.915	25.3	.605
DT entropy	.911	16.1	.602
RF Gini	.727	25.2	.350
RF entropy	.727	16.2	.350

The accuracy here fares better than under the previous data but not as well as the SAPHYRE and arithmetic data. ANN and SVM provide the best accuracy, but SVM comes at the expense of substantially higher runtime. Overall, decision trees appear to provide the best most optimal balance of accuracy, runtime, and F1 score.

4 Discussion

Following a thorough presentation of results in graphical and tabular forms, it is equally beneficial to observe more subjective characteristics in the latter layout. By doing so, we can view trends for all techniques in all datasets from a single view.

Table 5 presents a summary of the results from a subjective standpoint. The blank cells indicate that there is nothing noteworthy to report. The single letters after each ML technique denotes an abbreviation of the subjective experimental parameter used. One key observation here is that DT and RF have the best mix of results in most cases except on the DEAP data, in which case ANN does best. SVM tends the fare on the low end of most metrics, including runtime of the RBF kernel and most other metrics under the linear kernel. ANN typically produces a wide variety of results ranging from substantially high values to substantially low ones.

Table 5. Summary of subjective results for each dataset.

Technique	Arithmetic	DEAP	Visual matching	SAPHYRE
ANN S	Wide range results	Best in most	Best runtime	Widest results
ANN L	Wide range results	Best in most	–	Highest max time
SVM L	Worst in all	Worst in most	–	–
SVM R	Worst max time	Worst runtime	Worst time	–
KNN U	Overfitting	Worst max time	Overfitting	–
KNN D	Overfitting	2nd worst time	Overfitting	–
DT G	Best all metrics	–	Best in most	–
DT E	Best all but time	–	2nd best in most	Best time
RF G	2nd best in most	–	Overfitting	Best max acc
RF E	3rd best in most	–	Overfitting	Best in most

Table 6 provides a summary of the four datasets as well as most optimal machine learning technique recommended for each one. When observing the numerical characteristics of each set of data, it useful to rewrite and reorganize the results into a simpler form as portrayed next to the numbers in the table. As the results in this table indicate, there is a coincidentally balanced combination of attributes in each dataset, in which number of inputs and number of possible output classifications is either large or small. Out of the four datasets, each combination of attributes has either two large values, two small values, or a combination of each. Hence, this approach is highly beneficial in determining the thresholds in which to select a particular ML method. In this case, 20 appears to be a well-rounded threshold value that can effectively separate high values from low values. It should also be noted that SVM and KNN did not produce results noteworthy enough to be considered in this simplified summary.

Table 6. Simplified summary of results for each dataset.

Dataset	No. inputs	No. classifications	Best method
Arithmetic	3 (Lo)	5 (Lo)	DT
Visual matching	66 (Hi)	3 (Lo)	DT
DEAP	42 (Hi)	49 (Hi)	ANN
SAPHYRE	16 (Lo)	165 (Hi)	RF

5   Conclusions and Future Work

This paper presented a comprehensive analysis of the advantages and disadvantages of five candidate machine learning techniques on cognitive workload assessment. This was conducted on four relevant datasets under a variety of different experimental parameters. The results showed decision trees as the best overall technique for the arithmetic and visual matching datasets, random forest for SAPHYRE data, and artificial neural networks for the DEAP data.

These observations provide a significant first step in the overall scope of this research, as the different machine learning techniques can be analyzed and selected based on the cognitive workload data being given. By being able to dynamically decide the most accurate and most efficient technique at any given point in time, early detection of mental overload and underload given some physiological data can be done easily and quickly in order to most effectively optimize many human factors applications.

Acknowledgments. This research is supported by the Dayton Area Graduate Studies Institute (DAGSI) fellowship program for a project titled "Assessment of Team Dynamics Using Adaptive Modeling of Biometric Data." The authors wish to thank their DAGSI sponsor Dr. Gregory Funke for his continued guidance and support throughout the project.

References

1. Wilson, G.F., Russell, C.A.: Real-time assessment of mental workload using psychophysiological measures and artificial neural networks. Hum. Factors **45**(4), 635–644 (2016)
2. Baldwin, C.L., Penaranda, B.N.: Adaptive training using an artificial neural network and EEG metrics for within- and cross-task workload classification. NeuroImage **59**(1), 48–56 (2012)
3. Jin, L., et al.: Driver cognitive distraction detection using driving performance measures. In: Discrete Dynamics in Nature and Society (2012)
4. Son, J., Oh, H., Park, M.: Identification of driver cognitive workload using support vector machines with driving performance, physiology and eye movement in a driving simulator. Int. J. Precis. Eng. Manuf. **14**(8), 1321–1327 (2013)

5. Putze, F., Jarvis, J., Schultz, T.: Multimodal recognition of cognitive workload for multitasking in the car. In: 2010 20th International Conference on Pattern Recognition (ICPR), pp. 3748–3751. IEEE Press, New York (2010)
6. Liang, Y., Reyes, M., Lee, J.: Real-time detection of driver cognitive distraction using support vector machines. IEEE Trans. Intell. Transp. Syst. 8(2), 340–350 (2007)
7. Solovey, E., et al.: Classifying driver workload using physiological and driving performance data: two field studies. In: Proceedings of the SIGCHI Conference on Human Factors in Computing Systems, CHI 2014, pp. 4057–4066. ACM, New York (2014)
8. Berka, C., et al.: EEG correlates of task engagement and mental workload in vigilance, learning, and memory tasks. Aviat. Space Environ. Med. 78(5), B231–B234 (2007)
9. Niaz, M., Logie, R.H.: Working memory, mental capacity and science education: towards an understanding of the 'working memory overload hypothesis'. Oxford Rev. Educ. 19(4), 511–525 (1993)
10. Lenneman, J., Backs, R.: Cardiac autonomic control during simulated driving with a concurrent verbal working memory task. Hum. Factors 51(3), 404–418 (2009)
11. Reimer, B., Mehler, B.: The impact of cognitive workload on physiological arousal in young adult drivers: a field study and simulation validation. Ergonomics 54(10), 932–942 (2011)
12. Collet, C., et al.: Physiological and behavioural changes associated to the management of secondary tasks while driving. Appl. Ergon. 40(6), 1041–1046 (2009)
13. Elkin, C., Nittala, S., Devabhaktuni, V.: Fundamental cognitive workload assessment: a machine learning comparative approach. In: 8th International Conference on Applied Human Factors and Ergonomics (AHFE), pp. 275–284. Springer, Cham (2017)
14. Elkin, C., Devabhaktuni, V.: Analysis of alternatives for neural network training techniques in assessing cognitive workload. In: 9th International Conference on Applied Human Factors and Ergonomics (AHFE 2018), pp. 27–37. Springer, Cham (2018)
15. Pedregosa, F., et al.: Scikit-learn: machine learning in python. J. Mach. Learn. Res. 12, 2825–2830 (2011)
16. Natarajan, A., Xu, K.S., Eriksson, B.: Detecting divisions of the autonomic nervous system using wearables. In: IEEE 38th Annual International Conference of the Engineering in Medicine and Biology Society (EMBC), pp. 5761–5764. IEEE Press, New York (2016)
17. Deap: a dataset for emotion analysis using eeg, physiological and video signals. http://www.eecs.qmul.ac.uk/mmv/datasets/deap/
18. UCI machine learning repository, University of California, Irvine, School of Information and Computer Sciences. http://archive.ics.uci.edu/ml
19. Nittal, S., et al.: Pilot skill level and workload prediction for sliding-scale autonomy. In: 17th IEEE International Conference on Machine Learning Applications (ICMLA), pp. 1166–1173. IEEE Press, New York (2018)

Virtual Reality and Machine Learning
for Neuroergonomics

A Test Setting to Compare Spatial Awareness on Paper and in Virtual Reality Using EEG Signals

Sander Van Goethem, Kimberly Adema, Britt van Bergen,
Emilia Viaene, Eva Wenborn, and Stijn Verwulgen(✉)

Faculty of Design Science, Product Development,
University of Antwerp, Antwerp, Belgium
{sander.vangoethem, stijn.verwulgen}@uantwerpen.be

Abstract. Spatial awareness and the ability to analyze spatial objects, manipulate them and assess the effect thereof, is a key competence for industrial designers. Skills are gradually built up throughout most educational design programs, starting with exercises on technical drawings and reconstruction or classification of spatial objects from isometric projections and CAD practice. The accuracy in which spatial assignments are conducted and the amount of effort required to fulfill them, highly depend on individual insight, interests and persistence. Thus each individual has its own struggles and learning curve to master the structure of spatial objects in aesthetic and functional design. Virtual reality (VR) is a promising tool to expose subjects to objects with complex spatial structure, and even manipulate and design spatial characteristics of such objects. The advantage of displaying spatial objects in VR, compared to representations by projecting them on a screen or paper, could be that subjects could more accurately assess spatial properties of and object and its full geometrical and/or mechanical complexity, when exposed to that object in VR. Immersive experience of spatial objects, could not only result in faster acquiring spatial insights, but also potentially with less effort. We propose that acquiring spatial insight in VR could leverage individual differences in skills and talents and that under this proposition VR can be used as a promising tool in design education. A first step in underpinning this hypothesis, is acquisition of cognitive workload that can be used and compared both in VR and in a classical teaching context. We use electroencephalography (EEG) to assess brain activity through wearable plug and play headset (Wearable Sensing-DSI 7). This equipment is combined with VR (Oculus). We use QStates classification software to compare brain waves when conducting spatial assessments on paper and in VR. This gives us a measure of cognitive workload, as a ratio of a resulting from subject records with a presumed 'high' workload. A total number of eight records of subjects were suited for comparison. No significant difference was found between EEG signals (paried t-test, p = 0.57). However the assessment of cognitive workload was successfully validated through a questionnaire. The method could be used to set up reliable constructs for learning techniques for spatial insights.

Keywords: Spatial awareness · Virtual reality · EEG ·
Brain-computer interface · Platonic solids · Cognitive workload

© Springer Nature Switzerland AG 2020
H. Ayaz (Ed.): AHFE 2019, AISC 953, pp. 199–208, 2020.
https://doi.org/10.1007/978-3-030-20473-0_20

1   Introduction

Spatial perception or spatial awareness constitutes a significant aspect of reality and virtual reality (VR) [1]. It can be interpreted as the capacity to create a cognitive representation of an object, allowing people to identify an object from different angles [2]. The difference in spatial awareness in reality versus VR is that in reality the human being is surrounded by spatial information, while in virtual reality there is only the illusion of being surrounded by spatial information.

According to Eisenberg, spatial perception comprises: (1) recognizing a three-dimensional object from different angles, (2) perceiving the internal structure of a given spatial configuration, (3) determining the spatial relationships between the context and the individual (orientation) [2]. Spatial perception is thus directly related to any kind of inter-action of senses with the environment. However, only vision provides highly accurate and detailed spatial information about three-dimensional properties of external objects; it is used to guide spatial judgments in other modalities as well, vision therefore plays a dominant role in spatial perception [3].

Micro-expressions are also important in training spatial perception, because they translate spontaneous emotions into facial expressions; e.g. when interpreting a spatial environment. These can be measured by facial expression software, in a fixed and controlled context but not easily measured in situ due to the changing environment and behavior of the user. In particular limited input of light hampers measurement when wearing a VR headset. Alternatively, EEG is a promising counterpart to solve these problems [4]. Due to the increased accessibility and portability brain computer interface (BCI) headsets, they are also accessible for research, education and commercial organizations and other sources of interest to finally measure spontaneous emotions [5]. In this study, a dry electrode EEG headset is used to measure the necessary level of cognitive workload to create spatial awareness.

VR and BCI are emerging as accessible methods to measurement of a user's perception [6, 7]; it seems useful to research spatial perception and experience using these tools. Spatial ability can be improved by VR [2, 8, 9], while there have been no definitive results for improving spatial ability with augmented reality (AR) [10, 11].

Platonic solids, also known as the regular polyhedra, are fundamental structures in three-dimensional space. They can be described in terms of the features, sizes, positions, orientations and - most relevant in this study - rotational symmetries of their individual edges and surfaces [12]. This test is conducted in the realm of a design educational program at Bachelor level (Architectural design) aimed at learning the rotational symmetries of Platonic solids. The research question we address is whether and to what extend rotational symmetry are easier and faster acquired in VR than in classical lecturing, using traditional two-dimensional tools.

2   Materials and Methods

Various hardware devices and their software programs were acquired and synchronized. Besides this core equipment, a setting in reality similar to the one in virtual reality was constructed. The entire process was filmed in order to assure that no important or unusual emotional responses were missed.

Subjects were assessed with QUASAR DSI-7 dry electrode EEG headset and the Oculus VR Rift VR headset. The information from this hardware was recorded with provided dedicated data acquisition software; DSI-streamer. Data analysis was done with QUASAR's software program QStates.

The headset hardware [14] consists of seven individually placed sensors positioned to the F3, F4, C3, C4, Pz, P3, P4 locations corresponding international 10–20 system [13]. It includes two linked-ears ear clip mounted active sensors. The EEG headset is easy to place on the participants head and is ready to record EEG in under one minute with wired connection. The EEG headset is worn by the participant during the test in addition to the oculus rift VR headset.

The Oculus headset comprises a head-mounted display that can visually transfer the user into an alternate, digital landscape and two controllers which can be used to interact with this digital world. The Oculus VR Rift Headset is used during part of the test in combination with the QUASAR DSI-7 dry electrode EEG Headset. First, the VR Headset is positioned on the participant's head, the EEG Headset is consecutively placed over the VR Headset. The sensors on the EEG Headset can then be positioned correctly on the participant's head according to the corresponding International 10–20 System.

For this test an empty room was created with a white table positioned in the middle of this room with Oculus VR Software. Ten 3D Platonic solids were presented on this table (Fig. 1). Using the Oculus Touch-controller, these solids can be picked up, rotated and viewed from every angle. The solids are adapted to mimic normal real life conditions, they are affected by gravity and can be stacked on top of each other.

Fig. 1. The environment with platonic solids seen as in Virtual Reality.

EEG signals were recorded, filtered and saved in CSV format with DSI-streamer data acquisition software from QUASAR's DSI-7.

QStates was used to analyze and translate the recorded signals from the DSI-streamer into easily interpretable data output [14]. QStates offers the possibility to create subject specific EEG patterns under controlled external cognitive inputs. To make a training model, at least one minute of EEG data for both a high and low state conditions is needed [19]. In this study, the training inputs are tuned to represent a high workload versus a low workload.

The test was conducted with staff and students at the department Product Development (University of Antwerp). All subjects gave their informed consent. A total of 12 participants were enrolled in this study. Seven male and five female students participated in this study. Their ages range between 20 and 30. These students all have a similar profile. Eight participants produced useful EEF data. Four participants were excluded due to problems with the calibration of the EEG headset. All participants are expected to have a strong spatial ability, due to their nature of field study. The participants were asked to report any nausea or dizziness if experienced during the tests, which could be caused by entering virtual reality.

The tests were conducted in the usability lab at the department of Product Development at the University of Antwerp. All distracting sensory input was excluded from the room; isolating the user from irrelevant influences during the test. Figure 2 shows the floor plan of the test environment combined with the chronology of the experiments and study design. The setup consisted of four tables, four chairs, room dividers, two laptops, a dry electrode EEG headset and a VR headset. In the VR environment and in reality the walls were either white or grey and the tables were both light grey. After measuring the high and the low state of each participant, we tested the case in reality (2D) and in VR (3D). The setting in reality was highly similar to the setting in VR. Subjects were presented a table with ten Platonic solids in VR in 3D and ten Platonic solids printed on paper (represented in parallel projection).

In Fig. 5, the dotted line indicates the path that was taken by the participants. Each participant was welcomed outside of the test environment and introduced to the subject matter of the study. This introduction includes an explanation and short training session on Platonic solids and how to discern their rotational symmetries, for which three-dimensional Platonic solids were used. The test consists of four steps and for each step, at least one minute of EEG data was captured.

While the participant was informed about the first step of the test, the EEG headset was set up on the participant's head. When the signals were considered reliable as indicated by the software, the test commenced. This step intends to measure the "low" state for the QStates training model. The participant was asked to stare blankly ahead at a white wall and relax, while avoiding stressful or complex thoughts. After having measured one minute of the brain activity, the participant moved on to the next step.

The second step intends to capture the 'high' state for the QStates training model. Each participant was presented with 15 different 3D solids and 15 sheets of paper on which the corresponding top- and front view of these solids were printed (Fig. 3). These assignment was adapted from [15]. The participants were asked to match every 3D solids with the corresponding top- and front view. The participants were informed not to talk or not to excessively move during the test. The participant was asked to raise

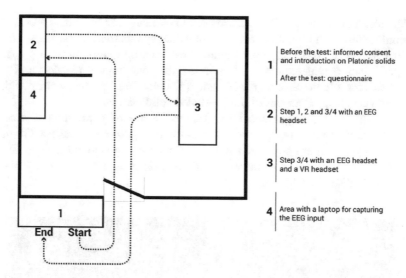

Fig. 2. Floor plan of the test environment and mapping of the subsequent steps in the experiment and study design.

their hand in case the test was completed before the minute was over or if something was unclear. This gesture minimizes interference with the brain measurement using the EEG-software. It was then possible to pause the EEG recorder and restart after helping the participant to proceed with the test.

Fig. 3. 3D solids & top and front view example, used for the "high" setting in the QStates training model.

After creating references for highs and lows, the participants start working with the platonic solids in the third and the fourth step. The order in which these steps take place were alternated for every other participant to prevent priming-effects to influence the results. Both of these steps start with a similar task being presented to the participant:

(1) select all Platonic solids with two, four and five rotational symmetries and (2) select all Platonic solids with two, three and five rotational symmetries. These questions were alternated for every consecutive participant. Two back-up questions were prepared in case a participant completed the first task in less than one minute; (3) elect all Platonic solids with four rotational symmetries and (4) select all Platonic solids with five rotational symmetries. Four out of ten presented solids are correct.

Participants were also presented ten different sheets of paper on the table with Platonic solids in parallel perspective. Due to the transparency of each solid, participants were able to discern any ribs behind the front view or side views (Fig. 4). They were asked to choose the Platonic solids that contain a certain number of rotational symmetries, as discussed before.

Fig. 4. The ten different 3D platonic solids printed on paper, used in step three of the experiment.

The order of VR and paper exposure was aternated between subjects. In step four (or three, depending on the participant), the EEG headset had to be removed from the participants head, to make room for the VR headset. When comfortable, the EEG headset was remounted on top of the VR headset. While ensuring a correct functioning of both devices, the participant had some time to get used to the controls of the Oculus Rift (Fig. 5). The participant was then introduced to the final task. Similarly to the previous step, they had to select the Platonic solids with certain given rotational symmetries.

Finally, the participants were asked to fill in a short questionnaire in which they were questioned about their experiences during the test.

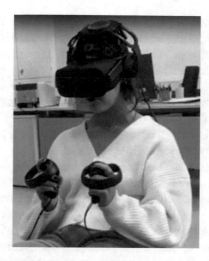

Fig. 5. EEG headset and VR headset mounted onto the participant's head

3 Results

Of the twelve participants, eight produced useable EEG data. Of participant number four, the data was not usable because of inconsistencies. Although the accuracy level for the model measured 100%, the test did not produce realistic results for this participant. Of three remaining participants with unsuccessful data, the EEG did not calibrate successfully throughout the entire test and the procedure was aborted.

This subsequent test measured the discrepancy in experience - more specifically, cognitive workload- between an assignment on two-dimensional sheets of paper versus 3D objects in virtual reality, as described above. The two datasets for this test can be found in Table 1.

Table 1. EEG results through QStates analysis and questionnaire results

Two-dimensional		Virtual reality			Question.: easiest	
1	M	,9148	91,48%	,5680	56,80%	VR
2	V	–	–	–	–	–
3	V	,9870	98,70%	,5610	56,10%	VR
4	V	–	–	–	–	–
5	M	,7281	72,81%	,4864	48,64%	VR
6	V	,6415	64,15%	,7285	72,85%	2D
7	V	,7761	77,61%	,9183	91,83%	2D
8	M	,4454	44,54%	,8168	81,68%	2D
9	V	,6034	60,34%	,3821	38,21%	VR
10	V	,9720	97,20%	,9948	99,48%	VR
11	M	–	–	–	–	–
12	M	–	–	–	–	–

To statistically process these datasets, a Paired Sample T-test was conducted to compare the means of cognitive workload for the test on sheets of paper (2D) versus in virtual reality (VR). Mean values are 0,758537 (75,8537%) and 0,681988 (68,1988%) for the results in 2D and VR respectively. A two-tailed confidence interval of 90%, with a one-tailed significance level of 5% shows that there is no significant difference between the two means, with a two-tailed significance value of 0,457 and thus a one-tailed value of 0,228 (> 0,05). The null hypothesis, that both means are the same, cannot be rejected. It can thus not be stated that the mean of the results in VR is significantly lower from the mean of the results in 2D.

To check whether the results of the test match with the experience of the participant, a short questionnaire was given to the participants afterwards. The most important question was whether the participant experienced the assignment as less difficult in 2D or in VR.

For seven out of eight participants, the EEG results of the test matched their perceived experience. The results of one participant (number ten) did not exactly match the experience. However, it needs to be mentioned that the results, measured with the EEG headset, for the assignment in VR and in 2D, are very close together. In addition, when answering the questionnaire, this participant also doubted which assignment type she experienced to be the least difficult one.

4 Discussion and Conclusion

Four out of eight of the successful tests show that the cognitive workload value for the assignment in virtual reality is lower than the value in the two-dimensional assignment. This would mean that these four participants experienced a lower cognitive workload to understand the Platonic solids to complete the assignment given, which was thus more easily solved. For the remaining four participants that successfully completed the test, this is the other way around.

The participant with the lowest cognitive workload value for the assignment in virtual reality compared to the value for two dimensions is participant number three, with a value for the 2D assignment of 0,9870 (98,70%) and a value for the assignment in virtual reality. This could indicate that this participant needed a relatively similar amount of cognitive workload to complete the test in both 2D as VR.

Before the test started, participants were introduced to the relevant principles of Platonic solids (rotational symmetry and order) so that they could perform the assignments in the test successfully. However, these principles were taught for the first time, which made the tasks more complicated for the. Therefore, an increased level of workload could be possible for each participant who did not know about the symmetries in Platonic solids.

In addition, the EEG headset did not perform properly with every participant. For three of the twelve participants, the headset did not perform well enough to be able to finish the test. The output signals from the headset did not calibrate properly, displaying flat or chaotic output lines in the streamer monitor. For this reason, it was not possible to proceed with the tests using this output, and thus these tests were aborted. It is suggested that the factor responsible for this malfunction is the structure of the hair of the participants, as a thick hair structure could affect and prohibit the connection with the dry EEG sensors. An additional difficulty is that the EEG headset had to be placed over the VR headset for one part of the test. The EEG headset first had to be removed after having completed previous tests and then placed over the VR headset. After this process the EEG signals had to calibrate again, which becomes even more difficult with participants for whom the headset does not calibrate properly even without the VR headset. When processing the data, it also appeared that the results of one participant were not usable. Even though the accuracy level of the model showed a measured value of 100%, the test did not produce realistic results. The result for the VR assignment was 0%, this would suggest that the output for this test was not properly processed. The output for this participant is not included in the results.

It is important to point out that the following assumptions were made to enable valuable conclusions:

- All disruptions caused while measuring brain activity with an EEG headset are filtered out with the software programs QStates and DSI-streamer.
- The VR experience does not influence the results. We give our participants time to get accustomed to the VR environment and the first test each participant takes is alternated, being either the one in VR or the one in 2D.

Finally, it is important to be critical on the ground truth of assessed output values. It was assumed that the measured 'high' and 'low' states were respectively a higher and lower representation of cognitive workload of spatial perception. This, however, has not been proven. QStates is a software program that relies on personal decision and judgment about the 'high' and 'low' states, and does not define why certain changes in experiences happen. QStates is therefore a software program that is partially subjective, and which therefore produces partially subjective results. By cross checking the out-comes with the results of the questionnaire, however, it seemed sufficient to consider the outcomes as reliable for this study.

Furthermore, there is no closure about whether the tests are a proper representation of cognitive workload for spatial perception. These tests could encompass more than only cognitive workload of spatial perception, and could relatively be composed of other, greater or smaller influences. Consequently, it would be of interest to define which variables define the spatial ability in practice, and how these practices differ. Combination of wearable EEG with VR and/or AR could be used to set up reliable constructs for learning techniques for spatial insights. Future research could therefore be conducted to define variables of the spatial ability.

References

1. Henry, D., Furness, T.: Spatial perception in virtual environments: evaluating an architectural application. In: IEEE Virtual Reality Annual International Symposium, pp. 33–40 (1993)
2. Morán, S., Rubio, R., Gallego, R., Suárez, J., Martín, S.: Proposal of interactive applications to enhance student's spatial perception. Comput. Educ. **50**, 772–786 (2008)

3. Eimer, M.: Multisensory integration: how visual experience shapes spatial perception. Curr. Biol. **14**, R115–R117 (2004)

4. Benlamine, M.S., Chaouachi, M., Frasson, C., Dufresne, A.: Physiology-based recognition of Facial Micro-expressions using EEG and Identification of the relevant sensors by emotion. In: PhyCS, pp. 130–137 (2016)

5. Ekandem, J.I., Davis, T.A., Alvarez, I., James, M.T., Gilbert, J.E.: Evaluating the ergonomics of BCI devices for research and experimentation. Ergonomics **55**, 592–598 (2012)

6. Loomis, J.M., Philbeck, J.W.: Measuring spatial perception with spatial updating and action. In: Carnegie Symposium on Cognition, 2006, Pittsburgh, PA, US (2008)

7. Moridis, C.N., Terzis, V., Economides, A.A., Karlovasitou, A., Karabatakis, V.E.: Using EEG frontal asymmetry to predict IT user's perceptions regarding usefulness, ease of use and playfulness. Appl. Psychophysiol. Biofeedback **43**, 1–11 (2018)

8. Creem-Regehr, S.H., Willemsen, P., Gooch, A.A., Thompson, W.B.: The influence of restricted viewing conditions on egocentric distance perception: implications for real and virtual indoor environments. Perception **34**, 191–204 (2005)

9. Faria, R.R.A., Zuffo, M.K., Zuffo, J.A. Improving spatial perception through sound field simulation in VR. In: Proceedings of the 2005 IEEE International Conference Virtual Environments, Human-Computer Interfaces and Measurement Systems, p. 6 (2005)

10. Kaufmann, H., Steinbügl, K., Dünser, A., Glück, J.: Improving spatial abilities by geometry education in augmented reality-application and evaluation design. In: Proceedings of the Virtual Reality International Conference (VRIC), pp. 25–34 (2005)

11. Franz, G., von der Heyde, M., Bülthoff, H.H.: The influence of the horizon height on spatial perception in VR. Horizon **47**, 66 (2004)

12. Pani, J.R., Zhou, H.-G., Friend, S.M.: Perceiving and imagining Plato's solids: the generalized cylinder in spatial organization of 3D structures. Vis. Cogn. **4**, 225–264 (1997)

13. Homan, R.W., Herman, J., Purdy, P.: Cerebral location of international 10–20 system electrode placement. Electroencephalogr. Clin. Neurophysiol. **66**, 376–382 (1987)

14. McDonald, N.J., Soussou, W.: Quasar's Qstates cognitive gauge performance in the cognitive state assessment competition 2011. In: 2011 Annual International Conference of the IEEE Engineering in Medicine and Biology Society, EMBC, pp. 6542–6546 (2011)

15. Corremans, J., Maes, W., Feys, S.: Basisvaardigheden in ruimtelijk denken: meer dan 350 oefeningen om het ruimtelijk inzicht te oefenen en het denken in twee en drie dimensies te bevorderen (2006)

Multimodal fNIRS-EEG Classification Using Deep Learning Algorithms for Brain-Computer Interfaces Purposes

Marjan Saadati[1]([⊠]), Jill Nelson[1], and Hasan Ayaz[2]

[1] Department of Electrical and Computer Enginnering,
George Mason University, Fairfax, VA, USA
{msaadati, jnelson}@gmu.edu
[2] School of Biomedical Engineering, Science and Health Systems,
Drexel University, Philadelphia, PA, USA
hasan.ayaz@drexel.edu

Abstract. The development of brain-computer interface (BCI) systems has received considerable attention from neuroscientists in recent years. BCIs can serve as a means of communication and for the restoration of motor function for patients with motor disorders. An essential part of the design of a BCI is correctly classifying the brain signals, historically collected using electroencephalography (EEG). However, recent studies have shown more robust classification results when EEG is combined with other neuroimaging methods such as fNIRS. Conventional classification methods need a priori feature preprocessing to train the model; such feature selection is a difficult and heavily studied problem. By using deep neural networks (DNN), in which recordings can be fed directly to the algorithm for training, we avoid the need for feature selection. In this study, the capabilities of DNNs in the classification of the hybrid EEG-fNIRS recordings of motor imagery (MI) and mental workload (MWL) tasks are investigated. A five-layer fully connected network is used for classification. This study makes use of two open-source meta-datasets collected at the Technische Universitat Berlin. The first dataset includes brain activity recordings of 26 healthy participants during three cognitive tasks: (1) n-back (0-, 2- and 3-back), (2) discrimination/selection response task (DSR) and (3) word generation (WG) tasks. The second dataset, motor imagery, consists of left and right-hand motor imagery tasks, each for 29 healthy participants. Our results show that classification accuracy is considerably higher for multimodal recordings when compared to EEG or fNIRS recordings alone. The proposed algorithm improves classification performance relative to a conventional support vector machine (SVM), reaching 90% average accuracy for both tasks, 8% higher than SVM performance. These results demonstrate the feasibility of achieving strong classification performance using multimodal BCI and deep learning.

Keywords: EEG · fNIRS · Deep neural networks · Deep learning · Brain imaging · Brain computer interfaces · Human machine interfaces

© Springer Nature Switzerland AG 2020
H. Ayaz (Ed.): AHFE 2019, AISC 953, pp. 209–220, 2020.
https://doi.org/10.1007/978-3-030-20473-0_21

1 Background

The development of brain-computer interface (BCI) systems has received considerable attention from neuroscientists in recent years. BCIs can serve as a means for communication, as a neurorehabilitation tool, and for restoration of motor function for people with motor disorders such as amyotrophic lateral sclerosis (ALS) and spinal cord injury, and/or people in the persistent locked-in state (LIS) [1]. Until recently, the dream of being able to control one's environment through thoughts was out of reach for these populations. However, new progress in BCI technology allows them use the electrical or hemodynamic signals collected from their scalp during a brain activity to connect with, maneuver in, or change their surroundings.

The primary purpose of a BCI is to detect and decode brain signals that indicate the users intentions and translate them into device commands that accomplish the users intent, usually a motor intent. These commands, or Motor Imagery (MI), are mainly from the primary motor cortex and consist of a range of upper or lower limb movements. To achieve the necessary decoding, a BCI system needs to perform a set of signal processing operations and classify the signals into the corresponding commands.

Standalone electroencephalography (EEG) and functional Near Infrared Spectroscopy (fNIRS) have been used for brain imaging extensively [2]. In the recent years, however, combination of these methods have shown promising results in neuroimaging, which is due to their complimentary resolution for accurate recognition of the brain activities in time and location; the high temporal resolution of the EEG and high spatial resolution of the fNIRS [3–5]. However, the classification of cognitive tasks based on fNIRS and EEG signals involves tackling complicated pattern recognition problems. Conventional classification methods need a priori feature selection and preprocessing which are not always guaranteed to result into the optimum classification. Deep Neural Networks (DNN) can overcome this challenge; DNNs directly extract the features from the fNIRS and EEG signals and they require a minimum feature preprocessing. Furthermore, with recent increases in computational power, deep neural networks (DNN) have become a practical option for high-accuracy classification.

In EEG based BCIs, many machine learning methods have been used to classify brain activity and have shown high classification accuracy, as summarized in [6]. The earliest studies on the capability of deep learning for brain-signal translation are described in [6, 7]. A deep belief network was used by Hajinoroozi et al. [8] for EEG-based classification of a driver's cognitive state. A motor imagery classification (left versus right) was investigated by using a few EEG recording channels and a DNN classifier by An et al. [9], with an average recorded accuracy was approximately 80%. On fNIRS based BCIs' classification, various types of fNIRS experiments for BCI applications have been investigated, including using mental workload [10, 11], motor imagery [12], motor execution [13, 14], and other approaches. Deep learning, however, has been applied in only a few studies. Mental workload task classification using a DNN classifier was investigated by Abibullaev et al. [14]. The proposed approach reached an accuracy of 94% for a four-class experiment. Hennrich et al. [15] studied the DNN classification efficiency of three mental tasks; accuracies were similar to conventional classification algorithms such as Linear Discriminant Analysis (LDA) and

SVM. Nguyen et al. [16] applied DNN on a tapping motor imagery task classification (left versus right) with an average accuracy of 85%. Most recently, in [17], the application of DNNs and Convolutional Neural Networks (CNNs) to fNIRS signals in the classification of a three-class motor execution task was investigated. Shallow convolutional neural networks with various numbers of filters in each layer were studied, and the results were compared to Artificial Neural Networks (ANNs) and SVMs.

The hybrid neuroimaging method is in its infancy, and to the best of our knowledge, few publications address classification using the combination of EEG and fNIRS recordings. In a study in 2017, Croce et al. [18] compared hybrid EEG-fNIRS to standalone EEG and reported an improvement in terms of classification accuracy; other scientists [1, 10, 13, 19] have published similar results. In the realm of machine learning algorithms, Fazli et al. [13] and Ma et al. [20] employed LDA and a Gaussian radial-basis kernel SVM, respectively, to classify motor imagery tasks. LDA was also used by Lee et al. [21] and Buccino et al. [22] for hybrid EEG and fNIRS classification for three conditions (left-hand, right-hand, and rest) motor imagery; average classification accuracy was reported as approximately 65% and 75%, respectively for the two methods. Khan et al. [10] and Khan and Hong [1] used EEG peak amplitudes of selected motor area channels and mean values of hemodynamic response as the feature set and achieved accuracy ranging from 80% to 95%.

Within deep learning, DNN has been successfully applied by Jirayucharoensak et al. [23] to classify three levels of valence and arousal based on EEG power spectral density features, achieving an accuracy of about 50%. In the most recent study of the classification of EEG + fNIRS motor imagery tasks (3 classes), a 5 layer DNN was introduced and applied by Chiarelli et al. [19]. A 10% improvement in accuracy compared to SVM and LDA methods was achieved.

2   Dataset and Signal Processing

This study makes use of two open-source meta-datasets collected at the Technische Universitat Berlin by Jaeyoung Shin et al. in 2016 and 2017 [24, 25]. The first dataset includes simultaneous EEG (30 sensors) and fNIRS (36 channels) recordings of the scalp for mental workload. Included are brain activity recordings of 26 healthy participants during three cognitive tasks: (1) n-back (0-, 2- and 3-back), (2) discrimination/selection response task (DSR) and 3) word generation (WG) tasks. These tasks are classified into four (0-, 2- and 3-back tasks and rest), two (target vs. non-target), and two (baseline and word generation) classes, respectively. The second dataset, motor imagery, consists of left and right-hand motor imagery tasks, each for 29 healthy participants.

EEG active electrodes were distributed on the entire scalp according to the 5–10 system. In the MWL experiment, 16 sources and 16 detectors were placed at frontal (16 channels), motor (4 channels), parietal (4 channels), and occipital (4 channels) areas. In the MI experiment, 15 sources and 16 fNIRS detectors were placed at frontal (9 channels), motor (12 channels), and visual (3 channels) areas. Figures of the locations of the EEG and fNIRS channels on the scalp for the first and second dataset can be found in [24] and [25]. The specifications of the datasets are described briefly below.

Dataset 1.A: n-back - The dataset consisted of 9 series of 0-, 2-, and 3-back tasks in a counterbalanced order. A single series included a 2 s. instruction showing the type of task (0-, 2- or 3-back), a 40 s. task period, and a 20 s. rest period [25].

Dataset 1.B: DSR - The dataset included 9 series of DSR in a counterbalanced order. A single trial involved a 2 s. instruction, a 40 s. task period, and a 20 s. rest period. In the task period, the symbol O and the symbol X were given every 2 s. in a random order. Volunteers were asked to press a button (target) when symbol O appeared and another button (non-target) when symbol X appeared [25].

Dataset 1.C: Word Generation - Each session included 20 random order of Word Generation (WG) and Baseline (BL) trials (10 each). For WG, the participants were instructed to keep thinking of words beginning with a given letter in the 2 s. instruction, whereas in BL they were asked to relax and gaze at the fixation cross for a low cognitive load [25].

Dataset 2.A: Left Hand MI vs. Right Hand MI - The experiment consisted of 3 sessions of left and right hand MI. Each session included a 1 min pre-experiment resting period, 20 repetitions of the given task, and a 1 min post-experiment resting period. The task started with 2 s. of a visual introduction of the task followed by 10 s. of a task period and a resting period with random length from 15 to 17 s. The subjects were asked to imagine hand gripping (opening and closing their hands) with a 1 Hz pace [25].

A summary of the number of the trials, number of series, and number of sessions of the experiments for all of the datasets is presented in 1.

2.1 Signal Processing

The EEG sampling frequency was 200 Hz; EEG data was stored over windows with lengths varying from 2 to 5 s., and bandpass filtered from 1 to 40 Hz using a second order Butterworth filter. Multiple window lengths are considered to investigate the impact of this hyperparameter on classification accuracy. Event-related desynchronization and synchronization analysis, ERD/ERS, was performed for all datasets. ERD/ERS are conventional indicators of cortical activation/deactivation and have been extensively used for various motor and workload memory tasks. ERD represents a short-lasting and localized amplitude decrease of rhythmic activity prior to the actual movement, and conversely, ESD exhibits an increase in amplitude with respect to the rest [26, 27]. To compute ERD, the signal was squared and averaged over one second to obtain a power estimate, i.e., ERD/ERS = $(A-R)/R$, where A is the average power over the time window during the task and R is the average power in a 2 s. window prior to the task onset (the rest).

The fNIRS sampling frequency is 10 Hz. Deoxy- and oxyhemoglobin data, HbR and HbO, were originally computed based on the modified Beer-Lambert law (mBLL) from the fNIRS optical density [24, 25]. As with the EEG data, several window lengths, from 1 to 5 s., are considered to evaluate the impact of window length on classification accuracy. Windowed recordings were filtered from 0 and 0.08 Hz with a second order Butterworth to remove physiological noises due to motion artifact, heart pulsation, and respiration. The signals HbR and HbO were normalized by dividing them by the maximum value of the window samples. To calculate the HbR and HbO changes relative to rest, the difference between the HbR and HbO of the window and the mean of the HbR and HbO of the rest period (base) prior to the window are calculated and used to build the features to feed to the DNN.

3 Deep Neural Network and Input Features

The proposed Deep Neural Network consists of four hidden layers, shown in Fig. 1, each with 60 neurons. The number of the neurons in the input layer is the sum of the HbO/HbR and EEG channels. The output layer consists of two, three, or four units, depending on the number of classes in the experiment, with a soft-max activation representing a probability distribution over the classes. This architecture was selected after experimenting with several different architectures, both shallower and wider.

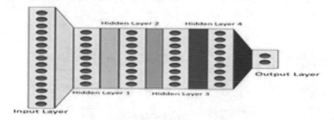

Fig. 1. Specifications for the proposed DNN. The DNN consists of four hidden layers with 60 neurons for each layer. The output layer consists of two, three or four units, depending on the number of classes in the experiment, with a soft-max activation representing a probability distribution over the classes.

The input features to the DNN are defined as N samples, each of which contains values for ERD, Δ_{HbO}, and Δ_{HbR} averaged over the chosen time window for all channels. The value of N depends on the length of the time window. Due to the low sampling rate of fNIRS, the dataset is relatively small, and the window must be chosen carefully. The chosen window length should neither result in N small enough that overfitting occurs, nor should the window be so large as to compromise accuracy. To assure this, a window length of 3 s is selected. The number of the features of the input varies depending on which recording is used. For full hybrid fNIR-EEG, a sample has 102 (36 + 36 + 30) features. In order to compare the classification accuracy of the full hybrid fNIRS-EEG system with the standalone fNIRS and EEG, four other cases are tested: combined EEG and Δ_{HbO}, standalone EEG, combination of Δ_{HbO} and Δ_{HbR}, and standalone Δ_{HbO}. The number of the features corresponding to each of these cases is shown in Table 2.

The classification process flow is shown in Fig. 2. After filtering, fNIRS-derived features are squared and then averaged over the window to compute Δ_{HbR} and Δ_{HbO}. Also after filtering, EEG-derived features are averaged over the window, and ERD/ERS analysis is performed on the result. All features are then fed to the DNN algorithm to train the network.

Table 1. Summary of experiment specifications: Number of trials, number of series, and number sessions of the experiments for all datasets

Experiment	n-back	WG	DSR	MI
Number of participants	26	26	26	29
Total number of trials	180	180	60	90
Number of trials per session	20	20	10	30
Number of series	3	3	1	3
Number of sessions	3	3	3	3
Classification	Level of difficulty	WG vs. BL	T vs. NT	LH vs. RH

Table 2. Number of features for each of the scenarios investigated

Methodologies	EEG	$\Delta_{HbO}-\Delta_{HbR}$	EEG$-\Delta_{HbO}$	Δ_{HbO}	Full hybrid EEG-fNIRS
Image width	30	72	66	36	102

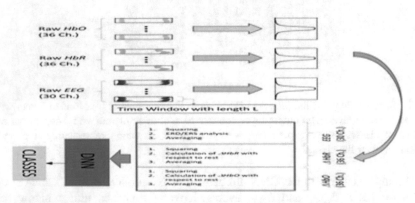

Fig. 2. Classification process flow: After filtering, fNIRS-derived features are squared and averaged over the window to compute Δ_{HbO} and Δ_{HbR}. Also after filtering, EEG-derived features are averaged over the window, and ERD/ERS analysis is performed on the result. All features are then fed to the DNN algorithm to train the network.

4   Results

Single subject accuracy of DNN-based classification and SVM-based classification for all the examined combinations of recordings - EEG, fNIRS ($\Delta_{HbO}-\Delta_{HbR}$), fNIRS ($\Delta_{HbO}$), EEG + fNIRS ($\Delta_{HbO}$), and EEG + fNIRS ($\Delta_{HbO}-\Delta_{HbR}$) - are reported in Tables 3–6 for n-back, DSR, WG, and MI tasks, respectively. The average accuracy is computed among the 26 subjects for mental workload tasks and 29 participants for MI tasks.

Table 3. Single subject and average classification performance, n-back task

Participants	1	2	3	4	5	6	7	8	9
fNIRS DNN	0.77	0.7	0.78	0.8	0.82	0.81	0.83	0.75	0.83
OXY DNN	0.68	0.42	0.71	0.57	0.84	0.61	0.77	0.79	0.76
EEG DNN	0.83	0.66	0.67	0.38	0.59	0.53	0.61	0.66	0.8
OXY-EEG DNN	0.71	0.7	0.71	0.7	0.79	0.7	0.75	0.83	0.81
Hybrid SVM	0.82	0.8	0.76	0.76	0.87	0.71	0.8	0.84	0.8
Hybrid DNN	0.91	0.7	0.82	0.86	0.83	0.81	0.96	0.92	0.85
Participants	10	11	12	13	14	15	16	17	18
fNIRS DNN	0.88	0.79	0.74	0.74	0.82	0.81	0.9	0.89	0.86
OXY DNN	0.8	0.68	0.79	0.77	0.72	0.61	0.79	0.61	0.77
EEG DNN	0.6	0.57	0.79	0.79	0.57	0.75	0.68	0.71	0.7
OXY-EEG DNN	0.74	0.55	0.8	0.66	0.78	0.86	0.75	0.76	0.79
Hybrid SVM	0.91	0.78	0.91	0.67	0.82	0.87	0.93	0.78	0.84
Hybrid DNN	0.89	0.89	0.89	0.77	0.93	0.83	0.95	0.83	0.85
Participants	19	20	21	22	23	24	25	26	**Avg.**
fNIRS DNN	0.86	0.86	0.88	0.6	0.62	0.81	0.89	0.81	**0.80**
OXY DNN	0.78	0.7	0.7	0.69	0.77	0.7	0.69	0.69	**0.70**
EEG DNN	0.7	0.73	0.7	0.63	0.74	0.76	0.7	0.73	**0.67**
OXY-EEG DNN	0.77	0.79	0.77	0.79	0.7	0.73	0.72	0.79	**0.74**
Hybrid SVM	0.84	0.91	0.8	0.87	0.78	0.93	0.84	0.89	**0.82**
Hybrid DNN	0.84	0.93	0.91	0.92	0.89	0.9	0.89	0.9	**0.87**

For all events, the full hybrid EEG-fNIRS has the highest average accuracy when compared to standalone methods. Applying DNN as the classifier for all of the experiments demonstrated higher performance than SVM, the conventional classifier. The highest improvement, 8%, is observed for the DSR task. Overall, average classification accuracy is higher in the 2-class experiments (MI, DSR, and WG) than the 4-class n-back test, as expected due to the larger number of possible classes in the n-back test. The proposed DNN classifier achieves the highest accuracy for the WG task 92% for DNN compared to 86% for SVM.

The ELU activation function was used to generate the results shown in the tables. To investigate effect of the activation function on classification performance, we applied two of the most recently introduced activation functions and compared their performance in this application. The activation functions applied were the rectified linear unit (ReLU) function, which has been proven to dampen the vanishing gradient problem due to its non-saturating properties, and exponential linear unit (ELU), which exhibits fast cost convergence to zero [6]. When the ELU activation function was applied, classification accuracy was an average of 2% higher than when ReLU was used.

Table 4. Single subject and average classification performance, DSR.

Participants	1	2	3	4	5	6	7	8	9
fNIRS DNN	0.83	0.73	0.81	0.85	0.85	0.85	0.87	0.81	0.87
OXY DNN	0.74	0.46	0.76	0.61	0.9	0.65	0.82	0.83	0.79
EEG DNN	0.87	0.7	0.7	0.42	0.65	0.59	0.64	0.72	0.85
OXY-EEG DNN	0.76	0.72	0.74	0.74	0.82	0.74	0.79	0.86	0.85
Hybrid SVM	0.83	0.82	0.78	0.77	0.9	0.73	0.82	0.87	0.82
Hybrid DNN	0.94	0.74	0.85	0.89	0.86	0.85	0.99	0.96	0.87
Participants	10	11	12	13	14	15	16	17	18
fNIRS DNN	0.92	0.85	0.78	0.78	0.86	0.84	0.92	0.93	0.92
OXY DNN	0.84	0.73	0.81	0.79	0.74	0.65	0.83	0.65	0.83
EEG DNN	0.64	0.59	0.81	0.81	0.61	0.79	0.7	0.74	0.75
OXY-EEG DNN	0.78	0.61	0.83	0.72	0.81	0.89	0.81	0.81	0.83
Hybrid SVM	0.93	0.81	0.94	0.69	0.85	0.88	0.95	0.8	0.86
Hybrid DNN	0.95	0.91	0.94	0.81	0.96	0.89	0.99	0.88	0.87
Participants	19	20	21	22	23	24	25	26	**Avg.**
fNIRS DNN	0.91	0.88	0.93	0.63	0.67	0.86	0.94	0.84	**0.84**
OXY DNN	0.83	0.74	0.75	0.72	0.81	0.75	0.72	0.72	**0.74**
EEG DNN	0.74	0.77	0.73	0.67	0.79	0.79	0.73	0.78	**0.71**
OXY-EEG DNN	0.81	0.83	0.81	0.85	0.74	0.78	0.76	0.82	**0.78**
Hybrid SVM	0.85	0.94	0.82	0.89	0.8	0.94	0.87	0.91	**0.84**
Hybrid DNN	0.89	0.95	0.95	0.96	0.93	0.93	0.94	0.93	**0.91**

The confusion matrices for all of the experiments is shown in Fig. 3. The rows of these matrices stand for predicted condition labels, while the columns represent reference labels. For all of the experiments, significant performances are achieved in all individual classes and they were dominantly and correctly identified in the confusion matrices (diagonal elements with dark brown color). In the n-back test, the 3-back class has the highest true positive classification relative to the other classes, and the 2-back class has the lowest. 3-back and 2-back trials exhibit the most misclassification, with 3-back being classified as 2-back and vice-versa. The patterns of misclassifications show that when items are misclassified they are more likely to be confused with items with close level of difficulty. In the MI test, better decoding performance is obtained for right hand movements and left hand movements show the smallest true positives classification. Classification performed on two-class experiments produces higher overall accuracies (a) than the 4-class experiment. In WG and DSR, both classes have similar rates of true positive classification and misclassifications are almost evenly happen in both classes.

Table 5. Single subject and average classification performance, WG.

Participants	1	2	3	4	5	6	7	8	9
fNIRS DNN	0.83	0.75	0.84	0.84	0.87	0.85	0.87	0.82	0.88
OXY DNN	0.75	0.47	0.75	0.63	0.91	0.66	0.82	0.84	0.82
EEG DNN	0.87	0.71	0.73	0.43	0.64	0.58	0.65	0.72	0.85
OXY-EEG DNN	0.76	0.75	0.76	0.75	0.84	0.76	0.82	0.88	0.86
Hybrid SVM	0.85	0.84	0.79	0.78	0.9	0.75	0.83	0.86	0.84
Hybrid DNN	0.94	0.75	0.88	0.9	0.89	0.87	0.99	0.96	0.89
Participants	10	11	12	13	14	15	16	17	18
fNIRS DNN	0.93	0.85	0.78	0.79	0.86	0.85	0.94	0.93	0.93
OXY DNN	0.84	0.73	0.83	0.81	0.76	0.65	0.85	0.66	0.84
EEG DNN	0.65	0.61	0.84	0.84	0.62	0.78	0.71	0.76	0.75
OXY-EEG DNN	0.78	0.62	0.84	0.71	0.84	0.91	0.81	0.8	0.82
Hybrid SVM	0.94	0.82	0.95	0.71	0.84	0.89	0.96	0.82	0.88
Participants	19	20	21	22	23	24	25	26	**Avg.**
fNIRS DNN	0.91	0.89	0.95	0.65	0.66	0.87	0.95	0.84	**0.85**
OXY DNN	0.82	0.77	0.75	0.75	0.8	0.75	0.75	0.75	**0.76**
EEG DNN	0.76	0.78	0.75	0.68	0.79	0.81	0.74	0.78	**0.72**
OXY-EEG DNN	0.82	0.84	0.84	0.86	0.74	0.79	0.76	0.83	**0.79**
Hybrid SVM	0.86	0.94	0.83	0.9	0.82	0.97	0.87	0.91	**0.86**
Hybrid DNN	0.89	0.96	0.97	0.96	0.94	0.96	0.95	0.95	**0.92**

Table 6. Single subject and average classification performance, MI.

Participants	1	2	3	4	5	6	7	8	9	10
fNIRS DNN	0.79	0.69	0.78	0.79	0.81	0.8	0.82	0.77	0.82	0.82
OXY DNN	0.7	0.43	0.72	0.58	0.85	0.6	0.77	0.79	0.77	0.77
EEG DNN	0.83	0.67	0.66	0.39	0.61	0.55	0.59	0.67	0.79	0.79
OXY-EEG DNN	0.71	0.68	0.7	0.7	0.78	0.72	0.76	0.82	0.82	0.82
Hybrid SVM	0.8	0.77	0.74	0.75	0.84	0.69	0.79	0.82	0.77	0.77
Hybrid DNN	0.91	0.7	0.82	0.85	0.82	0.82	0.97	0.93	0.83	0.83
Participants	11	12	13	14	15	16	17	18	19	20
fNIRS DNN	0.88	0.8	0.74	0.75	0.82	0.8	0.89	0.89	0.86	0.86
OXY DNN	0.81	0.7	0.78	0.75	0.71	0.59	0.81	0.59	0.79	0.79
EEG DNN	0.6	0.56	0.78	0.78	0.57	0.73	0.68	0.7	0.7	0.7
OXY-EEG DNN	0.74	0.56	0.79	0.68	0.79	0.85	0.75	0.75	0.77	0.77
Hybrid SVM	0.91	0.78	0.91	0.67	0.82	0.87	0.93	0.78	0.67	0.84
Hybrid DNN	0.9	0.89	0.9	0.76	0.91	0.84	0.94	0.85	0.85	0.85
Participants	21	22	23	24	25	26	27	28	29	Avg.
fNIRS DNN	0.88	0.85	0.9	0.61	0.63	0.82	0.88	0.8	0.83	**0.83**
OXY DNN	0.79	0.7	0.71	0.7	0.75	0.7	0.69	0.69	0.72	**0.77**
EEG DNN	0.72	0.73	0.69	0.62	0.74	0.75	0.7	0.74	0.77	**0.73**
OXY-EEG DNN	0.78	0.79	0.77	0.81	0.68	0.75	0.7	0.79	0.73	**0.80**
Hybrid SVM	0.84	0.91	0.8	0.87	0.78	0.93	0.84	0.89	0.9	**0.85**
Hybrid DNN	0.85	0.91	0.93	0.91	0.91	0.91	0.89	0.9	0.91	**0.91**

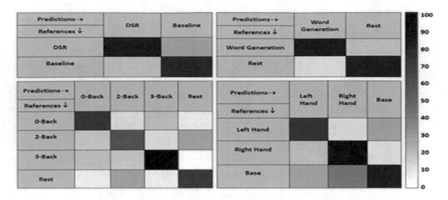

Fig. 3. Confusion matrices for all of the experiments.

5 Conclusion

We investigated the performance of a DNN structure in the classification of hybrid EEG-fNIRS recordings of motor imagery (MI) and mental workload (MWL) tasks. Our results show that classification accuracy is considerably higher for multimodal recordings in comparison to EEG or fNIRS recordings alone. The proposed algorithm significantly improves classification performance relative to a conventional support vector machine (SVM), reaching 90% average accuracy for the MI tasks, 8% higher than SVM performance. For n-back tasks, the pro-posed DNN achieves 87% accuracy compared to 82% accuracy using an SVM classifier. The ELU activation function is chosen because of its higher accuracy relative to ReLU. The results presented in this paper demonstrate the feasibility of achieving strong classification performance using multimodal BCI and a fully-connected DNN.

References

1. Hong, K.-S. Naseer, N., Kim, Y.-H.: Classification of prefrontal and motor cortex signals for three-class fNIRS-BCI. Neurosci. Lett. **587**, 87–92 (2015)
2. Ayaz, H., Dehais, F.: Neuroergonomics: The Brain at Work and Everyday Life, 1st edn. Elsevier, Academic Press, Cambridge (2019)
3. Ahn, S., Jun, S.C.: Multi-modal integration of EEG-fNIRS for brain-computer interfaces - current limitations and future directions. Front. Hum. Neurosci. **11**, 503 (2017). https://doi.org/10.3389/fnhum.2017.00503
4. Liu, Y., Ayaz, H., Shewokis, P.A.: Mental workload classification with concurrent electroencephalography and functional near-infrared spectroscopy. Brain Comput. Interfaces **4**, 1–11 (2017). https://doi.org/10.1080/2326263x.2017.1304020
5. Dehais, F., Duprès, A., Di Flumeri, G., Verdière, K.J., Borghini, G., Babiloni, F., Roy, R.N.: Monitoring pilot's cognitive fatigue with engagement features in simulated and actual flight conditions using an hybrid fNIRS-EEG passive BCI. In: IEEE SMC (2018)

6. Schirrmeister, R.T., Springenberg, J.T., Fiederer, L.D.J., Glasstetter, M., Eggensperger, K., Tangermann, M., Hutter, F., Burgard, W., Ball, T.: Deep learning with convolutional neural networks for EEG decoding and visualization. Hum. Brain Mapp. **38**(11), 5391–5420 (2017)
7. Wang, Z., Lyu, S., Schalk, G., Ji, Q.: Deep feature learning using target priors with applications in ECOG signal decoding for BCI. In: IJCAI, pp. 1785–1791 (2013)
8. Hajinoroozi, M., Mao, Z., Jung, T.-P., Lin, C.-T., Huang, Y.: EEG-based prediction of driver's cognitive performance by deep convolutional neural network. Sig. Process. Image Commun. **47**, 549–555 (2016)
9. An, X., Kuang, D., Guo, X., Zhao, Y., He, L.: A deep learning method for classification of EEG data based on motor imagery. In: Huang, D.S., Han, K., Gromiha, M. (eds.) Intelligent Computing in Bioinformatics. Lecture Notes in Computer Science, vol. 8590 (2014)
10. Hong, K.-S., Naseer, N., Kim, Y.-H.: Classification of prefrontal and motor cortex signals for three-class fNIRS-BCI. Neurosci. Lett. **587**, 87–92 (2015)
11. Khan, M.J., Hong, M.J., Hong, K.-S.: Decoding of four movement directions using hybrid NIRS-EEG brain-computer interface. Front. Hum. Neurosci. **8**, 244 (2014)
12. Coyle, S.M., Ward, T.E., Markham, C.M.: Brain computer interface using a simplified functional near-infrared spectroscopy system. J. Neural Eng. **4**(3), 219 (2007)
13. Fazli, S., Mehnert, J., Steinbrink, J., Curio, G., Villringer, A., Muller, K.-R., Blankertz, B.: Enhanced performance by a hybrid NIRS-EEG brain computer interface. Neuroimage **59**(1), 519–529 (2012)
14. Abibullaev, B., An, J., Moon, J.-I.: Neural network classification of brain hemodynamic responses from four mental tasks. Int. J. Optomechatronics **5**(4), 340–359 (2011)
15. Hennrich, J., Her, C., Heger, D., Schultz, T.: Investigating deep learning for fNIRS based BCI. In: EMBC, pp. 2844–2847 (2015)
16. Nguyen, H.T., Ngo, C.Q., Truong Quang Dang, K., Vo, V.T.: Temporal hemodynamic classification of two hands tapping using functional near-infrared spectroscopy Front. Hum. Neurosci. **7**, 516 (2013)
17. Trakoolwilaiwan, T., Behboodi, B., Lee, J., Kim, K., Choi, J.-W.: Convolutional neural network for high-accuracy functional near-infrared spectroscopy in a brain-computer interface: three-class classification of rest, right-, and left-Hand motor execution. Neurophotonics **5** (2007)
18. Croce, P., Zappasodi, F., Merla, A., Chiarelli, M.: Exploiting neurovascular coupling: a Bayesian sequential Monte Carlo approach applied to simulated EEG fNIRS data. J. Neural Eng. **14**(4) (2017)
19. Chiarelli, A.M., Zappasodi, F., Di Pompeo, F., Merla, A.: Simultaneous functional near-infrared spectroscopy and electroencephalography for monitoring of human brain activity and oxygenation: a review. Neurophotonics **4**(4) (2017)
20. Ma, L., Zhang, L., Wang, L., Xu, M., Qi, H., Wan, B., Ming, D., Hu, Y.: A hybrid brain-computer interface combining the EEG and NIRS. In: 2012 IEEE International Conference Virtual Environments Human-Computer Interfaces and Measurement Systems (VECIMS), pp. 159–162 (2012)
21. Lee, M.-H., Fazli, S., Mehnert, J., Lee, S.-W.: Hybrid brain-computer interface based on EEG and NIRS modalities. In: 2014 International Winter Workshop on Brain-Computer Interface (BCI), (2014)
22. Buccino, A.P., Keles, H.O., Omurtag, A.: Hybrid EEG-fNIRS asynchronous brain computer interface for multiple motor tasks. PloS ONE **11**(1) (2016)
23. Jirayucharoensak, S., Pan-Ngum, S., Israsena, P.: EEG-based emotion recognition using deep learning network with principal component based covariate shift adaptation. Sci. World J. (2014)

24. Shin, J., Von Luhmann, A., Kim, D.-W., Mehnert, J., Hwang, H.-J., Muller, K.-R.: Simultaneous acquisition of EEG and NIRS during cognitive tasks for an open access dataset. In: Generic Research Data (2018)
25. Shin, J., Von Lhmann, A., Kim, D.-W., Mehnert, J., Hwang, H.-J., Muller, K.-R.: Simultaneous aquisition of EEG and NIRS during cognitive tasks for an open access dataset. In: Scientific Data, vol. 5 (2018)
26. Pfurtscheller, G.: Functional brain imaging based on ERD/ERS. Vis. Res. **41**(10–11), 1257–1260 (2001)
27. Pourshafi, A., Saniei, M., Saeedian, A., Saadati, M.: Optimal reactive power compensation in a deregulated distribution network. In: 44th International Universities Power Engineering Conference (UPEC), pp. 1–6 (2009)
28. Neuper, C., Pfurtscheller, G.: Event-related dynamics of cortical rhythms: frequency-specific features and functional correlates. Int. J. Psychophysiol. **43**(1), 41–58 (2001)
29. Saadati, M., Nelson, J.K.: Multiple transmitter localization using clustering by likelihood of transmitter proximity. In: 51st Asilomar Conference on Signals, Systems, and Computers, pp. 1769–1773 (2017)

Convolutional Neural Network for Hybrid fNIRS-EEG Mental Workload Classification

Marjan Saadati[1]($\boxtimes$), Jill Nelson[1], and Hasan Ayaz[2]

[1] Department of Electrical and Computer Enginnering,
George Mason University, Fairfax, VA, USA
{msaadati,jnelson}@gmu.edu
[2] School of Biomedical Engineering, Science and Health Systems,
Drexel University, Philadelphia, PA, USA
hasan.ayaz@drexel.edu

Abstract. The classification of workload memory tasks based on fNIRS and EEG signals requires solving high-dimensional pattern classification problems with a relatively small number of training patterns. In the use of conventional machine learning algorithms, feature selection is a fundamental difficulty given the large number of possible features and the small amount of available data. In this study, we bypass the challenges of feature selection and investigate the use of Convolutional Neural Networks (CNNs) for classifying workload memory tasks. CNNs are well suited for learning from the raw data without any a priori feature selection. CNNs take as input two-dimensional images, which differ in structure from the neural time series obtained on the scalp surface using EEG and fNIRS. Therefore, both the existing CNN architectures and fNIRS-EEG input must be adapted to allow fNIRS-EEG input to a CNN. In this work, we describe this adaptation, evaluate the performance of CNN classification of mental workload tasks. This study makes use of an open-source meta-dataset collected at the Technische Universität Berlin; including simultaneous EEG and fNIRS recordings of 26 healthy participants during n-back tests. A CNN with three convolution layers and two fully connected layers is adapted to suit the given dataset. ReLU and ELU activation functions are employed to take advantage of their better dampening property in the vanishing gradient problem, fast convergence, and higher accuracy. The results achieved with the two activation functions are compared to select the best performing function. The proposed CNN approach achieves a considerable average improvement relative to conventional methods such as Support Vector Machines. The results across differences in time window length, activation functions, and other hyperparameters are benchmarked for each task. The best result is obtained with a three-second window and the ELU activation function, for which the CNN yields 89% correct classification, while the SVM achieves only 82% correct classification.

Keywords: EEG · fNIRS · Convolutional neural networks · Deep learning · Brain imaging · Brain computer interfaces · Human machine interfaces

© Springer Nature Switzerland AG 2020
H. Ayaz (Ed.): AHFE 2019, AISC 953, pp. 221–232, 2020.
https://doi.org/10.1007/978-3-030-20473-0_22

1   Introduction

Human-Machine-Interfaces (HMI) are widely used in everyday life, and their navigation imposes high cognitive demands on the operator's brain. This mental workload (MWL) affects the operator's interaction with computers and other devices. MWL may compromise the user's functioning, and sometimes safety, by increasing fault rates and reaction times, introducing fatigue [1], and causing the neglect of critical information known as cognitive tunneling [2]. Therefore, taking into consideration the operator's cognitive characteristics and condition is an integral step in improving the design of HMIs through installing adaptive features that can adjust to MWL changes [1]. We consider two essential aspects of a protocol for measuring MWL: selecting the data to be acquired and analyzed, and developing an algorithm for event-related classification.

To measure MWL, brain activity is typically recorded using the conventional method of Electroencephalography (EEG), or more recently using functional Near Infrared Spectroscopy (fNIRS) [3]. EEG measures voltage fluctuations resulting from ionic current within the neurons of the brain. EEG signals are spontaneous relative to the brain activity through electrical measurement that provide high temporal resolution, but these measurements are highly susceptible to electrical noise and motion artifacts, e.g., eye-blinking or muscle movement. fNIRS, on the other hand, monitors the hemodynamic responses that follow neuronal activity. fNIRS is less susceptible to electrical noise and movement artifacts yet has lower temporal resolution and low depth resolution. However, the higher spatial resolution offered by fNIRS provides a better indication of which part of the cortex is activated. Integration of EEG and fNIRS provides us with two different sources of data that are associated to the same neuronal activities but are sensitive to different types of events. We can leverage the complementary temporal and spatial resolution of the two data types; while EEG identifies the cortical reaction to a specific stimulus with relatively fine temporal resolution, fNIRS measures hemodynamic changes that arise from neuronal activity to determine the location of the reaction.

The classification of cognitive events based on fNIRS and EEG signals requires solving high-dimensional pattern classification problems with a relatively small number of training patterns. Conventional classification methods require that a priori feature selection be performed before the model can be trained [4, 5]. Such feature selection is a difficult and heavily studied problem; the large number of possible features and relatively small amount of available data introduce the curse of dimensionality and the possibility of overfitting. Deep Neural Networks (DNNs), and in particular Convolutional Neural Networks (CNNs), allow for classifiers that bypass the need for feature selection, since both are well suited for learning from raw data and hence recordings can be fed directly to the algorithms for training.

CNNs take as input two-dimensional images, which differ in structure from the neural time series obtained on the scalp surface using EEG and fNIRS. Therefore, both the existing CNN architectures and fNIRS-EEG input must be adapted to allow fNIRS-EEG input to a CNN. Furthermore, since the number of samples in the fNIRS-EEG recordings is relatively small, avoiding underfitting or overfitting is a primary challenge. In this study, we aim to focus on how the CNNs of different architectures can be designed and trained for end-to-end learning from fNIRS (or fNIRS-EEG) signals recorded in human subjects. An input adaptation method, an efficient network structure, and methods to overcome the overfitting are introduced. The performance of a CNN classifier for a 4-class mental workload task classification is evaluated, and the impact of design choices (e.g., the overall network architecture and type of nonlinearity used) on classification accuracy are investigated.

The remainder of the paper is organized as follows. Section 2 provides background on CNNs and on published results related to the work presented here. After a description of the dataset specifications, the preprocessing and input adaptation methods are explained in Sects. 3 and 4. A description of the deep neural networks and proposed Convolutional Neural Network including the method spec, the network architecture and hyperparameters configuration, are described in Sect. 5. Finally, classification results and discussion conclude the paper in Sect. 6.

2 Background and Literature Review

2.1 Convolutional Neural Networks

CNNs are an effective classification approach derived from Artificial Neural Networks (ANNs). A CNN applies convolutional filters to the input data and optimizes the gradient descent problem with a smaller number of the weights than an ANN. A CNN consists of several layers: the input layer, the convolutional layer(s), fully connected hidden layer(s), and the output layer(s). Our proposed CNN, represented in Fig. 1, demonstrates a sample structure of a CNN.

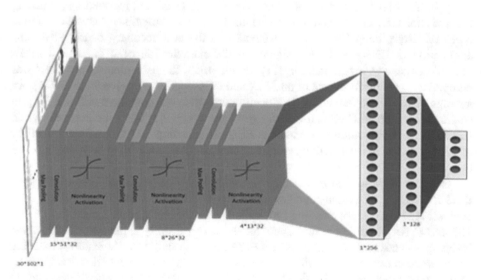

Fig. 1. The proposed network spec: A CNN with three convolution layers (including max-pooling and dropout layers after each convolutional layer), two fully connected layers and a Softmax output layer.

In the convolutional layer, a series of learnable convolution filters (Kernels) are convolved across the raw pixel data of an image to extract and learn higher-level features. These higher-level features are called activation maps. In these layers, a nonlinear activation function is applied to the output to introduce nonlinearities into the model. Stacking the activation maps for all filters along the depth dimension forms the full output volume of the convolution layer [6]. The pooling layer downsamples the image data extracted by the convolutional layers to reduce the dimensionality of the feature map in order to decrease processing time. In addition to making the input representations smaller and more manageable, the pooling layer reduces the number of parameters and computations in the network, thereby controlling overfitting [6].

The fully connected layer performs classification on these features based on the training dataset. Every node in a fully connected layer is connected to every node in the previous layer. Finally, the output layer contains a single node for each target class in the model with a Softmax activation function to compute the probability of each class. The Softmax activation function ensures that the final outputs fulfill the constraints of a probability density. The activation function introduces nonlinearity in the node's output, making it capable of learning more complex tasks [6].

2.2   Related Work

In EEG classification, many studies have investigated the performance of deep learning for brain-signal translation; a review can be found in [7–9]. In 2016, Bashivan et al. trained a CNN using EEG power in three different frequency bands of interest. They reported a best-performance accuracy of approximately 92% [7]. The authors represent the EEG as a time series of topographically organized images. In a comprehensive study in 2017 [10], a number of CNNs were investigated and reported as promising methods for both classification and visualization. In a comparison between different types of CNNs, deep CNNs have demonstrated the best accuracy, nearly 98%. The application of CNNs to fNIRS signals in the classification of a three-class motor execution task was investigated in [11]. In this paper, the performance of a CNN was comparted to the performance of an ANN and of a support vector machine (SVM). An average of 6% improvement in classification was observed compared to SVM [11]. In studies on combined EEG and fNIRS experiments, Croce et al. [12] compared hybrid EEG-fNIRS classification to standalone EEG classification and showed the feasibility of improvements in brain activity reconstruction. Similar results have been published in [13–15].

In terms of evaluation of machine learning algorithms for classification of hybrid EEG-fNIRS systems, many studies have employed SVMs and linear discriminant analysis (LDA) to classify various types of cognitive or motor cortical tasks [1, 13, 16–18]. In the deep learning realm, DNN has been successfully applied to classify three levels of valence and arousal based on EEG power spectral density features, reaching an accuracy of roughly 50% [19]. In the most recent study of the classification of EEG-fNIRS motor imagery tasks (3 classes), a 5-layer DNN was applied [14], and an improvement of 10% in accuracy compared to SVM and LDA methods was achieved. To the best of our knowledge, there are no studies on the application of CNNs in hybrid fNIRS-EEG BCI classification, which makes this an excellent opportunity to focus on unique aspects of using CNN methods in this field.

3   Dataset and Preprocessing

This study makes use of a dataset collected at the Technische Universität Berlin by Shin et al. [20, 21] in 2017. The dataset includes simultaneous EEG (30 sensors) and fNIRS (36 channels) recordings of the scalp for mental workload during n-back (0-, 2- and 3-back) tasks. These tasks are classified into four possibilities: 0-, 2- and 3-back tasks, and rest. The study included twenty-six right-handed healthy participants. The dataset consists of three sessions, where each session contained three series of 0-, 2-, and 3-back tasks. Hence, nine series of n-back tasks were performed for each participant. A single series included a 2 s. instruction showing the type of the task (0-, 2- or 3-back), a 40 s. task period, and a 20 s. rest period. A total of 180 trials were performed for each n-back task (20 trials, 3 series, 3 sessions). All EEG and fNIRS signals were recorded simultaneously.

Thirty EEG active electrodes were distributed on the entire scalp according to the 5-10 system. Sixteen sources and sixteen detectors were placed at frontal, motor, parietal, and occipital areas. The locations of the EEG and fNIRS channels and details of the sequence of the tasks can be found in [20].

3.1   EEG Preprocessing

EEG data were sampled at 200 Hz and stored over windows with lengths varying from 2 to 5 s. Data were bandpass filtered to between 1 and 40 Hz (second order digital Butterworth filter). The range of window lengths is designed to investigate the impact of this hyperparameter on classification accuracy. Event-related desynchronization and synchronization analysis, ERD/ERS, was done for all the dataset. ERD/ERS are conventional indicators of cortical activation/deactivation and have been extensively used for various motor and workload memory tasks. ERD represents a short-lasting and localized amplitude decrease of rhythmic activity prior to the actual movement, and conversely, ESD exhibits an increase in amplitude with respect to the rest [22, 23]. After bandpass filtering to retain the mu and beta band, the signal was squared and averaged over the second to obtain a power estimate to compute the ERD, i.e., ERD/ERS = $(A-R)/R$, where A is the average power over the window time during the task and R is the average power in a 2 s. window prior to the task onset (the rest).

3.2   fNIRS Preprocessing

The sampling frequency is 10 Hz. Deoxy- and oxyhemoglobin data, HbR and HbO, were originally computed based on the modified Beer-Lambert law (mBLL) from the fNIRS optical density [17]. Similar to EEG, several window lengths, from 2 to 4 s., are considered to evaluate the impact of window length on accuracy. The recording within each time window is filtered from 0 and 0.08 Hz using a second-order digital Butterworth filter to remove the physiological noises due to motion artifact, heart pulsation, and respiration. The signals HbR and HbO are normalized by dividing them by the maximum value of the window samples. To calculate the HbR and HbO changes relative to rest, the deference between the HbR and HbO of the window and the mean of the HbR and HbO of the rest period (base) prior to the window are calculated and adapted to build images to feed to the CNN, as below:

$$\Delta_{HbO} = HbO(Window) - HbO(Base) \tag{1}$$

$$\Delta_{HbR} = HbR(Window) - HbR(Base) \tag{2}$$

4   Input Image Structure

CNN, like other supervised machine learning algorithms, require training data in the form of labeled examples. All recordings in the dataset described above are labeled as one of four classes. The EEG recordings are downsampled to 10 Hz to match the sampling frequency of the fNIRS data. Data matrices form a set of N input images; the length of the image is equal to the number of samples in the time window, and N depends on the window length. Due to the low sampling rate of the fNIRS data, the dataset is relatively small, and the length of the window must be chosen carefully. The length of the windows should not result in such a small N that overfitting occurs (in case of windows with large length), nor compromise the accuracy (in case of small windows). The image width varies depending on which recordings are used. For the full hybrid fNIRS-EEG data, the width is 102 (36 + 36 + 30) columns which include the average ERD/ERS of the EEG recordings, and Δ_{HbO} and Δ_{HbR} of the fNIRS recordings in the desired time window. In order to compare the classification accuracy of the full hybrid fNIRS-EEG system to that of standalone fNIRS and EEG, four other cases are tested, including; the combined EEG and Δ_{HbO}, standalone EEG, combined of Δ_{HbO} and Δ_{HbR}, and standalone Δ_{HbO}. The widths of the images corresponding to each of these cases are presented in Table 1. The input image shape for the full hybrid EEG-fNIRS system is represented in Fig. 1.

Table 1. Dimensions of the input images for the scenarios investigated

Methodologies:	EEG	$\Delta_{HbO}-\Delta_{HbR}$	EEG$-\Delta_{HbO}$	Δ_{HbO}	Full hybrid EEG-fNIRS
Image width	30	72	66	36	102

5   Proposed Convolutional Neural Network

A CNN with three convolutional layers (including max-pooling and dropout layers after each convolutional layer) and two fully connected layers is adapted to suit the given dataset. Its structure is shown in Fig. 1. The output layer consists of four units, with a Softmax activation representing a probability distribution over the classes. Dropout, as a regularization technique to improve generalization, randomly sets the output of units in the network to zero during training, preventing those units from affecting the output or the gradient of the loss function for an update step [6]. This ensures that the network does not overfit by depending on specific hidden units. There are 32 3 * 3 kernels in each layer. The two fully connected layers have 256 and 128 neurons, respectively. This architecture is selected after experimenting with many different architectures, both shallower and deeper. Changes in image size as the data passes through the CNN is depicted in Fig. 1. For the full hybrid EEG-fNIRS classification, the image size is 30 * 112; after three convolutional layers with a filter size of 3 * 3, the final image is 4 * 13.

We apply two of the most recently introduced activation function and compare their performance in this application. The activation functions we consider are the rectified

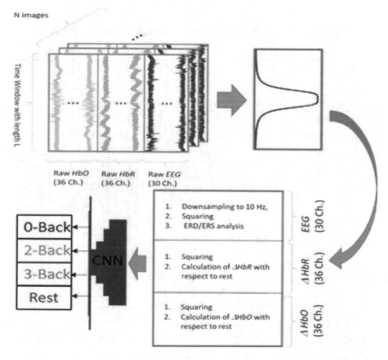

Fig. 2. Processing of the data: After filtering, fNIRS data is squared, then Δ_{HbR} and Δ_{HbO} are calculated. EEG data is downsampled to 10 Hz, and ERD/ERS analysis is performed on the downsampled data. Finally, data are fed to the CNN algorithm to train the network.

linear unit (ReLU) function, which was proven to dampen the vanishing gradient problem due to its non-saturating properties [10] and exponential linear unit (ELU) because of its fast cost convergence to zero and more accurate results.

The entire classification process ow is shown in Fig. 2. After filtering, fNIRS data is squared, then Δ_{HbR} and Δ_{HbO} are calculated. EEG data is downsampled to 10 Hz, and ERD/ERS analysis is performed on the downsampled data. Finally, data are fed to the CNN algorithm to train the network.

6 Classification Performance Results

Single subject classification accuracy for all the examined combinations of recordings (EEG, fNIRS ($\Delta_{HbO}-\Delta_{HbR}$), fNIRS ($\Delta_{HbO}$), EEG + fNIRS ($\Delta_{HbO}-\Delta_{HbR}$), EEG + fNIRS ($\Delta_{HbO}$)) and classifiers (SVM and CNN) are reported in Table 2. Average accuracy is computed across the 26 subjects. Similar to previous investigations comparing classification performance using standalone fNIRS, standalone EEG, and hybrid EEG-fNIRS with other machine learning methods such as SVM and LDA, a clear increase in CNN accuracy can be seen when fNIRS is employed with EEG when compared to standalone modalities (average EEG-CNN accuracy 69%, average fNIRS + CNN accuracy 82%, average fNIRS + EEG-CNN accuracy 89% at iteration

Table 2. Summary of classification accuracy results, n-back tasks.

Participant	1	2	3	4	5	6	7	8	9
fNIRS CNN	0.8	0.71	0.8	0.82	0.84	0.82	0.84	0.78	0.84
OXY CNN	0.71	0.44	0.73	0.6	0.87	0.62	0.8	0.82	0.78
EEG CNN	0.84	0.69	0.69	0.4	0.62	0.56	0.62	0.69	0.82
OXY-EEG CNN	0.73	0.71	0.73	0.71	0.8	0.73	0.78	0.84	0.84
Hybrid SVM	0.82	0.8	0.76	0.76	0.87	0.71	0.8	0.84	0.8
Hybrid CNN	0.92	0.73	0.84	0.88	0.85	0.84	0.98	0.94	0.86
Participant	10	11	12	13	14	15	16	17	18
fNIRS CNN	0.89	0.82	0.76	0.76	0.84	0.82	0.91	0.91	0.89
OXY CNN	0.82	0.71	0.8	0.78	0.73	0.62	0.82	0.62	0.8
EEG CNN	0.62	0.58	0.8	0.8	0.58	0.76	0.69	0.73	0.73
OXY-EEG CNN	0.76	0.58	0.82	0.69	0.8	0.87	0.78	0.78	0.8
Hybrid SVM	0.91	0.78	0.91	0.67	0.82	0.87	0.93	0.78	0.84
Hybrid CNN	0.92	0.9	0.92	0.79	0.94	0.86	0.96	0.86	0.86
Participant	19	20	21	22	23	24	25	26	Avg.
fNIRS CNN	0.89	0.87	0.91	0.62	0.64	0.84	0.91	0.82	0.81
OXY CNN	0.8	0.73	0.73	0.71	0.78	0.73	0.71	0.71	0.71
EEG CNN	0.73	0.76	0.71	0.64	0.76	0.78	0.71	0.76	0.66
OXY-EEG CNN	0.8	0.82	0.8	0.82	0.71	0.76	0.73	0.8	0.76
Hybrid SVM	0.84	0.91	0.8	0.87	0.78	0.93	0.84	0.89	0.82
Hybrid CNN	0.86	0.94	0.94	0.94	0.92	0.92	0.92	0.92	0.89

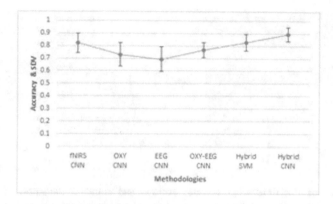

Fig. 3. Average and standard deviation of classification accuracy for each scenario considered.

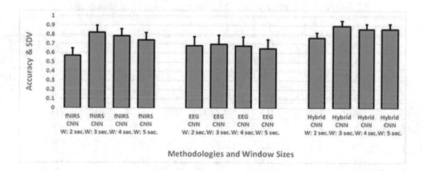

Fig. 4. Average recall (a) and classification accuracy (b) in the training process for ELU and ReLU activation functions.

1900). Using both Δ_{HbO} and Δ_{HbR} yields a higher average accuracy than only Δ_{HbO} (81% compared to 71%). fNIRS shows a higher average accuracy than EEG and than the combination of EEG and Δ_{HbO}. The last two columns of the table compare the classification accuracy for SVM to CNN for the full hybrid scenario. CNN achieves better classification performance than SVM, reaching an average 89% accuracy which shows a 7% improvement. Figure 3 summarizes the average accuracy and standard deviation for all of the modalities considered. EEG shows the highest standard deviation, and full hybrid modality shows the lowest. Applying CNN on the fNIRS data and SVM on the hybrid system yield very similar average accuracies.

To investigate the impact of the time window size on classification accuracy, we calculated the average and standard deviation of the accuracy for three methodologies with the highest performance. We considered four window lengths ranging from 2 to 5 s; the results is shown in a bar chart in Fig. 5. A 2 s. window yields the lowest accuracy for all three modalities, likely because the small dimensions of the window. The 3 s. window yields the highest accuracy and is used for the rest of the analysis. The effect of the window is negligible for EEG data, since the higher sampling frequency results in a large number of samples even in a small window.

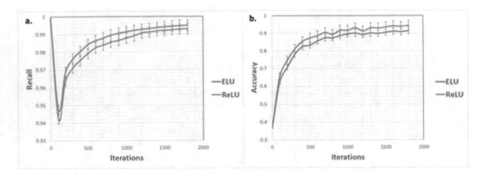

Fig. 5. Impact of the time window size; average and standard deviation of classification accuracy for the three methodologies with the best classification performance.

Table 3. Summary of the CNN results for hybrid method

Participants	1	2	3	4	5	6	7	8	9
Min	0.85	0.66	0.73	0.82	0.74	0.77	0.87	0.83	0.77
Max	0.94	0.74	0.87	0.92	0.86	0.88	1	0.97	0.87
Average	0.92	0.73	0.84	0.88	0.85	0.84	0.98	0.94	0.86
SDV	0.05	0.04	0.07	0.05	0.06	0.05	0.06	0.07	0.05
Recall	0.86	0.83	0.8	0.77	0.86	0.78	0.9	0.79	0.8
Participants	10	11	12	13	14	15	16	17	18
Min	0.83	0.83	0.84	0.71	0.88	0.8	0.87	0.79	0.78
Max	0.94	0.93	0.93	0.8	0.95	0.88	0.99	0.89	0.89
Average	0.92	0.9	0.92	0.79	0.94	0.86	0.96	0.86	0.86
SDV	0.05	0.05	0.04	0.05	0.03	0.04	0.06	0.05	0.05
Recall	0.83	0.8	0.76	0.84	0.86	0.83	0.82	0.83	0.84
Participants	19	20	21	22	23	24	25	26	Avg.
Min	0.78	0.88	0.86	0.88	0.85	0.83	0.85	0.81	0.81
Max	0.88	0.97	0.98	0.97	0.93	0.93	0.96	0.96	0.91
Average	0.86	0.94	0.94	0.94	0.92	0.92	0.92	0.92	0.89
SDV	0.05	0.04	0.06	0.05	0.04	0.05	0.05	0.07	0.09
Recall	0.86	0.84	0.88	0.89	0.78	0.84	0.86	0.76	0.82

Minimum, maximum, standard deviation, and recall for classification accuracy are computed for each participant for the CNN method and full hybrid system, and results appear in Table 3. The lowest Min single subject accuracy is 71%, while the average Min is 81%. The Max single subject accuracy reaches 99%, while the average Max is 91%. The average recall is 82%, similar to the recall of the SVM method.

In order to study the effects of the activation functions and number of iteration, we computed the average recall and classification accuracy in the training process. The results are shown in Fig. 4. Figure 4(b) shows the average training accuracy. Note that that the highest accuracy is achieved in 1900 iterations, and application of ELU yields

2% better performance than ReLU. The recall graph in Fig. 4(a) confirms this result, as well.

In conclusion, the CNN based classifier introduced in this paper shows promising results in comparison with a more conventional SVM approach. The CNN based classification demonstrates the highest accuracy when the hybrid EEG-fNIRS recordings are used, in comparison with the standalone methods. The best result were obtained with a three-second window and the ELU activation function, for which the CNN yields 89% correct classification, while the SVM achieves only 82% correct classification. The ELU activation function shows higher accuracy and recall in comparison with the ReLU. For the future work, we plan to investigate different image structures and performance of the other CNN structures such as ResNet in HMI and BCI applications.

References

1. Aghajani, H., Garbey, M., Omurtag, A.: Measuring mental workload with EEG + fNIRS. Front. Hum. Neurosci. **11**, 359 (2017)
2. Jarmasz, J., Herdman, C.M., Johannsdottir, K.R.: Object-based attention and cognitive tunneling. J. Exp. Psychol. Appl. **11**(1), 3 (2005)
3. Ayaz, H., Dehais, F.: Neuroergonomics: The Brain at Work and Everyday Life, 1st edn. Elsevier, Academic Press, Cambridge (2019)
4. Wang, L., Curtin, A., Ayaz, H.: Comparison of machine learning approaches for motor imagery based optical brain computer interface. In: Ayaz, H., Mazur, L. (eds.) Advances in Neuroergonomics and Cognitive Engineering, vol. 775, pp. 124–134. Springer, Cham (2019)
5. Liu, Y., Ayaz, H., Shewokis, P.A.: Multisubject "learning" for mental workload classification using concurrent EEG, fNIRS, and physiological measures. Front. Hum. Neurosci. **11**(389) (2017). https://doi.org/10.3389/fnhum.2017.00389
6. Schmidhuber, J.: Deep learning in neural networks: an overview. Neural Networks **61**, 85–117 (2015)
7. Bashivan, P., Rish, I., Yeasin, M., Codella, N.: Learning representations from EEG with deep recurrent-convolutional neural networks. In: arXiv preprint arXiv:1511.06448 (2015)
8. Hajinoroozi, M., Mao, Z., Jung, T.-P., Lin, C.-T., Huang, Y.: EEG-based prediction of driver's cognitive performance by deep convolutional neural network. Sig. Process. Image Commun. **47**, 549–555 (2016)
9. Thodoro, P., Pineau, J., Lim, A.: Learning robust features using deep learning for automatic seizure detection. In: Machine Learning for Healthcare Conference, pp. 178–190 (2016)
10. Schirrmeister, R.T., Springenberg, J.T., Fiederer, L.D.J., Glasstetter, M., Eggensperger, K., Tangermann, M., Hutter, F., Burgard, W., Ball, T.: Deep learning with convolutional neural networks for EEG decoding and visualization. Hum. Brain Mapp. **38**(11), 5391–5420 (2017)
11. Trakoolwilaiwan, T., Behboodi, B., Lee, J., Kim, K., Choi, J.-W.: Convolutional neural network for high-accuracy functional near-infrared spectroscopy in a brain-computer interface: three-class classification of rest, right-, and left-hand motor execution. Neurophotonics **5**(1) (2007)
12. Croce, P., Zappasodi, F., Merla, A., Chiarelli, A.M.: Exploiting neurovascular coupling: a Bayesian sequential Monte Carlo approach applied to simulated EEG fNIRS Data. J. Neural Eng. **14**(4) (2017)

13. Hong, K.-S., Naseer, N., Kim, Y.-H.: Classification of prefrontal and motor cortex signals for three-class fNIRS-BCI. Neurosci. Lett. **587**, 87–92 (2015)
14. Chiarelli, A.M., Zappasodi, F., Di Pompeo, F., Merla, A.: Simultaneous functional near-Infrared spectroscopy and electroencephalography for monitoring of human brain activity and oxygenation: a review. Neurophotonics **4**(4) (2017)
15. Fazli, S., Mehnert, J., Steinbrink, J., Curio, G., Villringer, A., Muller, K.-R., Blankertz, B.: Enhanced performance by a hybrid NIRS-EEG brain computer interface. Neuroimage **59**(1), 519–529 (2012)
16. Saadati, M., Nelson, J.K.: Multiple transmitter localization using clustering by likelihood of transmitter proximity. In: 51st Asilomar Conference on Signals, Systems, and Computers, pp. 1769–1773 (2017)
17. Khan, M.J., Hong, M.J., Hong, K.-S.: Decoding of four movement directions using hybrid NIRS-EEG brain-computer interface. Front. Hum. Neurosci. **8**, 244 (2014)
18. Lee, M.-H., Fazli, S., Mehnert, J., Lee, S.-W.: Hybrid brain-computer interface based on EEG and NIRS modalities. In: Brain-Computer Interface (BCI), 2014 International Winter Workshop (2014)
19. Jirayucharoensak, S., Pan-Ngum, S., Israsena, P.: EEG-based emotion recognition using deep learning network with principal component based covariate shift adaptation. Sci. World J. (2014)
20. Shin, J., Von Luhmann, A., Kim, D.-W., Mehnert, J., Hwang, H.-J., Muller, K.-R.: Simultaneous acquisition of EEG and NIRS during cognitive tasks for an open access dataset. In: Generic Research Data (2018)
21. Shin, J., Von Luhmann, A., Kim, D.-W., Mehnert, J., Hwang, H.-J., Muller, K.-R.: Simultaneous aquisition of EEG and NIRS during cognitive tasks for an open access dataset. In: Scientific Data, vol. 5 (2018)
22. Pfurtscheller, G.: Functional brain imaging based on ERD/ERS. Vis. Res. **41**(10–11), 1257–1260 (2001)
23. Neuper, C., Pfurtscheller, G.: Event-related dynamics of cortical rhythms: frequency-specific features and functional correlates. Int. J. Psychophysiol. **43**(1), 41–58 (2001)
24. Pourshafi, A., Saniei, M., Saeedian, A. Saadati, M.: Optimal reactive power compensation in a deregulated distribution network. In: 44th International Universities Power Engineering Conference (UPEC), pp. 1–6 (2009)
25. Ma, L., Zhang, L., Wang, L., Xu, M., Qi, H., Wan, B., Ming, D., Hu, Y.: A hybrid brain-computer interface combining the EEG and NIRS. In: 2012 IEEE International Conference on Virtual Environments Human-Computer Interfaces and Measurement Systems (VECIMS), pp. 159–162 (2012)
26. Saadati, M., Mortazavi, S., Pourshafi, A., Saeedian, A.: Comparing two modified method for harmonic and flicker measurement based on RMS, p. R1 (2009). https://rms.scu.ac.ir/Files/Articles/Conferences/Abstract/v56-27.pdf2009102231223218.pdf
27. Schirrmeister, R.T., Springenberg, J.T., Fiederer, L.D.J., Glasstetter, M., Eggensperger, K., Tangermann, M., Hutter, F., Burgard, W., Ball, T.: Deep learning with convolutional neural networks for EEG decoding and visualization. In: Hum. Brainmapping **38**(11), 5391–5420 (2017)

Impacts of the Time Interval on the Choice Blindness Persistence: A Visual Cognition Test-Based Study

Qiuzhu Zhang[1], Yi Lu[2], Huayan Huangfu[2,3]([✉]), and Shan Fu[2]

[1] School of Life Science and Technology,
University of Electronic Science and Technology of China, No. 4, Section 2,
Jianshe North Road, Chengdu 610054, China
`201821140602@std.uestc.edu.cn`
[2] Department of Automation, School of Electronic and Electrical Engineering,
Shanghai Jiao Tong University, 800 Dongchuan Road, Shanghai 200240, China
`hyhuangfu@sjtu.edu.cn`
[3] Counseling and Support Services, Shanghai Jiao Tong University,
800 Dongchuan Road, Shanghai 200240, China

Abstract. Choice blindness presented as a person's failure to notice a mismatch between one's preference and task decision. We studied the persistence of choice blindness through time interval based on visual cognitive tests. We designed a 2 (high and low similarity pictures) × 2 (one-day and one-week time interval) mixed design. 20 pairs of scenery pictures with different similarities were used as materials and 52 adults were recruited as participants. The results verified the existence of choice blindness. It was found that no significant difference existed in the perception of false feedback between the first day and the first week in the first experiment. However, there was significant difference in the second experiment between one-day later and one-week later. In other words, time intervals had an effect on the persistence of choice blindness, especially the time interval of one-day.

Keywords: Choice blindness · Time interval · The false feedback

1 Introduction

People make choices frequently and they usually think they are aware of the reasons for their choices and can easily detect the mismatch between their choices and the results. However, more and more studies have found that this is not the case. Sometimes even the simplest choice task, people may not know what their real choice is. This phenomenon is called choice blindness [1]. Choice blindness is a robust, repeatable and dramatic phenomenon, which is ubiquitous in choice preference and has important influence on individual decision. Choice blindness plays an important role in decision because of its universality and uniqueness. Therefore, the study of choice blindness has great theoretical significance and application value. An important feature of choice blindness is persistence [2], and time interval is the important measurement index. In theory, the study on the impact of time interval on the persistence of choice blindness

© Springer Nature Switzerland AG 2020
H. Ayaz (Ed.): AHFE 2019, AISC 953, pp. 233–242, 2020.
https://doi.org/10.1007/978-3-030-20473-0_23

can provide a deeper understanding of choice blindness, deepen the study on the factors, and provide a theoretical basis for the blindness study of later researchers. In practice, choice blindness is closely related to decision-making, which is common in consumer psychology, attitude formation, moral judgment and other fields. This study can help individuals make better decisions.

Choice blindness always presents as a person's failure to notice a mismatch between one's preference and task decision, that is, "Participants cannot find that their real choices are manipulated" [3]. Choice blindness is an important evidence to prove the change of individual choice preference. In recent years, the researchers have paid more attention to it. They first proposed the concept of choice blindness and verified the existence of the phenomenon in laboratory experiments and field experiments [3]. The experiment became a classical experimental paradigm. In this experiment, participants were told that they were taking part in an experiment on the attractiveness of female faces. Then participants were presented with pairs of female face pictures and were required to make a choice between the two pictures based on attractiveness. After making a choice, the picture they chose was presented again and they were asked to explain why it was more attractive. But in fact, the participants were given false feedback in partial selection feedback. That is, the participants were presented with another photo they did not choose at first. The final results showed that most of the participants did not detect that the photos presented to them were not the original photos in the false feedback.

Some others studied the choice blindness. They mainly focus on three aspects, the first one is the scene of the choice blindness, the second one is the effect of the choice blindness' persistence and the third one is the different channels of the choice blindness [4, 5]. Whether the impact of choice blindness on selection is short or long-term, the current research results are still inconclusive. In a choices blindness study all the pictures were randomly presented to the participants for 15 min after the completion of the entire experimental task, and the participants were asked to rate their preference for pictures [6]. Participants' score of the false preference picture was still higher than the real preference picture. So researchers argued that the participants' preference for the false feedback pictures did not change after 15 min, which indicated that the choice blindness had a lasting effect on the change of preference. Based on the classical paradigm which used with female faces as experimental materials, Johansson [7] added ranking. It was found that the participants' preference was affected by the experimenter's manipulation, and they rated the unselected option higher. The results shown that the choice blindness effect was long-lasting and can affect future choices, but the duration time was short. A choice blindness paradigm with female faces as experimental materials and Singaporean as participants found that the effect of choice blindness on selection preference manipulation was short [8]. This manipulation effect only lasted for a few minutes without lasting effects of days or even weeks.

Some trails changed the materials based on the classic paradigm to explore the choice blindness. For example, a choice blindness task used vacation destinations as materials and asked participants to rate vacation destinations [9]. The results showed that the participants rated their unselected vacation destinations more highly. When the researchers repeated the trail after 3 years, they found that the participants still chose the vacation destinations they had not chosen in the first place. That is, the influence of choice blindness on selection preference manipulation still existed after 3 years [10].

Zhang [1] explored the psychological mechanism of choice blindness from the perspective of memory representation and recruited Chinese as participants. Although the choice blindness was persistent, the lasting time wasn't long. The memory representation theory is the cognition perspective to explain the choice blindness. Both the representation failure and the extraction failure may be the occurrence mechanism of the choice blindness, but the representation failure is likely to the basis. By studying cognitive style and sensory channel on choice blindness [11], and finding that sensory channel was an important factor of choice blindness. The participants (from china) were more likely to suffer from choice blindness under the auditory condition than the visual condition. There was also a comprehensive overview of the current studies about choice blindness was conducted and it discussed the stability of choice blindness [2].

In short, although choice blindness involves many fields, it is still a relatively new research content. At present, the researchers have paid more attention to the choice blindness. But the research on the influencing factors of choice blindness is still in the preliminary stage.

2 Method

2.1 Participants

52 healthy adults (half male and half female, age range 18–22 years) were recruited to participants in this study, and were randomly divided into two groups (each group of 26 participants, half male and half female). All participants had normal or corrected to normal vision. And they had never participated in the choice blindness experiment to ensure that they did not know the purpose of the experiment. They also had basic computer operation ability.

Material. 20 pairs of scenery pictures with different similarities (10 pairs of high similarity pictures and 10 pairs of low similarity pictures) were used as materials. Eight of the twenty pairs were chosen as target pairs. Four of these eight pairs were false feedback pictures (2 pairs of high similarity pictures and 2 pairs of low similarity pictures), that is, the pictures presented to the participants was the non-preference pictures that the participants did not choose at the beginning. For other four of eight pictures were true feedback pairs (2 pairs of high similarity pictures and 2 pairs of low similarity pictures), that is, the pictures presented to the participants was the preference pictures that the participants chosen at the beginning.

2.2 Experiment Design

We designed a 2 (high and low similarity pictures) × 2 (one-day and one-week time interval) mixed design based on the classical selection of blind paradigm. The independent variables were time interval (one-day and one-week) and picture similarity (high similarity and low similarity). In independent variables, time interval was the variable between subjects and picture similarity was the variable within subjects. The dependent variables were the reaction after participants receiving the false feedback scenery pictures. 20 pairs of scenery pictures were presented 4 pairs were veridical feedback pictures, 4 pairs were false feedback, and the remaining 12 pairs were no feedback. The order of veridical and false feedback pairs and pictures' similarity was presented by ABBA design.

2.3   Experiment Procedure

First step, all participants completed the trail on the computer. Before the formal experiment begins, the participants will be informed that "This is an experiment on scenic attraction. Please take a deep breath and relax. This experiment will not have adverse effect on you. Next, we will start the experiment, please follow the instructions to do every step."

Second step, we used the E-prime software programming to complete the experiment (Fig. 1). A pair of scenery pictures were displayed on the left and right side of the computer screen. Participants were asked to choose the more attractive scenery picture between presented pairs. And the choice process was free. After chosen, the selected pictures will appear again in the center of the computer screen. And participants were asked to write down the reasons of the chosen the preferred picture on the experimental record chart. The process was also no limited time.

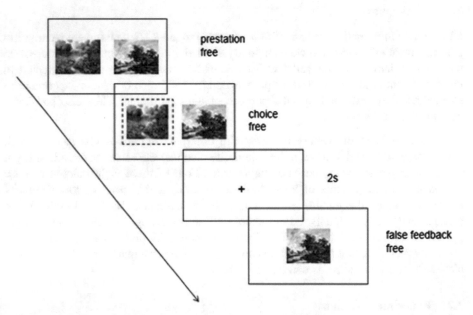

Fig. 1. The process of false feedback. First the participants were shown two scenery pictures and were asked to choose which one he or she found more attractive. The picture in the imaginary line was chosen (the imaginary line didn't appear in trails). Then a fixation appeared in the screen. After 2 s, the chosen picture was shown in the center. At the same time, the participants need to write down their reasons of choosing the scenery pictures in the experimental record chart.

Third step, the one group repeated the experiment on the second day after completing the experiment, and the other group repeated the experiment after one week.

2.4   Data Recording and Processing

We used E-prime and SPSS 20.0 software to process the results. The participants' unawareness of the false feedback pictures was scored as 0, their detections of once was scored as 1, and twice was scored as 2. We recorded whether the pictures chosen in the first day (or the first week) and the second day (or the second week) were changed or not. If the pictures were not changed, we marked as 0. Otherwise it was marked as 1. We also recorded the reasons for the participants to choose the preferred pictures and the reaction time. The reasons were divided into four types: wholeness, detail, part, and intuition. The detail reason was smaller area in the scenery pictures than part reason.

3   Results

3.1   Choice Blindness

For verifying the existence of choice blindness, we analyzed the detection rate of choice blindness by participants' reports and used SPSS20.0 software to taking statistical analysis. There were the 208 false feedback in total. In the first experiment, the detection rate of false feedback was 40.87%, and it was 56.73% in the second experiment. In other words, nearly half of the participants in both experiments were not aware of the false feedback. The results showed the choice blindness existed really on Chinese adults.

3.2   Impacts of the Time Interval on the Choice Blindness Persistence

To examine the effect of time interval on chosen-scenery pictures preferences, we used the chi-square test to compare the difference in number of detections for false feedback pictures. In the first experiment, there was no significant difference in the detections of false feedback pictures between the two groups ($\chi^2 = 0.85$, $p = 0.66$; Table 1). And there was no significant difference between low similarity and high similarity pictures (in the low-similarity, $\chi^2 = 1.58$, $P = 0.46$; in the high similarity, $\chi^2 = 0.10$, $P = 0.95$).

Table 1. The chi-square test for detecting false feedback pictures was tested for the first experiment in both groups (one-day and one-week)

Groups		0	1	2
One-day	Low similarity	12	9	5
	High similarity	12	8	6
One-week	Low similarity	10	7	9
	High similarity	11	9	6

We used the chi-square test to compare the difference in number of detections trail for the second experiment. The results shown that there was significant difference in the detections of false feedback pictures between the two groups ($\chi^2 = 6.41$, $P = 0.04 < 0.05$; Table 2). However, there was no significant difference between the two groups in the false feedback of low and high similarity pictures (in the low-similarity, $\chi^2 = 2.32$, $P = 0.31$; in the high similarity, $\chi^2 = 4.12$, $P = 0.13$).

Table 2. The chi-square test for detecting false feedback pictures was tested for the second experiment in both groups (one-day and one-week)

Groups		0	1	2
One-day	Low similarity	10	2	14
	High similarity	13	3	10
One-week	Low similarity	6	5	15
	High similarity	7	8	11

We also compared the results of the two groups before and after the two experiments. Participants both in the one-day and one-week time interval group noticed more false feedback scenery pictures in the second experiment than the first experiment (Tables 3 and 4). By using chi-square to examine the number of their detections of false feedback, it shown that there was significant difference between the twice experiments on the one-day time interval group and indicated that the false feedback was more easily detected in the second experiment ($\chi^2 = 11.40$, $P = 0.003 < 0.01$; Table 3). But there was no significant difference between the twice experiments on the one-week time interval group ($\chi^2 = 5.14$, $P = 0.08$; Table 4).

Table 3. The chi-square test of participants' detections false feedback in the one-day time interval group

Experiment	0	1	2
The first experiment	24	17	11
The second experiment	23	5	24

In the experiment we saw difference between females and males, so we used chi-square test to identify the gender difference in the detections of false feedback scenery pictures. There was no significant difference in the detections of false feedback between females and males on the one-day time interval group ($\chi^2 = 2.37$, $P = 0.31$; Table 5). But chi-square test showed that it was significant different of the one-day time interval group' females and males ($\chi^2 = 12.85$, $P = 0.002 < 0.01$; Table 5).

The average detection value of false feedback for one-week time interval group males was 0.82, while that of female participants was 0.42. Therefore, the male participants were more likely to detect the false feedback than the females in one-week time interval group.

Table 4. The chi-square test of participants' detections false feedback in the one-week time interval group

Experiment	0	1	2
The first experiment	21	16	15
The second experiment	13	13	26

Table 5. Gender difference detections of false feedback chi-square test

Gender	Groups	0	1	2
Female	One-day	23	14	15
	One-week	22	16	14
Male	One-day	24	8	20
	One-week	12	13	37

We recorded the participants' reasons for choosing preferences and divided four part (i.e. wholeness, detail, part, and intuition). The two groups participants were focus on scenery pictures' wholeness in two experiments (e.g. picture's layout, structure). And there was no significant difference for reasons ($\chi^2 = 3.96$, $P = 0.27$; Table 6).

Table 6. Reasons chi-square analysis of false feedback received by both groups (one-day and one-week)

Groups	Experiment	Wholeness	Intuition	Part	Detail
One-day	First	37	11	7	7
	Second	18	6	5	9
One-week	First	40	6	7	16
	Second	32	5	6	10

For investigating whether false preferences affect preferences, we compared each participant's choice of preferences scenery pictures in twice experiments. The results shown that the number of unchanged choice preference was significantly more than the number of changing (Table 7). The consistency of choice preference between the twice experiments in the one-day time interval group was 78%, and the one-week time interval group was 70.2%. There was no significant difference for choice preference ($\chi^2 = 1.60$, $P = 0.21$; Table 7).

Table 7. Chi-square analysis for choice preference of false feedback

	Preference unchanged	Preference change
One-day	81	23
One-week	73	31

In our experiments, males spent more time on choosing their more preference scenery pictures than females in the one-week time interval group (Table 8). That is, males' react time was different with females and their react time was longer. And the difference was significant by independent-sample t test (t(24) = −2.48, p = 0.02<0.05; Table 8).

Table 8. Gender difference of react time to false feedback pictures in one-week time interval group (ms)

Gender	M ± SD	t	p
Female	5627 ± 3385	−2.48	0.02
Male	9567 ± 4613	N/A	N/A

4 Discussion

In this study, the effect of time interval on the choice blindness' persistence was discussed by comparing the detections response to false feedback in twice trials at different time interval.

In extant literature on choice blindness, most of them believe that the choice blindness' persistence is transient (e.g., Taya et al.) [6]. We found that in the first experiment there was no significant difference in the detections of false feedback between the two groups. However, in the second experiment there was significant difference in the detections of false feedback between the two groups. In other words, there was no significant difference in the perception of false feedback between the first day and the first week in the first experiment. However, there was a significant difference in the detections of false feedback in the second experiment between one-day later and one-week later. That is, time interval had an effect on the persistence of choice blindness, especially the one-day time interval. The reason for this phenomenon may be related to the participants' own emotion, situation and picture presentation time. Besides, as the time of one-day time interval is short, when the participants repeated the same experiment next day, which may produce practice effect to detections of false feedback.

In the classical choice blindness experiment, the researchers used the faces pictures with different similarity as materials and found that the difference in similarity had no significant difference in the detections rate of choice blindness. That is, similarity was not the influencing factor of choice blindness [3]. In the second experiment, although there was significant difference in the detections of false feedback scenery between the two groups. But the difference of detections in low and high similarity scenery pictures was no significant. It also indicated that pictures' similarity has no salient influence on the detections of false feedback.

When participants received the feedback of pictures (including veridical and false feedback), their reasons for the preference pictures were focused on the wholeness. No matter how many times the experiment was carried out, it can be clearly seen that the overall description was the main reason. This was consistent with previous studies.

Some researchers pointed out in the cross-cultural study of attention that the overall processing mode of Oriental people involved a larger area of attention [12]. In exploring the psychological mechanism of choice blindness, it was found that the overall feature description was more than the specific feature description when the participants gave causal descriptions to the pictures with veridical or false feedback [1].

Some researchers found that the choice blindness may affect the selection preference [5, 13]. In our experiment, we found that whether the one-day time interval group or the one-week time interval group, the participants' choice preference for false feedback scenery pictures did not change much. That is to say, after receiving false feedback from the first experiment, most of the preference pictures selected in the second experiment were still consistent with the first experiment. But our results showed that there was no significant difference between the pictures selected in the twice experiments and indicated that the choice preference in the second experiment was not significantly affected by the choice blindness.

In the study of the response to the false feedback scenery pictures, it was found that males chosen the preferred pictures about 4 s longer than females in the one-week time interval group. And there was significant difference in gender's react time to the false feedback pictures by t-test. As for the reasons for false feedback pictures, we found that although both males and females paid more attention to the picture' wholeness, and females focused on the details of the picture was 13.4%. While males paid attention to the details of the picture was 24.5%. This suggested that men pay more attention to detail than women, which may explain why males who were tested in one-week time interval group spent significantly more time than females. Therefore, male participants in one-week time interval group were more likely to detect false feedback.

Future work should increase the number of participants to enhance the sample representativeness. The selection of experimental materials also needs to be more careful and considerable. At the same time, a more detailed study can be carried out on the variable of time interval. The time interval in our experiment were one-day and one-week time interval. So three or five days-time interval can be added to further explore the influence of time interval on the choice blindness' persistence.

5   Conclusions

We studied the persistence of choice blindness through time interval based on visual cognitive tests. Draw conclusions under research conditions: in the first experiment there was no significant difference in the detections of false feedback between the two groups, $P > 0.05$; in the second experiment there was a significant difference in the detections of false feedback between the two groups, $P < 0.05$. Time intervals have an effect on the choice blindness' persistence, especially the time interval of one-day.

References

1. Zhang, H., Xu, F.M., Xu, M.B.: Choice blindness: did you really know what you have chosen? Adv. Psychol. Sci. **22**(8), 1312–1318 (2014). (in Chinese)
2. Ding, S.W.: Choice blindness: how stable is your choice blindness? Guide Sci. Educ. **3**, 177–178 (2017). (in Chinese)
3. Johansson, P., Hall, L., Sikstrm, S., Olsson, P.A.: Failure to detect mismatches between and outcome in a simple decision task. Science **310**(5745), 116–119 (2005)
4. Hall, L., Johansson, P., Taring, B., Sikstrom, S., Deutgen, P.: Magic at the marketplace: choice blindness for the taste of jam and smell of tea. Cognition **117**(1), 1–3 (2010)
5. Tang, Y.: A review of foreign studies on choice blindness. Psychol. Doctor **23**(4), 1–3 (2017). (in Chinese)
6. Hall, L., Johansson, P., Chater, N.: Preference change through choice, neuroscience of preference and choice: cognitive and neural mechanisms. Acad. Press **9**(5), 121–138 (2012)
7. Johansson, P., Hall, L., Taring, B., Sikstrom, S., Chater, N.: Choice blindness and preference change: you will like this paper better if you (believe you) chose to read it. J. Behav. Decis. Making **27**(3), 281–289 (2014)
8. Taya, F., Gupta, S., Farber, I., O'Dhaniel, P.: A manipulation detection and preference alterations in a choice blindness paradigm. PLoS ONE **9**(9), e108515 (2014)
9. Sharot, T., Velasquez, C.M., Dolan, R.J.: Do decision shape preference? Evidence from blind choice. Psychol. Sci. **21**(9), 1231–1235 (2010)
10. Sharot, T., Fleming, S.M., Yu, X., Koster, R., Dolan, R.J.: Is choice-induced preference change long lasting. Psychol. Sci. **3**(10), 1123–1129 (2012)
11. Zen, K., Duan, J.Y., Tian, X.M.: The impact of cognitive styles and sensory modalities to choice blindness. Chin. J. Ergon. **22**(4), 5–9 (2016). (in Chinese)
12. Liu, S., Wang, H., Peng, K.: Cross-cultural research on attention and its implications. Adv. Psychol. Sci. **21**(1), 37–47 (2013). (in Chinese)
13. Johansson, P., Hall, L., Sikstrm, S.: From change blindness to choice blindness. Psychologia **51**(2), 142–155 (2008)

Systemic-Structural Activity Theory

Self-regulation Model of Decision-Making

Alexander Yemelyanov[✉]

Department of Computer Science, Georgia Southwestern State University,
800 Georgia Southwestern State University Drive, Americus, GA 31709, USA
Alexander.Yemelyanov@gsw.edu

Abstract. The paper proposes the self-regulation model (SRM) of decision-making, which is based on the self-regulation model of the thinking process developed within the systemic-structural activity theory. SRM includes two sub-models: formation of mental model (FMM), which is executed by the divide and concur algorithm, and formation of the level of motivation (FLM), which is executed by the dynamic programming algorithm, as well as the regulation of their interaction by using feedback and feedforward controls. Feedback control is regulated by the factor of difficulty and feedforward control is regulated by the factor of significance. These two factors determine four general criteria of success in evaluating and regulating the level of motivation. The paper formulates primary rules of self-regulation in decision-making in which the factors of significance and difficulty are designated the leading role. In a real-life example with a Facebook friend request we demonstrate how these rules were implemented in Performance Evaluation Process, which relies on Express Decision, a mobile web application for supporting an individual in making quick decisions in complex problems.

Keywords: Decision-making · Mental model · Level of motivation ·
Self-regulation · Feedback and feedforward controls ·
Systemic-structural activity theory · Factors of significance and difficulty ·
Mobile web application

1 Introduction

Decision-making is considered a self-regulative thinking process driven by a motivation to attain a goal [1]. The foundation for decision-making is a continuous reformulation of a problem, and the development of such a mental model that can determine the level of motivation for selecting the best of the available alternatives based on the expected outcomes. The initial mental model of a problem often cannot facilitate the achievement of desired outcomes. In this situation, after understanding the problem at hand, the decision-maker divides a problem into sub-problems. The decision-maker then begins creating sub-problems by formulating various hypotheses. Each hypothesis has its own potential goal. Based on the comparison and evaluation of such hypotheses, the decision-maker selects one and formulates the first sub-goal associated with the selected hypothesis. Comparing a new sub-goal with an existing mental model of the problem allows to transform the original mental model into a new one that is adequate for the new sub-problem. If the problem is evaluated positively (positive feedback), the

© Springer Nature Switzerland AG 2020
H. Ayaz (Ed.): AHFE 2019, AISC 953, pp. 245–255, 2020.
https://doi.org/10.1007/978-3-030-20473-0_24

thinking process cycle is complete. However, if the result is evaluated negatively (negative feedback), internal or external information should be added to continue this process on a lower level (or to search for other alternatives of solving the problem). In SSAT, unlike cognitive psychology, the decision-maker is able to regulate his behavior not only externally, but also internally by using the inner mental plane. This process is not straightforward and includes a continuous cycle of updating the mental model with feedback and feedforward controls. It is necessary to note that in traditional dynamic decision-making only external feedback from the environment is considered [2]. The description of the suggested self-regulation model of decision-making is presented below. In this model, a key role is assigned to the factors of significance and difficulty. These factors pass through the entire process of decision-making. The factors themselves and their interactions determine the rational mechanism of any choices including the ones in which the emotional component has a decisive impact. In the present work we demonstrate the role of these factors in determining the primary rules of self-regulation.

2 Formation of Mental Model and the Level of Motivation

The Self-Regulation Model (SRM) of decision-making includes two sub-models: Formation of Mental Model (FMM) and Formation of the Level of Motivation (FLM). Execution of SRM is driven by the motivation to attain a goal and includes the execution of FMM and FLM, as well as the regulation of their interaction by using feedback and feedforward controls.

The Design strategy for FMM implements a divide-and-conquer algorithm (D&C) to construct a Decision Tree. The divide-and-conquer technique [3] uses a recursive breakdown approach in decision-making: decompose the problem into smaller sub-problems, solve them, and then recombine their results to solve the bigger problem. This division of the problem into sub-problems may span several levels deep until a basic (ad hoc) level of certainty will be reached, at which point the problem can be positively evaluated based on IL-Frame within the process of FLM. In other words, the problem will contain only those outcomes for which the decision-maker will be able to determine their respective positive (or negative) intensity and likelihood, which, in turn, will allow to determine the positive (or negative) motivational level (preference). It should be noted that the efficiency of the divide-and-conquer algorithm increases when people apply hypotheses and split the problem into two mutually exclusive hypotheses. In FMM, feedback control is used to verify whether the current state of the individual's mental model is capable of either evaluating the problem based on IL-Frame or choosing the best alternative. The feedback is *positive* (+fb_FMM) when the individual can perform the verification and *negative* (−fb_FMM) when the individual cannot perform it. Feedforward control leads to an upgrade of the existing mental model. With this purpose, by considering various hypothetical situations and alternative solutions, the problem is divided into sub-problems with corresponding sub-goals.

The design strategy for FLM implements a dynamic programing algorithm (DP). This algorithm determines the level of an alternative's motivation by evaluating its outcomes in IL-Frame and aggregating results with the help of K-Rules. IL-Frame is used as a template to evaluate outcomes, according to four primary criteria of success: *positive significance* (which is presented by positive intensity $I+$), *positive component of difficulty* (positive likelihood $L+$), *negative significance* (negative intensity $I-$), and *negative component of difficulty* (negative likelihood $L-$). These are four significance-difficulty criteria of success in the evaluation of outcomes (further abbreviated as 4-SD Criteria). K-Rules are relations between these four criteria which have been experimentally determined by Kotik [4, 5] with the purpose to define positive $S+$ $(I+, L+)$, negative $S-(I-, L-)$, and cumulative $P = S(P+, P-) = S(S+ (I+, L+), S-(I-, L-))$ levels of preference of outcomes, as well as to combine these levels into a cumulative level of preference (motivation) for an alternative. The dynamic programing technique [3] is used to solve a problem by breaking it down into smaller and simpler sub-problems. This method is applied to solve problems that have the properties of overlapping sub-problems and that fulfill the principle of optimality. The first requirement is satisfied because the divide-and-conquer method splits problems into sub-problems. The principle of optimality is also satisfied, in the case that the formation of the level of motivation indicates that this level can accomplish the individual's goal: if the level of motivation for solving a problem accomplishes the individual's goal, then the level of motivation for solving sub-problems will also accomplish the individual's corresponding sub-goals. In FLM, feedback control verifies whether the level of motivation for choosing an alternative is created. Feedback is *positive* (+fb_FLM) when the level of motivation is created, but *negative* (−fb_FLM) otherwise. Feedforward control allows to predict the level of the alternative's motivation after changing verbal characteristics (manipulated control inputs) in IL-Frame.

Feedback Control provides a connection between FMM and FLM models. It is regulated by the factor of *difficulty*, which determines the individual's self-efficacy [6] in attaining the goal. Feedback control is corrective, connected to the individual's past experience, and provides robustness and error elimination in formation of the mental model and the level of motivation [7].

Feedforward Control produces upgrades in FMM and FLM. It is regulated by the factor of *significance*, which determines the directness to the goal [1]. Feedforward control is predictive and leads to an upgrade in the existing mental model or level of motivation in order to enhance its ability to solve the problem and obtain the desired outcomes.

Within SRM, both FMM and FLM, there are two concurrently running processes that are self-regulated by feedback and feedforward controls and driven by motivation to attain the goal. FMM implements the divide-and-conquer algorithm for mental model formation, while FLM implements the dynamic programing algorithm for motivation level formation. The level of motivation for selecting an alternative forms dynamically, according to the mental model that forms recursively. It is worth noting that using the dynamic programming algorithm to determine the level of motivation demonstrates just how powerful motivation really is. Many combinatorial decision problems that typically require exponential time and space to be solved by standard algorithms can be solved in polynomial time and space using dynamic programming. Therefore, self-regulation in decision-making optimizes the use of time and memory resources. It is worth noting that in the dynamic programming of the level of motivation, the *power of motivation* itself is demonstrated.

Kotik [8] illustrated that self-regulation increases the reliability of the goal-directed activity. He described two types of self-regulation: self-regulation in the area of information processes (*information-based self-regulation*) and self-regulation in the area of energy (emotional) processes (*energy-based self-regulation*). The complexity in decision-making is compensated by the intensification of information processes, as well as by the energy reaction of the brain. All of this generate energy mobilization of the organism, aimed at bringing it into readiness for intensive spending forces and overcoming the difficulties. Zarakovsky and Pavlov [9] showed that emotions have an inducing function that provides energy for switching, reinforcing, compensating, and organizing functions in self-regulation.

According to behavioral and neuroscientific research, cognition and rational decision-making are not entirely the product of rational information processing and symbol manipulation, but instead require the involvement of emotion. Damasio's somatic marker hypothesis [10] suggests a mechanism by which emotional processes can self-regulate decision-making. When making decisions, these somatic markers and their induced emotions are consciously or unconsciously associated with prior positive and negative outcomes that allow to quickly evaluate these outcomes based on past experiences. This is a demonstration of energy-based self-regulation in which the formation of the level of motivation (FLM) is activated based on internal positive feedback from the formation of the mental model (FMM). This means that entirely emotionally-driven decisions can be advanced in a purely rational way by comparing positive and negative outcomes of each of the hypothetical decision options.

The conducted analysis allows to formulate the following four primary rules of self-regulation (further abbreviated as 4-SR Rules), where the factors of significance and difficulty are assigned a key role.

SR1: +fb_FMM ⇒ FLM. Positive feedback on formation of the mental model activates formation of the level of motivation (D&C algorithm in FMM calls DP algorithm in FLM). This means that, if the difficulty of evaluation of outcomes in the IL-Frame is adequate, then the level of motivation is formed.

SR2: −fb_FMM ⇒ ff_FLM. Negative feedback on formation of the mental model activates feedforward control of formation of the level of motivation (D&C algorithm in FMM upgrades DP algorithm in FLM). This means that if the difficulty of evaluation of outcomes in the IL-Frame is not adequate, then the significance of the problem determines whether this problem should be split into sub-problems (hypotheses) for their further evaluation in the Il-Frame, or if the decision-making process should be terminated.

SR3: +fb_FLM ⇒ FMM. Positive feedback on formation of the level of motivation activates formation of the mental model (DP algorithm in FLM calls D&C algorithm in FMM). This means that if the difficulty of evaluation of the level of motivation is adequate, then the mental model is formed and a decision can be made.

SR4: −fb_FLM ⇒ ff_FMM. Negative feedback on formation of the level of motivation activates feedforward control of formation of the mental model (DP algorithm in

FLM upgrades D&C algorithm in FMM). This means that if the difficulty of evaluation of the level of motivation is not adequate, then significance determines whether the goal and/or alternatives of the problem should be modified and the decision-making process repeated or completely terminated.

Figure 1 presents a model of the self-regulation process of decision-making in which the factors of significance and difficulty are assigned the leading role. This model includes construction of a *Decision Tree*, based on evaluating outcomes in *IL-Frame* according to 4-SD Criteria and aggregating the results of this evaluation, based on the K-Rules. All of this is done to determine the level of motivation (preference) for choosing alternatives. Consistent interaction between the Decision Tree and IL-Frame is regulated by the primary 4-SR Rules of self-regulation.

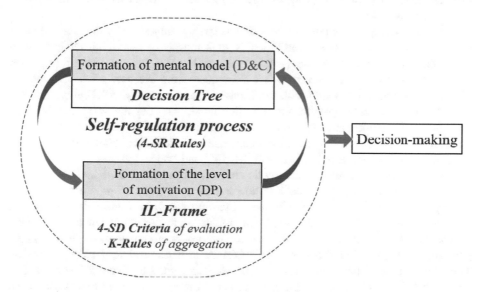

Fig. 1. Self-regulation model of decision-making

In the following example, we present a scenario with a Facebook friend request in which we demonstrate how the rules of self-regulation were implemented in Performance Evaluation Process [5], which relies on Express Decision [11], a web application for supporting an individual in making quick decisions in complex problems.

3 Applying Rules of Self-regulation to Make a Decision About Facebook Friend Request

Gena has a decision-making problem regarding accepting, rejecting, or considering a Facebook friend request. However, instead of rushing to solve it on the fly, she has decided to turn to the assistance of Express Decision (ED) to help her make the decision that is most appropriate to her particular situation. A couple months ago, Gena was accepted for a position as an analyst in a mid-size marketing company. Gena was able to establish professional friendly relations with her boss, Robert. As a side gig, Gena is a photographer at music shows and posts her photos to her Facebook page, which is only visible to her friends. Her boss in the marketing company, Robert, recently came across some of the photos that Gena has taken that had been shared on others' Facebook pages; he is an ardent music lover and becomes interested to see more of her photography. Robert sends Gena a friend request with the hopes that she will accept it, so that he will have access to all her photos and related work. However, Gena is concerned that if she accepts his request, he will also gain access to all her private information, along with some graphic content that he might find inappropriate or offensive, and this could negatively affect her career prospects. Gena tries to weigh the pros and cons of either accepting or rejecting Roberts's friend request, and wonders what she should do.

The uncertainty of such decision-making is largely influenced by the fact that Gena cannot predict how a new friend's access to her account could impact her career prospects if their friendly business relationship might come to an end in the future. On one hand, Gena does not want to delete the request from her boss for fear of jeopardizing her current relationship with him, since friendly relations with him would only contribute to her career growth. On the other hand, she does not want any personal information he could retrieve from her account to be used against her, in case the relationship with her boss somehow deteriorates or does not work out over time. Evidently, the correct decision between "accept" and "reject" is very important (significant) for Gena because it is directly tied to her professional growth and the future of her career. At the same time, each option has its own share of pros and cons, and finding a reasonable compromise between them proves to be a difficult task for her. This determined the fact that she was highly motivated to carefully evaluate each of the alternatives for the subsequent selection of the best one among them. In the given situation, Gena, worried about making a mistake in her selection, decided to rely on more than her intuition alone. This is why she has turned to Express Decision. She knows that Express Decision is a mobile web application no more difficult to use than a standard calculator, and that it can quickly assist anyone with their mixed feelings regarding evaluating and selecting the best solution for them. The only requirement is a general understanding or awareness (in broad terms, at the very least) of what it is that you ultimately want to accomplish with your selection – you need to have a general awareness of the *goal* you wish to realize. For Gena, the ultimate goal is continued *career growth*, which is why she chose to use Express Decision with this specific goal in mind, since it will help her solve the problem of whether or not she should *accept* her boss's friend request or *reject* it. We will denote this for brevity as *Problem (Decision-making: accept friend request vs. reject friend request)*. According to the Performance Evaluation Process used in ED, Gena must first name the alternatives (or options) available to her. Then, each alternative should be evaluated in the IL-Frame by four verbal characteristics: *positive intensity* and *its likelihood*, and *negative intensity* and *its likelihood*. In terms of this framework, subjective intensity is measured on a fuzzy verbal scale from *extremely weak* to *extremely strong,* while subjective likelihood is measured on the scale from *extremely seldom* to *extremely often.* Assume that these characteristics were determined by the following reasoning for each alternative:

Problem 1: Evaluation of Alternative 1 (Accept Friend Request). Pros (+): Gena knew that accepting the request could *strongly* (*I5*) contribute to her career growth, since her boss also uses Facebook for business purposes, but she rated her chances as *seldom* (*L3*), since she uses Facebook relatively infrequently and often doesn't log into it for days.

Cons (–): On the other hand, it is not entirely unlikely (*L4* – *not seldom-not often*) that her photography and private information could be used against her – if not now, then sometime down the line – and this could *very strongly* (*I6*) negatively impact her career prospects.

Because Gena does not experience any difficulties in selecting adequate verbal characteristics for the IL-Frame, *Alternative 1* is considered to be positively evaluated. This means that the *initial* mental model state allows Gena to evaluate *Alternative 1* (*+fb_FMM₁*), which in turn allows ED to support Gena in the formation of the level of motivation for choosing this alternative (*SR1: +fb_FMM₁ ⇒ FLM₁*).

- Level of positive motivation (preference) is 52% ("middle"): $S+ (I5, L3) = 0.52$.
- Level of negative motivation (preference) is 71% ("high"): $S-(I6, L4) = 0.71$.
- Level of motivation is *37%* ("*low*"): $S(S+ (I5, L3), S-(I6, L4)) = 0.37$.

This indicates that for Gena, accepting Robert's friend request has a few select pros, but there are many more significant cons involved, which ultimately allows for ED to determine that Gena's level of motivation for selecting alternative 1 is 37%, which is "low." Gena accepts the level of motivation for choosing *Alternative 1* (*+fb_FLM₁*) – since she has always firmly believed that mixing work with pleasure is likely to severely jeopardize professional relations – which in turn upgrades her mental model state (*SR3: +fb_FLM₁ ⇒ FMM₁*). This is why she decides against correcting the IL-Frame and instead turns to the evaluation of *Alternative 2*.

Problem 2: Evaluation of Alternative 2 (Reject Friend Request). Pros (+): Gena assumes that if she rejects the request, thus denying Robert the right to gain access to her private account information, this could *often* (*L5*) yet *very weakly* (*I2*) contribute to her career growth.

Cons (–): Gena only associates the cons to her career with the potential consequences of a rift in friendly relations with her boss. However, these consequences are *not apparent* to Gena, since her current mental model state does not allow her to evaluate Alternative 2 (*−fb_FMM₂*). Because of this, Gena has difficulty measuring the negative consequences from Robert's perspective and therefore, *Alternative 2* is considered to be negatively evaluated. At the same time, because evaluation of alternative 2 is very significant for Gena, which is why she is highly motivated to evaluate it, the new challenges that have arisen in light of this situation do not deter her.

Gena simplifies Problem 2 by splitting it into two sub-problems, 2.1 and 2.2, which relate to two possible and mutually-exclusive *hypothetical situations* (Robert could either *maintain* or *refuse* his friendly relations with Gena) with the purpose of determining their levels of motivation (*SR2: −fb_FMM₁ ⇒ ff_FLM₂₁_FLM₂₂*). Problem 2.1 evaluates *Alternative 2* assuming the first hypothetical situation, in which Robert continues to *maintain friendly relations* with Gena after having his friend request rejected by her, while Problem 2.2 evaluates *Alternative 2* assuming the second hypothetical situation, in which Robert *refuses to maintain friendly relations* with Gena after having his friend request rejected by her.

Problem 2.1: Evaluation of Alternative 2.1 (Reject Friend Request and Maintain Friendly Relations). Pros (+): Gena realizes that there is a chance to preserve friendly relations with Robert even after rejecting his friend request, since they still both have a strong mutual interest in music. This provides a relatively good (*L4*) chance for her to be able to maintain strong (*I5*) positive prospects beneficial to her career growth.

Cons (−): Gena understands that negative outcomes could occur not only from a falling out with her boss, but also from maintaining friendly relations with him. She recalls a scenario with one of her friends, in which the friend's boss received a rejection to his invitation for the friend to attend a concert with him. Although outwardly, the boss seemed to maintain good relations with this friend, he was still inwardly upset with her rejection, which ultimately ended up affecting the friend's annual evaluation, albeit weakly. Due to this, Gena decided to analyze as *seldom (L3)* the likelihood of a *weak (I3)* negative impact on her career growth, if she were to reject the friend request from Robert, although he would still appear to maintain friendly relations with her. Here we will note that initially, Gena only associated the cons to her career with the consequences of a potential rift in friendly relations with her boss. Using ED enabled Gena to also see the potential disadvantages to her career in the case that she is able to maintain friendly relations with her boss.

Because Gena does not experience any difficulties in selecting adequate verbal characteristics for the IL-Frame, *Alternative 2.1* is considered to be positively evaluated. The current mental model state allows Gena to evaluate *Alternative 2.1* (+*fb_FMM$_{21}$*), which in turn allows ED to support Gena in formation of the level of motivation for choosing this alternative (*SR1: +fb_FMM$_{21}$* $\Rightarrow$ *FLM$_{21}$*).

- Level of positive motivation (preference) is 57% ("middle"): *S+ (I4, L5) = 0.57.*
- Level of negative motivation (preference) is 38% ("low"): *S−(I3, L3) = 0.38.*
- Level of motivation (preference) is 58% ("middle"): *S(S+ (I4, L5), S−(I3, L3)) = 0.58.*

Gena agreed with this evaluation of her level of motivation (+*fb_FLM$_{21}$*), since she's keen on staying professional and maintaining some distance with her boss, which in turn upgrades her mental model (*SR3: +fb_FLM$_{21}$* $\Rightarrow$ *FMM$_{21}$*). This is why she decides against correcting the IL-Frame and instead turns to the evaluation of *Alternative 2.2*.

Problem 2.2: Evaluation of Alternative 2.2 (Reject Friend Request and Refuse to Maintain Friendly Relations). Pros (+): Gena understands that abruptly ending friendly work relations with Robert would likely cause him to evaluate her professional abilities from a more objective, and perhaps even slightly unfavorable, position. Moreover, Robert would have far less of a justification to try to advance to a more personal level in their relations, beyond what would be comfortable for Gena and acceptable professionally. With all this, Gena associated *very realistic (L5)*, although *relatively insignificant (I3)*, prospective career growth.

Cons (−): At the same time, Gena has not discounted the *likely (L4)* possibility of a *very strong (I6)* negative impact on her career growth if she were to reject the friend request from Robert, and he consequently ends up refusing to maintain friendly relations with her.

Because Gena has not experience any difficulties in selecting adequate verbal characteristics for *Alternative 2.2* ($+fb-FMM_{22}$), this alternative is positively evaluated, which in turn allows ED ($SR1: +fb\_FMM_{22} \Rightarrow FLM_{22}\_FLM_2$) to

(1) form the level of motivation for choosing Alternative 2.2 (FLM_{22}):

- Level of positive motivation (preference) is 52% ("middle"): $S+$ *(I3, L5)* = *0.52*.
- Level of negative motivation (preference) is *71%* ("high"): $S-$*(I6, L4)* = *0.71*.
- Level of motivation (preference) is 37% ("low"): $S(S+$ *(I3, L5)*, $S-$*(I6, L4))* = *0.37*.

(2) form the level of motivation for choosing *Alternative 2* (FLM_2) by aggregating the preference scores of *Alternatives 2.1* and *2.2*:

- Level of positive motivation (preference) is 52% ("middle").
- Level of negative motivation (preference) is 52% ("middle").
- Level of motivation (preference) is 47% ("middle").

In this way, the conducted analysis indicates that Gena was more motivated to reject her boss's friend request (47%, "middle" level) than to accept it (37%, "low" level). Although Gena had initially leaned towards *Alternative 2 ("reject")*, as it seemed more preferable than *Alternative 1 ("accept")*, she was surprised to learn that ED established a relatively low (47%) level of motivation for *Alternative 2*, which she did not consider sufficient for its selection ($-fb\_FLM_2$). In relation to this, Gena decides to re-evaluate *Alternative 2.2* by changing the potential for cons from *"very strong"* (*I6*) to *"strong"* (*I5*) ($SR4: -fb\_FLM_2 \Rightarrow ff\_FMM_2$). The current mental model state allows Gena to re-evaluate *Alternative 2* ($+fb\_FMM_2$), which in turn allows ED to form a new level of motivation for choosing *Alternative 2* ($SR1: +fb\_FMM_2 \Rightarrow FLM_2$).

Although ED reflected this change by increasing the level of motivation for *Alternative 2* from 47% to 53% ("middle"), for Gena this increase was not sufficient for making a motivated selection in favor of *"reject"* ($-fb\_FLM_2$).

In the resulting situation, when Gena is unable to make a motivated decision in favor of either alternative, *"accept"* or *"reject"*, Gena decides not to terminate the decision-making process (because the right choice was significant for her), and instead to turn to evaluate her third and final option, *Alternative 3* (wait and consider the friend request) ($SR4: -fb\_FLM_2 \Longrightarrow ff\_FMM_3$).

Problem 3: Evaluation of Alternative 3 (Wait and Consider Friend Request). Pros (+): Gena believes that waiting and considering the request *strongly* (*I5*) and *very often* (*L6*) contributes to her career growth, since the cost of a mistake in the case of an incorrect choice between "accept" and "reject" is relatively great. If anything, Robert could resend his friend request if he so desires – if he does not forget about it altogether.

Cons (−): On the other hand, this alternative, although *seldom* (*L3*), is nonetheless able to cause *tangible* (*I4*) negative effects, since Gena's apparent laxity in her personal correspondence could contribute to Robert perceiving her as being similarly irresponsible from a professional standpoint.

Because Gena does not experience any difficulties in selecting adequate verbal characteristics for the IL-Frame ($+fb\_FMM_3$), *Alternative 3* is positively evaluated, which in turn allows ED to support Gena in the formation of the level of motivation for choosing this alternative ($SR1: +fb\_FMM_3 \Rightarrow FLM_3$).

- Level of positive motivation (preference) is 76% ("very high").
- Level of negative motivation (preference) is 48% ("middle").
- Level of motivation (preference) is 63% ("high").

Gena accepts the level of motivation (63%) for choosing *Alternative 3* ($+fb-FLM_3$), which in turn upgrades her mental model state ($SR3: +fb\_FLM_3 \Rightarrow FMM_3$). With the help of ED ($SR1: +fb\_FMM_3 \Rightarrow FLM$), this *final* mental model state allows Gena ($+fb\_FMM_3$) to compare all alternatives and make a decision in favor of *Alternative 3*.

Indeed, there is a significant positive to the uncertainty present in *Alternative 3* (waiting to either "accept" or "reject" her boss's friend request), which puts Gena in a more neutral position, allowing her to buy some time and better consider both sides of the situation – i.e. the different possible pros and cons of accepting versus rejecting her boss's friend request. Therefore, Gena has decided for herself that she is content with her selection of *Alternative 3*, which involves waiting and considering Robert's friend request.

In this way, Express Decision has enabled Gena not to rush with making a final decision relying on her intuition alone. Instead, ED has let her make a well thought-out decision closely aligned with her own values and preferences within the span of just a few minutes. Express Decision helped Gina *build a mental model* to solve a decision-making problem by providing her with an option that was the best fit for her circumstances.

More specifically, using ED has allowed Gena to *clarify her goal* and *the criteria* of her decision problem. For example, as a result of splitting Problem 2 into two sub-problems – Problem 2.1 and Problem 2.2 – Gena was able to find additional pros in the situation, when her boss decides to refuse to maintain friendly relations, as well as pinpoint additional cons in the situation when her boss decides to maintain friendly relations; both of these insights helped Gena to better understand the ultimate goal, i.e. her continued career growth. ED also enabled Gena to correct the criteria of success in her decision-making. At the beginning of solving her problem, Gena had a specific criteria of success, such as the fact that she associated her career growth with maintaining friendly relations with her boss. In the process of decision-making with the help of ED, Gena was forced to change this criterion, since she came to understand the complexity of the association between friendly work relations and career growth.

References

1. Bedny, G., Karwowski, W., Bedny, I.: Applying Systemic-Structural Activity Theory to Design of Human-Computer Interaction Systems. CRC Press and Taylor & Francis Group, Boca Raton and London (2015)
2. Brehmer, B.: Dynamic decision making: human control of complex systems. Acta Physiol. **81**(3), 211–241 (1992)

3. Levitin, A.: Introduction to the Design and Analysis of Algorithms. Pearson, London (2011)
4. Kotik, M.A.: Developing applications of "field theory" in mass studies. J. Russ. East Eur. Psychol. 2(4), 38–52 (1994)
5. Yemelyanov, A.M.: Decision support of mental model formation in the self-regulation process of goal-directed decision-making under risk and uncertainty. In: Ayaz, H., Mazur, L. (eds.) Advances in Neuroergonomics and Cognitive Engineering (Advances in Intelligent Systems and Computing), vol. 775, pp. 225–236. Springer International Publishing (2018)
6. Bandura, A.: Self-Efficacy. The Exercise of Control. W. H. Freeman, New York (1997)
7. Yemelyanov, A.M., Yemelyanov, A.A.: Applying SSAT in computer-based analysis and investigation of operator errors. In: Baldwin, C. (ed.) Advances in Neuroergonomics and Cognitive Engineering (Advances in Intelligent Systems and Computing), vol. 586, pp. 331–339. Springer International Publishing (2017)
8. Kotik, M.A.: Self-regulation and Reliability of Operator. Valgus, Tallinn (1974)
9. Zarakovsky, G.M., Pavlov, V.V.: Laws of Functioning Man-Machine Systems. Soviet Radio, Moscow (1987)
10. Bechara, A., Damasio, A.R.: The somatic marker hypothesis: a neural theory of economic decision. Games Econ. Behav. 52, 336–372 (2005)
11. Yemelyanov, A.M., Baev, S., Yemelyanov, A.A.: Express decision – mobile application for quick decisions: an overview of implementations. In: Ayaz, H., Mazur L. (eds.) Advances in Neuroergonomics and Cognitive Engineering (Advances in Intelligent Systems and Computing), vol. 775, pp. 255–264. Springer International Publishing (2018)

Application of Systemic-Structural Activity Theory to Evaluation of Innovations

Inna Bedny[✉]

United Parcel Service, 3 Nuthatcher Court, Wayne, NJ 07470, USA
innabedny@gmail.com

Abstract. This paper discusses methodological issues related to the innovation process. Generally, the approach to the determination of innovation efficiency consists of comparing the effect of innovative activities and the cost of innovations which provide this effect. Here we examine the innovation efficiency from the perspective of human performance. Systemic-structural activity theory offers analytical methods for studying how innovations impact productivity, complexity and reliability of task performance. Systemic-structural activity theory approach concentrates on reducing the complexity of work, enhancing reliability of task performance, evaluating innovations, determining the most efficient versions of software or equipment and on developing the efficient strategies of performance. The results of this study suggest that SSAT offers methods of evaluation of efficiency of innovation at the design level. The use of these methods allows to determine if innovations increase productivity and reliability of performance, and reduce the rate of human errors without costly development process.

Keywords: Innovation · Efficiency · Complexity · Reliability ·
Systemic-structural activity theory

1 Introduction

At a microeconomic level viewpoint deems participation in the global market demands and continuous improvement in technology and business processes as vital to economic prosperity, thereby providing a strong incentive to invest in innovation [1].

Many researchers explored the patterns of adoption of innovations across industries, but only some have come up with the quantitative methods of evaluation of innovations.

Innovation efficiency is simply an output produced by invested resources. Increasing the former while reducing the latter increases efficiency. There is, quite rightly, a strong focus on the effectiveness of innovation. But what is the human effect of innovation? How do you ensure repeatability?

It is possible to develop efficient methods of task performance, evaluate the efficiency of training, increase productivity, and improve the user experience in general based on analysis of the structure of activity.

Continuous progress of technology and business processes is vital to economic prosperity, thereby providing strong incentives for investing in innovations.

© Springer Nature Switzerland AG 2020
H. Ayaz (Ed.): AHFE 2019, AISC 953, pp. 256–265, 2020.
https://doi.org/10.1007/978-3-030-20473-0_25

The vital role of innovation in national competitiveness is recognized by most nations [2].

In this paper we consider the following questions: Does the innovation lead to increased productivity? Does it reduce the rate of human errors? Does it positively affect reliability of performance? Is the memory workload reduced?

The methods of evaluation of efficiency of innovation at the design level have been created within the SSAT framework. These methods allow to evaluate the innovations and their effect on productivity and reliability of performance, on reduction of human errors at the design stage.

SSAT produces beneficial results when examining if innovation would be effective and allows to make an informed decision for implementation of innovations.

Here we demonstrate how to build human algorithm of the task performance considering two versions of the same task. One - before the implementation of the software enhancement and the other one after its implementation. By using these algorithms and the measures of complexity of the task developed in the framework of SSAT, we calculate these measures and compare the performance time of the task and its complexity before and after the improvement.

The suggested method can be also applied when choosing the most efficient version of the website or the software.

The analytical principles of task analysis presented here allow to reduce the costly cycles of continuous enhancements of software and/or equipment and to make informed decisions about the efficiency of the proposed innovations and about their effect on human performance.

2   Analysis of the Existing Version of the Task

For this study we choose the very widely performed task of receiving inventory in the warehouse. This task includes manual work and also involves using the software for the warehouse management in order to register the incoming inventory.

After conducting the qualitative analysis of this task, we have depicted task performance as an algorithm. This human algorithm is presented in Table 1. The operator performs the task of receiving an individual item that includes choosing the item number form the list on the screen and evaluating the quantity of the item that has been delivered to the warehouse after being ordered from the vendor. If the ordered and received quantity are the same the operator receives the item, if it's not the same the operator should recall the rules and make the decision accordingly. The detailed description of the task and the full version of this table can be found in the book by Bedny and Bedny published in 2018 [3].

In this table O^ε depicts performance of motor actions, O^α identifies perceptual actions, O^μ involves memory functions and O^{th} describes thinking actions, while the l symbols are associated with decision-making.

Analysis of this algorithm shows that I this short fragment of the task the operator has to make 3 decisions, keep information in working memory and recall the rules. Considering that this task is performed hundreds of times during the shift and the decision-making and memory workload lead to numerous errors, we proposed the following innovations.

Table 1. Algorithmic description of the inventory receiving task before innovation.

Members of algorithm	Description of elements of task	Time (s)
	Individual item processing	
$\overset{17(2)}{\downarrow}\,\overset{11(2)}{\downarrow}\,\overset{7(2)}{\downarrow}O^{\varepsilon}_{17}$	Take an item out of the box, and return to the computer area again (while a worker makes several steps his/her right hand releases the item and worker holds the item only by left hand)	5.7
$O^{\alpha\mu}_{18}$	Look at the item number on the physical item and compare it with the item numbers on the screen	Average time ≈ 6
$l_4 \overset{4}{\uparrow}$	If the item number is on the first page, go to O^{ε}_{20}. If the item number is not on the first page, go to O^{ε}_{19}	0.3
O^{ε}_{19}	Hit arrow key (repeat if required)	0.62
$\downarrow O^{\varepsilon}_{20}$	Put cursor on the selected line and hit ENTER to go to the screen with a detailed item information	2.64
$O^{\alpha\mu}_{21}$	*Compare received quantity with PO (purchase order) quantity*	1.64
$l_5 \overset{5}{\uparrow}$	If received quantity and ordered quantity are the same, go to O^{ε}_{29} (This output is performed in other subtasks). If received quantity is greater or less than ordered quantity, go to O^{ε}_{22}	*0.4*
O^{ε}_{22}	Type the received quantity and press ENTER to get a question at the bottom of the screen	1.7
O^{ε}_{23}	Read the statement: THE RECEIVED QUANTITY AND ORDERED QUANTITY DO NOT MATCH. DO YOU ACCEPT? (YES/NO). Scan and read $\approx$ four words	1.14
$O^{\mu th}_{23}$	Recall instructions and perform required calculation and estimation (relationship between ordered and received quantity)	4.2
$l_6 \overset{6}{\uparrow}$	If quantity is not accepted (computer defaults to 'N') go to O^{ε}_{25}. Otherwise, go to O^{ε}_{28} (This output is performed in another version of the task)	1.5
O^{ε}_{25}	Press ENTER (default is conformed)	0.84
O^{ε}_{26}	Put rejected item in the Put-Aside Area, Figure 2, area 10. Return to the base unit 5. O^{ε}_{26} includes: (1) the left hand grasps the item; (2) worker turns body and make approximately 6 steps to the Put-Aside area; move the left hand and releases the item; (3) worker turns body and make approximately 4 steps to the base unit for unpacking	12.5
Total performance time	*Task of receiving an item ends here*	**39.18**

3 Suggested Innovation

As it can be seen the operator has to read the item number from the label and brows the purchase order list of items on the computer screen to find the matching number. That involves making a number of decisions and keeping information in working memory.

We proposed to use the handheld scanner that is connected to the computer. When the operator scans the barcode label on the item the corresponding item number is highlighted on the screen and operator just hits ENTER to confirm the match.

Comparison of the human algorithms of the task performance before and after implementation of the innovation should demonstrate the efficiency the proposed innovation.

4 Analysis of the Task After Implementation of the Innovation

Table 2 describes the algorithm of the task performance after the proposed innovation has been implemented.

The number of decisions has been reduced as well as the performance time of the task. Memory workload related to memorizing the item number is also eliminated.

5 Comparison of the Quantitative Measures of Task Performance Before and After Innovation

Let us evaluate the task performance before and after innovation. For this purpose, we select the most relevant measures of complexity for the considered task. The comparative analysis of the complexity of the task before and after improvement will give us an idea of the efficiency of the implemented innovations based on quantitative criteria. The times are taken from the corresponding algorithmic descriptions of the task performance.

We determine the duration of the separate components of activity that are classified as perceptual components, thinking components, the total performance time for keeping information in working memory, the performance time for decision-making components (logical conditions), the performance time for executive components of activity (motor components), and the performance time of all cognitive components of activity. There are the following absolute measures of complexity: T_{ex} - the performance time of executive (motor) components of activity; T_α - the performance time of perceptual components; L_g - the performance time of logical conditions (decision making); T_{wm} - the time for keeping information in working memory; T_{th} - the time for performing thinking operations.

This data allows calculation of the fraction of time for all the above-listed components (relative measures of task complexity). All these measures are presented in Table 3. These measures facilitate quantitative evaluation of the complexity of the task performance before and after implementation of innovation.

Table 2. Algorithmic description of inventory receiving task after innovation

Members of algorithm	Description of elements of task	Time (s)
	Individual item processing	
$\overset{8(2)}{\downarrow}\overset{5(2)}{\downarrow}O^{\varepsilon}_{11}$	Take an item out of the box, and return to the computer area again (while a worker makes several steps his/her right hand releases the item and worker holds the item only by left hand)	5.7
O^{ε}_{12}	Take the barcode scanner and scan the item number. (The matching item is highlighted)	2.9
O^{ε}_{13}	Hit ENTER to go to the screen with a detailed item information	0.9
$O^{\alpha\mu}_{14}$	*Compare received quantity with PO (purchase order) quantity*	1.64
$\overset{3}{l_3\uparrow}$	If received quantity and ordered quantity are the same, go to O^{ε}_{22} (P = 0.9). If received quantity is greater or less than ordered quantity, go to O^{ε}_{15}	*0.4*
O^{ε}_{15}	Type the received quantity and press ENTER to get a question at the bottom of the screen	1.7
O^{ε}_{16}	Read the statement: THE RECEIVED QUANTITY AND ORDERED QUANTITY DO NOT MATCH. DO YOU ACCEPT? (YES/NO). Scan and read	1.14
$O^{\mu th}_{17}$	Recall instructions and perform required calculation and estimation (relationship between ordered and received quantity)	≈4.2
$\overset{4}{l_4\uparrow}$	If quantity is not accepted (computer defaults to 'N'), go to O^{ε}_{18}. Otherwise, go to O^{ε}_{21}	1.5
O^{ε}_{18}	Press ENTER (default is conformed and after performing O^{ε}_{19} start to work with new item)	0.84
O^{ε}_{19}	Put rejected item in the Put-Aside Area. Return to the base unit O^{ε}_{19} includes: (1) the left hand grasps the item; (2) worker turns body and makes approximately 6 steps to the put aside area; moves the left hand and releases item; (3) worker turns body and make approximately 4 steps to the base unit	12.5
Total performance time	*Task of receiving an individual item ends here*	*33.42*

The duration of perceptual components of activity is determined utilizing the formula:

$$T\alpha = O^{\alpha\mu}_{18} + O^{\alpha\mu}_{18} + O^{\alpha}_{23} = 6 + 1.64 + 1.14 = 8.78\,\text{s}, \qquad (1)$$

where $O^{\alpha\mu}_{18}, O^{\alpha\mu}_{21}, O^{\alpha}_{23}$ are members of algorithms that are involved in receiving information.

There is only one thinking member of the algorithm, $O^{\mu th}_{24}$, and therefore $T_{th} = 4.2$ s.

The duration of keeping information in working memory can be determined by summarizing the performance time of the members of the algorithm presented in the below formula:

Table 3. Measures of complexity of the task performance.

Measure number	Measures	Before innovation (s)	After innovation (s)	Psychological meaning
1	Algorithm (task) execution time (T). $T = \Sigma P_i\, t_i$	39.18	33.42	Duration of task performance
2	Total of the performance time of all afferent operators (T_α). $T\alpha = \Sigma P_\alpha\, t_\alpha$	8.78	2.78	Duration of perceptual components of task
3	Total of the performance time of all thinking operators (T_{th}). $Tth = \Sigma P_{th}\, t_{th}$	4.2	4.2	Duration of thinking components of task
4	Total of the performance time of all operators that requires keeping information in working memory (including mnemonic operations that combined with other components of activity) (T_{wm}). $Twm = \Sigma P_{wm}\, t_{wm}$	11.84	5.84	Time for retaining current information in working memory
5	Total of the performance time of all logical conditions (L_g). $L_g = \Sigma P_l\, t_l$	2.2	1.9	Duration of decision-making components of task
6	Total of the performance time of all efferent operators (T_{ex}). $T_{ex} = \Sigma P_j\, t_j$	24	24.54	Duration of executive components of task
7	Total of the performance time of all cognitive components including perceptual activity $(Twm$ is not considered as independent component of cognitive activity in this formula; $Twm = 0$), $(Tcog)$ $Tcog = T\alpha + Tth + L_g$	15.18	8.88	Duration of all cognitive components of task
8	Fraction of time for afferent operators in the time for entire task performance $(N\alpha)$. $N\alpha = T\alpha/T$	0.22	0.08	Perceptual workload in the task performance
9	Fraction of time for thinking operators in the time for the entire task performance (Tth). $Nth = T_{th}/T$	0.11	0.13	Problem solving workload in the task performance
10	Fraction of time for logical conditions in the time for the entire task performance (N_l). $N_l = L_g/T$	0.06	0.06	Decision-making workload in the task performance

(continued)

Table 3. (*continued*)

Measure number	Measures	Before innovation (s)	After innovation (s)	Psychological meaning
11	Fraction of time for performance of cognitive components of task to the entire task performance, including perceptual components (N*cog*) (combined in time with other components of activity mnemonic operations are not considered). '*Ncog = Tcog/T*	0.39	0.27	General cognitive workload in the task performance
12	Fraction of time for logical conditions, which largely depends on information selected from long term memory rather than from external sources of information, in the entire time for logical conditions performance. ('N$_{ltm\ l}$ = l_μ/L$_g$)	0	0	Level of memory workload and complexity of the decision-making process
13	Fraction of time for retaining current information in working memory in the time for the entire task performance (N*wm*). N*wm* = T*wm*/T	0.3	0.17	Memory workload in the task performance
14	Fraction of time for performance of all efferent operators in the time for the entire task performance '*Nbeh = Tex/T*	0.61	0.73	Behavioral or motor workload in the task performance

$$T_{wm} = O_{18}^{\alpha\mu} + O_{21}^{\alpha\mu} + O_{24}^{\alpha\mu} = 6 + 1.64 + 4.2 = 11.84\,\text{s} \qquad (2)$$

The duration of the decision-making components (logical conditions) in task performance is:

$$L_g = l_4 + l_5 + l_6 = 0.3 + 0.4 + 1.5 = 2.2\,\text{s} \qquad (3)$$

The duration of the executive components of activity (the performance time of efferent operators) is:

$$T_{ex} = O_{17}^{\varepsilon} + O_{19}^{\varepsilon} + O_{20}^{\varepsilon} + O_{22}^{\varepsilon} + O_{25}^{\varepsilon} + O_{26}^{\varepsilon}$$
$$= 5.7 + 0.62 + 2.64 + 1.7 + 0.84 + 12.5 = 24 \, s \tag{4}$$

The duration of all cognitive components of the task are determined by utilizing the formula below:

$$T_{cog} = T_{\alpha} + T_{th} + L_g = 8.78 + 4.2 + 2.2 = 15.18 \, s \tag{5}$$

If we know total the task performance time and duration of its separate components, we can then determine the fractions of time for the above-listed components in the total task performance time. These fractions are presented in Table 3.

We have completed the analysis of the quantitative measures of complexity of the task performance before improvement. After introduction of the barcode scanner, the quantity of members of the algorithm reduced.

All measures after improvement have been obtained the same way as for the version of the task before improvement. Let us now compare the obtained data for measures of complexity before and after improvement.

The first measure (total task execution time) demonstrates that before improvement, the task performance time was 39.18 s and after improvement, this time was 33.42 s. Hence, the performance time for the entire task has been reduced by 5.66 s (14%). The perceptual workload has been significantly reduced (see measures 2 and 8). Measure 2 (T_{α}) for the perceptual components of activity before improvement was 8.78 s, while the performance time of the whole task is 39.18 s. Therefore, $N_{\alpha} - T_{\alpha}/T = 0.22$ before innovation, and $N_{\alpha} = T_{\alpha}/T = 2.78/33.42 = 0.08$ after improvement which demonstrates the significant reduction of the portion of time spent on perceptual components of activity.

The memory workload after implementation of the innovation has significantly decreased. Before innovation, the time for retaining information in working memory was 11.84 s and after innovation it was 5.84 s (measure 4). The relative value of the memory workload was 0.3 before and 0.17 after innovation (measure 13). The duration of decision-making processes has been slightly reduced after innovation, but the relative value of the decision-making workload has not changed (measures 5 and 10). The duration of all cognitive components in the task performance after innovation has been significantly reduced (see measure 7). Measure 11 also demonstrates that the general cognitive workload decreased. The duration of motor components in the task performance is practically the same (measure 6). The difference, especially for motor activity, is negligible. However, the fraction of behavioral or motor components in the task performance increased slightly (measure 14). Such changes are due to the fact that the time for the motor activity in both versions of the task are practically unchanged because using the barcode scanner after innovation involves some additional motor actions. Before innovation, the worker can often perform some additional motor actions for scanning an item (see the member of the algorithm O_{20}^{ε} "repeat if required").

The time for cognitive components of activity and their complexity are reduced after introduction of the handheld scanner. Usage of the handheld scanner eliminated the following members of the algorithm before improvement in the task performance after innovation (see $O_{18}^{\alpha\mu} - 6$ s; $l_4 - 0.3$ s).

We can observe the following changes in motor activity. Before improvement, after decision-making (l_4), an operator performs O_{19}^{ε} (press arrow key, repeat if required). The performance time of this action is 0.62 s. O_{20}^{ε} has a complicated structure, and its performance time is 2.64 s (Table 1). In the version of the task after innovation, O_{19}^{ε} is eliminated and O_{13}^{ε} has the same purpose as the member of the algorithm O_{20}^{ε} before innovation. At the same time, O_{23}^{ε} after innovation requires 0.9 s (Table 2).

It is interesting to compare members of the algorithm O_{19}^{ε} and O_{20}^{ε} before innovation with O_{12}^{ε} and O_{13}^{ε} after innovation. The duration of the motor activity after innovation for the last two members of the algorithm has slightly increased. The performance time of O_{12}^{ε} and O_{13}^{ε} after innovation is 3.8 s. The performance time of O_{19}^{ε} and O_{20}^{ε} before innovation was 3.26 s (difference 0.54 s). An operator has to press the arrow key in order to move the cursor to the required line. If the item at hand is not on the list on the first page of the screen, an operator has to browse the item list and press the down key more times. Such browsing requires perceptual actions and motor actions. This can increase the duration of the motor activity, and the difference in 0.54 s might be eliminated. As a result of the innovations, the cognitive activity and its complexity are reduced, and the behavioral components of activity are shorter on average.

When comparing the complexity of the logical conditions before and after improvement one can see that before innovation there were three logical conditions: l_4, l_5, and l_6. The first two conditions l_4, l_5, belong to the third category of complexity and l_6 to the fourth category of complexity. The performance time of l_4 and l_5 is 0.7 s, and the performance time of l_6 is 1.5 s. The total performance time for logical conditions is 2.2 s. The fraction of time for the logical conditions of the fourth category is $1.5/2.2 = 0.68$.

After improvement, there are only two logical conditions, l_3 and l_4. The first one requires 0.4 s and belongs to the third category of complexity. The second one requires 1.5 s and belong to the fourth category of complexity. Therefore, the fraction of time for logical conditions of the fourth category of complexity is 0.79, and 0.21 is of the third category of complexity. Hence, the performance time for the decision making L_g after innovation has been reduced from 2.2 to 1.9 s.

Based on the analysis of various measures of complexity, we can make the following conclusions. After implementation of the innovation, the complexity of the task is reduced in general due to the optimization of the cognitive components of activity. The time of the task performance is also reduced. The cognitive workload, and especially the absolute value of the measures, has reduced significantly.

6 Conclusion

This paper offers an insight into the quantitative methods of evaluation of the efficiency of the innovations suggested within the framework of SSAT.

Deep understanding of the problems of innovativeness demonstrates not only what has to be done, but also an explanation of why it is worth doing.

Some innovations are very costly. It is important to consider the tangible and intangible benefits that the innovation can provide.

Developed in SSAT methods allow to evaluate the proposed innovations quantitatively at the design stage, save money and allow to make the right decisions before the implementation of the proposed innovations.

The discussed here method can be utilized for the comparison of different versions of the innovation.

The considered in this paper example has been used to demonstrate the method.

References

1. Drucker, P.: Innovation and Entrepreneurship: Practice and Principles, pp. 133–140. Heinemann, London (1985)
2. Nasierowski, W., Arcelus, F.J.: Procedia Soc. Behav. Sci. **58**, 792–801 (2012)
3. Bedny, G.Z., Bedny, I.S.: Work Activity Studies Within the Framework of Ergonomics, Psychology, and Economics. CRC Press and Taylor and Francis Group, Boca Raton and Florida (2018)

Goal-Classification and the Influence of Activity-Goal-Formation on Individuals' Systemic-Consideration of Activity-Strategies and Decision-Outcomes

Mohammed-Aminu Sanda[1,2(✉)]

[1] University of Ghana Business School, P. O. Box LG 78, Legon, Accra, Ghana
masanda@ug.edu.gh, mohami@ltu.se
[2] Luleå University of Technology, 97187 Luleå, Sweden

Abstract. The influencing role of students' activity goal formation informed by their goal classification (i.e. highest or best) in their cognitive considerations of both activity strategies and decision outcomes for a pending task is examined in this study. Using data from a sample of 300 Graduate students preparing for an end-of-semester examination and the systemic structural activity analytical approach, it is found that actors' cognitive classification of goals for pending activity as "highest" or "best" has no significant effect on the students' goal formulation and the dynamic influence it has on their cognitive considerations for both activity strategy and decision outcome. Irrespective of goal classification, the students' cognitive process of activity goal formation is found to significantly influence both their cognitive considerations of activity strategies and decision outcomes. It is concluded that the cognitive classification of goal has no direct significance on an students' Goal formation process for a pending activity.

Keywords: Goal classification · Activity goal formation · Highest goal · Best-goal · Activity strategy · Decision outcome

1 Introduction

In every workplace, workers are known to perform variety of activities. In the performance of such activities, several actions, each of which entail different but complementary operations are undertaken. In this respect, individuals performing tasks are cognitively affected by their psychosocial interaction with their work contexts. Such interaction starts from the goal formulation process and runs through to the use of their cognition in strategizing for the pending task and considering possible task-outcome decisions. Thus, activities that individuals engage in at the workplace, as [1] intimated can be viewed in macro-contexts. Such macro-contexts, according to [1], provide commonalities of action by outlining the micro-context. Therefore, the interaction between the micro and macro contexts makes visible the uncertainties associated with the activity [2] and [3] as well as provide the avenue for formulating practice that is adaptive [1] and [4]. Since uncertainty in a task is indicative of the task's difficulty and complexity [2] and [3], a task characterized as complex can be evaluated differently by

© Springer Nature Switzerland AG 2020
H. Ayaz (Ed.): AHFE 2019, AISC 953, pp. 266–278, 2020.
https://doi.org/10.1007/978-3-030-20473-0_26

different persons [2]. While some can evaluate a complex task as very difficult, some might view the same task as not difficult [2]. As it is argued by [2] and [5], the differences in such evaluations by the different persons, is informed by the differences in their levels of cognition [2], skills and individual features [2] and [5]. These observations by [2] and [5] was empirically tested by [6] in a study that examined the notions of complexity and difficulty by graduate students preparing for examination activity. In the study, [6] established that the conscious goal-directed processes of the students while preparing for their pending examination were influenced by their past experiences in the numerous examinations they have written, and also their subjective perceptions of the difficulty level of the impending examination. The findings by [6] showed that when the students were preparing for their pending examination, as an organizational activity, they experienced different cognitive-oriented activities. The cognitive dynamics of the transition process in such examination preparation activity, as argued by [6], is influenced by the psychological characterization of the activity's goal. In this regard, [6] concluded that when a student is readying for an examination activity, underscored by his/her setting of either a best-goal or highest-goal, such goal formulation will affect the influence that the student's consideration of activity strategies has on his/her considerations of decision outcomes. Therefore, the purpose of this study is to empirically identify the dominant goal-characteristic entailed in graduate students' goal-formulation process when preparing for examinations, and how the characterized-goal influences the students' considerations of their activity-strategies and decision-outcome. This is to address the gap on whether students' characterization of goals at the workplace realistically affect their activity performances. In this respect, the Fig. 1 below is conceptualized to guide the study.

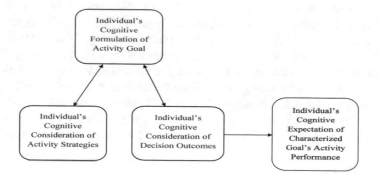

Fig. 1. Conceptual model of the influence of an individual's cognitive goal formulation on his/her considerations of activity strategies and decision outcomes

2   Literature Review

It has been established in the extant literature that when designing tasks for individuals, an important issue that requires consideration is the need to identify and distinguish the complexities associated with the task. In the views of [5] and [7], such complexities should include the cognitive attributions of the task in question, which is informed by the specificity of task's information processing [5], as well as the task's emotional-motivational attribution, which is informed by energetic aspects of the task [7]. This therefore, requires that all attributions of complexity associated with any task be made visible [5] and [7]. This is to provide persons concerned with designing the tasks of other individuals an understanding of practice enhancing strategies [5] and [7] that could be used to navigate the cognitive difficulties and emotional-motivational challenges individuals encounter in their task performances [5] and [7]. This, according to [5], is because, an individual's activity cannot be adequately interpreted based on the assumption such activity is designed around a single, neatly identifiable goal [5]. Thus, arguing from the perspectives of [8], goal-setting is as a determinant of an individual's self-regulation activity. Therefore, using the perspectives of [9] as a point of departure, it could be argued that individuals mostly try to accomplish task with set-goals that are manifestations of the task's objective. This objective is interpreted by [9] as similar in meaning to the concepts of purpose and intent. As such, activities of individual's entail multiple goals, which in the views of [5] and [7] are often in interaction and sometimes in conflict. As it is posited by [6], it is imperative for individuals to choose appropriate strategies when preparing for complex tasks. Such individuals, according to [10], must be viewed as having the ability to strategized towards attaining their set-goals. In this vein, it could be argued from the perspectives of [10] that work strategy development by individuals is motivated by their set-goals. This is because, the strategy development mechanism is cognitive [10], since it entails individual creative problem-solving and skill development capabilities. As it is noted by [10], when the goals of individuals are expressed as specific intentions to take a certain action rather than as vague intentions of trying hard [10] or as subjective estimates of task difficulty [10], the goals mostly end up helping in the regulation of such individuals' performances. Similarly, [11] argued that when individuals set for themselves goals that are challenging, they mostly end up posting higher outputs than when the goals are vague. Thus, using the explanation by [10] as a point of departure, individuals whose tasks are not goal-bonded will try to do their task well up to their possible abilities. The rational for this, as argued by [8], is that when individuals are confronted with task bonded by set-goals, they automatically use their inherent skills and knowledge to facilitate the attainment of such goal. In this vein, it is important to also underline the position by [12] that when the task with set-goals is new to individuals, such individuals will likely engage in deliberate planning in order to develop strategies that will facilitate their abilities of attaining the set-goals [12].

3 Methodology

The study is guided by the quantitative method of task complexity evaluation. In this method, as it is explained by [2], units of measurement and measurement procedures that permit the comparison of different elements of activity are requirement [2]. This, according to [2], implies that while task complexity can be evaluated both experimentally and theoretically [2], expert judgments, such as the use of a five-point scale can also be used for the subjective evaluation of the task's complexity [2] and [5], goal formation, strategy formulation and outcome decision-making [5]. In this study therefore, the quantified-qualitative philosophical approach was used to guide the capture of the graduate students' subjective opinions on their exam's preparation.

3.1 Data Collection

The well-established knowledge that activities of individuals are realized by goal-directed actions [1] was used to guide the data collection method. These actions, as explained by [7], are informed by individuals' mental processes. Based on these perspectives, an experimentation on the graduate students' individual differences of maximization in examination activity goal formation and decision outcomes was conducted. A self-administering questionnaire prepared from the maximization measurement scale by [13] was used as the data collection tool. The data was collected from 338 graduate students, comprised of I86 (55%) females and 152 (45%) males. All the respondents were graduate students enrolled in Master-level Degree programs at the University of Ghana Business School, in Accra, Ghana. As graduate students, all the respondents are have completed 4-years undergraduate tertiary programs and have written examinations severally and for so many years as they progressed on their educational developments. At the time of data collection, the respondents were preparing to write their end of second semester examinations between May-June, 2018. The questionnaire included a synopsis that explained the research purpose.

3.2 Data Analysis

In the data analysis, the stepwise approach was used. The data was collated and firstly analyzed descriptively. Secondly, the data was analyzed inferentially using the structural equation modelling (SEM) approach. In this approach, the analysis of moment structures (AMOS) was used as the analytical technique as applied by [14]. This analytic procedure, according to [14] and [15] has the advantage of maximizing the validity of the estimates. The AMOS graphics statistical software was used as the analytical tool. The tool facilitates the conduct of analyses for multiple levels of variables using a range of in-built statistical techniques [14]. In testing the predictability of measured individual factors that constitute the various components of the conceptual model shown in Fig. 1, path analysis was conducted. The analysis was started by loading each of the latent variables in the conceptual model (i.e. Fig. 1) in the AMOS software in order to assess the model-fitness of their respective measurable indicators (factors). The model fit was interpreted using the path coefficient benchmark value of 0.7 or higher as proposed by [16] in order to ensure robustness. The criteria used to determine model fit are the Chi Square (CMIN) value, which is the absolute test of model fit and the Comparative Fit Index (CFI) value. A CMIN value below 0.05 implies model acceptance. For the CFI, a value close to 1.0 indicates a very good model fit.

4 Results and Analysis

All the three hundred and thirty-eight (338) questionnaires administered were retrieved. Prior to the analysis, thirty-six (36) of the retrieved questionnaires were found to be incomplete and by implication, unusable and were thus rejected. Thus, the analysis was conducted using the remaining 302 usable questionnaires, which comprised of I83 (54.1%) females and 119 (35.2%) males.

4.1 Descriptive Analysis of Students' Perceptual Linkage of Goal Formulation to Strategy and Decision Outcome

A summary of the descriptive comparison of the students' perceptions of how the characteristic of their goal formulations influence their abilities to link their strategies to their decision of what-to-do in a pending examination is shown in Fig. 2 below.

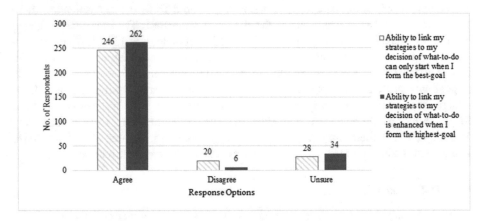

Fig. 2. Comparison of respondents' perceptions of the influence of goal characteristic on abilities to link strategies to decision of what-to-do

As highlighted in Fig. 2 above, a very significant number of the respondents hold the perceptions that their abilities to link their strategies to their decision of what-to-do in pending examinations is strongly influenced when they characterized their goal-formulation as either "highest" (86.75%) or "best" (81.46%). This descriptive outcome is indicative that the respondents do not perceive the characterization of goals as having any specific overriding influence on their cognitive formulation of goals in pending activity. This outcome is further examined inferentially to check its viability relative to the outlined purpose of the study.

4.2 Structural Analysis of Model for Best Goal Formulation

Figure 3 below shows the AMOS-generated standardized path diagram highlighting the standardized path coefficients in the structural model underlined by best-goal for-mulation. As highlighted in the figure, five (5) predictive indicators for four (4) latent variables (cognitive processes) were tested in the graduate students' preparation towards examination (as an organizational activity), and minimum was achieved for the model. The goodness of fit statistics shows that the overall model fit is quite good, because the estimated CMIN ($\chi 2$) value of 62.947 (df = 5) has probability level of 0.000, which is lesser than the 0.05 used by convention. This implied that the acceptance of the null hypothesis that the model fits the data. Additionally, the esti-mated CFI value of 0.740 is close to 1.0, indicating an acceptance of the null hypothesis of a good fit for the tested model. Table 1 below gives a summary of the indicator variables and their path coefficients or standardized regression weights (R) for the respective latent variables (cognitive processes). As it is established by convention, the indicators are interpreted as significant predictors of their latent variables only if the values of their estimated loadings are 0.7 or higher as proposed by [16].

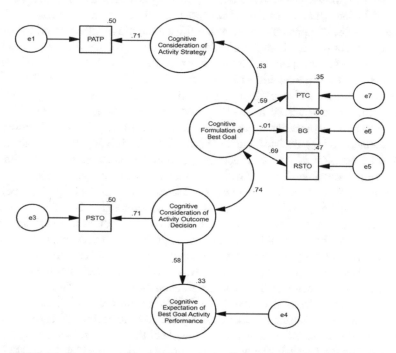

Fig. 3. AMOS graphics generated path diagram showing standardized path coefficients in the structural model for best goal formulation

Table 1. Path coefficients of indicators predictive of unobserved variables for best goals

Unobservable variables	Indicators	Label in model	R	R^2
Cognitive formulation of best goal	Perception of task difficulty	PTC	0.59	0.35
	Perception of task experience	RSTO	0.69	0.48
	Perception of best goal	BG	−0.01	0.0
Cognitive consideration of activity strategy	Assessment of activity performance options	PATP	0.71	0.50
Cognitive consideration of activity outcome decision	Decision outcome expectation	PSTO	0.71	0.50
Cognitive expectation of best goal activity performance	Cognitive consideration of activity outcome decision		0.58	0.34

The results in Table 1 above shows that only the students' assessment of activity performance options (PATP) and decision outcome expectations (PSTO) have factor loadings above the 0.7 threshold. The factor loading of the students' perception of task experience (RSTO) has a factor loading of 0.69, which is approximated to 0.7. The students' formulation of best goals for the task (BG) has a factor loading of −0.01 which is far below the 0.7 threshold. Therefore, three indicators have factor loadings or standardized regression estimates (R) of approximately 0.7 and higher, and as such are predictive of their unobserved variables (cognitive processes) as proposed by [16]. The implications of these results are that, in preparing towards an examination activity underlined by the students' notion of best goal-setting, the characterization of the goal formulation as "best" appears to have no significant cognitive influence on the students. Rather, the students' cognitive formulation of the best goal for the activity is largely underlined by their perceptions of their task experience ($R = 0.69$, $R^2 = 0.48$), and to a lesser extent by their perception of the task's difficulty ($R = 0.59$, $R^2 = 0.35$). On the other hand, the students' cognitive consideration of their activity strategies is under-scored by their assessment of the activity performance options ($R = 0.71$, $R^2 = 0.50$), while their cognitive consideration of decisions for the activity outcome is informed by their expectation of the decision outcome ($R = 0.71$, $R^2 = 0.50$). Finally, the students' cognitive expectation of best goal activity performance is to an extent influenced by their cognitive consideration of activity outcome decision ($R = 0.58$, $R^2 = 0.34$). Correlation analysis of the latent variables in the structural model (Fig. 3 above), show that the standardized correlation weights (α) for the association between the students' cognitive formulation of best goal and their cognitive consideration of activity outcome decision is above the threshold value of 0.7, while that with their cognitive consider-ation of activity strategy is below the threshold value of 0.7. This implies that when students are faced with pending activities, their cognitive processes in the activity-goal formulation will be such that their characterization of the goal as "best" will strongly influence their consideration of the activity outcome ($\alpha = 0.74$), while such charac-terization will have minimal influence on their consideration of activity strategy ($\alpha = 0.53$). Thus, the conclusion by [6] to the effect that when students readying for a best-goal-oriented task, their characterization of the goal as "best" in their activity goal formulation mediates the influences that their consideration of activity strategies has on their considerations of decision outcomes does not hold. By implication, the charac-terization of the goal as "best" appears to have no effect on the influence that students' cognitive processes in goal formulation has on the transition of their cognitive con-sideration of the activity strategy and the cognitive consideration of decisions for the activity outcome. Rather, the best-goal characterization appears to have significant positive influence on the students' cognitive consideration of activity strategy. The sense in this finding is underscored by [8] and [17]'s argument that when individuals are confronted with a task that appears complex, urging them to do their best some-times leads to their development of better strategies. The opposite appears to be the case when a specific difficult goal is set for the individual.

4.3   Structural Analysis of Model for Highest Goal Formulation

Figure 4 below shows the AMOS-generated standardized path diagram highlighting the standardized path coefficients in the structural model underlined by best-goal for-mulation. As highlighted in the figure, five (5) predictive indicators for four (4) latent variables (cognitive processes) were tested in the graduate students' preparation towards examination (as an organizational activity), and minimum was achieved for the model. The goodness of fit statistics shows that the overall model fit is quite good, because the estimated CMIN ($\chi2$) value of 58.227 (df = 5) has probability level of 0.000 which is smaller than the 0.05 used by convention. This implied that the acceptance of the null hypothesis that the model fits the data. Additionally, the estimated CFI value of 0.757 is close to 1.0, indicating an acceptance of the null hypothesis of a good fit for the tested model. Table 2 below gives a summary of the indicator variables and their path coefficients or standardized regression weights (R) for the respective latent variables (cognitive processes). As it is established by convention, the indicators are interpreted as significant predictors of their latent variables only if the values of their estimated loadings are 0.7 or higher as proposed by [16].

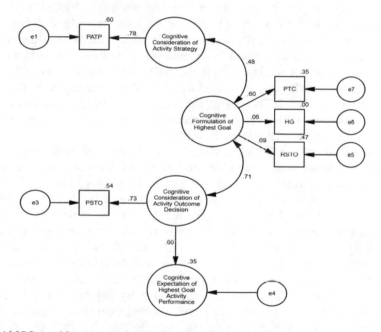

Fig. 4. AMOS graphics generated path diagram showing standardized path coefficients in the structural model for highest goal formulation

Table 2. Path coefficients of indicators predictive of unobservable variables for highest goals

Unobservable variables	Indicators	Label in model	R	R^2
Cognitive formulation of highest goal	Perception of task difficulty	PTC	0.60	0.36
	Perception of task experience	RSTO	0.69	0.48
	Perception of highest goal	HG	−0.06	0.00
Cognitive consideration of activity strategy	Assessment of activity performance options	PATP	0.78	0.61
Cognitive consideration of activity outcome decision	Decision outcome expectation	PSTO	0.73	0.53
Cognitive expectation of highest goal activity performance	Cognitive consideration of activity outcome decision		0.60	0.36

Table 2 above shows that only the students' assessment of activity performance options (PATP) and decision outcome expectations (PSTO) have factor loadings above the threshold level of 0.7. The students' perception of task experience (RSTO) has a factor loading of 0.69, which is approximated to 0.7. Only the students' formulation of highest goals for the task (HG) that has a factor loading of 0.06, which is far less than the 0.7 threshold. As such, three indicators have factor loadings (R) of approximately 0.7 and higher, and as such are predictive of their unobserved variables (cognitive processes) as recommended by [16]. The sense of these results is that, in preparing towards an examination activity underlined by the students' notion of highest goal-setting, the characterization of the goal formulation as "highest" appears to have no significant cognitive influence on the students. Rather, the students' cognitive formulation of the highest goal for the activity is largely underlined by their perception of their task experience ($R = 0.69$, $R^2 = 0.48$), and to a lesser extent by their perception of the task's difficulty ($R = 0.60$, $R^2 = 0.36$). On the other hand, the students' cognitive consideration of the activity strategy is underscored by their assessment of the activity performance options ($R = 0.78$, $R^2 = 0.61$), while their cognitive consideration of decisions for the activity outcome is informed by their expectation of the decision outcome ($R = 0.73$, $R^2 = 0.53$). Finally, the students' cognitive expectation of highest goal activity performance is to an extent influenced by their cognitive consideration of activity outcome decision ($R = 0.60$, $R^2 = 0.36$).

Correlation analysis of the latent variables in the structural model (Fig. 4 above), show that the standardized correlation weights (α) for the association between the students' cognitive formulation of highest goal and their cognitive consideration of activity outcome decision is above the threshold value of 0.7, while that with their cognitive consideration of activity strategy is below the threshold value of 0.7. This implies that when students are faced with a pending activity, their cognitive processes in the activity-goal formulation will be such that their characterization of the goal as "highest" will strongly influence their consideration of the activity outcome ($\alpha = 0.71$), while such characterization will have minimal influence on their consideration of activity strategy ($\alpha = 0.48$). Thus, the conclusion by [6] that when students are readying for a highest-goal-oriented task, the characterization of the goal as "highest" in their activity goal formulation moderates the influences that their consideration of activity strategies has on their considerations of decision outcomes does not hold. By implication, the characterization of the goal as "highest" has no effect on the influence that the students' cognitive processes in the goal formulation has on the transition of their cognitive consideration of the activity strategy and their cognitive consideration of decisions for the activity outcome. These finding underscores [8] and [18] 's observations that when individuals are trained in proper strategies, those given specific high-performance goals are more likely to use those strategies than those given other types of goals.

4.4   Confirmatory Structural Analysis of Model with Uncharacterized Goal

A confirmatory structural analysis was conducted to affirm the findings that the students' characterization of goal formulation as best or highest have no significant influence on their goal formulation cognitive process. when preparing towards an examination activity. The AMOS-generated standardized path diagram showing the standardized indicator loadings of the respective latent variables (cognitive processes) in the re-tested structural model is shown in Fig. 5 below. Based on the goodness of fit statistics, it is evident that the overall model fit appears quite good. This is because the estimated CMIN ($\chi 2$) value of 57.69 (df = 2) has probability level of 0.00 which is lesser than the 0.05 used by convention. Therefore, the null hypothesis that the model fits the data is accepted. The model has a Comparative Fit Index (CFI) value of 0.75, which is close to 1.0, and as such, indicated an acceptance of the null hypothesis that the tested model has a good fit.

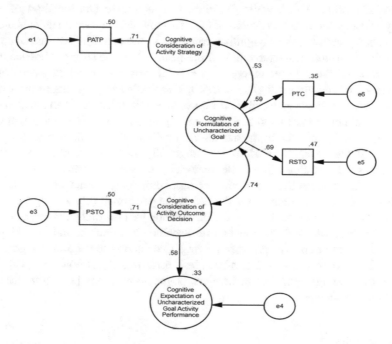

Fig. 5. AMOS-generated path diagram showing standardized indicator loadings in the structural model for uncharacterized goal formulation

The results as highlighted in Fig. 1 above showed that it is only the students' assessment of activity performance options (PATP) and decision outcome expectations (PSTO) that have factor loadings whose values are greater than 0.7 threshold. The students' perception of task experience (RSTO) has a factor loading of 0.69, which value is approximated to 0.7. In this regard, the three indicators (i.e. PATP, PSTO and RSTO) have factor loadings (R) which values are approximately 0.7 and higher. Therefore, using the recommendations of [16], the three factors are predictive of their unobserved variables (cognitive processes). This implies that when preparing towards an examination activity with uncharacterized goal formulation, the students' cognitive formulation of a goal for the activity is largely underlined by their perception of their task experience ($R = 0.69$, $R^2 = 0.48$), and to a lesser extent by their perception of the task's difficulty ($R = 0.60$, $R^2 = 0.36$). On the other hand, the students' cognitive consideration of the activity strategy is underscored by their assessment of the activity performance options ($R = 0.71$, $R^2 = 0.50$), while their cognitive consideration of decisions for the activity outcome is informed by their expectation of the decision outcome ($R = 0.71$, $R^2 = 0.50$). Finally, the students' cognitive expectation of an activity performance with an uncharacterized goal, is to an extent influenced by their cognitive consideration of activity outcome decision ($R = 0.59$, $R^2 = 0.35$). Correlation analysis of the latent variables in the structural model show that the standardized correlation weights (α) for the association between the students' cognitive formulation of highest goal and their cognitive consideration of activity outcome decision is above the threshold value of 0.7, while that with their cognitive consideration of activity strategy is below the threshold value of 0.7. This implies that when students are faced with a pending activity, their cognitive processes in the activity-goal formulation will be such that their non-characterization of the goal will have a strong influence on their consideration of the activity outcome ($\alpha = 0.74$), but a minimal influence on their consideration of activity strategy ($\alpha = 0.63$). This finding implies that in preparing for complex tasks, students must choose appropriate strategies [6], and must have the ability to attain or at least approach their goals [10]. This is well situated within [6]'s postulation to the effect that in the transition of individuals' cognitive goal formulation processes to the emergence of their thoughtfully mastered learning activity, the dichotomous interrelationship of the historicity of the individuals' self-regulation activity and their subjective perception of task complexity influences only their consideration of decision-outcomes for the task to be performed (i.e. writing examination in this study). The findings relate to the observation by [10] that strategy development is motivated by goals, with the mechanism itself being cognitive, and involving either skill development or creative problem-solving. Considering the notion that goal-setting works, it is relevant to ask how it affects task performance, as noted by [10]. This is because, even though goal-setting involves cognitive elements [10], it is primarily a motivational mechanism.

5   Conclusion

This study has provided understanding on the dynamics of students' cognitive formulation of goals and its influence on the mutual effect of students' cognitive consideration of activity strategies and activity outcome decision. Arguing from the perspective that students' subjective perception of task complexity does not influence his/her activity goal formation, as well as their consideration of activity strategies, the findings show that cognitively characterizing the goal of a pending activity as either "highest" or "best" has no significant effect on the influence that students' goal formulation cognitive processes has on their cognitive considerations of the activity strategy, and their cognitive consideration of decisions for the activity outcome. In this regard, it is firstly concluded that the cognitive processes of goal formulation must be viewed as the aim of the activity, and as such, the activity can be formulated without characterizing the goal. It is also concluded that the cognitive characterization of task-goal in terms of "best" or "high" as indicative of the cognitive measures of students goal formation for an activity has no significant influence on the dichotomous interrelationship that exists between the historicity (experience) of their self-regulation activity and their subjective perception of task complexity (difficulty), both of which are engraved in the conscious goal-directed processes of the individual student. The students' cognitive processes of activity goal formation tend to have enormous influence on their cognitive considerations of activity strategies when compared to their cognitive considerations of decision outcomes.

References

1. Sanda, M.A.: Cognitive and emotional-motivational implications in the job design of digitized production drilling in deep mines. In: Hale, K.S., Stanney, K.M. (eds.) Advances in Neuroergonomics and Cognitive Engineering, Advances in Intelligent Systems and Computing, vol. 488, pp. 211–222. Springer, Switzerland (2016)
2. Bedny, G.Z., Karwowski, W.: A Systemic-Structural Theory of Activity: Applications to Human Performance and Work Design. Taylor and Francis, Boca Raton (2007)
3. Sanda, M.A., Johansson, J., Johansson, B., Abrahamsson, L.: Using systemic structural activity approach in identifying strategies enhancing human performance in mining production drilling activity. Theor. Issues Ergon. Sci. 15(3), 262–282 (2014)
4. Jarzabkowski, P.: Strategy as social practice: an activity theory perspective on continuity and change. J. Manage. Stud. 40(1), 23–55 (2003)
5. Sanda, M.A.: Mediating subjective task complexity in job design: a critical reflection of historicity in self-regulatory activity. In: Carryl, B. (ed.) Advances in Neuroergonomics and Cognitive Engineering, pp. 340–350. Springer, Cham (2017)
6. Sanda, M.A.: Dichotomy of historicity and subjective perception of complexity in individuals' activity goal formation and decision outcomes. In: Ayaz, H., Mazur, L. (eds.) Advances in Neuroergonomics and Cognitive Engineering, pp. 265–277. Springer, Cham (2019)
7. Sanda, M.A., Johansson, J., Johansson, B., Abrahamsson, L.: Using systemic approach to identify performance enhancing strategies of rock drilling activity in deep mines. In: Hale, K. S., Stanney, K.M. (eds.) Advances in Neuroergonomics and Cognitive Engineering, Advances in Intelligent Systems and Computing, vol. 488, pp. 135–144. CRC Press, Boca Raton (2012)
8. Locke, E.A., Latham, G.P.: Building a practically useful theory of goal setting and task motivation: a 35-year odyssey. Am. Psychol. 57(9), 705–717 (2002)
9. Locke, E.A.: Purpose without consciousness: a contradiction. Psychol. Rep. 25, 991–1009 (1969)
10. Locke, E.A., Saari, L.M., Shaw, K.N., Latham, G.P.: Goal setting and task performance: 1969–1980. Psychol. Bull. 90(1), 125–152 (1981)
11. Locke, E.A.: Toward a theory of task motivation and incentives. Organ. Behav. Hum. Perform. 3, 157–189 (1968)
12. Smith, K., Locke, E., Barry, D.: Goal setting, planning and organizational performance: an experimental simulation. Organ. Behav. Hum. 46, 118–134 (1990)
13. Dalal, D.K., Diab, D.L., Zhu, X.S., Hwang, T.: Understanding the construct of maximizing tendency: a theoretical and empirical evaluation. J. Behav. Decis. Making 28, 437–450 (2015)

14. Sanda, M.A., Kuada, J.: Influencing dynamics of culture and employee factors on retail banks' performances in a developing country context. Manage. Res. Rev. **39**(5), 599–628 (2016)
15. Di Stefano, C., Zhu, M., Mîndrilă, D.: Understanding and using factor scores: considerations for the applied researcher. Pract. Assess. Res. Eval. **14**(20) (2009). http://pareonline.net/getvn.asp?v=14&n=20
16. Schumacker, R.E., Lomax, R.G.: A Beginner's Guide to Structural Equation Modeling. Lawrence Erlbaum, Mahwah (2004)
17. Earley, P.C., Connolly, T., Ekegren, G.: Goals, strategy development and task performance: some limits on the efficacy of goal setting. J. Appl. Psychol. **74**, 24–33 (1989)
18. Earley, P.C., Perry, B.: Work plan availability and performance: an assessment of task strategy priming on subsequent task completion. Organ. Behav. Hum. **39**, 279–302 (1987)

Performance Evaluation Process and Express Decision

Alexander Yemelyanov[✉] and Alla Yemelyanov

Department of Computer Science, Georgia Southwestern State University,
800 Georgia Southwestern State University Drive, Americus, GA 31709, USA
{Alexander.Yemelyanov, Alla.Yemelyanov}@gsw.edu

Abstract. The Performance Evaluation Process (PEP) is a structured technique specifically designed to support an individual with making personal and quick decisions in complex problems, in which the factors of uncertainty and risk are both present, and in which individual biases and emotional factors may be an important influence. PEP is based on a self-regulation model of a decision-making process developed within the systemic-structural activity theory. PEP regulates two concurrently running processes in decision-making – the formation of a mental model and the formation of the level of motivation by using two general regulators: the factor of significance and factor of difficulty. The factor of difficulty provides feedback control, while the factor of significance provides feedforward control. Formation of a mental model is provided dynamically by constructing a decision hierarchy. Formation of the level of motivation is provided recursively by using the IL-Frame. When evaluating outcomes, the IL-Frame operates with four general criteria of success. Express Decision (ED), a progressive mobile web application, is guided by PEP to provide decision-making support.

Keywords: Decision-making · Emotions · Quick decisions ·
Systemic-structural activity theory · Self-regulation ·
Feedback and feedforward controls · Performance Evaluation Process ·
Decision support · Progressive mobile web application

1 Introduction

Decision-making is no longer viewed simply as a rational process in which logical thinking determines the best means to achieve a goal. Researchers from different fields of cognitive science have shown that human decisions and actions are influenced to a far greater degree by intuition and emotional reactions than was previously thought. Neurological research demonstrates that not only are information and emotional processes inherently connected, but also that without emotion, logical reasoning is impossible and decisions cannot be made. Antonio Damasio determined that formerly well-adapted individuals who experienced damage to their ventromedial cortex could no longer recognize what they really wanted to do in situations when there was no obviously superior choice. These patients remained overall intelligent and sensible, yet they became unable to make decisions preferable to them regarding personal issues.

© Springer Nature Switzerland AG 2020
H. Ayaz (Ed.): AHFE 2019, AISC 953, pp. 279–287, 2020.
https://doi.org/10.1007/978-3-030-20473-0_27

For instance, when requested to decide whether to schedule an appointment for Tuesday or Wednesday, one individual simply went back and forth for half of hour, listing reasons for and against each day. Ultimately, he simply could not decide because his ability to "feel" was impaired, and therefore, he had no emotionally-based preference system to help guide his decision-making [1]. According to Damasio's somatic marker hypothesis, decision-making is guided largely by images of future positive and negative outcomes; this hypothesis refers to somatic markers as a "special instance of feelings" of outcomes (their valence, strengths, and frequencies), generated from secondary emotions. Based on the somatic marker hypothesis, Bechara and Damasio [2] suggested a neutral model which demonstrated how feedback from the body (body loop) may regulate successful economic decisions in situations of complexity and uncertainty. Dunn et al. [3] noted that even the somatic marker hypothesis accurately identified some of the brain regions involved in decision-making, emotion, and body-state representation, but exactly how they all interact at the psychological level remains somewhat unclear. It should be noted that some decisions made by a person are completely emotionally driven. Slovic et al. [4] consider affect heuristic, a mental shortcut that allows people to make decisions quickly and efficiently, in which current emotion – fear, pleasure, surprise, etc. – influences decisions. Affective responses occur rapidly and automatically and usually refer to simple solutions.

In the work, we consider the complex problems in solving what might be of particular interest to an individual. In such solutions, there is a goal and a complex interweaving of emotional and rational factors. Difficult decisions such as buying a house, divorcing a partner, or investing money cannot only be made on the basis of situational emotions, such as affect heuristic. They are driven by both current and distant goals. The presence of a goal contributes both a rational and emotional component to decision-making. Moreover, the emotional component is determined not only by the subjective value of the expected outcomes, but also by the subjective difficulty of obtaining or avoiding them. Therefore, we turned to the systemic-structural activity theory (SSAT) [5], which builds a psychologically-based model of thinking processes, taking into account the goal, as well as the factors of difficulty and significance associated with its achievement. These factors are involved in complex self-regulation processes of forming a mental model of a problem to be solved and forming a level of motivation for choosing the best alternative.

2   Self-regulation Model of Thinking Process

A distinctive feature of this model is the mandatory presence of the goal and an emphasis on its achievement, which is important for mental activity associated with decision-making. It is important to note here that not only does the goal divide the outcomes into positive and negative categories [6] and assign them a marker of subjective importance, but, together with the conditions of the problem (problem = goal + conditions for its attaining), it also provides another marker for the outcomes: the difficulty of obtaining them (for positive outcomes) and the difficulty of avoiding them (for negative outcomes). Both of these markers form an integral marker for the results – the level of motivation.

According to SSAT [7], at the start of making a decision, the goal may have a very general form. Only during the process of problem solving does the goal gradually become clearer and more specific, and it may even be corrected if necessary during the course of activity. The problem solving process begins when the *initial mental model* of the problem is created. However, the initial mental model of the problem often is unable to facilitate the attainment of a desired result on its own. Therefore, after understanding the problem at hand, the subject divides the problem into *sub-problems*. He/she initiates the formation of sub-problems by formulating various *hypotheses*. Each hypothesis has its own potential goal. Based on the comparison and evaluation of such hypotheses, a subject selects one and formulates the first sub-goal associated with the selected hypothesis. Problem solving includes the continuous *reformulation (disaggregation)* of a problem and the development of its corresponding mental model. If the received data is evaluated *positively (positive feedback)*, the thinking process cycle is complete. However, if the result is evaluated *negatively (negative feedback)*, internal or external information should be added to continue this process on a lower level (or to search for other alternatives of solving the problem). In other words, in SSAT, unlike cognitive psychology, the decision-maker is able to regulate his/her behavior not only externally, but also internally by using the inner mental plane.

The levels of motivation for solving problems are formed based upon the *imaginative, verbally-logical*, and *emotionally-evaluative analyses,* along with a collection of appropriate external and internal information. When the evaluation process for each positively evaluated sub-problem is finished, the aggregation process begins. Within this process, all determined levels of alternatives' motivations in sub-problems will be aggregated backward according to the previously created structural hierarchy (problem, sub-problems, sub-sub-problems, etc.) within the disaggregation process. The aggregation process results in the formation of the final mental model of the problem, with the levels of motivation available for all alternatives. This allows the individual to make a decision concerning whether the alternative with the highest level of motivation will be executed, or if its level of motivation is not high enough, which would then require new alternatives to be added for consideration.

The level of positive motivation is determined by the level of significance of positive outcomes and the level of difficulty in obtaining them, and the level of negative motivation is determined by the level of significance of negative outcomes and the level of difficulty to avoid them. With this in mind, the evaluation of significance of positive (negative) outcomes reflects the level (intensity) of their positive (negative) importance for the individual. Evaluation of difficulty depends on the valence of outcomes and for positive outcomes reflects the level of subjective possibility to attain these outcomes and for negative outcomes – level of subjective possibility to avoid them. Since the decision takes place in the conditions of uncertainty of the outcomes, we assume that subjective possibility can be reflected by subjective perception/feeling of their likelihood.

In relation to this, the level of *positive motivation* is determined by *intensity (I+)* of positive outcomes (*positive intensity*) and *likelihood (L+)* of attaining them (*positive component of likelihood*), while the level of *negative motivation* is determined by *intensity (I−)* of negative outcomes (*negative intensity*) and *likelihood (L−)* of avoiding them (*negative component of likelihood*).

These assumptions allow the use of studies by Kotik [8, 9], in which he measured the attractiveness (preference) of emotion-inducing events based on the intensity and likelihood of their outcomes. Essentially, this measured positive (negative) preference reflects the level of positive (negative) motivation to attain positive (and avoid negative) outcomes, which will be demonstrated below. In these studies, likelihood (L) and intensity (I) were measured on the verbal fuzzy scales of "weak-strong" and "seldom-often," respectively, and each of these scales contains nine levels. In the future, it will be convenient for us to present the results obtained by Kotik in the following mathematically stricter form [10]. *Positive preference (P+)* is considered a composition of *positive intensity (I+)* (subjective value of the positive outcomes in the context of attaining the goal) and *positive component of likelihood* or shortly – *positive likelihood* (*L+*) (subjective expectations of attaining *I+* in the context of existing conditions). *Negative preference (P−)* is considered a composition of *negative intensity (I−)* (subjective value of the negative outcomes in the context of attaining the goal) and *negative component of likelihood* or shortly – *negative likelihood (L−)* (subjective expectations of avoiding *I−* in the context of existing conditions). *Preference (P)* is considered to be a function of marginal utility functions $P+$ and $P−$: $P = S(P+, P−)$.

Kotik's results provide a constructive mechanism for measuring the level of motivation through measuring feelings of significance and feelings of difficulty of positive and negative outcomes on the terminal level of the decision tree – in other words, at the level when these feelings may be recognized and expressed verbally. The role of the IL-Frame – a special template for detecting and processing these feelings – will be demonstrated below.

3 Self-regulation Model of Decision-Making

The proposed self-regulation model (SRM) of decision-making includes two sub-models: Formation of mental model (FMM) and formation of the level of motivation (FLM). Execution of SRM is driven by the motivation to attain a goal and includes the execution of FMM and FLM, as well as the regulation of their interaction by using feedback and feedforward controls.

The Design strategy for FMM implements a divide-and-conquer algorithm (D&C). The divide-and-conquer technique uses a recursive breakdown approach in decision-making: decompose the problem into smaller sub-problems, solve them, and then recombine their results to solve the bigger problem. This division of the problem into sub-problems may span several levels deep until a basic (ad hoc) level of certainty will be reached, at which point the problem can be positively evaluated based on IL-Frame. In FMM, feedback control is used to verify whether the current state of the individual's mental model is capable of either evaluating the problem based on IL-Frame or choosing the best alternative. The feedback is *positive* (+fb_FMM) when the individual can perform the verification and *negative* (−fb_FMM) when the individual cannot perform it. Feedforward control leads to an upgrade of the existing mental model. With this purpose, by considering various hypothetical situations and alternative solutions, the problem is divided into sub-problems with corresponding sub-goals.

The design strategy for FLM implements a dynamic programing algorithm (DP). The dynamic programing technique is used to solve a problem by breaking it down into smaller and simpler sub-problems. This method is applied to solve problems that have the properties of overlapping sub-problems and that fulfill the principle of optimality. In FLM, feedback control verifies whether the level of motivation for choosing an alternative is created. Feedback is *positive* (+fb_FLM) when the level of motivation is created, but *negative* (−fb_FLM) otherwise. Feedforward control allows to predict the level of the alternative's motivation after changing verbal characteristics (manipulated control inputs) in IL-Frame.

Feedback Control provides a connection between FMM and FLM models. It is regulated by the factor of difficulty, which determines the individual's self-efficacy in attaining the goal. Feedback control is corrective, connected to the individual's past experience, and provides robustness and error elimination in formation of the mental model and the level of motivation.

Feedforward Control produces upgrades in FMM and FLM. It is regulated by the factor of significance, which determines the directness to the goal. Feedforward control is predictive and leads to an upgrade in the existing mental model or level of motivation in order to enhance its ability to solve the problem and obtain the desired outcomes.

Within SRM, both FMM and FLM, there are two concurrently-running processes that are self-regulated by feedback and feedforward controls and driven by motivation to attain the goal. FMM implements the divide-and-conquer algorithm for mental model formation, while FLM implements the dynamic programing algorithm for motivation level formation. The level of motivation for selecting an alternative forms dynamically, according to the mental model that forms recursively.

4 Performance Evaluation Process

PEP implements the self-regulation model of decision making. It takes into account both emotions and logical reasoning that is important for complex decisions. PEP regulates two concurrently running processes in decision-making – formation of mental model (FMM) and formation of the level of motivation (FLM) by using *two general regulators*: *factor of significance* and *factor of difficulty*; factor of difficulty provides feedback control (fb) and factor of significance – feedforward control (ff). Formation of mental model (i.e. decision hierarchy) is provided dynamically by using decision tree. Formation of the level of motivation (i.e. alternatives' evaluation) is provided recursively by using IL-Frame. IL-Frame operates with *four general criteria of success*, when evaluating outcomes: positive intensity $(L+)$ and its likelihood $(L+)$ and negative intensity $(I-)$ and its likelihood $(L-)$ (see Fig. 1).

The Performance Evaluation Process includes the following three stages.

1. *Problem decomposition and formation of mental model.* The decomposition process starts from the evaluation of the decision-making problem by applying the IL-Frame. If, for each alternative, the intensities and likelihoods of the positive and negative outcomes can be determined, this means that the problem can be positively

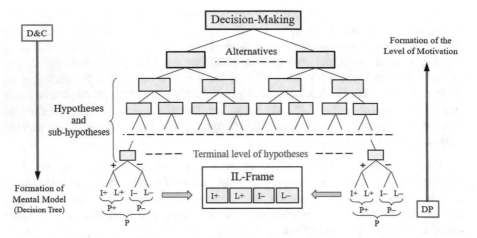

Fig. 1. Performance Evaluation Process: decision hierarchy (decision tree) with IL-Frame.

evaluated (+fb_FMM – positive feedback on FMM), does not need further decomposition of alternatives, and so the next stage, aggregation, will be applied (FLM). At the *aggregation* stage, the level of preference for each alternative will be measured with the IL-Frame, which allows to make a decision regarding which alternative is more preferable in the final *decision-making* stage. If the problem is unable to be positively evaluated (−fb_FMM – negative feedback on FLM) – for example, when the intensities of positive outcomes cannot be determined – the problem should then be divided into sub-problems with corresponding sub-goals (ff_FLM – feedforward control of FLM). For a decision-making problem, this means that the decomposition process is required at the alternatives' levels, with further evaluation and measurement of each alternative by using the IL-Frame. As a result of the decomposition, the problem should be divided into sub-problems, until the point that such a level of detail is reached that the positive and negative outcomes will be assessed by their levels of intensity and likelihood – so that in the aggregation stage, this will allow to determine (measure) the problem's preference level. For this purpose, hypotheses are used. Each hypothesis creates such a sub-problem that the alternatives are evaluated from the position of the outcomes (positive or negative), which are more specific (less uncertain) than the previous group of outcomes. If for such outcomes, their intensity and likelihood can be evaluated on the given scale, this signifies that further specification of the sub-problem and its outcomes is not required (+fb_FLM – positive feedback on FLM). Otherwise (−fb_FLM – negative feedback on FLM), a new hypothesis should be considered with even more specified outcomes (ff_FMM – feedforward control of FMM). The hypothesis allows to assess the intensity(s) and likelihood(s) of the specific outcome(s), which in turn allow to determine the level of a problem's preference, according to the IL-Frame. The choice of hypotheses is determined by the decision-maker and helps him to construct a mental model of the problem being solved. Hypotheses are typically considered for mutually exclusive situations (i.e. the occurrence of some outcomes and their non-occurrence) in order to simplify the assessment of outcomes' degrees of influence on the higher preference levels of sub-problems or alternatives. As a result of decomposition, a *decision tree* (DT) is established for each alternative, and this reflects the functional structure of all the sub-problems examined. In this case, nonterminal vertices determine the sub-problems examined, and terminal vertices contain estimates of intensity and like-lihood for positively evaluated sub-problems.

2. *Problem aggregation and formation of the level of motivation.* In the process of aggregation, a preference level will be determined (measured) for each alternative, which reflects the level of motivation for choosing a particular alternative. Aggregation includes *three stages.* In the *first stage,* by using the IL-Frame, the preference levels of all positively evaluated sub-problems should be determined. In the *second stage,* it is necessary to separately determine the positive and negative preference levels for each alternative. In order to do this, we use special rules for aggregation that present hierarchical and timing dependences of outcomes. The timing dependence of outcomes is determined by hypotheses that are usually considered for two mutually exclusive cases – the occurrence of an event (for example, an accident with severe consequences) and the non-occurrence of this event. In the *third stage,* by using the IL-Frame, the positive and negative prefer-ence levels of each alternative are combined into its cumulative level of preference. Thus, the result of problem aggregation is the construction of a *goal-motivational decision tree* (GMDT) for each alternative [11]. This tree differs from *decision tree* in that all sub-problems are assigned their respective measured levels of preference. The level of preference of the entire alternative aggregates the levels of preference of the examined sub-problems. The level of preference of the entire alternative aggregates the levels of preference of the examined sub-problems.

3. *Making a decision.* In this process, the preference levels of each of the alternatives are compared, and a decision is made regarding selecting an alternative for exe-cution. An alternative with a sufficient level of preference for execution is selected. A low level of preference can occur when the problem is too complicated or not significant enough. The specification of the goal and the addition of new alterna-tives make it possible to arrive at an acceptable solution. If both alternatives have a sufficiently high level of preference, but their preference levels are equal or very close to each other (−fb_FLM – negative feedback on FLM), then positive and negative preference levels should both be evaluated (ff_FMM – feedforward control of FMM). If the levels of motivation for choosing all alternatives are low (−fb_FLM – negative feedback on FLM), then a new alternative should be considered (ff_FMM – feedforward control of FMM).

5 Express Decision

Express Decision (ED), a mobile web application, is guided by Performance Evaluation Process to provide an individual with decision-making support in making quick decisions in complex problems. ED supports the decision-maker by transforming his/her mental model of the decision-making problem from the initial to the final state.

By implementing the IL-Frame, ED helps to evaluate the problem and then measure positive, negative, and resulting levels of preferences for alternatives in order to make a final decision.

ED supports the following processes.

1. Formation of mental model by generating hypotheses and constructing the decision tree.
2. Formation of the level of motivation by using the IL-Frame for collecting relevant data and evaluating alternatives and related hypotheses. ED uses IL-Frame as a template for assessing on a verbal scale the intensity of expected positive (I+) outcomes and their likelihood (L+), as well as the intensity of expected negative (I−) outcomes and their likelihood (L−). These four verbal characteristics operate as general criteria of success when evaluating outcomes: positive (I+) and negative (I−) significances and their positive (L+) and negative (L−) components of difficulty. ED uses experimentally determined relations P = S(P+, P−) = S(S+(I+ , L+), S−(I−, L−)) to combine levels of positive and negative preferences of outcomes into a cumulative level of preference (motivation) for an alternative.
3. Interaction between FMM and FLM by providing feedback and feedforward controls according to special self-regulation rules.

ED is not required to formulate the goal, alternatives, and criteria of success in advance; they can be formulated within the process of decision-making. ED checks completeness of collected data and their consistency with the constructed mental model and then proceeds to conduct sensitivity and what-if analyses.

Express Decision is developed as a progressive web application (PWA) by using a modern Java script framework Vue.js (https://vuejs.org) (frontend) and Google Firebase framework (https://firebase.google.com) (backend). Compared to other web applications, PWA is fast, reliable, and engaging [11]. The core functionality of the app utilizes the ED algorithm augmented with an intuitive user interface that allows to run the app on both *mobile* and *desktop* platforms. The PWA nature of the app also supports *offline workflow*. The app makes it possible for multiple users to work simultaneously while providing each user with an ability to save/restore their session in the cloud. The application can also be used in *guest mode* for quick prototyping of a decision-making workflow without the ability to save results.

Express Decision has a variety of applications [12, 13] and using it is no more difficult than using a regular calculator.

References

1. Damasio, A.R.: Descartes' Error: Emotion, Reason, and the Human Brain. Grosset/Putnam, New York (1994)
2. Bechara, A., Damasio, A.R.: The somatic marker hypothesis: a neural theory of economic decision. Games Econ. Behav. **52**, 336–372 (2005)
3. Dunn, B.D., Dalgleish, T., Lawrence, A.D.: The somatic marker hypothesis: a critical evaluation. Neurosci. Biobehav. Rev. **30**(2), 239–271 (2005)

4. Slovic, P., Finucane, M., Peters, E., MacGregor, D.: The affect heuristic. Eur. J. Oper. Res. **177**, 1333–1352 (2007)
5. Bedny, G.: Application of Systemic-Structural Activity Theory to Design and Training. CRC Press and Taylor & Francis Group, Boca Raton and London (2015)
6. Heath, C., Larrick, R.P., Wu, G.: Goals as reference points. Cogn. Psychol. **38**, 79–109 (1999)
7. Bedny, G., Karwowski, W., Bedny, I.: Applying Systemic-Structural Activity Theory to Design of Human-Computer Interaction Systems. CRC Press and Taylor & Francis Group, Boca Raton and London (2015)
8. Kotik, M.A., Yemelyanov, A.M.: A method for evaluating the significance-as-anxiety of information. In: Studies in Artificial Intelligence. Scientific Notes of Tartu University, vol. 654, pp. 111–129. Tartu University Press, Tartu (1983)
9. Kotik, M.A.: A method for evaluating the significance-as-value of information. In: Studies in Artificial Intelligence. Scientific Notes of Tartu University, vol. 688, pp. 86–102. Tartu University Press, Tartu (1984)
10. Yemelyanov, A.M.: Decision support of mental model formation in the self-regulation process of goal-directed decision-making under risk and uncertainty. In: Ayaz, H., Mazur, L. (eds.) Advances in Neuroergonomics and Cognitive Engineering. Advances in Intelligent Systems and Computing, vol. 775, pp. 225–236. Springer International Publishing (2018)
11. Sheppard, D.: Beginning Progressive Web App Development. Creating a Native App Experience on the Web. Apress, Berkeley (2017)
12. Yemelyanov, A.M., Baev, S., Yemelyanov, A.A.: Express decision – mobile application for quick decisions: an overview of implementations. In: Ayaz, H., Mazur, L. (eds.) Advances in Neuroergonomics and Cognitive Engineering. Advances in Intelligent Systems and Computing, vol. 775, pp. 255–264. Springer International Publishing (2018)
13. Yemelyanov, A.M.: The model of the factor of significance of goal-directed decision making. In: Baldwin, C. (ed.) Advances in Neuroergonomics and Cognitive Engineering. Advances in Intelligent Systems and Computing, vol. 586, pp. 319–330. Springer International Publishing (2017)

Estimation of Emotional Processes in Regulation of the Structural Afferentation of Varying Contrast by Means of Visual Evoked Potentials

Sergey Lytaev[1,2(✉)], Mikhail Aleksandrov[2], and Mikhail Lytaev[3]

[1] Saint Petersburg State Pediatric Medical University, Litovskaya. 2,
194100 Saint Petersburg, Russia
salytaev@gmail.com
[2] Almazov National Medical Research Centre, Saint Petersburg, Russia
[3] The Bonch-Bruevich State University of Telecommunications,
Saint Petersburg, Russia
mikelytaev@gmail.com

Abstract. We studied central mechanisms for regulation of incoming structural afferentation of varying contrast of mentally diseased by means of multichannel registration of visual evoked potentials (VEPs) and brain topographic mapping with subsequent correlation with behavioral data. The first type of VEPs responses by direct relationship between contrast and VEPs amplitude was characterized. In second case, initially there was VEPs amplitude increase, but later on the relationship became the reverse one, manifested by marked decrease of the amplitude. Comparison of behavioral investigation results with electrophysiological data enabled to find that the first variant of reaction was characterized by lower degree of visual images identification, under condition hampering their identification, irrespective of test objects modality.

Keywords: Visual evoked potentials · Augmentation · Reduction · Emotional status · Contrast · Reversible pattern

1 Introduction

The impact of dynamic factors of the external and internal environment contributes to the formation of adaptive rearrangements in the central, peripheral nervous and sensory systems. As a result, this is reflected in the realization of attention, memory, consciousness and other mental functions. The universal mechanism explaining dynamic shifts is the law of force and, as its stage of development, the ultimate protective inhibition [1]. The mechanism of protective inhibition, which makes it possible to explain compensatory reorganizations in pathology, leaves a number of questions in ensuring mental functions under conditions of extreme influences. In the study of reflex activity, the hypothesis of "preventive" inhibition, which arises under the action of weak stimuli on a healthy organism, was proposed as a solution to the question posed

© Springer Nature Switzerland AG 2020
H. Ayaz (Ed.): AHFE 2019, AISC 953, pp. 288–298, 2020.
https://doi.org/10.1007/978-3-030-20473-0_28

[2]. In sensory physiology, this question has received the most comprehensive analysis in the study of the phenomenon of augmentation/reduction (A/R) [3–6].

One of the most known neuroscience theories of personality is the reinforcement sensitivity theory (RST) [7–9]. The original formulation of the RST set emphasis upon only two neurobehavioural systems, the behavioural inhibition system (BIS) and the behavioural approach system (BAS). There is a hidden complexity in and between these systems and this is captured in the revised RST (rRST) which postulates three major neuropsychological systems, two for defensive behaviours (the fight-flight-freeze system, FFFS; and the BIS), and one for approach behaviours (the BAS). By origin, the RST is associated with the kinesthetic psychological phenomenon, named augmentation/reduction (A/R). In such situations, some persons mostly tended to augment virtual weight of an object, and others – to reduce. The former were named "augmentors", and the latter "reducers" accordingly.

The purpose of present research was aimed for estimation of emotional processes in regulation of the structural afferentation of varying contrast by means of visual evoked potentials (VEPs) recording, analysis of person factors and behavioural data.

2   Methods

The research of 52 healthy men (age 20–22 years) and 19 psychiatric patients (males: mean age 30.5, range 21–43) without active productive pathology was performed. An estimation of emotional activity was executed by two factors C (emotional stability) and Q4 (relaxation – tension) of Cattell's test. VEPs using 19 monopolar points system 10/20 with neuromapper were recorded. Stimulation was carried out with eyes open by presenting reversible patterns on the display screen, which are black and white chess squares, replacing each other every 1.0 s. 3 series of stimuli differing in degree of contrast of checkerboard field elements – low, medium and high were used.

VEPs were analyzed at the time period of 400.0 ms since the instance of stimulus application, using data of brain topographic mapping, as well as estimation of time-spatial characteristics of evoked potentials (EPs). Averaged VEPs were divided into intervals of 120–200 and 200–250 ms, measuring amplitude of the N1/P1 complex from peak to peak, as well as peak latency (PL) of the N1 wave. For selection of more informative indices there was used a stepwise discriminant analysis, where parameters assessment was made by criterion of F-statistics of single-factor variance analysis. Differences were considered as significant in case of $F > 4.0$.

Behavioural research consisted in testing vision system by screen tachistoscopic technique, presenting images. The set of 32 complex geometrical Perret's figures for testing short time visual memory, independent of language abilities was used [10]. Fragments of these figures, which when put together make complete image, after proper briefing were presented one after another to observers at interval and exposition of 500.0 ms with further search for initial figure in the table.

Nine images with incomplete set of signs, consisting of familiar objects – key, spectacles, anchor, balance, nippers, scissors, tea-kettle, electric lamp and trumpet – were presented to examinees under conditions of time deficit (exposition of 4.0–3000.0 ms).

3   Outcomes

Character of response reaction to increase of checkerboard contrast patients and healthy examinees divided into two groups. In the first variant (11 patients and 29 healthy) there was registered direct relationship between contrast degree and VEPs amplitude (Fig. 1). In the second variant (8 patients and 23 healthy persons) initially there was observed an increase of VEPs amplitude, but later on contrast degree increase was accompanied by decrease of evoked responses amplitude (Fig. 2).

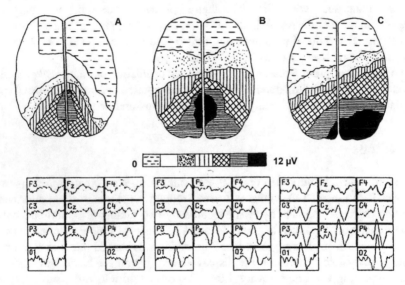

Fig. 1. VEPs topomaps by reversal pattern stimulation of min (A), middle (B) and max (C) contrast from augmentation.

Analysis of spatial distribution of VEPs on cortex and assessment of brain topographic maps allows to trace the above-mentioned changes by typical cases in the first (Fig. 1) and the second (Fig. 2) groups.

It is characteristic that in the first group, when comparing VEPs parameters under conditions of minimal and maximal contrast, meaningful differences were observed actually in all sites concerning both amplitude and PL ($F > 4.0$). At the same time the response to reversion of the medium contrast checker-board differed to far lesser degree from the VEPs parameters of maximum contrast stimulation. In last case meaningful differences were observed only in the right occipital site ($F = 8.2$). Besides attention was arrested by decrease of values of the N1 wave PL when contrast rose, the former being especially manifested in frontal and middle sites.

In the second group PL changes were of less distinct character and at the right occipital area there was even observed PL increase. For this group there were characteristic also topographic changes of the N1/P1 complex amplitude (Fig. 2). In spite of, common for this group, tendency to initial increase of amplitude with it's following decrease, meaningful differences in occipital sites were not observed ($F < 4.0$). At the

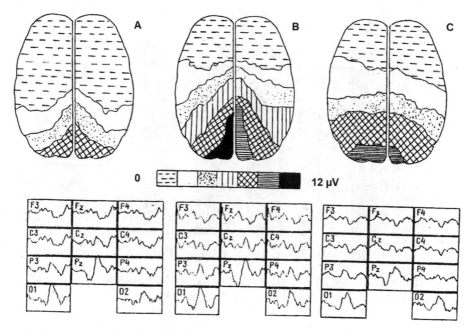

Fig. 2. VEPs topomaps by reversal pattern stimulation of min (A), middle (B) and max (C) contrast from reducing.

same time given situation was manifested more meaningfully at parietal, central and frontal areas (F > 4.0).

Groups formed based on evoked response by checkerboard contrast changing compared the results of tachistoscopic testing. It was characteristic that, in general, images identification degree irrespective of VEPs response reaction character and test objects modality was higher with healthy examinees. However, among psychopathological groups under examination there some differences were observed. So, in the second group the identification of the Perret's figures consequent parts (Fig. 3) came near the normal one, whereas in the first group the number of correct answers was significantly lower (P < 0.05). Resembling correlations between groups under examination also in identifying images with incomplete set of signs were observed (Fig. 3).

For all this under conditions of low test exposition (4.0–30.0 ms) differences of the results from the controls were more meaningful in the first group (P < 0.01), than in the second one (P < 0.05). Increase of test images exposition was followed by increase of number of correct answers, especially seen in the second group (P > 0.05).

The estimation of profiles from factors C and Q4 allowed characterizing the inspected persons. The values of the factor "relaxation – tension" in range from 1 to 6 balls were distributed. Parameters of emotional stability laid within the limits 5–9 balls. For definition of relation of influencing of human emotional state into character of the brain EPs, all persons on 2 groups were divided. 31 persons compounded a category of people with high-performance tensions (4–6 balls) and 21 men had low estimations of the given parameter (1–3 balls). It is necessary to take attention that higher estimations

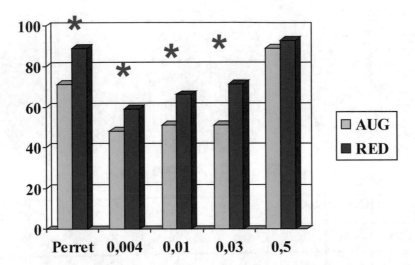

Fig. 3. The values of recognition (%) of Perret's figures and images with incomplete set of signs during presentation: 0.004, 0.01, 0.03 and 0.5 s. * – P < 0.05.

(4–6) of factors Q4 corresponded to lower values of balls of the factor C (5–6). Vice-versa, for low estimations of tension (1–3) high values of emotional stability (7–9) have corresponded.

The averaged values of the amplitudes of the VEPs by groups reflect the characteristics of the dynamics of the N70 and N150 components (Figs. 4 and 5). It is characteristic that in both cases the spatial parameters of N70 differ slightly from each other. At the same time, the minimum amplitude values are noted during the period of exposure of the least contrasting chess field. An increase in the degree of contrast is accompanied in all subjects by an increase in the amplitude of N70. The similarity of the spatial dynamics of the N70 wave in both groups also draws attention. At the same time, the maximum amplitudes are noted in the occipital-parietal leads with predominance in the right occipital area. The maximum values of the amplitude factors also are recorded here, which indicates a low dispersion of these indicators. The variance of the amplitudes of the frontal leads, and especially in the group of subjects with low scores of factor Q4, is higher (low ordinal numbers of factors). In addition, in these sections, in this group, an increase in the amplitude of N70 is observed when exposed to an average degree of contrast in comparison with the maximum (Fig. 4).

Significant differences among the studied groups for the amplitude of the N150 wave of VEPs are registered. If individuals with high Q4 values have a direct relationship between the degree of contrast of the chess field and the magnitude of the amplitude N150 (Fig. 5), then the opposite group of subjects has a pronounced inversion of answers (Fig. 5). It is characteristic that in individuals with a low score of factor Q4, in response to stimulation with patterns of high contrast in the occipital-parietal sites, minimal values of the amplitudes N150 are observed. In turn, in the same areas, minimally contrast stimulation causes the most pronounced N150 waves with a noticeable predominance of amplitude in the right hemisphere. N150 takes an

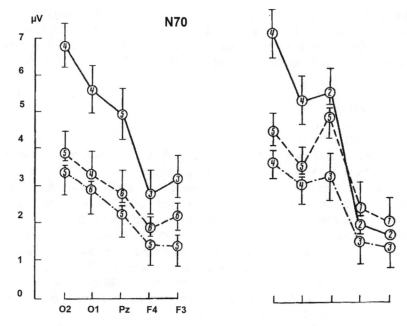

Fig. 4. The factor analyses of the values of amplitudes of the wave N70 VEPs from min, middle and max contrast. Left – augmentation, right – reduction. On the ordinate scale – μV, abscissa – sites on 10/20 system. The numbers in the circles are the ordinal numbers of the factors.

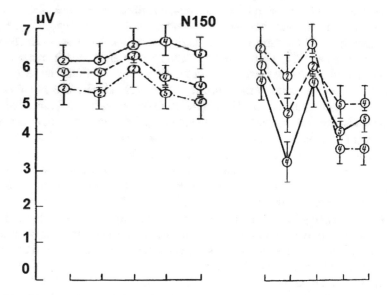

Fig. 5. The factor analyses of the values of amplitudes of the wave N150 VEPs from min, middle and max contrast. Notes see Fig. 4.

intermediate position in response to medium-contrast patterns. However, in the frontal regions, the amplitude of this wave becomes maximal. Characteristically, in general, in both groups, the amplitude values of the N150 are more stable in the frontal cortex, as evidenced by the higher values of the factors (Figs. 4 and 5).

4 Discussion

As a rule in works connected with investigation of mechanisms for regulation of sensory signals of various intensity entering the brain, there was made a connection with stimulation/inhibition processes in the central nervous system [11–13]. Where it was assumed that character of response depended upon threshold sensitivity. Presence of low sensitivity, that is of initially high threshold of absolute sensory sensitivity, increased extent of response to rising intensity of stimulant. On the contrary, system with high sensitivity, that is with low absolute sensory threshold, triggers a "program", protecting against "overloading" and in spite of increase of stimulus intensity, reduced evoked response is obtained. Accordingly, the examinees in the first case were customarily called "augmentors", but in the second case – "reducers" [14, 15].

But in all previous works on research of A/R mechanism simple signals of varying intensity were used as adequate stimulants, where, as a matter of fact, was only gradual variation of stimulus physical characteristics. Employment of reversible checkerboard patterns as a stimulant have certain principal peculiarities, and, in particular, that energy effect upon visual system was insignificant comparing with informational effect, the one a potential was evoked by [15].

The results of given research have indicated, that initially response of all examinees to contrast degree increase was the same and consisted in increase of the N1/P1 complex amplitude. At the same time, in the first group this fact was registered mainly at the occipital areas (Fig. 1), but in the second group amplitude increase appeared more pronounced at the parietal and central sites (Fig. 2). Indicated circumstance enabled to make conclusion that contrast increase at initial stage is of sufficient informational meaningfulness for all examinees. In one case, it was manifested mainly at the projection zone of vision system, whereas in another case the answers prevailed at the areas of neocortex. Such response to contrast increase, manifested by further increase of VEPs amplitude all over the brain surface in the first group, and vice versa, the amplitude reduction in the second group, had plenty in common with previously suggested conception of "points of switching" [16, 17]. The main point of the latter consisted in nervous system beginning to work in search mode continuously controlling sensitivity, when stimulation process approaches switching points. From the point of view of the Pavlovian law of force [1, 2], switching mechanism could be considered as a transfer process, where in some cases response follows the law of force directly, but in others an adaptive preventive mechanism is working.

In number of works on evoked activity research, it was stated that increase of stimulus intensity was accompanied by reduction of EPs peak latency [18–20]. Actually, we observed similar trend in the first group practically in all sites. However, in the second group, this dependence practically was not manifested. Apparently dynamics of the N1/P1 complex peak with changing contrast degree were not specific ones, the same being proved by some published data demonstrating reduction of LP under conditions of stimulant intensity increase for single EPs only, while time characteristics of averaged EPs followed another laws [6, 15].

In publications about A/R phenomenon in Psychiatry clinic there was met sufficient number of descriptions of effecting correlations of electrophysiological data with results of biochemical investigations [6, 13, 17], data of PET-scan [18], as well as with results of psychological experiments [5–7, 12]. So, a number of authors, for instance, considered that Cz and Fz middle sites were the most informative for EPs registration, and also believed that reducers, by integral character of nervous activity, should be regarded as extroverts.

Based on existing conception of functional stratification [16], we considered the A/R phenomenon from position of integration of "strata" of functional activity in the brain activities. At the bottom of this scale of ranks there were located biochemical and biophysical processes. The neurophysiological level was the next with it's indices being electrographic correlates (EPs). The top level was represented by various forms of higher psychic (behavioral) functions.

Accepting a task of investigation of the A/R mechanism in vision system during perception of structural afferentation, detection of relations of electrophysiological data with capabilities of higher vision function was of obvious interest for us. Our results demonstrated, first, reduced characteristics of identification ability in groups of mentally diseased, and secondly, indicated higher abilities of perception functions in the second group of patients in comparison with the first one, irrespectively of employed sensory modalities (Figs. 3 and 4). Mentioned facts, on the one hand, confirmed nonspecific character of sensory failures of schizophrenic patients, detected under conditions hampering identification. On the other hand, they enabled to effect correlation with neurophysiological manifestations of the aug/red and in this connection to evaluate specific failures of sensory identification mechanisms.

Therefore, identification of the Perret's figures triggers at least three consecutive processes – picking out of signs, keeping of characteristics set of every fragment in short-time memory and, finally, spatial synthesis with following identification. It is characteristically that in the second group of patients reduction of identification indications differ a little from the control group (P > 0.05), while indications of the first group are characterized by much smaller number of correct answers (Fig. 3; P < 0.05). Similar tendency, but more pronounced is observed in recognition of images with incomplete set of signs (Fig. 3), which are identified mainly by operation of images invariant identification mechanism.

Thus, it becomes apparent that efficiency of functioning of specific mechanisms of sensory identification – invariant evaluation of signals, short-time visual memory, spatial analysis and synthesis – is in certain relation with electrophysiological characteristics of perception of structural afferentation with varying contrast. It could be supposed that more pronounced response in associative cortex to intermediate, by it's intensity, stimulant, correlating with higher images identification in the second group of schizophrenic patients, tells about more synchronous work of brain hemispheres when processing meaningful information. The adaptivity mechanism, engaged later on, in response to contrast increase, manifested by EPs amplitude reduction, probably, tells about reduction of biological meaningfulness of the given informative stimulant. On the contrary, absence of switching adaptivity mechanism, probably, is a reflection of more inert processes of brain, the latter being indicated also by rise of images sensory identification thresholds.

The structural substrate of the switching mechanisms is mainly the parietal and occipital cortex. Data in favor of asymmetry with an increase of amplitude in the right occipital lobe correlate with the previously obtained results when registering the VEP in response to the reversion of the chess field, both in normal [13] and in pathology [20].

As established, the emotional status to a largely extent determines the relationship between the degree of contrast of the chess field and the amplitude N150 VEPs. The results of this study show that, initially, to increase the degree of contrast, the nature of the response in all subjects was partially similar and was accompanied by an increase in the amplitude of N150 VEPs in all leads in persons of the first group, and in individuals of the second group in frontal areas. The noted facts allow us to conclude that an increase in contrast at the initial stage is of sufficient informational significance for all subjects. The subsequent course of the response to an increase in the degree of contrast, manifested by a further increase in the N150 amplitude over the entire brain surface in subjects with a high degree of tension and, on the contrary, its decrease in subjects with the opposite characteristic of the emotional sphere has much in common with some previously proposed mechanisms for the regulation of increasing afferentation.

The neurophysiological basis of the switching mechanism of the A/R phenomenon is most often associated with biochemical changes. In particular, the meaning has the concentration of catecholamines, platelet monoamine oxidase, level of endorphins, etc. [16, 17]. Of certain interest are also some correlations of changes in the bioelectric activity induced with the data of psychological testing. According to some authors, frontal sites are the most informative [6, 7, 12]. The Q4 factor (from the Cattell questionnaire) has a direct connection with both the EP components having a PL of about 150 ms and 300 ms was established. There are observations that, in terms of the integral nature of nervous activity, the inclusion of a switching mechanism with a low intensity of stimulation is more characteristic of extroverts [15].

It is characteristic that the individual characteristics of the behavior of man and higher animals in recent times are largely associated with the above-noted bio-chemical indicators [16–18]. An established point of view is that emotional processes by reticulo-frontal interaction with the direct participation of a number of subcortical formations are determined. In this case, the frontal cortex plays the role of the information component of emotions [13].

Thus, the results of this study indicate the presence of a switching mechanism when changing the contrast of structural afferentation, which is dependent on the human emotional state and mainly by the frontal informational neocortex is provided.

5   Conclusion

The increase of reversible checkerboard patterns contrast degree enabled to find two types of VEPs responses. The first variant by direct relationship between contrast and VEPs amplitude was characterized. In the second case, initially there was VEPs amplitude increase, but later on, the relationship became the reverse one, manifested by marked decrease of the amplitude. Comparison of behavioral investigation results with electrophysiological data enabled to find that the first variant of reaction was characterized by lower degree of visual images identification, under condition hampering their identification, irrespective of test objects modality.

Thus, it becomes apparent that efficiency of functioning of specific mechanisms of sensory identification – invariant evaluation of signals, short-time visual memory, spatial analysis and synthesis – is in certain relation with electrophysiological characteristics of perception of structural afferentation with varying contrast. It could be supposed that more pronounced response in associative cortex to intermediate, by it's intensity, stimulant, correlating with higher images identification in the second group of schizophrenic patients, tells about more synchronous work of brain hemispheres when processing meaningful information. The adaptivity mechanism, engaged later on, in response to contrast increase, manifested by EPs amplitude reduction, probably, tells about reduction of biological meaningfulness of the given informative stimulant. On the contrary, absence of switching adaptivity mechanism, probably, is a reflection of more inert processes of brain, the latter being indicated also by rise of images sensory identification thresholds.

- An increase in the intensity of the contrast of a reversible chess field, regardless of the emotional sphere is accompanied by an increase in the amplitude of the N70 VEP component. A similar relationship for the amplitude of the N150 wave in persons with a high degree of tension and reduced emotional stability was traced.
- For persons with a low degree of tension and high emotional stability in the occipital-parietal leads, an inverse relationship between the magnitude of contrast and amplitude N150 is noted. At the same time, in the frontal areas, the switching mechanism of regulation is observed, which after the initial increase of the N150 amplitude is accompanied by its further decrease with increasing degree of contrast.

References

1. McNaughton, N., Corr, P.J.: The neuropsychology of fear and anxiety: a foundation for reinforcement sensitivity theory. In: The Reinforcement Sensitivity Theory of Personality, pp. 44–94. Cambridge University Press, Cambridge (2008). http://dx.doi.org/10.1017/CBO9780511819384.003
2. Lytaev, S., Aleksandrov, M., Ulitin, A.: Psychophysiological and intraoperative AEPs and SEPs monitoring for perception, attention and cognition. Commun. Comput. Inf. Sci. **713**, 229–236 (2017). https://doi.org/10.1007/978-3-319-58750-9_33
3. Prescott, J., Connolly, J.F., Gruzelier, J.H.: The augmenting/reducing phenomenon in the auditory evoked potential. Biol. Psychol. **19**(1), 31–44 (1984)
4. Yakovenko, I.A., Tcheremoushkin, E.A., Lytaev, S.A.: Reflecting the augmenting – reducing phenomenon in the spatio-temporal EEG characteristics. Zhurnal Vysshei Nervnoi Deyatelnosti Imeni I.P. Pavlova **44**, 25–32 (1994)
5. Carrillo-de-la-Peña, M.T.: ERP augmenting/reducing and sensation seeking: a critical review. Int. J. Psychophysiol. **12**(3), 211–220 (1992)

6. Pascalis, V., Fracasso, F., Corr, P.J.: Personality and augmenting/reducing (A/R) in auditory event-related potentials (ERPs) during emotional visual stimulation. Sci. Rep. **7**, Article number: 41588 (2017). https://doi.org/10.1038/srep41588

7. Corr, P.J., DeYoung, C.G., McNaughton, N.: Motivation and personality: a neuropsychological perspective. Soc. Pers. Psychol. Compass **7**, 158–175 (2013). https://doi.org/10.1111/spc3.12016

8. Corr, P.J.: Reinforcement sensitivity theory of personality questionnaires: structural survey with recommendations. Personality Individ. Differ. **89**, 60–64 (2016). https://doi.org/10.1016/j.paid.2015.09.045

9. Corr, P.J., McNaughton, N.: Neuroscience and approach/avoidance personality traits: a two stage (valuation–motivation) approach. Neurosci. Biobehav. Rev. **36**, 2339–2354 (2012). https://doi.org/10.1016/j.neubiorev.2012.09.013

10. Perret, E.: Gehirn und Verhalten. Neuropsychologie des Menschen. Huber, Stuttgart (1973)

11. Shostak, V.I., Lytaev, S.A., Golubeva, L.V.: Topography of afferent and efferent flows in the mechanisms of auditory selective attention. Neurosci. Behav. Physiol. **25**(5), 378–385 (1995). https://doi.org/10.1007/BF02359594

12. Sylvers, P., Lilienfeld, S.O., LaPrairie, J.L.: Differences between trait fear and trait anxiety: implications for psychopathology. Clin. Psychol. Rev. **31**, 122–137 (2011). https://doi.org/10.1016/j.cpr.2010.08.004

13. Gray, J., McNaughton, N.: The Neuropsychology of Anxiety: An Enquiry into the Functions of the Septo-Hippocampal System, 2nd edn. Oxford University Press, Oxford (2000)

14. Petrie, A.: Individuality in Pain and Suffering. University of Chicago Press, Chicago (1967)

15. Lytaev, S.A., Shostak, V.I.: The significance of emotional processes in man in the mechanisms of analyzing the effect of varying contrast stimulation. Zhurnal Vysshei Nervnoi Deyatelnosti Imeni I.P. Pavlova **43**(6), 1067–1074 (1993)

16. Von Knorring, L., Monakhov, K.K., Perris, C.: Switching adaptive mechanism for regulation of signals coming into central nervous system. Hum. Physiol. **7**, 784–795 (1981)

17. von Knorring, L., Perris, C.: Biochemistry of the augmenting-reducing response in visual evoked potentials. Neuropsychobiology **7**, 1–8 (1981). https://doi.org/10.1159/000117825

18. Lytaev, S., Aleksandrov, M., Popovich, T., Lytaev, M.: Auditory evoked potentials and PET-scan: early and late mechanisms of selective attention. Adv. Intell. Syst. Comput. **775**, 169–178 (2019). https://doi.org/10.1007/978-3-319-94866-9_17

19. Hensch, T., Herold, U., Diers, K., Armbruster, D., Brocke, B.: Reliability of intensity dependence of auditory-evoked potentials. Clin. Neurophysiol. **119**, 224–236 (2008). https://doi.org/10.1016/j.clinph.2007.09.127

20. Lytaev, S.A., Belskaya, K.A.: Integration and disintegration of auditory images perception. Lect. Notes Comput. Sci. **9183**, 470–480 (2015). https://doi.org/10.1007/978-3-319-20816-9_45

Cognitive Computing

Modelling an Adjustable Autonomous Multi-agent Internet of Things System for Elderly Smart Home

Salama A. Mostafa[1]([✉]), Saraswathy Shamini Gunasekaran[2],
Aida Mustapha[1], Mazin Abed Mohammed[3],
and Wafaa Mustafa Abduallah[4]

[1] Faculty of Computer Science and Information Technology,
Universiti Tun Hussein Onn Malaysia, 86400 Johor, Malaysia
{salama, aidam}@uthm.edu.my
[2] Department of Computer Science and Informatics, Universiti Tenaga Nasional,
43000 Kajang, Selangor, Malaysia
sshamini@uniten.edu.my
[3] Faculty of Computer Science and Information Technology,
University of Anbar, Anbar 31001, Iraq
mazinalshujeary@uoanbar.edu.iq
[4] Faculty of Computers and IT, Nawroz University, Duhok 42001, Iraq
wafaa.mustafa@nawroz.edu.krd

Abstract. Internet of Things (IoT) introduces many intelligent applications that are closely attached to humans' daily activities. This advanced technology attempts to bridge the gap between the information world and the physical world. Recent studies investigate efficient, flexible, scalable and reliable IoT systems that not only control things and devices on behalf of humans but adaptable to humans' preferences. However, the autonomous control of the IoT in a smart home or healthcare environment subjects to many factors such as human health, time and date. For example, peoples' needs and behaviours during workdays differ from weekends or a young person needs and behaviours differs from an elderly person. Hence, the practical setting of a smart home entails flexible management to the autonomous control of IoT systems. This paper proposes an architecture of Adjustable-Autonomous Multi-agent IoT (AAMA-IoT) system to resolve a number of the IoT management of control and application interface challenges. The AAMA-IoT is applied in an elderly smart home simulation in which autonomous agents control passive things such as a chair or door and active things such as a television or an air conditioner. The test results show that the AAMA-IoT system controls 14 things with average activities recognition accuracy of 96.97%.

Keywords: Autonomous agents · Adjustable autonomy ·
Human-agent interaction · Internet of things · Smart home · Elderly user

© Springer Nature Switzerland AG 2020
H. Ayaz (Ed.): AHFE 2019, AISC 953, pp. 301–311, 2020.
https://doi.org/10.1007/978-3-030-20473-0_29

1   Introduction

The Internet of Things (IoT) research and development are growing fast due to its importance and impact on facilitating humans' life. According to the ABI research, more than 30 billion devices will serve under IoT by 2020 [1]. The research direction of autonomous systems in the current years focuses on incorporating multi-agent systems' technologies within the IoT to formulate advance ambient intelligence environments. The agents are embodied in things to manifest dynamic and distributed problem-solving capabilities [2–5].

The literature presents several attempts to manage autonomous control of multi-agent systems based on users' preferences by utilizing the concept of adjustable autonomy [6]. Adjustable autonomy is built upon grading the autonomy of operators to a particular range or extent so that it can be changed over time [7, 8]. It concerns with controlling the autonomy of agents to improve their performance, increase satisfaction, reduce workload and adapt preferences of users [8, 9]. Adjustable autonomy application covers a wide range of domains including unmanned system, robotics, smart home, organizational coordination, electronic commerce, and even space mission [6, 10]. However, the proposed models of adjustable autonomous systems are meant for a limited number of operators or operations and not IoT like environments [11, 12].

Several attempts to integrate agents with IoT are visible in the literature. To the best of our knowledge, that there is no available modelling of an adjustable autonomy for multi-agent system in IoT environment [8, 13, 14]. This research attempts to propose an Adjustable Autonomous Multi-Agent IoT (AAMA-IoT) system that efficiently and reliably satisfies the preferences of elderly users leaving in a smart home. The contribution of the research is adding the global control sub-layer in the application layer of the IoT. The work is simulated in an elderly smart home environment that provides various autonomous functions and tracks the elderly daily activities.

The rest of the paper is organized in five sections. Section 2 presents the related work in modelling multi-agent IoT systems. Section 3 views the theoretical representation of the AAMA-IoT system. Section 4 demonstrates the implementation and discusses the obtained results. Section 5 concludes the work and suggests future work.

2   Related Work

Autonomous agents and multi-agent systems are widely adopted in IoT research and applications. It is clearly stated in the literature that a multi-agent approach mitigates the complexity of modelling IoT systems [15–17]. There are different deployments to agents in the IoT systems and they simply can be categorized based on the IoT architecture to physical layer as in [2]; network layer as in [15] and [18]; application layer as in [16] and more than one layer as in [3]. The following is a review summary of the related work that shows the agents' role in the IoT systems.

Elkhodr et al. [3] propose a Dynamic Disclosure-Control Method (DDCM) that deploys agents to improve the security and privacy of the IoT systems. They assume that the activities of the IoT system users are instantly recorded. The agents manage and protect the tracing of the IoT users' locations when they are interacting with and using the system by analyzing the context of the requested tracing services. The agents respond to the requests based on a user privacy setting and the analysis process. The agents are given the absolute autonomous control to provide precise, imprecise or fake locations based on the analysis results. This approach implies that things can provide incorrect or faulty information. A better approach can be defining privacy protocols to prevent untrusted things or third-party systems from abusing users' data [19].

Godfrey et al. [18] propose a framework that uses mobile agents to connect things in the IoT environment. The mobile agents proactively provide resources to things and services to users of the environment. The mobile agents are fully autonomous and have global control over the things. The human user only commands a particular service using a mobile phone application and the mobile agents pursue and process the command. The framework is implemented in a robot path planning and control scenario of an IoT environment. The things are network nodes that collect data about the robot's movements and the surrounding then the mobile agents use the data to track and guide the robot to complete particular tasks and avoid obstacles.

Mzahm et al. [2] propose the concept of Agents of Things (AoT) as a complement to the IoT concept. The AoT formally embedded reasoning and intelligence on things instead of including them as a layer or as a part of a layer (most common at the interface part of the application layer). The architecture of the AoT classifies the agents' abilities according to the devices that they are embedded in via a spectrum of intelligence. The spectrum ranges from passive to highly active intelligent agents. The AoT architecture theoretically improves the autonomy, reasoning ability, and the scalability of the things. However, this architecture demands a distributed computing in which each active thing has its own microprocessor and memory while the traditional IoT architecture utilizes cloud computing instead. Additionally, human interaction with the system to enable global control over the things is challenging.

3   Modelling

This paper proposes an architecture for adjustable autonomous multi-agent IoT (AAMA-IoT) systems. This architecture aims to resolve a number of the conventional IoT challenges of global control and human interaction. It has a basic IoT architecture that consists of a physical (perception) layer, the network (transport) layer and the application layer [2, 16]. It operates based on the adjustable autonomy of a multi-agent system. Figure 1 shows the architecture of the AAMA-IoT system.

The physical layer includes things/devices, sensors and actuators. It might contain RFIDs, tags, tag readers, sensors (temperature and humidity), GPS and digital camera. The network layer includes the internet gateway for all the things/devices that need to interact with, and a data centre or cloud server to pre-process, manage, and store the data. Finally, the application layer (e.g., mobile application) includes operational algorithms such as artificial intelligence algorithms to process the data and decide on the conditions and actions of the IoT. It further includes an interface for human

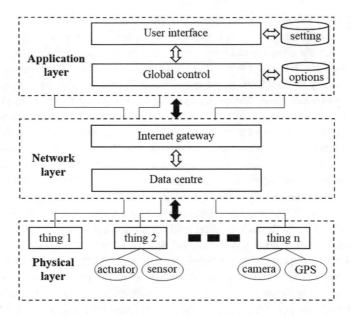

Fig. 1. The architecture of the AAMA-IoT system

interaction with the IoT system. The only modification to the AAMA-IoT architecture is in the application layer in which it has a user interface and a global control sub-layers along with the things setting and control options databases. The global control sub-layer forms the main contribution of this work. The global control includes three parties, which are humans, autonomous agents and an IoT control centre of the IoT service provider. The global control sub-layer aims to synergize the control of the three parties. Figure 2 shows the global control sub-layer of the AAMA-IoT architecture.

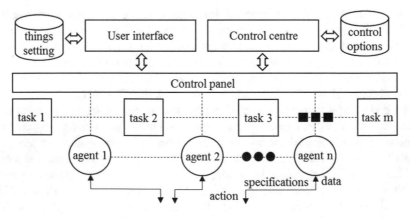

Fig. 2. The global control sub-layer

The application domain of the AAMA-IoT is an elderly smart home. The user interface enables humans to set their preferences using the things setting and control options databases. The control centre is the party that manages the end user application including setting and updating policies, protocols and services. Autonomous agents mainly concern with controlling the functionality of things. The agents have a decentralized multi-agent architecture. This architecture provides self-organization, flexibility, and scalability to the IoT system. Some agents control passive things such as a chair or sofa and others control active things such as a television or an air conditioner.

Basically, in AAMA-IoT, an individual agent has five operational modes and an adjustable autonomy mechanism. These modes are *not-active, inactive, reactive, active* and *proactive*. The adjustable autonomy mechanism has five autonomy levels that correspond to and control the five operational modes of the agent. Each of these operational modes is equipped with a course of actions. The execution of the actions leads to assist in achieving the task of a particular thing. Hence, this mechanism enables the agent to show different types of autonomous behaviours. It implies the adjustment of the autonomy levels changes the operational mode of the agents, which changes the autonomous behaviour of controlling the corresponding things.

Moreover, in the AAMA-IoT, things have functional, autonomy and safety properties. These properties are defined in the databases of the IoT. As mentioned earlier, the agents' architecture provides five operational modes. These modes are designed to operate according to the properties of the things. Hence, the operational modes of the agents are relatively constrained by the properties of the things and the autonomy level options. For example, an agent of a chair is considered as *inactive* based on the properties of the chair while an agent of an electric wheelchair has a range of *inactive*, *reactive* and *active* operational modes based on the properties of the electric wheelchair. The following table shows an abstract association between the five operational modes of agents and five adjustable autonomy levels based on the functional properties of things.

Table 1. The configuration options

Mode	Functional properties	Autonomy
Not-active	No sensing and no actions	Level 1
Inactive	Sensing but no actions	Level 2
Reactive	Sensing and reactive actions	Level 3
Active	Sensing and active actions	Level 4
Proactive	Sensing and proactive actions	Level 5

Based on Table 1, the *not-active* mode is associated with an autonomy level 1, l_1. This agent mode controls things that have no sensing and no action capabilities. It is useful when a thing that the agent controls is removed or not functioning. The *inactive* mode is associated with an autonomy level 2, l_2. This agent mode controls things with sensing but no actions functional properties or has a pending or hold action state. The *reactive* mode is associated with an autonomy level 3, l_3. This agent mode controls things with sensing and reactive functional properties in which the agent can response to specific situations or events. The *active* mode is associated with an autonomy level 4, l_4. This agent mode controls things with sensing and active functional properties in which the agent performs actions that cause changes to situations or events. The *proactive* mode is associated with an autonomy level 5, l_5. This agent mode controls things with sensing and proactive functional properties in which the agent performs actions that invoke or revoke situations or events. An essential difference between the *active* and *proactive* modes is that the active agent only controls particular things while the proactive agent interferes with or influences other agents of lower autonomy levels. For example, the *proactive* mode of an agent enables the agent to change the operational mode of other agents in some security or emergency conditions.

Let an AAMA-IoT control unit U of a smart home is defined as a tuple of things' specifications S, agents A, tasks T, data sources D, control policies C, autonomy levels L and user preferences P.

$$U = <S, A, T, D, C, L, P> \tag{1}$$

Where $S = \{s_1, s_2, \ldots, s_n\}; A = \{a_1, a_2, \ldots, a_n\}; T = \{t_1, t_2, \ldots, t_m\};$ $D = \{d_1, d_2, \ldots, d_n\}; C = \{c_1, c_2, \ldots, c_n\}; L = \{l_1, l_2, \ldots, l_5\}; P = \{p_1, p_2, \ldots, p_n\};$

Then the priority of global control has the distribution of $C > L > P$. The agent updates its autonomy level by mapping the control polices of the things, current autonomy level options and user preferences as shown in (2):

$$l \leftarrow update(a, (c, l, p)) \tag{2}$$

The agent then achieves t according to l by using s and d to generate a new d.

$$d \leftarrow execute(a, (s, d, (achieve(a, (t, l))))) \tag{3}$$

The resulted d of (3) includes operational instructions, actions and data of a particular thing. It is transmitted to the thing using wired or wirelessly connectors of the network layer and the thing operates according to the d. The thing transmits back its d to the AAMA-IoT control unit. Figure 3 shows an abstract AAMA-IoT control unite of a smart home.

A basic AAMA-IoT control unit has general properties of things, agents, tasks, data sources, control policies and autonomy levels. The specific properties of the control unit are configurable according to the application domain and users' preferences. The Fig. 3 setting of the AAMA-IoT control unit considers five rows of abstract autonomy levels and an undefined number of agents. The agents of a particular row are assumed to share the same abstract autonomy setting and associated with things that have similar functional properties.

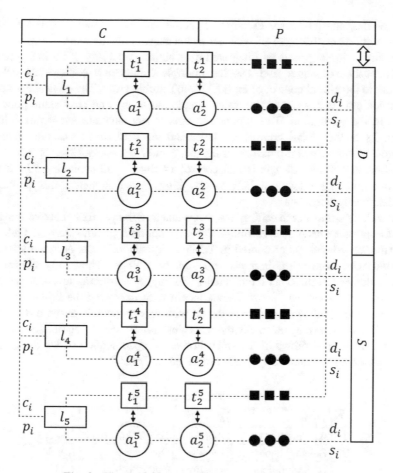

Fig. 3. The AAMA-IoT control unit of a smart home

4 Simulation and Results

The application domain of the AAMA-IoT system is an automated elderly smart home simulation in which autonomous agents control things of an Elderly house including furniture and devices. The elderly smart home makes a good test environment due to the complexity of smart home tasks achievement and elderly preference satisfaction. The application or simulation entails accessing sensor data of various things and deciding based on the pattern of the data [20]. The data is analyzed and processed by the AAMA-IoT modules to comprehend its context and perform responses.

This work adopts the activity recognition dataset of [21]. This dataset consists of real-world scenarios of an IoT-based smart home. It is used in [22] for monitoring and recognizing the activities of a human in a smart home IoT environment. The dataset has 1319 instances of 14 things that are recorded for 28 days. This work uses the dataset to determine the smart home things, the order and the time of the elderly usage to the things and the pattern of seven elderly daily activities. Four of these things are carefully changed with other things to fit with the test scenarios of this work.

The simulation supports an integrated distribution capability of an IoT system that simultaneously exchanges and handles multiple inputs and changes its behaviour according to the global control of an AAMA-IoT architecture. The control centre of the IoT service provider is out of the tested simulation scope and only static policies are applied in the simulation. The simulation aims to demonstrate the applicability and functionally of the global control sub-layer and the performance of the agents. It is implemented in the Java programming language and Java Agent Development (JADE) framework. It includes 14 agents that control 14 things and achieving 17 tasks. The agents determine the interaction between the things based on their corresponding tasks and elderly daily activities.

The elderly smart home setting has seven inactive things, four reactive things, two active things and one proactive thing. The corresponding agents have default operational autonomy levels as presented in Table 1. Agents with low autonomy levels are associated with *not-active*, *inactive* and *reactive* things. They only concern with operating things. On the other hand, agents with higher autonomy levels are associated with *active* and *proactive* things. They concern with operating the things and tracing elderly activities. The active agents interact with other agents to assist in the tracking process. The proactive agent tracks the activities and directs the agents to assist in the tracking process. The things of the smart home, the specification of the things and the default autonomy levels of the agents are given in Table 2.

Table 2. The simulation setting

ID	Thing	S	T	L
1	Air-conditioner	*Active*	On/off time, track	$l_1 - l_4$
2	Balcony door	*Inactive*	Open/close	$l_1 \vee l_2$
3	Bathroom door	*Inactive*	Open/close	$l_1 \vee l_2$
4	Bedroom door	*Inactive*	Open/close	$l_1 \vee l_2$
5	Cups cupboard	*Inactive*	Open/close	$l_1 \vee l_2$
6	Freezer	*Active*	Replenish, track	$l_1 - l_4$
7	Fridge	*Proactive*	Replenish, track	$l_1 - l_5$
8	Front door	*Inactive*	Open/close	$l_1 \vee l_2$
9	Groceries cupboard	*Inactive*	Open/close	$l_1 \vee l_2$
10	Microwave	*Reactive*	On/off time	$l_1 - l_3$
11	Plates cupboard	*Inactive*	Open/close	$l_1 \vee l_2$
12	Television	*Reactive*	On/off time	$l_1 - l_3$
13	Toilet flush	*Reactive*	On/off time	$l_1 - l_3$
14	Washing machine	*Reactive*	On/off time	$l_1 - l_3$

Table 3 shows the simulation results of the default autonomy setting in which no adjustment has been made to the autonomy levels. It presents the things' ID, the achieved tasks, the average consumed time to achieve the tasks and the interaction between the agents that control the things.

Table 3. The preliminary results of the simulation

ID	t_1	t_2	Time (s)	Average (s)	Send	Receive
1	15	242	3518	14.08	519	1557
2	141	–	685	5.25	577	–
3	257	–	120554	469.08	1013	–
4	216	–	755785	6515.38	831	–
5	50	–	613	12.26	475	–
6	40	217	565	2.19	514	398
7	191	257	965739	3758.13	448	1033
8	85	–	399	5.09	617	–
9	75	–	35227	470.19	592	–
10	32	–	1321	41.28	284	326
11	65	–	1143	17.58	549	–
12	52	–	51722	995.05	451	416
13	142	–	147	1.035	176	213
14	13	–	312	25.15	195	288

The agents complete 2090 tasks with average activities recognition accuracy of 96.97% in which 1374 tasks belong to t_1 that are used for operating the things and 716 tasks belong to t_2 that are used for tracking the elderly activities. The consumed time to achieve each task is related to the functional properties of the things and the tracking process. The send and receive functions are performed according to the things functional properties, autonomy levels and the needs of tasks achievement.

5 Conclusion and Future Work

In IoT environments such as smart home, smart things are autonomously working together to achieve common goals. The global aim of such autonomous environments is to provide means that facilitate humans' life. The smart home systems raise some usability and safety concerns as humans are essential parts of these systems. These concerns impose the need for global control. Amending the performance of smart things can reduce such concerns, increase confidence and reset preferences. It requires modelling a flexible IoT architecture that facilitates global control over multiple things without disturbance to the system functions. The literature proposes a self-regulated multi-agent system or conventional human-agent interaction methods to maintain global control. In this paper, we use the concept of adjustable autonomous to model an AAMA-IoT system's architecture that resolves a number of IoT global control challenges. The system is simulated and tested in an elderly smart home environment. This result validates the applicability of the proposed architecture in which the agents successfully operate the things. The average activities recognition accuracy score of the simulated system is 96.97% which is considered very high. Future work considers including elderly healthcare settings in the system's functionality. Elderly usually follow a certain daily routine and they do not change this routine unless there are a physical or mantel needs such as disability. The simulation should periodically change the autonomy setting according to the elderly health conditions. Additional work can be including adjustable security functions of a smart home based on some security conditions.

Acknowledgements. This project is partially sponsored by University Tenaga Nasional (UNITEN) under the UNIIG Grant Scheme No. J510050772. It is also supported by Universiti Tun Hussein Onn Malaysia (UTHM) under the Postdoctoral D004 grant.

References

1. Allied Business Intelligence: More Than 30 Billion Devices Will Wirelessly Connect to the Internet of Everything in 2020. Allied Business Intelligence (ABI) Research, New York (2013). Accessed 2 July 2017
2. Mzahm, A.M., Ahmad, M.S., Tang, A.Y.: Agents of things (AoT): An intelligent operational concept of the internet of things (IoT). In: 2013 13th International Conference on Intelligent Systems Design and Applications (ISDA), pp. 159–164. IEEE (2013)
3. Elkhodr, M., Shahrestani, S., Cheung, H.: A contextual-adaptive location disclosure agent for general devices in the internet of things. In: 2013 IEEE 38th Conference on Local Computer Networks Workshops (LCN Workshops), October 2013, pp. 848–855. IEEE
4. Mostafa, S.A., Ahmad, M.S., Tang, A.Y., Ahmad, A., Annamalai, M., Mustapha, A.: Agent's autonomy adjustment via situation awareness. In: Intelligent Information and Database Systems, pp. 443–453. Springer International Publishing, Switzerland (2014)
5. Mostafa, S.A., Ahmad, M.S., Ahmad, A., Annamalai, M., Gunasekaran, S.S.: An autonomy viability assessment matrix for agent-based autonomous systems. In: 2015 International Symposium on Agents, Multi-Agent Systems and Robotics (ISAMSR), IEEE, pp. 53–58, August 2015
6. Mostafa, S.A., Ahmad, M.S., Mustapha, A.: Adjustable autonomy: a systematic literature review. Artif. Intell. Rev 1–38 (2017)
7. Mostafa, S.A., Mustapha, A., Mohammed, M.A., Ahmad, M.S., Mahmoud, M.A.: A fuzzy logic control in adjustable autonomy of a multi-agent system for an automated elderly movement monitoring application. Int. J. Med. Inform. **112**, 173–184 (2018)
8. Flemisch, F., Heesen, M., Hesse, T., Kelsch, J., Schieben, A., Beller, J.: Towards a dynamic balance between humans and automation: authority, ability, responsibility and control in shared and cooperative control situations. Cogn. Technol. Work **14**(1), 3–18 (2012)
9. Mostafa, S.A., Ahmad, M.S., Mustapha, A., Mohammed, M.A.: Formulating layered adjustable autonomy for unmanned aerial vehicles. Int. J. Intell. Comput. Cybern. **10**(4), 430–450 (2017)
10. Mostafa, S.A., Mustapha, A., Hazeem, A.A., Khaleefah, S.H., Mohammed, M.A.: An agent-based inference engine for efficient and reliable automated car failure diagnosis assistance. IEEE Access **6**, 8322–8331 (2018)
11. Mostafa, S.A., Ahmad, M.S., Annamalai, M., Ahmad, A., Basheer, G.S.: A layered adjustable autonomy approach for dynamic autonomy distribution. In: Frontiers in Artificial Intelligence and Applications, pp. 335–345. IOS Press (2013)
12. Mostafa, S.A., Ahmad, M.S., Ahmad, A., Annamalai, M., Gunasekaran, S.S.: A flexible human-agent interaction model for supervised autonomous systems. In: 2016 2nd International Symposium on Agent, Multi-Agent Systems and Robotics (ISAMSR), August 2016, pp. 106–111. IEEE

13. Mostafa, S.A., Mustapha, A., Ahmad, M.S., Mahmoud, M.A.: An adjustable autonomy management module for multi-agent systems. Procedia Comput. Sci. **124**, 125–133 (2017)
14. Mostafa, S.A., Darman, R., Khaleefah, S.H., Mustapha, A., Abdullah, N., Hafit, H.: A general framework for formulating adjustable autonomy of multi-agent systems by fuzzy logic. In: Smart Innovation, Systems and Technologies, pp. 23–33. Springer, Cham (2018)
15. Laghari, S., Niazi, M.A.: Modeling the internet of things, self-organizing and other complex adaptive communication networks: a cognitive agent-based computing approach. PLoS ONE **11**(1), e0146760 (2016)
16. Jie, Y., Pei, J.Y., Jun, L., Yun, G., Wei, X.: Smart home system based on IoT technologies. In: 2013 Fifth International Conference on Computational and Information Sciences (ICCIS), pp. 1789–1791. IEEE (2013)
17. Abbas, H., Shaheen, S., Elhoseny, M., Singh, A.K., Alkhambashi, M.: Systems thinking for developing sustainable complex smart cities based on self-regulated agent systems and fog computing. Sustain. Comput. Inform. Syst. **19**, 204–213 (2018)
18. Godfrey, W.W., Jha, S.S., Nair, S.B.: On a mobile agent framework for an internet of things. In: 2013 International Conference on Communication Systems and Network Technologies (CSNT), pp. 345–350. IEEE (2013)
19. Al-Fuqaha, A., Guizani, M., Mohammadi, M., Aledhari, M., Ayyash, M.: Internet of things: a survey on enabling technologies, protocols, and applications. IEEE Commun. Surv. Tutorials **17**(4), 2347–2376 (2015)
20. Synnott, J., Nugent, C., Jeffers, P.: Simulation of smart home activity datasets. Sensors **15**(6), 14162–14179 (2015)
21. Van Kasteren, T.: Datasets for activity recognition. https://sites.google.com/site/tim0306/datasets. Accessed June 2017
22. Van Kasteren, T., Noulas, A., Englebienne, G., Kröse, B.: Accurate activity recognition in a home setting. In: Proceedings of the 10th International Conference on Ubiquitous Computing, pp. 1–9. ACM (2008)

Energy-Efficient Analysis in Wireless Sensor Networks Applied to Routing Techniques for Internet of Things

Carolina Del-Valle-Soto[✉], Gabriela Durán-Aguilar,
Fabiola Cortes-Chavez, and Alberto Rossa-Sierra

Facultad de Ingeniería, Universidad Panamericana, Álvaro del Portillo 49,
45010 Zapopan, Jalisco, México
{cvalle, gaduran, fcortes, lurosa}@up.edu.mx

Abstract. Imagine being able to connect to the work network from a public park and then meet a friend for coffee or shopping. Imagine finding everything a tourist needs, such as bus schedules, nearby restaurants and other entertainment options in touch screen kiosks conveniently located throughout the city. Most applications of IoT depend on a battery for their operation and they are designed to reduce or even avoid the human intervention in the sensing process. Most IoT projects are motivated by a need to reduce operating energy costs or increase revenue. This paper presents and analyses the energy model of a wireless sensor using four different routing protocols: Multi-Parent Hierarchical (MPH), On Demand Distance Vector (AODV), Dynamic Source Routing (DSR) and Zig-Bee Tree Routing (ZTR). In these applications, the energy consumption is a key factor, sensors can be located in remote zones difficult to access, so it is not possible to replace the battery continuously. Due to the limitations of battery life, the nodes are designed to save as much energy as possible, and most of the time they are in sleep mode (low power consumption). Finding energy sources for difficult-to-connect device has become a priority for technology, in large part due to the rise of the IoT concept. This is why the network itself must provide energy saving mechanisms and a good solution could be in charge of the packets administration in the network. required format.

Keywords: Wireless Sensor Networks · Energy consumption ·
Performance metrics · Routing protocol · Internet of things

1 Introduction

Within the automation processes, sensors play a fundamental role given that they are responsible for providing information of the environment. That will be used for decision making, taking into account this, it is vitally important that sensor measurements must be reliable, and all of them are considered the external elements that may affect the quality of measurements. Wireless Sensor Networks (WSNs) are based on low-cost devices (nodes) that are able to get information from their environment, process it locally, and communicate via wireless links to a central coordinator node. Additionally, the coordinator node might also send control commands to the nodes [1].

© Springer Nature Switzerland AG 2020
H. Ayaz (Ed.): AHFE 2019, AISC 953, pp. 312–321, 2020.
https://doi.org/10.1007/978-3-030-20473-0_30

WSNs may not rely on a predetermined structure and require the capacity of self-organization in order to deal with communications impairments, mobility and node failures. Moreover, it is important to study the scalability and adaptation methods of the network in the face of topology changes and packet transmission failures in the wireless medium. Our study presents scenarios sharing information among devices within wireless environment, it defines and analyses different kind of performance metrics, such as: received or lost packets, retransmissions, overhead, battery life, network lifetime and, so on. In addition, it describes a specific atmosphere with time and space conditions to be monitored under a given statistical model for sending packets in four protocols under study: two of them proactive and the others reactive. Our proposal is to classify the metrics to suggest the relationships among them and study their influence on routing protocols and how impact them on energy. Furthermore, we intend to analyze a real scenario with another energy model in order to prove our best two proactive protocols. These tests are given to observe energy in our simulator and a real scenario with CC2650 sensors in the laboratory with an error of only 2% of difference between our tool and the scenario.

Figure 1 describes the possible sensors with which the human being could interact in their daily life, such as reading their pulse when exercising through a smart watch, monitoring their health with a smart phone, as well as using printers or computers as a tool for work or leisure by means of electronic tablets. Moreover, scenarios of wireless sensor networks are proposed under different configurations of topology arrays. The aim is to contrast the performance of the sensor network under three widely known protocols in the literature: AODV [2], DSR [3] and MPH [4]. The latter was designed and implemented in the reference cited in [4]. In this study, AODV, DSR, ZTR [5] and MPH are compared based on various efficiency metrics and how they optimize routing protocols through energy. There are several schemes to find the best routes in the shortest possible time. In terms of hierarchy algorithms, such as ZTR, it has a simple and fast routing, which reduces overload in the network, is reliable and has a distributed addressing scheme that does not require nodes to have routing tables. Results from the work [4] shows that for the single sink scenario, the MPH protocol has an energy saving of 35% against AODV and DSR protocols and 8% compared with ZTR. MPH has 27% less overhead compared with AODV and DSR. Additionally, MPH presents a 10% increase in packet delivery compared with AODV, DSR and ZTR. Resilience is better in MPH for 26% regarding AODV and DSR.

2 Routing Protocols in WSNs

In communication networks, there are routing protocols classified into two groups: proactive routing protocols and reactive routing protocols. When nodes are under a reactive protocol, they ask for a route only when it is needed. This involves high latency for the first packet and some independence among routes. The AODV routing protocol is based on routing efficiency of wireless ad-hoc networks with a huge number of nodes and it uses a route discovery mechanism in broadcast mode. It is considered as a reactive protocol: the routes are created only when they are needed, on demand. AODV uses the bandwidth efficiently and responds to the network changes in a very

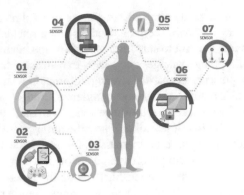

Fig. 1. Distribution of sensors in a real scenario.

quick mode, preventing network loops [2]. Some advantages of AODV include more reliability and less cost in bandwidth. However, there are some disadvantages, as follows: more complexity and computing, more cost in memory, and this protocol was designed to work in a network where there are no malicious nodes. In conclusion, it is not a secure protocol. The DSR protocol is a reactive protocol. Furthermore, the protocol adapts itself quickly to routing changes when a node is frequently moving, and finally, this protocol decreases the overhead in the network [6]. ZTR is a simple protocol which establishes parent-child links and the nodes always carry information to their parent. It has a tree topology and is easy to implement. Some of the advantages of ZTR are that in the algorithm implemented in the network layer, there is a balance between cost per unit, battery expense, complexity of implementation to achieve a proper cost-performance relation to the application [7]. MPH creates a hierarchical network logical topology where the hierarchy of the nodes is given by its location level, which is proactive. It works like a hierarchical tree: nodes establish parent and child links that constitute the possible routes. This protocol takes advantage of the controlled maintenance of routes of the proactive nature, but combines the agility that allows to have more than one route for a node. This makes it more versatile and adaptable to different topologies [4].

3   Study of Energy in WSNs

Wireless Sensor Networks differ from other wireless networks because they are formed by low-cost and low-processing sensors that send information to a collector or base station node. Due to the small size of the nodes in WSNs, saving energy consumption is vital, as it is very difficult to recharge and aims to achieve maximum efficiency in the delivery of information. The problem of energy consumption is not the same way in all network nodes. When there are one or several collectors nodes, and close to these nodes there are other nodes that forward the whole network traffic, they are more likely to exhaust their energy faster. This problem is known as energy hole problem. Uneven depletion of energy causes the expiration time unexpectedly creating network information loss. One of the possible solutions to this problem that is mentioned in the literature is the creation of clusters in order to promote the network scalability and zoning problem. Nevertheless, there are different definitions and concepts regarding lifetime [8]. The analysis of areas showing the network is distributed in concentric rings to pace traffic across nodes as they approach their final destination. There are local and global parameters to dis-aggregate energy consumption. Global parameters show what energy cost is throughout the entire network, taking into account each one of the types of energy for each specific activity. On the other hand, the local aspect refers to the total energy consumption in a single node. This total energy depends on the location of the node in the topology, if it is close or far to the coordinator node, and how much traffic is forwarded through the node. We make the analysis of the proposed energy model for our studied protocols: AODV, DSR, ZTR and MPH.

4 Performance Metrics Study

Metrics of the network layer are very important because they show the performance and usefulness of a routing protocol. Each routing protocol is designed for specific applications and certain scenarios. These metrics indicate how the use of bandwidth is affected by the overhead of the routing protocol in use. In addition, the availability of effective routes and the ability of the network for self-configuration show the capacity of the protocol to recover from topology changes. Recovery times have an impact on the latency in the network and even though the networks conform with different technologies, it is highly important to understand and evaluate the performance metrics as shown in [9]. This analysis allows customization of the network to improve its different aspects and provide better communication. Our event-driven simulator, programmed under C++ language, has three important parts: Physical layer, MAC layer, Network layer, and energy model. Furthermore, for the energy model we take into account the behavior of CC2530 [10] sensors, from Texas Instruments. Then, for a real scenario, we test another energy model applied to CC2650 [11] sensors, form Texas Instruments in order to compare the MPH performance regarding the other afore-mentioned protocols.

5 Analysis of Performance Metric Results Under a Grid Topology

WSNs are multi-functional, low-cost and low-power networks, and rely on communications among nodes or from sensor nodes to one or more sink nodes. Sink nodes, sometimes called coordinator nodes or root nodes, may be more robust and have larger processing capacity than the other nodes. Sensor networks can be widely used in various environments, sometimes hostile. Some of the many applications of WSNs are in the medical field, agriculture, monitoring and detection, automation and data mining. We compare the performance of the MPH protocol with that of commonly used protocols in sensor networks, such as AODV, DSR and the well-known algorithm, ZTR.

We consider the following important metrics that are indicative of network performance and they are tested under the topology described in Fig. 2.

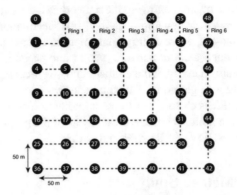

Fig. 2. Network topology.

Conversely, Table 1 describes the parameters of the simulations.

Table 1. Simulation and real network parameters. CSMA/CA, carrier sense multiple access with collision avoidance [12].

Parameter	Value
Physical Layer Parameters	
Sensitivity threshold receiver	−94 dBm
Transmission power	4.5 dBm
MAC Layer Parameters	
Maximum retransmission number	3
Maximum retry number	5
Maximum number of tries to reach a node from the collector	9
Packet error rate	1%
Average error rate	22 bytes
Maximum number of backoffs	4
MAC protocol	IEEE 802.15.4
MAC layer	CSMA/CA
Network Layer Parameters	
Number of nodes	49
Maximum data rate	250 kbps
Scenario	Static nodes

The energy model implemented for the studied protocols is presented in Table 2.
Next we present Table 3, that is a qualitative analysis of how impact the performance metrics previously mentioned in energy for each of the four routing protocols.

Table 2. Energy model.

	Voltage (mV)	Current (mA)	Time (ms)
Start-up mode	120	12	0.2
MCU running on 32-MHz clock	75	7.5	1.7
CSMA/CA algorithm	270	27	1.068
Switch from RX to TX	140	14	0.2
Switch from TX to RX	250	25	0.2
Radio in RX mode (processing and waiting)	250	25	4.1915
Radio in TX mode	320	32	0.58
Shutdown mode	75	7.5	2.5

This analysis allows to visualize which metrics have a greater or lower impact on energy, according to each protocol. For example, the response time is higher in AODV and DSR, if we take an average between the 4 protocols. The response time of AODV and DSR would be above this average, but the measure of the AODV is higher than DSR. Now, the energy impact of this performance metric in MPH is lower than the average, which is good, but it is even better (much lower) in ZTR. This shows that the characteristics of MPH and ZTR are quite similar, with the difference that MPH is a complete protocol with better redundancy and routing than ZTR, but with minimal energy consumption. We observe that ZTR and MPH have a low impact of metrics on energy, however, ZTR has 78% of the metrics at the minimum impact. Moreover, MPH is very close to this value with 67% of metrics with minimal impact on energy. This is good news because MPH is a complete protocol, with redundancy and hierarchy in its design, which favors the response reduction times and less information loss.

Table 3. Energy behavior per performance metric.

Metric	Energy			
	AODV	DSR	ZTR	MPH
Response time	↑↑↑	↑↑	↓↓	↓
Network lifetime	↑↑↑	↑↑	↓↓	↓↓
Overhead	↑↑↑	↑↑↑	↓	↓↓
Battery life	↑↑↑	↑↑↑	↓	↓
Complexity and cost	↑↑↑	↑↑↑	↓	↓↓
Channel retries	↑↑↑	↑↑↑	↓	↓
Packet retransmissions	↑↑↑	↑↑↑	↓	↓
Packet loss	↑↑	↑↑	↓	↓
Resilience	↑↑	↑↑	↓	↓

↑↑↑: Very above of energy average.
↑↑: Above of energy average.
↓↓: Very below of energy average.
↓: Below of energy average.

Figure 3 shows the results of the four routing protocols using nine different metrics. The main results are summarized below: The MPH and ZTR protocols have a similar response time throughout the simulation; their values vary between 0.2 s and 0.3 s. This is three times less than the protocol with the highest response time, the AODV protocol. This indicates that packets among the different nodes are sent three times faster in ZTR and MPH than in AODV. This can be caused due to the design simplicity and hierarchical topology that ZTR and MPH have. Similar to the response time, when overhead is analyzed, ZTR and MPH have a similar behavior with an average overhead of 21% and 27% respectively. On the other hand, DSR and AODV protocols have an average overhead of 41% and 43%, this means that DSR and AODV protocols send more packets to keep the network running, that is, more control packets. This feature might cause more collisions with traffic packets. One of the most important parameters is the energy consumption; this can be analyzed using two metrics: network lifetime and battery life. On one hand, network lifetime indicates the useful lifetime of the entire network, so at second 100 the AODV and DSR protocols have almost spent all their lifetime with 80% used, the double, compared with the MPH and ZTR. On the other hand, the results show an average battery percentage used by each node in the network, again, the AODV and DSR protocols have a similar behavior, however the nodes of these networks spend twice as much battery life as compared with the protocol with the best performance, ZTR.

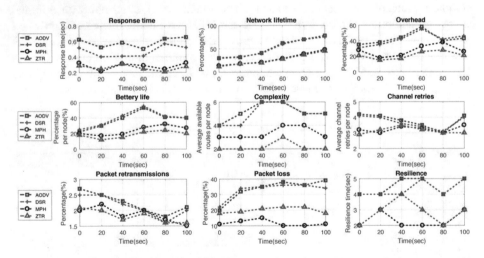

Fig. 3. Performance metrics for proactive and reactive protocols.

6   Comparison with a Real Scenario

We implemented a real scenario shown in Fig. 4, on the left we have the proposed implementation, and on the right we observe our real scenario of sensors. This scenario consists of 4 nodes located in star topology and a coordinator (5 nodes in total).

Regarding the sensors we implemented CC2650 from Texas Instruments. We ran a simulation of 100 s with the conditions specified in the Table 1.

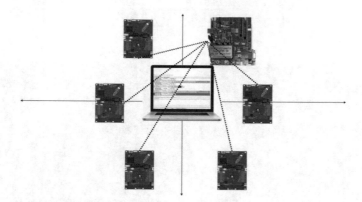

Fig. 4. Proposed implementation with the mentioned sensors.

The energy model [13] implemented for the four protocols is presented in Table 5. We design an energy model based on the main modes of operation of the sensors with respect to the activities that a node performs in the network (Table 4).

Table 4. CC2650 energy model with a DC power supply of 2 V.

	Current (mA)	Time (µs)
Start-up mode	3.26	1160
MCU running on 32-MHz clock	1.45	5
Radio preparation	4.30	101
Switch from RX to TX	3.43	370
Switch from TX to RX	4.66	112
Radio in RX mode	6.47	184
Radio in TX mode	7.47	168
Shut down mode	2.45	685

The model described in Table 5 was applied to the simulator with the same number of sensors described in scenario shown in Figs. 4 and 5 and was tested under the MPH and ZTR protocols. Subsequently, the same simulation scenario is actually implemented with CC2650 sensors and code conditions were modified in order to implement the MPH protocol and the Zigbee algorithm, which would represent ZTR.

The results show that the MPH protocol has an energy waste very similar to the Zigbee algorithm, which is a simple and easy algorithm to implement, but with the disadvantage that it has little redundancy and scalability. Therefore, the fact that ZTR spends only 6% less energy than MPH is a big advance, because MPH is a complete

Table 5. Energy comparison.

	ZTR energy (J)	MPH energy (J)
Simulation	0.009279	0.009834
Real scenario	0.009165	0.009745

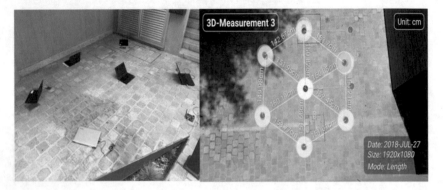

Fig. 5. Our real implemented scenario of sensors

routing protocol with greater redundancy and scalability among other feature to control packet loss.

7   Conclusion

IoT enables physical devices or sensors to measure, perform a defined task, use the cloud for storage and to actuate the alert system automatically in case of an emergency situation with the aid of Internet as its underlying technology. Thus, IoT transforms these traditional devices to work in a smart way by using various deriving technologies such as pervasive computing, embedded devices, various communication standards and technologies, Internet protocols, and various application services. The results for MPH protocol are encouraging because this protocol has good performance in terms of processing, fast and efficient information delivery and energy conservation. Protocols such as AODV and DSR are very efficient in terms of backup routes and connectivity from any node to any node in the network. ZTR is a simple and low energy cost algorithm, but it is not very reliable in adverse network conditions or failure on the links. The combination of a hierarchical topology with self-configuration and maintenance mechanisms of the MPH protocol makes the nodes optimize network processes, reduce delays, take short routes to the destination and decrease network overhead. All this is reflected in the successful delivery of information and lower energy consumption.

References

1. Yigitel, M.A., Incel, O.D., Ersoy, C.: QoS-aware MAC protocols for Wireless Sensor Networks: a survey. Comput. Netw. **55**(8), 1982–2004 (2011)
2. Perkins, C.E., Royer, E.M.: Ad-hoc on-demand distance vector routing. In: Proceedings of 2nd IEEE Workshop on Mobile Computing Systems and Applications, WMCSA 1999, pp. 90–100 (1999). Cited By: 5922. www.scopus.com
3. Maltz, D.A., Broch, J., Jetcheva, J., Johnson, D.B.: Effects of on-demand behavior in routing protocols for multihop wireless ad hoc networks. IEEE J. Sel. Areas Commun. **17**(8), 1439–1453 (1999)
4. Del-Valle Soto, C., Mex Perera, C., Orozco Lugo, A., Galvan Tejada, G.M., Olmedo, O., Lara, M.: An efficient Multi-Parent Hierarchical routing protocol for WSNs. In: Wireless Telecommunications Symposium (2014). Cited By: 5
5. Alliance, Z.: ZigBee specification (document 053474r17), vol. 21, January 2008
6. Johnson, D.B., Maltz, D.A.: Dynamic source routing in ad hoc wireless networks. In: Mobile Computing, pp. 153–181. Springer (1996)
7. Wadhwa, L., Deshpande, R.S., Priye, V.: Extended shortcut tree routing for zigbee based wireless sensor network. Ad Hoc Netw. **37**, 295–300 (2016)
8. Han, G., Liu, L., Jiang, J., Shu, L., Hancke, G.: Analysis of energy-efficient connected target coverage algorithms for industrial wireless sensor networks. IEEE Trans. Industr. Inf. **13**(1), 135–143 (2017)
9. Fauzia, S., Fatima, K.: Performance evaluation of AODV routing protocol for free space optical mobile Ad-Hoc networks. In: Advances in Intelligent Systems and Computing, vol. 683 (2018)
10. Texas Instruments: CC2530 data sheet (2009)
11. Texas Instruments: Multi-standard CC2650 SensorTag Design Guide. Texas Instruments Incorporated, Dallas (2015)
12. IEEE: Wireless medium access control (MAC) and physical layer (PHY) specifications for low-rate wireless personal area networks (WPANs). IEEE Standard 802.15.4-2006. IEEE Computer Society, New York (Revision of IEEE Standard 802.15.4-2003) (2006)
13. Kamath, S., Lindh, J.: Measuring bluetooth low energy power consumption. Texas instruments application note AN092, Dallas (2010)

Internet of Things: Analyzing the Impact on Businesses and Customers

Mandeep Pannu[✉] and Iain Kay

Department of Computer Science and Information Technology,
Kwantlen Polytechnic University, Surrey, BC, Canada
{mandeep.pannu,iain.kay}@kpu.ca

Abstract. The Internet of Things (IoT) refers to intelligently connected smart devices and systems using embedded technology software and sensors to communicate, collect and exchange data with one another. According to analyst firm Gartner, there will be 20 billion connected devices to be deployed by 2020. The IoT encompasses sensor, actuators, electronic processing, microcontrollers, embedded software, communications services and information services associated with the things. The focus of this research is to find out the impact of IoT on businesses and their customers and design a prototype system that will assist users to make informed decision and protect their IoT devices. IoT is a collection of fragmented and complicated data from different devices. Data are collected from and shared between devices to gather information for various uses, such as improving customer relationships and identifying customer segments. This research will use quantitative research methods to collect data from selected companies that utilize IoT. By analyzing the data collected, we can identify and define the requirements to design a prototype system. The prototype system will collect information from IoT devices, perform critical data analysis, and provide insightful information to help businesses make decision about IoT future.

Keywords: Internet of Things (IoT) · Security+

1 Introduction

The Internet of Things (henceforth known as IoT) refers to intelligently connected smart devices and systems using embedded technology, software, and sensors to communicate, collect, and exchange data with the user and one another. According to the Gartner analyst firm, there will be 20 billion connected devices to be deployed by 2020 [1, 2]. However, these increasingly connected devices bring security vulnerabilities to aspects of the home and business that have never before experienced digital attacks. These could be IoT connected door locks that leak network passwords or IoT coffee makers that can then be set to make coffee from outside one's home network.

These problems arise mainly from earlier IoT devices where security was either an afterthought, or simply not thought of at all. There exists a large number of legacy IoT devices that have little to no security. In addition, these legacy IoT devices can be thought of as incapable of having security implemented due to limited amounts of

© Springer Nature Switzerland AG 2020
H. Ayaz (Ed.): AHFE 2019, AISC 953, pp. 322–327, 2020.
https://doi.org/10.1007/978-3-030-20473-0_31

storage, memory, and processing power [3]. This research will use quantitative research methods to collect data from selected companies that utilize IoT. By analyzing the data collected, we can identify and define the requirements of a prototype system and create it. This prototype system will be connected to a local network, sitting in-between network IoT devices and the network, to allow it to encrypt otherwise unencrypted information meant for outside world before tunneling it to an external Content Delivery Network (henceforth known as "CDN") for forwarding. It will also allow the user to control access to the IoT device from the local network. This system will allow otherwise unsecured legacy IoT devices to have secure external communication independent of software updates and hardware that may or may not be provided by the manufacturer.

In layman's terms, most smart appliances that connect to the internet, and other devices, are unsecure, opening up home networks to attack from something as simple as someone remotely connecting to a refrigerator. By placing a buffer between the smart fridge and both the internet and other devices, users can ensure the security of their networks. The buffer would be an IoT secure hub that acts as a firewall for unsecured devices to connect to the wider internet without opening themselves to infiltration (Fig. 1).

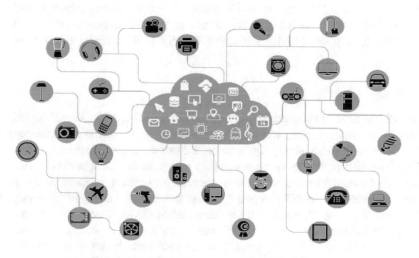

Fig. 1. An illustration of common IoT devices

2 Background

When we talk about IoT security, we're really talking about minimizing the vulnerable attack surfaces. In this instance, it is the network communication that presents the largest attack surface for all modern and legacy IoT devices. This network communication can further be broken down into two types: internal and external. This research will focus on the securing of both parts of this network communication, as in legacy IoT devices, this represents the bulk of existing attacks and exposure. An example would be the Haier SmartCare Device which runs a telnet server open to the local network [4]. Indeed, the popular hacker search engine, Shodan, not only indexes all of these unsecured IoT devices, but makes that data available for anyone to see. This service works because these external connections are available with either little, or no security. This problem is hard to solve for an end user. Normally fixing such a vulnerability would require a software update to the IoT devices, changing the way it encrypts its data, or provides its services externally.

These threats are not just hypothesized, but already being seen in the wild. In 2017, a casino was successfully hacked, with over 10 gigabytes of private information being leaked out through an IoT connected fish tank [5]. IoT devices are also used to overwhelm internet services. According to NexusGuard report, the average and maximum size of attacks coming from the IoT's have gone from 4.10 Gbps and 63.70 Gbps respectively in Q2 2017 to 26 Gbps and 359 Gbps respectively in Q2 2018, which is approximately a 500% increase [6].

3   Proposed Design

The proposed design is characterized by two domains. "Local" being the private network on which the IoT is located, and "external" being the network through which you access the IoT through the internet. For simplicity, we will assume local network is the home network of a user, and the external network is the internet beyond the home network.

3.1   Local

Local IoT security has three primary concerns. These are network infiltration, using IoT devices as an attack vector, and tracking user activity. Unsecure network communication can be monitored locally and used to infiltrate the local network using tools such as Wireshark if a malicious actor has gained access to the local network. This information can be used to track all network requests and their responses, and even send unauthorized requests to IoT devices. IoTs can also be used as an attack vector. They can be used to spy on the users that are connected to the local network. The malicious actor can then gather intel based on IoT sensor data for use in unscrupulous activities.

To solve these problems, we would provide a network device which acts as a pass-through and be the central hub to access the IoT devices. The IoT's will connect to this device.

To be able to use and access the IoT from the local network, users will have to allow their devices in the Access Control List (ACL) which is set on incoming connections to the IoT. This allows us to create an extra layer of security when accessing the IoT. When an unauthorized third-party try to access an IoT, the ACL triggers an alert which will then send a notification to the owner of the IoT device.

Suppose a user wants to connect to his IoT device from his home computer. First, it would check to ensure that his computer is approved on the Access Control List, then it would forward the traffic to the IoT device normally (Fig. 2).

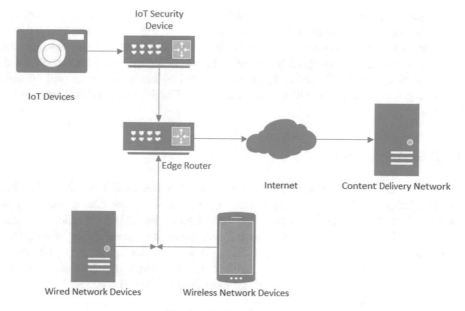

IoT Security
Device

IoT Devices

Edge Router

Internet   Content Delivery Network

Wired Network Devices   Wireless Network Devices

Fig. 2. Proposed approach

3.2   External

IoT's have the same security concerns as local networks. These apply to IoT's that are connected to the internet and send unencrypted data, or even encrypted data in some cases, over the internet. As IoT's are low-powered devices, the encrypted, when applied to external communications, is sometimes not strong enough and can be broken easily.

These problems can be solved by using the same network device in the same manner. Any communication with the IoT destined for the internet is intercepted by the network device and sent securely to a CDN. The routing of the external traffic proceeds normally after it has reached the CDN.

Suppose an IoT device is attempting to push sensor data or publish a camera feed outside the home network. This data would be intercepted by the network device, encrypted, and tunneled out of the network to the CDN for final delivery. Any device or service that wishes to access this data or view the camera feed would need to connect to the CDN through whatever security suite is being used to secure the connection with the IoT, such as an Access Control List, Country Block, or Web Application Firewall.

3.3   General Features

While these internet and external network communications are occurring, the network device also works to track traffic statistics, as well as sending notifications when it see's unauthorized access attempts or new MAC addresses attempting communication. These services will be available through a simple web interface, allowing the user to view stats, manage the Access Control List, and setup encryption keys.

Through this system we will enable modern security features on legacy IoT devices by routing their otherwise unsecured traffic though our custom security layer. All that is required is for end users to connect their IoT devices to the network through the network device access point, whitelist their devices on the internal access control list, and subscribe to a CDN service to secure external connections. Users will be able effectively track, monitor and block access to their IoT inside and outside their home or business.

4   Recommendations

The biggest obstacle faced is that of user apathy [7]. While an IoT secure device hub would almost certainly guarantee device security, most users aren't interested in securing their devices until it's too late. According to the official McAfee Business twitter [1], 43% of users claim that device security is the responsibility of manufacturers, with an additional 26% who believe device security is the responsibility of someone other than themselves. To ensure adequate utilization and dissemination, ISP providers should embed the device within their product from the outset.

5   Conclusion

IoT security plays an increasingly important role in today's world. Unsecured IoT devices are being used for diverse malicious activities including DDoS attacks, infiltration, and unauthorized user tracking.

Traditionally, it has been hard to secure legacy IoT devices, which amount to a huge number of the total active IoT devices, as they lack tough encryption and strict security policies.

The proposed network device provides an approach to secure unencrypted communications and also limit access to the connected IoT devices. It puts the control of IoT devices in the owner's hands.

As such, the development and implementation of a system, such as the one described herein, adds a level of security previously unavailable in this area of computing. The development of this system will be well rewarded through the increased security of the IoT devices, which in turn will be a big deterrent for any malicious agent.

References

1. McAfee Business whose job do you think it is to ensure they're properly secured (2019). https://twitter.com/McAfee_Business/status/1096448932899483648
2. Van der Meulen, R.: Gartner Says 8.4 Billion Connected "Things" (2017). https://www.gartner.com/en/newsroom/press-releases/2017-02-07-gartner-says-8-billion-connected-things-will-be-in-use-in-2017-up-31-percent-from-2016

3. Gerber, A.: Top 10 IoT Security Challenges (2017). https://developer.ibm.com/articles/iot-top-10-iot-security-challenges/. Accessed Mar 2019
4. Wurm, J., Hoang, K., Arias, O., Sadeghi, A. R., Jin, Y.: Security analysis on consumer and industrial IoT devices. In: Report Published in 21st Asia and South Pacific Design Automation Conference (2016)
5. Schiffer, A.: How a fish tank helped hack a Casino (2017). The Washington Post. https://www.washingtonpost.com/news/innovations/wp/2017/07/21/how-a-fish-tank-helped-hack-a-casino/?utm_term=.62a1ae94d829. Accessed Mar 2019
6. Miu, T., Yeung, R., Li, D.: DDoS Threat Report Q2 2018. Report published by NexusGuard (2018)
7. Rayome, A.D.: CES 2019: 58% of Consumers Don't Secure Their Personal Devices (2019). https://www.techrepublic.com/article/ces-2019-58-of-consumers-dont-secure-their-personal-devices/

Brain - Machine Interface and Artificial Intelligence Systems

An Application of Extended Kalman Filter for the Pressure Estimation in Minimally Invasive Surgery

Van-Muot Nguyen[✉], Eike Christian Smolinski,
Alexander Benkmann, Wolfgang Drewelow, and Torsten Jeinsch

Institute of Automation, University of Rostock, Richard-Wagner-Str. 31/Haus 8,
18119 Rostock, Germany
{van.nguyen, eike.smolinski, alexander.benkmann,
wolfgang.drewelow, torsten.jeinsch}@uni-rostock.de

Abstract. The paper presents an application of the Extended Kalman Filter (EKF) as an observer method to estimate the pressure in the operation area for the controlled process of minimally invasive surgery (MIS). Via trocars and pipes, the inflow and outflow of the rinsing fluid at the operation area are controlled by a double roller pump (DRP). Additionally, those flows are affected by the inside pressure of the pipes. The pressure sensor in the operation area is not allowed to utilize for surgery on the real patients at the current stage. Therefore, it is necessary to reconstruct the state of pressure in the operation area of MIS. The estimated pressure from the EKF estimator is used to replace the measured feedback signal to control the pump. The EKF worked based on the input signals of the rinsing fluid flows and the observable signal from available pressure sensors at the double roller pump. The proposed method was successfully implemented on MATLAB Simulink. For the further verification, it was also applied on the real device simulator environment. The results from the research show that the estimated pressure gives a high precision. In addition, the noises from the measured states are effectively eliminated.

Keywords: Extended Kalman Filter · Minimally invasive surgery ·
Double roller pump · Estimator

1 Introduction

Minimally invasive surgery (MIS) technique provides advantageous solutions to the patients in diagnosing and curing problems in parts of the body. By using special instruments and tiny tools for treatment, this technique is performed with small wounds on the operation field. Therefore, the problems of trauma, blood depletion and recovery time are reduced significantly [1–3]. A crucial thing is to remain a clear visibility

---

Van-Muot Nguyen is currently a Ph.D. Candidate at the University of Rostock, Germany. He has worked at Can Tho University, Can Tho city, Vietnam (corresponding author's phone number: +49-152-172-00738; fax: +49-381-498-7702).

© Springer Nature Switzerland AG 2020
H. Ayaz (Ed.): AHFE 2019, AISC 953, pp. 331–343, 2020.
https://doi.org/10.1007/978-3-030-20473-0_32

during MIS. In arthroscopy, rinsing fluid is provided into the operation areas like the knee, shoulder or hip joints for the visibility and expansion [2, 4–6]. This paper presents an investigation of the knee arthroscopy as an example of MIS. Figure 1 describes an overview of the system operation in the knee arthroscopy [2]. The DRP has been used as a medical therapy device to control the inflow automatically according to the outflow for the desired pressure in MIS. In order to minimize the problems of haemorrhages and fluid depletion, the pressure in the surgical region needs to be controlled close to the blood pressure of the patient body. The controller design was successfully investigated in the citation of [3] by measuring the pressure inside the simulator of the operation area for the feedback.

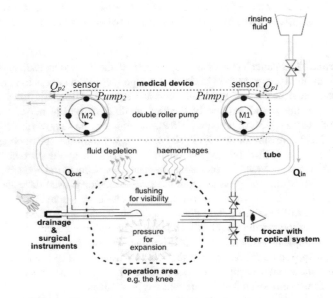

Fig. 1. System overview in minimally invasive surgery.

In the real-life operation of MIS, measuring the pressure inside the operation area is unacceptable for the safety of the patient. To deal with the problem of pressure control without using pressure sensor for the feedback, an algorithm of state estimation is proposed. The citation [7] presented the methods of Luenberger and basic Kalman Filter for the estimation. Those methods were implemented to the linearized process. However, the results from those estimators remained some error and noise which need to be more effectively reduced. Hence, with the nonlinear process of the knee arthroscopy simulator constructed for the research, a solution of Extended Kalman Filter (EKF) algorithm is implemented for the estimator. The EKF is functioned to calculate and update the filter gain vector recursively and infinitely. The filter gain vector is determined depending on both the state covariance matrix and the error noise covariance of the current measurement realized on the experiment [8]. The available pressure sensors at the DRP are used for the observation. The pressure in the operation area is estimated and compared to the desired pressure for evaluating the estimate error.

The estimate results are really meaningful for the combination of the observer as a feedback data to the sensorless controlled process in MIS.

The paper content is organized in the following sections. Section 2 describes an overview of the system and mathematical equations. Section 3 presents the EKF implementation. Section 4 shows the results and discussion. Section 5 is the conclusion.

2 System Overview

Figure 1 describes an overview of the operation in MIS, for example in the knee arthroscopy. Together with some other specialized surgical tools, trocars with fiber optical instruments are used to obtain the image transferation from the operational area of the body to the monitor. Rinsing fluid is transported from the tubes into the operation area for a suitable expansion and a clear visibility. The drainage instrument can be activated any time by a surgeon for the outflow of fluid during MIS.

The double roller pump is used to support the control of the rinsing fluid flows (called inflow Q_{in} and outflow Q_{out}) of the operation area. There are four roller wheels fixed on the shaft of each DC motor ($M1$ or $M2$) and the available pressure sensor at each roller pump of the DRP. One roller pump on the right (called pump1) with the motor $M1$ creates the pressure in the flexible tube for the direction of inflow to the operation area. The other roller pump on the left (called pump2) with the motor $M2$ produces the pressure for the outflow of the operation area. By changing the motor speed from the DC voltage input, the pressure at each roller pump is changed respectively. The change of pressure causes the flow of fluid in the tube.

2.1 Mathematical Equations of the System

The rotation speed of the motor $M1$ is defined as n_1; and n_2 is the rotation speed of the motor $M2$. Equations of flows at the DRP (Q_{p1} and Q_{p2}) are formed in (1) and (2). The relationship between the flow at the pump and motor speed is indicated in Fig. 2.

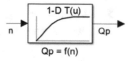

Fig. 2. The flow at the pump depending on the motor revolution

$$Q_{p1} = f(n_1). \tag{1}$$

$$Q_{p2} = f(n_2). \tag{2}$$

With the hydraulic capacity of the tube as well as the reaction of the current pressure inside the operation area, it contains a small difference of ΔQ_1 between the real flow of fluid (Q_{in}) into the operation area and the flow created at the pump1 (Q_{p1}). Similarly, in the inverse direction of the outflow, the difference of flow ΔQ_2 is created by the real value of Q_{out} and the flow at the pump2 Q_{p2}. In ideal case, the pressure in the surgical region is stable when the inflow Q_{in} is equal to the outflow Q_{out}.

The changes of the pressure ($\dot{p}_1$ and $\dot{p}_2$) in the tubes are defined in (3) and (4) whereas C_{tube} is represented to the hydraulic capacity of the tubes.

$$\dot{p}_1 = \frac{Q_{p1} - Q_{in}}{C_{tube}} = \frac{\Delta Q_1}{C_{tube}}. \tag{3}$$

$$\dot{p}_2 = \frac{Q_{p2} - Q_{out}}{C_{tube}} = \frac{\Delta Q_2}{C_{tube}}. \tag{4}$$

The process of the knee arthroscopy was proposed for the investigation. During the stage of validation and verification in experiments, the operation area was modelled by a simulator called the knee model. The pressure in the knee model is called p_{knee}. Any change of pressure $\dot{p}_{knee}$ in (5) depends on the flow of incompressible fluid inside the area, where C_{knee} is defined as a capacity of the knee model which describes the ability of the volume change via the pressure p_{knee}.

$$\dot{p}_{knee} = \frac{Q_{sum}}{C_{knee}}. \tag{5}$$

The value Q_{sum} in (6) describes the total capable flows of fluid in the knee model.

$$Q_{sum} = Q_{in} - Q_{out} + Q_{hae} - Q_{dep}. \tag{6}$$

Noting that Q_{hae} is represented to the flow of the haemorrhage when p_{knee} is lower than the blood pressure. Likewise, Q_{dep} is the flow of the fluid depletion which happens when p_{knee} is higher than the blood pressure. By controlling the pressure in the surgical area close to the blood pressure, the values of Q_{hae} and Q_{dep} can be minimized for the eliminations of the troubles in MIS. The value of $Q_{loss} = (Q_{dep} - Q_{hae})$ is considered to the loss of flow during MIS. Thus, the Eq. (6) can be performed as (7).

$$\begin{aligned} Q_{sum} &= Q_{in} - Q_{out} - Q_{loss} \\ &= (Q_{p1} - \Delta Q_1) - (Q_{p2} - \Delta Q_2) - Q_{loss} \\ &= Q_{p1} - Q_{p2} - (\Delta Q_1 - \Delta Q_2 + Q_{loss}). \end{aligned} \tag{7}$$

The total flow error Q_{err} defined in (8) is nonzero in the nonlinear process of MIS.

$$Q_{err} = \Delta Q_1 - \Delta Q_2 + Q_{loss}. \tag{8}$$

$$Q_{sum} = Q_{p1} - Q_{p2} - Q_{err}. \tag{9}$$

In addition, by following the ideal gas equation in (10), the product of the pressure p_{knee} and the volume of gas V_{gas} in the knee is proportional to the mass 'm' of the gas

$$p_{knee} V_{gas} = mRT. \tag{10}$$

$$p_{knee} = \frac{mRT}{V_{gas}}. \tag{11}$$

where R is the gas constant, and T is the temperature of gas.

By taking derivation on two sides, Eq. (10) can be rewritten as (12).

$$\dot{p}_{knee} V_{gas} + p_{knee} \dot{V}_{gas} = \dot{m}RT. \tag{12}$$

In the closed surgical area of MIS including some fluid and gas, it is considered to be no change of gas amount during MIS. This means that the mass 'm' of gas is a constant. The Eq. (12) can be thus rearranged into (13). This is clear that the change of pressure in the operation area is mutually depended on the change of gas volume, or the gas compression in the knee model.

$$\dot{p}_{knee} = -\frac{p_{knee}}{V_{gas}} \dot{V}_{gas}. \tag{13}$$

The gas volume is inversely proportional to the fluid volume indicated in (14).

$$V_{gas} = V_{total} - V_{fluid} = V_{total} - \left(V_0 + \int_0^t Q_{sum} dt \right). \tag{14}$$

where V_{total} and V_{fluid} represent to the total volume of the knee model and the current volume of fluid inside respectively. V_0 is the initial fluid volume in the knee model. V_{total} and V_0 are considered to constants. Hence, the derivation of gas volume is (15).

$$\dot{V}_{gas} = -Q_{sum}. \tag{15}$$

Substituting (15) to (13), the change of pressure in the knee model is shown in (16).

$$\dot{p}_{knee} = \frac{p_{knee}}{V_{gas}} Q_{sum}. \tag{16}$$

Comparing (16) to (5), it is obvious that the capacity of the knee model C_{knee} is equal to the ratio between the gas volume and the pressure in the operation area currently.

2.2   State Space Presentation

During MIS, the flow of the fluid into the surgical region Q_{in} is created by the difference pressure of $(p_1 - p_{knee})$ as in (17). Similarly, the outflow Q_{out} is presented in (18).

$$Q_{in} = \frac{1}{R_1}(p_1 - p_{knee}) = \frac{1}{R_1}\Delta p_1. \tag{17}$$

$$Q_{out} = \frac{1}{R_2}(p_2 - p_{knee}) = \frac{1}{R_2}\Delta p_2. \tag{18}$$

Assuming that R_1 and R_2 are the flow resistances of the tubes for the inflow and the outflow correspondingly. Equations (3) and (4) can be rewritten in (19) and (20).

$$\dot{p}_1 = \frac{1}{C_{tube}}\left[Q_{p1} - \frac{1}{R_1}(p_1 - p_{knee})\right]. \tag{19}$$

$$\dot{p}_2 = \frac{1}{C_{tube}}\left[Q_{p2} - \frac{1}{R_2}(p_2 - p_{knee})\right]. \tag{20}$$

From (8) and (15), the change of gas volume in the knee model is determined in (21).

$$\dot{V}_{gas} = -(Q_{p1} - Q_{p2} - Q_{err}). \tag{21}$$

By substituting p_{knee} from (11) into (19) and (20), rearranging the elements in these equations, a state space presentation in the nonlinear process is formulated in (22).

$$
\begin{bmatrix} \dot{p}_1 \\ \dot{p}_2 \\ \dot{V}_{gas} \end{bmatrix} =
\begin{bmatrix} -\frac{1}{R_1 C_{tube}}p_1 + \frac{mRT}{R_1 C_{tube}}\cdot\frac{1}{V_{gas}} \\ -\frac{1}{R_2 C_{tube}}p_2 + \frac{mRT}{R_2 C_{tube}}\cdot\frac{1}{V_{gas}} \\ 0 \end{bmatrix} +
\begin{bmatrix} \frac{1}{C_{tube}} \\ 0 \\ -1 \end{bmatrix} [Q_{p1}]
$$
$$
+ \begin{bmatrix} 0 \\ \frac{1}{C_{tube}} \\ 1 \end{bmatrix} [Q_{p2} + Q_{err}] = f(x, u, z) \tag{22}
$$

where $u = Q_{p1}$ is defined as the main input; extra input $z = (Q_{p2} + Q_{err})$ is defined as an input disturbance of the process; and vector of three state variables: $x = [p_1, p_2, V_{gas}]^T$.

The pressure p_{knee} is depended on the variable of the gas volume V_{gas} which should also be estimated for the replacement of the measured data feedback to the controller. When the state of V_{gas} is estimated, then the pressure p_{knee} can be determined by (11) and (13). This research contributes an application of the EKF as an estimator for the state of the pressure p_{knee}. Figure 3 is a general description of the observer design using

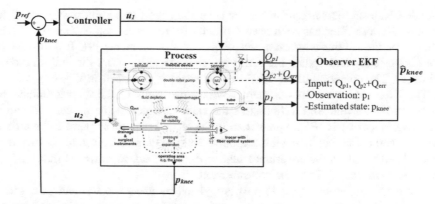

Fig. 3. An overview of observer design in the controlled process of MIS.

EKF for the process of MIS. The variable pressure p_1 at the *pump1* was selected for the measurement data to the EKF observer.

The details on the implementation of the Extended Kalman Filter in the MIS is described in Sect. 3.

3 Implementation of the Extended Kalman Filter in the MIS

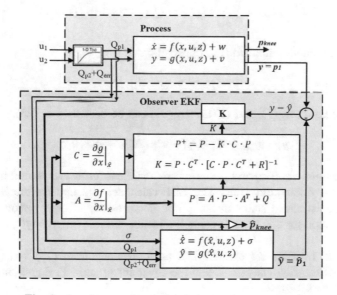

Fig. 4. An observer description in state space presentation.

Extended Kalman filter is crucial for the state estimation in nonlinear systems. While the basic Kalman filter has been used normally for the linearized process, the EKF is utilized for the nonlinear process and nonlinear measurements [9]. It is to linearize about a trajectory and update the estimated states resulting from the real time observation [8]. The operation of the EKF in state space is described in Fig. 4.

The EKF observer works sequently within two steps of the loop infinitely: prediction and correction. In the first step, the state variables of the process '*x*' and the state covariance matrix '*P*' are predicted. In the second step, these states and matrix are corrected and updated by using the computed Kalman gain called vector *K*. This vector is calculated based on the covariance matrix *P*, the linearized observed matrix *C*, and the noise covariance *R* from the measurement.

The correcting signal σ (in Fig. 4) is updated by using the Kalman gain and the current predicted error. The predicted error is the difference between the measurement and the estimated data. In this paper, the EKF is implemented in two different cases of state space presentations on the model of the knee arthroscopy. The first case is the usage one state variable equation, and the second case is using three state variables presentation of the process.

3.1   Case 1: One State Variable Presentation

In this case, p_{knee} is the estimated state; and the observable measurement is the pressure *p1* at the *pump1* of the DRP. From (8) and (16), the state space presentation is in (23).

$$\dot{x} = \dot{p}_{knee} = \frac{p_{knee}}{V_{gas}} \cdot (Q_{p1} - Q_{p2} - Q_{err}) = f(x, u, z). \tag{23}$$

The measurement output for the observation:

$$y = p_1 = p_{knee} + R_1 \cdot Q_{in}. \tag{24}$$

From the Eqs. (3), (7), and (17), then (24) can be described in (25).

$$y = p_{knee} + R_1 \cdot Q_{p1} - R_1 \cdot \Delta Q_1 = g(x, u, z). \tag{25}$$

Both functions *f(x, u, z)* and *g(x, u, z)* are nonlinear forms and should be linearized at each period of sampling time. The EKF calculates and updates the Kalman gain vector for the estimated states instantly and continuously at each period of working time. The solution for this work is using the Jacobi matrices based on *f(x, u, z)* and *g(x, u, z)*.

The Jacobi matrices *A1* and *C1* can be determined from (23) and (25) for the calculations of the Kalman gain *K* and the covariance matrix *P*. Equation (27) proves that the process is observable.

$$A_1 = \left[\frac{\partial f}{\partial x} \right]_{(x)} = \left[\frac{1}{V_{gas}} \cdot (Q_{p1} - Q_{p2} - Q_{err}) \right]_{(x)}. \tag{26}$$

$$C_1 = \left[\frac{\partial g}{\partial x}\right]_{(x)} = [1].\tag{27}$$

3.2   Case 2: Three State Variables Presentation

Three state variables presentation in state space is (22) and equivalent to the form (28).

$$\dot{x} = \begin{bmatrix} \dot{x}_1 \\ \dot{x}_2 \\ \dot{x}_3 \end{bmatrix} = \begin{bmatrix} \dot{p}_1 \\ \dot{p}_2 \\ \dot{V}_{gas} \end{bmatrix} = \begin{bmatrix} f_1(x,u,z) \\ f_2(x,u,z) \\ f_3(x,u,z) \end{bmatrix} = f(x,u,z).\tag{28}$$

The measurement output for the observation is (29).

$$y = p_1 = \begin{bmatrix} 1 & 0 & 0 \end{bmatrix} \begin{bmatrix} p_1 \\ p_2 \\ V_{gas} \end{bmatrix} = g(x,u,z).\tag{29}$$

From (22), (28), (29), the Jacobi matrices A_2 and C_2 are determined in (30) and (31).

$$A_2 = \begin{bmatrix} \frac{\partial f_1}{\partial p_1} & \frac{\partial f_1}{\partial p_2} & \frac{\partial f_1}{\partial V_{gas}} \\ \frac{\partial f_2}{\partial p_1} & \frac{\partial f_2}{\partial p_2} & \frac{\partial f_2}{\partial V_{gas}} \\ \frac{\partial f_3}{\partial p_1} & \frac{\partial f_3}{\partial p_2} & \frac{\partial f_3}{\partial V_{gas}} \end{bmatrix}_{(x)}$$

$$= \begin{bmatrix} \frac{-1}{R_1 C_{tube}} & 0 & \frac{-p_{knee}}{V_{gas}}\left(\frac{1}{R_1 C_{tube}}\right) \\ 0 & \frac{-1}{R_2 C_{tube}} & \frac{-p_{knee}}{V_{gas}}\left(\frac{1}{R_2 C_{tube}}\right) \\ 0 & 0 & 0 \end{bmatrix}_{(x)}.\tag{30}$$

$$C_2 = \begin{bmatrix} \frac{\partial g}{\partial p_1} & \frac{\partial g}{\partial p_2} & \frac{\partial g}{\partial V_{gas}} \end{bmatrix}_{(x)} = \begin{bmatrix} 1 & 0 & 0 \end{bmatrix}.\tag{31}$$

For the observation of the EKF, it is assumed that there are two mutually independent distributions of the white noises w and v. These noises are shown in (32) and (33) for the process noise and the measurement noise respectively.

$$\dot{x} = f(x,u,z) + w.\tag{32}$$

$$y = g(x,u,z) + v.\tag{33}$$

The expectations of the noise covariances are given in (34), (35) and (36). The matrix Q is the noise covariance of the process; R is the noise covariance of the measurement.

$$E[w(t)w^T(\tau)] = Q \cdot \delta(t - \tau); \text{ and } E[w(t)] = 0. \tag{34}$$

$$E[v(t)v^T(\tau)] = R \cdot \delta(t - \tau); \text{ and } E[v(t)] = 0. \tag{35}$$

$$E[w(t)v^T(\tau)] = 0, \forall t \text{ and } \tau. \tag{36}$$

$E[]$ is the expectation operator; δ is Kronecker delta function defined in (37) [10].

$$\delta(t - \tau) = \begin{cases} 1, & \text{if } t = \tau \\ 0, & \text{if } t \neq \tau \end{cases}. \tag{37}$$

In both two presented cases, the Jacobi matrices A_1 and A_2 are altered at each sampling time. Thus, it should be computed and updated instantly in discrete time. The Eqs. (32) and (33) are rewritten in discrete time as (38) and (39).

$$x_{(k+1)} = f(x_k, u_k, z_k) + w_k. \tag{38}$$

$$y_k = g(x_k, u_k, z_k) + v_k. \tag{39}$$

The algorithm of the two steps in the loop of the EKF is described briefly as follow:

a. Updating the prior data from the estimated states $\hat{x}_k$ and covariance matrix P_k^- at the time t_k. The loop is initialized with the provided conditions x_0 and P_0.
b. Predicting the state vector and the state covariance matrix.

$$\hat{x}_{k+1}^- = f(\hat{x}_k, u_k, z_k). \tag{40}$$

$$P_k^- = A_k \cdot P_k^- \cdot A_k^T + Q_k. \tag{41}$$

c. Calculating the Kalman gain vector K_k.

$$K_k = P_k \cdot C_k^T [C_k \cdot P_k \cdot C_k^T + R_k]^{-1}. \tag{42}$$

d. Correcting the predicted state vector of x and updating the covariance matrix P.

$$\hat{x}_{k+1} = \hat{x}_{k+1}^- + K_k [y_k - C_k \cdot \hat{x}_{k+1}^-]. \tag{43}$$

Table 1. Parameters used in the MIS simulator of the knee arthroscopy.

Name	Symbol	Value				Unit
Hydraulic capacity of the tubes	C_{tube}	9.75e−11				m³/Pa
Total volume of the knee model	V_{total}	0.0016				m³
Initial fluid volume in the knee model	V_0	0.001				m³
Ambient pressure (initial pressure state)	p_0	101325				Pa
Process noise covariance matrix (Case 1)	Q_k	[1e13]				
Process noise covariance matrix (Case 2)	Q_k	$\begin{bmatrix} 1e13 & 0 & 0; \\ 0 & 1e13 & 0; \\ 0 & 0 & 25e11 \end{bmatrix}$				
Measurement noise covariance (Case 1, Case 2)	R_k	[1]				

$$P_k^+ = P_k - K_k \cdot C_k \cdot P_k. \tag{44}$$

At each time of t_k, the EKF estimates the states and corrects the estimated states depending on the covariances of P_k, Q_k, and R_k.

Table 1 shows some initial conditions and parameters used in the MIS simulator of the knee arthroscopy. The results of the EKF implementation are shown in Sect. 4.

4   Results and Discussion

Figure 5 shows the results of estimation using the EKF observer with two cases of state space presentation implemented in the knee arthroscopy of MIS. The left column of Fig. 5 presents the estimation results on modelling with MATLAB Simulink. The right column of Fig. 5 demonstrates the results in real time simulator environment. The upper panel of Fig. 5 presents the results of the estimated pressure of the knee model (Fig. 5a and b). It is seen that the estimated pressure from the EKF with the case of one state was nearly the same to the estimated state in the case of three states presentation. The Fig. 5c and d (in the middle) display the errors of the estimated pressure compared to the measured pressure. In real time experiment with both two cases of state space presentation implemented, the estimation error as well as the noise from measured data was minimized significantly. Furthermore, the lower panel (Fig. 5e and f) shows the observed data from the EKF compared to the measurement of pressure at the roller pump1 p_1. The noise from the observed data was also reduced from the measurement.

Additionally, in real time device experiments, when the authors tried to ignore the loss of error flow Q_{err}, then the estimated pressure was incorrectly (indicated by the dotted signal in Fig. 5b). This proves that the EKF is more sensitive in nonlinear system with the measured data in real time device experiments.

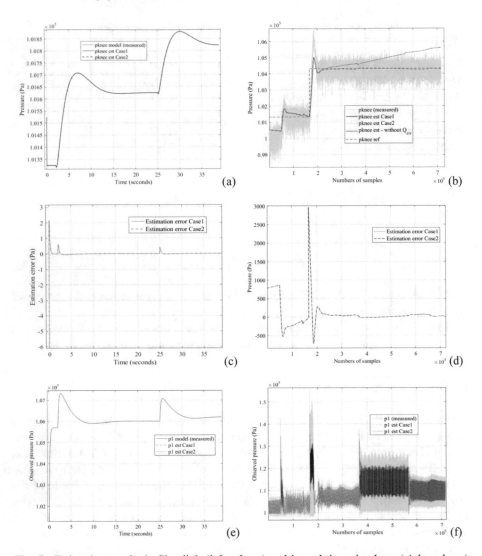

Fig. 5. Estimation results in Simulink (*left column*) and in real time simulator (*right column*).

5   Conclusion

The paper introduced an application of the Extended Kalman filter on the pressure estimation in the knee arthroscopy of MIS. The estimator worked based on the data from the flows of fluid and the available measured pressure at the double roller pump. Two different cases of choosing the numbers of state were applied to the process for the comparision of results. The results from the two cases were implemented in both MATLAB Simulink and real time device simulator of the knee arthroscopy. The results in Sect. 4 proved that the estimate error as well as the noises from measurement were minimized and eliminated effectively. This illustrates that the Extended Kalman filter estimates the pressure in the operation area of MIS more precisely compared to the results from the basic Kalman filter implementation in [7]. The estimated pressure is worthy helpful in using as a feedback data for the interaction to the sensorless controlled process. This solves the problem of unmeasurable pressure in the surgical area of the real patient during MIS.

Acknowledgments. Research funding: This research and development project is funded by the German Federal Ministry of Education and Research (BMBF) within the program "Medizintechnische Lösungen für die digitale Gesundheitsversorgung" under the project number 13GW0164B and managed by the Project Management Agency VDI Technologiezentrum GmbH. The author is responsible for the contents of this publication. Conflict of interest: Authors state no conflict of interest. Ethical approval: The conducted research is not related to either human or animals use.

References

1. Davies, B.: Robotics in minimally invasive surgery. In: IEE Colloquium on Through the Keyhole: Micro-engineering in Minimally Invasive Surgery, pp. 5/1–5/2 (1995)
2. Smolinski, E., Benkmann, A., Westerhoff, P., Nguyen, V.M., Drewelow, W., Jeinsch, T.: A hardware-in-the-loop simulator for the development of medical therapy devices. IFAC **50**, 15050–15055 (2017)
3. Nguyen, V.M., Jeinsch, T.: Pressure control in minimally invasive surgery. In: International Symposium on Automatic Control (2017)
4. Hsiao, M.S., Kusnezov, N., Sieg, R.N., Owens, B.D., Herzog, J.P.: Use of an irrigation pump system in arthroscopic procedures. Orthopedics **39**(3), 474–478 (2016)
5. Muellner, T., Menth-Chiari, W.A., Reihsner, W.A., Eberhardsteiner, R., Eberhardsteiner, J., Engebretsen, L.: Accuracy of pressure and flow capacities of four arthroscopic fluid management systems. Arthrosc. J. Arthrosc. Relat. Surg. Off. Publ. Arthrosc. Assoc. N. Am. Int. Arthrosc. Assoc. **17**, 760–764 (2001)
6. Bergstrom, R., Gillquist, J.: The use of an infusion pump in arthroscopy. Arthrosc. J. Arthrosc. Relat. Surg. **2**, 4–45 (1986)
7. Nguyen, V.-M., Smolinski, E.C., Benkmann, A., Drewelow, W., Jeinsch, T.: An application of pressure estimation in minimally invasive surgery. In: 4th International Conference on Green Technology and Sustainable Development, pp. 658–662. IEEE (2018)
8. Brown, R.G., Hwang, P.Y.C.: Introduction to Random Signals and Applied Kalman Filtering. Wiley, New York (2012)
9. Romaniuk, S., Ambroziak, L., Gosiewski, Z., Isto, P.: Real time localization system with Extended Kalman Filter for indoor applications. In: 21st International Conference on Methods and Models in Automation and Robotics, pp. 42–47. IEEE (2016)
10. Shyam, M.M., Naik, N., Gemson, R.M.O., Ananthasayanam, M.R.: Introduction to the Kalman Filter and tuning its statistics for near optimal estimates and Cramer Rao bound (2015)

Stress Measurement in Multi-tasking Decision Processes Using Executive Functions Analysis

Lucas Paletta[1(⊠)], Martin Pszeida[1], Bernhard Nauschnegg[1],
Thomas Haspl[2], and Raphael Marton[1]

[1] Digital - Institute for Information and Communication Technologies,
Joanneum Research Forschungsgesellschaft mbH,
Steyrergasse 17, 8010 Graz, Austria
{lucas.paletta,martin.pszeida,
bernhard.nauschnegg}@joanneum.at,
raphael.marton@student.tugraz.at
[2] Robotics - Institute for Robotics and Mechatronics,
Joanneum Research Forschungsgesellschaft mbH, Lakeside B08a,
9020 Klagenfurt am Wörthersee, EG, Austria
thomas.haspl@joanneum.at

Abstract. The presented work aimed to investigate how the impact of cognitive stress would affect the attentional processes, in particular, the performance of the executive functions that are involved in the coordination of multi-tasking processes. The study setup for the proof-of-concept involved a cognitive task as well as a visuomotor task, concretely, an eye-hand coordination task, in combination with an obstacle avoidance task that is characteristic in human-robot collaboration. The results provide the proof that increased stress conditions can actually be measured in a significantly correlated increase of an error distribution as consequence of the precision of the eye-hand coordination. The decrease of performance is a proof that the attentional processes are a product of executive function processes. The results confirm the dependency of executive functions and decision processes on stress conditions and will enable quantitative measurements of attention effects in multi-tasking configurations.

Keywords: Stress measurement · Gaze analysis · Multi-tasking ·
Executive functions · Human-robot collaboration · Spatial fixation distribution

1 Introduction

Future progress in human-robot collaboration will depend on the ability to coordinate decision processes in robot and human actions in an efficient manner, such as, for the management of assembly in manufacturing. While machines tend to apply deterministic, precise and rapid operations, research focuses on how human operators would integrate without stress in spatiotemporal requirements of fast collaboration. Executive functions represent mechanisms of cognitive control that enable objective oriented, flexible and efficient behaviors [1, 2] and directly relate to dynamic coordination of attention which is a fundamental factor in human-robot collaboration [3]. Executive functions are impacted by stress whenever task requirements surpass the regulatory

© Springer Nature Switzerland AG 2020
H. Ayaz (Ed.): AHFE 2019, AISC 953, pp. 344–356, 2020.
https://doi.org/10.1007/978-3-030-20473-0_33

capacity of the organism [4]. Human stress reactions increase saliency for immediate challenging recognition and at the same time decrease weight on more complex cognitive processes [5]. During a single task stress can even increase the selectivity of attention [6] but should decrease attention in multi-tasking. We argue that human attention in the context of grasping can provide a function of measurable indication about how well human executive functions are supporting visuomotor decision processes in a collaborative task.

In a multi-tasking scenario, the operator's activity is triggered by a primary cognitive task, defined by an n-back test [7] where a randomly triggered image sequence is depicted with two parametrizations. The human operator performs a 'peg-in-hole' task and places in a left (right) hole for any 1-back (2-back) case of repetition. Imprecision of attentional fixations was defined as dependent variable and the stress condition as independent variable. A robot arm was moved in a random motion around target holes hence the operator paid attention not to touch since with any pressure the robot arm halted and delayed the task.

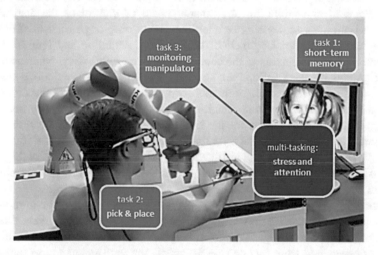

Fig. 1. Stress measurements in a multi-tasking study involving a cognitive short-term memory task, a pick and place task, and obstacle avoidance in the context of random robot arm motion.

The study was performed with 6 participants, performing the multi-tasking in the two stress conditions. The difference of the fixation distributions of the eye-hand coordination task correlated with the difference in mean arousal measurements, demonstrating the relation between stress, attention, and executive functions. The results confirm the dependency of executive functions and decision processes on stress conditions and will enable quantitative measurements of attention in multi-tasking stress related configurations in future work.

2   Background and Related Work

Stress is defined in terms of psychological and physical reactions of animate beings in response to external stimuli, i.e., by means of stressors. These reactions empower the animate to cope with specific challenges, and at the same time there is the impact of physical and mental workload. [9, 10] defined stress in terms of a physical state of load which is characterized as tension and resistance against external stimuli, i.e., stressors which refers to the general adaptation syndrome. Studies confirm that the selectivity of human attention is performing better when impacted by stress [6] which represents the 'narrowing attention' effect. According to this principle, attention processes would perform in general better with increasing stress impact.

Executive functions are related to the dynamic control of attention which refers to a fundamental component in human-robot collaboration [3]. Executive functions refer to mechanisms of cognitive control which enable a goal oriented, flexible and efficient behavior and cognition [1, 2], and from this the relevance for human everyday functioning is deduced. Following the definition of [11], executive functions above all include attention and inhibition, task management or control of the distribution of attention, planning or sequencing of attention, monitoring or permanent attribution of attention, and codification or functions of attention. Attention and inhibition require directional aspects of attention in the context of relevant information whereas irrelevant information and likewise actions are ignored. Executive functions are known to be impacted by stress once the requirements for regulatory capacity of the organism are surpassed [4, 12]. An early stress reaction is known to trigger a saliency network that mobilizes exceptional resources for the immediate recognition and reactions to given and in particular surprising threats. At the same time, this network is down-weighted for the purpose of executive control which should enable the use of higher-level cognitive processes [5]. This executive control network is particularly important for long-term strategies and it is the first that is impacted by stress [13].

In summary, impact of stress can support to focus the selective attention processes under single task conditions, however, in case of multiple task conditions it is known that stress impacts in a negative way the systematic distribution of attention onto several processes and from this negatively affects the performance of executive functions. Evaluation and preprevention of long-term stress in a production work cell is a specific application objective. Stress free work environments are more productive, provide more motivated co-workers and error rates are known to be lower than for stressed workers. The minimisation of interruption by insufficient coordination of subtasks is an important objective function which has executive functions analysis as well as impact by stress as important input variables.

3   Multi-tasking, Executive Functions and Attention

Multi-tasking activities are in indirect relation with executive functions. Following Baddeley's model of a central executive in the working memory [1] there is an inhibitor control and cognitive flexibility directly impacting the (i) multi-tasking, (ii) the change between task and memory functions, (iii) selective attention and inhibition functions.

The performance in relation to an activity becomes interrupted once there is a shift between one to the other ("task switch"). The difference between a task shift which particularly requires more cognitive resources, and a task repetition is referred to as 'switch cost' [14] and in particular of relevance if the task switch is more frequently, and also referring to the frequency of interruptions after which the operator has to continue the activity from a memorized point where the switch started. Task switches in any case involve numerous executive function processes, including shift of focus of attention, goal setting and tracking, resources for task set reconfiguration actions, and inhibition of previous task groups. Multi-Tasking and the related switch costs define and reflect the requirements for executive functions and from this the cognitive control of attention processes, which provides highly relevant human factors parameters for the evaluation of human-robot collaboration systems.

The impact of executive functions on decision processes was investigated in detail by [15]. Specifically they researched on aspects of applying decision rules based on inhibition and the consistency in the perception of risk in the context of 'task shifting'. The results verify that the capacity to focus and therefore to inhibit irrelevant stimuli represents a fundamental prerequisite for successful decision processes. At the same time, the change between different contexts of evaluation is essential for consistent estimation of risk.

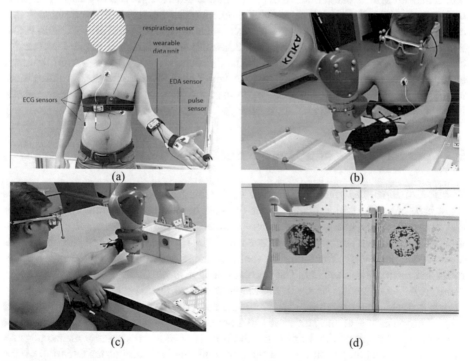

Fig. 2. Bio-sensor and eye tracking glasses setup. (a) Full bio-sensor setup with EDA sensor, HRV and breathing rate sensors. (b, c) Eye-hand coordination task interrupted by a robot arm movement. (d) Target areas with holes and gaze intersections centered around the hole center as landing point of eye-coordination task.

4 Measurements of Emotional and Cognitive Stress

Stress or activity leads to psychophysiological changes, as described by [16]. The body provides more energy to handle the situation; therefore, the hypothalamic-pituitary-adrenal (HPA) axis and the sympathetic nervous system are involved. Sympathetic activity can be measured from recordings of, for example, electro-dermal activity (EDA) and electrocardiogram (ECG). Acute stress leads to a higher skin conductance level (SCL) and also to more spontaneous skin conductance reactions (NS.SCR). [17] showed in his summary of EDA research that these two parameters are sensitive to stress reactions. In a further step [18] examined the reaction of stress by measuring EDA with a wearable device, during cognitive load, cognitive and emotional stress, and eventually classified the data with a Support Vector Machine (SVM) whether the person received stress. The authors conclude that the peak high and the instantaneous peak rate are sensitive predictors for the stress level of a person. The SVM was capable to classify this level correctly with an accuracy larger than 80%. The electrocardiogram shows a higher heart rate (HR) during stress conditions meanwhile the heart rate variability (HRV) is low [19], and the heart rate was found to be significantly higher during cognitive stress than in rest. [20] found in their survey that the HRV and the EDA are the most predictive parameters to classify stress. [21] used a cognitive task in combination with a mathematical task to induce stress. Eye activity can be a good indicator for stressful situations as well. Some studies report more fixations and sac-cades guided by shorter dwell time in stress, other studies reported fewer fixations and saccades but longer dwell time. Many studies use in addition the pupil diameter for classification of stress, such as reported. In the survey from [20], the pupil diameter achieved good results for detecting stress. A larger pupil diameter indicates more stress. In the study of [22], blink duration, fixations frequency and pupil diameter were the best predictors for work load classification.

5 Concept, Tasks and Human Factors of the Study

Study Concept. The overall objective of the multi-tasking study presented in this work is to prove the effect of an increase of stress on the executive functions in a multi-tasking study via an attentional outcome parameter. At the same time, we envision that vice versa attentional outcome parameters could be used to predict the level of stress as an outcome parameter. The complete multi-tasking scenario is embedded in a human-robot interaction environment. The human operator is involved in several tasks at the same time since she has to act in an adaptive way to the state of several other tasks. Figure 1 depicts the operator and the multiple tasks the operator is involved at the same time. The objective of the study was to involve the operator in multiple tasks at the same time and in particular, to measure the impact of induced stress by means of gaze. Task 1 highly challenges short-term memory and requires continuous attention on the screen. Task 2 is a pick-and-place task that requires physical but also visual attention with respect to the eye-hand coordination to pick an object and place it exactly at the goal position. Finally, task 3 requires the monitoring of the randomly controlled robot arm movement for the purpose of obstacle avoidance. All three tasks have to be performed concurrently most of the time by the human operator. The state of task 1 impacts the current or the next movement and its goal position in task 2 and 3 run in parallel, since the operator has to take care for the robot movement the whole time, the hand is moved toward the goal position (see Fig. 2 for details).

Gaze in principle is dependent on various contextual conditions and therefore careful setting of the gaze analysis is necessary. Gaze is in general dependent on 'psychophysiological' parameters. Gaze is dependent on cognitive parameters since executive functions impact the orientation of gaze by means of a meaningful distribution of attention on spatially distributed stimuli, gaze is impacted as well by affective state since gaze orientation dependent on the emotional value of a stimulus, and gaze is impacted by physical parameters since visual stabilization of balance of body movements is specifically supported by eye movements towards stimuli of the physical environment. Therefore, we investigate gaze under very specific conditions, i.e., in a hand-eye coordination task where gaze is strictly responding under visuomotor task specifications that are set in context with visual attention and from this provides an appropriate measure to quantify the impact of stress on attention.

5.1   Tasks, Interactions and Human Factors

Cognitive Task. Firstly, stress is induced via the workload of a cognitive task. The 'n-back test' [7] is known to impact very intensively the short-term memory of the participant. On the basis of this test the variable level of stress is induced onto the participant, since the n-back test can be easily configured by means of difficulty of task, i.e., by increasing the 'n', i.e., the number of events the participant has to track back in memory, as well as by changing the time period in which novel cues will become presented to the participant. In our study setting, the cues are represented by images from a standardized image database.

Visuomotor Task. On the basis of specific cognitive events, such as, either the occurrence of an immediate repetition (n equals 1) or a repetition with another image in between (n equals 2), the operator has to perform a visuomotor task that is dependent on the current state of the n-back test. The subsequent visuomotor task is a ,pick and place 'activity in which the operator has first to pick a piece out of a 'source' box and then, depending on the current outcome of the cognitive task, place that piece inside one of two other available 'target' boxes. Figure 1 shows the moment in which the operator actually picks a piece out of the source box. Figure 2b, c demonstrates the moment in which the operator is placing the piece inside a 'target' box. In case the operator detected an immediate repetition (n equals 1), the operator had to place the piece into the left (with respect to the operator's view) and in case of a repetition with another image in between (n equals 2) onto the right of the two boxes, respectively.

Movement of the Robot Arm. A specific condition on the motion of the hand and the complete arm of the operator is the motion of a robot arm that accidentally might be in the direct trajectory of the hand from the box to one of the two target holes or boxes. The movements of the robot arms have to be randomized in its spatiotemporal characteristics in order to urge the operator to watch the current state and predict the short term trajectory of the robot arm in order to make the movement of the own hand and arm without contact with the robot arm. The movement was specified by a professional and certified safety expert for human-robot interaction so that the movement could definitely and in no possible configuration harm the sanity of the operator. The robot arm is operated in its 'collaborative mode' and with constant velocity. The objective was solely to provide a dynamic obstacle in the process. Furthermore, it is typical that in human-robot collaboration collaborating robot arms might be in the way of a human movement and from this it is a conceptual way to represent the type of visuomotor challenges that arise from spatial collaboration in this way.

Hand-Eye Coordination Task. The 'place' part of the 'pick and place' task is a visuomotor task where the hand movement is coordinated by the location of the spatial goal that is represented by the hole of one of the two target boxes. Usually the piece is picked with two fingers and those fingers have to be maneuvered through the center of the hole and further a bit inside the hole and would then release the piece ins ide the box. The location of the spatial goal has to be fixated by the eyes of the operator at least once within the period of the visuomotor task in order to enable a sufficiently fine-tuned movement of the hand. The human gaze is known to be orientated towards the affordance of a spatial goal targeted movement, i.e., the position where the hand and finger movements are oriented to. [8] have identified in their seminal work that the human gaze usually performs a probing look towards the goal position of a manual task, such as, the tipping of a block with the finger or the positioning of a block towards a goal position, i.e., the 'landing point', in the physical environment. They found that the gaze fixates 150 ms on the landing point where the physical interaction will take place, and disengages 72 ms before the interaction, with a precision of ≈ 1 cm of the gaze position within the target coordinates of the actual landing point. Figure 2d depicts a view on the holes of the target boxes and overlaid the distribution of gaze during the eye coordination task of placing the piece into the hole. The points indicate the distribution of points where the gaze ray intersected with the virtual plane coinciding with the physical front plane of the target boxes with the hole.

5.2 Research Hypothesis

The research hypothesis assumes that the stress condition (no substantial versus substantial stress impact) would impact human visual attention and that this effect could be measured quantitatively in the measurement configuration. Furthermore, there is the assumption that the stress impact would take place under multi-tasking conditions which would argue for a deterioration of attention performance due to the increased challenge for successful performance of executive functions.

Concretely, it is expected that the fixation of the target position would deteriorate under the increased stress condition in terms of a less focused gaze behavior which would be represented by a wider spatial distribution of the gaze with respect to the goal position, i.e., the center coordinate of the hole in each of the target boxes.

The distribution with larger standard deviation would in particular be expected due to the fact that the activity of the manipulator would increasingly negatively impact the focus on the correct target position.

6 Experimental Results

Setup of the Study. Measurements were captured in the Human Factors Laboratory at the JOANNEUM RESEARCH Forschungsgesellschaft mbH in Graz, Austria. The study was performed with N = 6 participants, two of them female, four male, M_{age} = 42.3 and SD = 8.5 years. The participants were all technical experts from the research center, being naïve to the task. Each session took in total a maximum of 35 min, firstly sociodemographic data were captured, then the wearable sensing and eye tracking calibration was applied, then participants were introduced to the task and could test the required interaction for a maximum of 5 min, then two 'multi-tasking sessions' were applied as described in detail below.

Two sessions labelled 'NC' (normal condition) and 'SC' (stress condition) were performed by the participants (see Table 1). The duration of the two sessions, the parameters that define a more relaxed ('NC') and a more stressed ('SC') session are outlined in detail in Table 1. In general, the normal condition grants more time to watch individual images and more time between the image presentations as well as asking for less attention at the display since there are fewer 1-back and 2-back conditions which as a consequence triggers much less visuomotor tasks and hand-eye coordination and also fewer time for real multi-tasking, since the time sent purely for observing the display and waiting for the next repetition is actually spent as a single task. The robot arm was programmed to perform a sequence of robot movements from a current to a randomly selected target position. The target position was determined by randomized coordinate determination out of two predefined areas of interest, i.e., to the left and right of the central line separating the two target boxes, respectively, and with an area of about 30 × 30 cm.

Table 1. Parameters for the n-back task in terms of presentation of images on the screen.

Condition	tBetwIm	tImDis	imPSes	1StepPc	2StepPc	sTimSe
NC	1.4	2.0	150	10	30	8.5
SC	0.6	1.2	300	20	40	9.0

tBetwIm time between image presentations (sec.), tImDis time images are displayed (sec.), imPSes number of images per session, 1StepPc one step back chance (%), 2StepPc two step back chance (%), sTimSe total time per session (min.).

Participants were asked to perform with maximum possible speed and accuracy. After the sessions they were interviewed about their experience with the task and the robot arm. In general, there was no specific respect for the robot arm in terms of safety concerns, however, it was a challenge to find a rapid way around the moving arm. The majority of the participants reported learning effects and indicated that with on-going experience the task was subjectively performed in a more relaxing manner.

Hardware Configuration. To acquire video and eye-tracking data we used 'SMI Eye Tracking Glasses 2 Wireless', with 60 Hz sampling rate of gaze and a 1280×960 pixel resolution scene camera at 24 fps. In order to achieve the highest level of gaze estimation accuracy in the study setup, the dynamic position of the eye tracking glasses as well as the static position of the corners of the front plane of target boxes were tracked relative to the worker's environment The choice of vision based motion capturing system was OptiTrack [23] using 9 cameras of type 'Prime 41' with 4.1 MP resolution, a 100' tracking range, 180 fps resolution of motion capture, and 51° field of view. The accuracy of marker positioning was reported to be 0.06 mm by the Motive Software from OptiTrack. For the arousal measurements, a BIOPAC BN-PPGED system was used with EDA/GSR channel for Electro-dermal response, skin conductance activity/response that provides for indication of eccrine (skin sweating) activity, the unit interfaces with the MP150 basis platform for data acquisition and analysis. The robot arm was a Kuka iiwa LBR 7 R800 with standard control software for various safety modes. For the presentation of the n-back cognitive task a Microsoft Surface Pro 6 with 8 GB/128 GB RAM/SSD was used connected with a 17-in. TFT flat screen.

6.1   Descriptive Statistics

Arousal Measured with EDA. The raw data underlying the research and the descriptive as well as inferential statistics were collected from the EDA/SGR channel of the BIOPAC measuring unit, and from the gaze intersection with the front plane of the target box. From Fig. 4 it becomes obvious that the EDA activity in the SC sessions is of a higher level of arousal measured than during the NC sessions.

Impact on Gaze Distribution. Figure 4 depicts the precision in the eye-hand coordination tasks from the distribution of gaze intersecting with the frontal plane of the target boxes. Table 2 gives an overview about how the change in the electro-dermal activity due to stress conditions (NC, SC) is reflected in the precision of the eye-hand coordination in terms of the distribution of gaze on the frontal plane of the target boxes. The values λ correspond to the Eigenvalues of the Gaussian probability density as the main axes of the ellipsoid describing the error distribution.

6.2   Inferential Statistics

Eye-Hand Coordination Estimates the Stress Impact. The results demonstrate a high correlate between the stress measured by the bio-sensors, i.e., concretely by the EDA based arousal with the area of the ellipsoid spanned by the Eigenvectors of the Gaussian probability density which is a measure of the spread of the distribution or the size of the error in the precision of the eye-hand coordination towards the center of the hole in the frontal box planes. The Pearson correlation between the differences between the means in the arousal of NC and SC and the differences between the error distributions measured by area (see Table 2) for NC and SC is in total r = .790 for the box 2. The correlation in analogy for box 1 is only r = .703, however, that still represents a rather high value (Fig. 3).

Arousal and Eye-Hand Coordination. There is a lower but still substantial correlation between the SC based arousal values (EDA) and the area of the ellipsoid spanned by the Eigenvectors as described in detail above. The Pearson correlation is r = .547

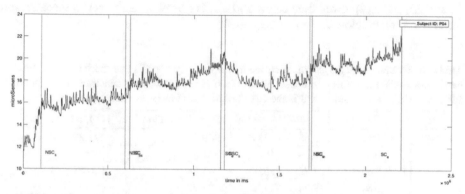

Fig. 3. Sample measurement curve of electro-dermal activity (EDA), participant 4. The data sampling starts with a 'NC' session (left), and then proceeds with an 'SC' stress condition, followed by another NC/SC session pair. From the EDA activity it becomes obvious that in the SC sessions there is a higher level of arousal measured than during the NC sessions.

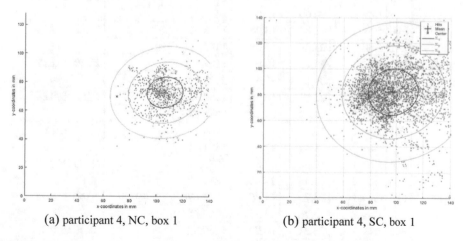

(a) participant 4, NC, box 1   (b) participant 4, SC, box 1

Fig. 4. Distribution of gaze under different stress conditions (NC, SC) measured during the eye-hand coordination task, the 'pick & place' task. (a, b) participant 4. It is obvious that the stress impact (SC) leads to a larger distribution of gaze hits on the frontal box plane, most probably due to the impact on the executive functions for the coordination of the multi-task study.

and therefore represents a direct dependency between these quantities. This proves that the stress has a direct impact to the precision of the attention based eye-hand coordination as predicted in the context of the executive function performance.

Stress Level Estimation. Eye movement based classification method, i.e., the 'stress level' estimation method, very well estimates the EDA arousal level, as well as the cardinal ranking for different levels of stress impact. The stress level estimation is computed by thresholds on the number of saccades and the mean dwell time during an observation window of 5000 ms. For level 2 the number of saccades should be between 15 and 25, and mean dwell time below 500 ms, For level 3, more than 25 saccades and mean dwell time below 250 ms is requested.

Table 2. Quantitative analysis of electrodermal activity ('arousal') and the impact on the eye-hand coordination in terms of the gaze hit distribution around the hole center of the target boxes (Box 1 is the left box and Box 2 is the right box in the view of the human operator).

		Condition	λ_1 [mm]	λ_2 [mm]	Error distribution [mm²]	M, arousal [µS]
NSC	P01	Box 1	23,58	13,08	969,14	21.33 ± 0.40
		Box 2	22,09	18,18	1261,61	
	P02	Box 1	22,26	13,46	941,58	20.57 ± 1.17
		Box 2	19,74	17,62	1092,77	
	P03	Box 1	17,92	15,03	846,09	5.8 ± 1.24
		Box 2	26,15	16,90	1388,53	
	P04	Box 1	13,85	10,64	462,87	17.12 ± 0.54
		Box 2	28,20	16,27	1441,24	
	P05	Box 1	27,11	18,76	1597,83	14.26 ± 1.12
		Box 2	24,60	18,84	1455,91	
	P06	Box 1	22,30	18,96	1328,57	23.86 ± 0.78
		Box 2	23,06	19,08	1382,47	
SC	P01	Box 1	24,22	15,61	1187,61	23.79 ± 1.12
		Box 2	29,88	22,65	2126,13	
	P02	Box 1	21,87	16,75	1150,48	20.50 ± 0.84
		Box 2	24,32	21,16	1616,72	
	P03	Box 1	19,78	15,52	964,46	6.43 ± 1.00
		Box 2	28,41	17,16	1531,44	
	P04	Box 1	19,38	17,84	1086,61	19.03 ± 0.67
		Box 2	29,48	19,65	1819,87	
	P05	Box 1	26,75	14,25	1197,67	10.56 ± 3.55
		Box 2	23,28	17,57	1285,45	
	P06	Box 1	28,21	20,99	1859,58	23.40 ± 0.92
		Box 2	26,05	20,39	1668,62	

λ_1, λ_2: *eigenvalues of the principal component analysis of the distribution of gaze hits on the frontal box plane centered at the hole center of the target boxes (B1, B2).*

7   Conclusions and Future Work

The presented work aimed to investigate how the impact of cognitive stress would affect the attentional processes, in particular, the performance of the executive functions that are involved in the coordination of multi-tasking processes. The study setup for the proof-of-concept involved a cognitive task as well as a visuomotor task, concretely, an eye-hand coordination task, in combination with an obstacle avoidance task that is characteristic in human-robot collaboration. The results provide the proof that increased stress conditions actually can be measured in a significantly correlated increase of an error distribution as consequence of the precision of the eye-hand coordination. The decrease of performance is a proof that the attentional processes are a product of executive function processes.

Future work will focus on using larger populations to specify the accuracy of the measurements, studies to determine the required frequency for robust analysis, and profound eye movement analysis on the generation of the error distribution in the eye-hand coordination task.

Acknowledgments. This work has been supported by the Austrian Ministry for Transport, Innovation and Technology (BMVIT) within the project framework CollRob (Collaborative Robotics), as well as by the Austrian BMVIT/FFG by projects MMASSIST II (No. 858623) and FLEXIFF (No. 861264). We would like to thank in particular Norah Neuhuber for her support in the interpretation of the bio-sensor data.

References

1. Baddeley, A.D., Hitch, G.: Working memory. In: Bower, G.A. (ed.) The Psychology of Learning and Motivation: Advances in Research and Theory, vol. 8, pp. 47–89 (1974)
2. Lezak, M.D.: Neuropsychological Assessment. Oxford University Press, New York (1995)
3. Bütepage, J., Kragic, D.: Human-robot collaboration: from psychology to social robotics arXiv:1705.10146 (2017)
4. Koolhaas, J.M., Bartolomucci, A., Buwalda, B., De Boer, S.F., Flügge, G., Korte, S.M., Meerlo, P., Murison, R., Olivier, B., Palanza, P., Richter-Levin, G.: Stress revisited: a critical evaluation of the stress concept. Neurosci. Biobehav. Rev. **35**, 1291–1301 (2011)
5. Hermans, E.J., Henckens, M.J.A.G., Joels, M., Fernandez, G.: Dynamic adaptation of large-scale brain networks in response to acute stressors. Trends Neurosci. **37**, 304–314 (2014)
6. Chajut, E., Algom, D.: Selective attention improves under stress: implications for theories of social cognition. J. Pers. Soc. Psychol. **85**, 231 (2003)
7. Jaeggi, S.M., Buschkuehl, M., Perrig, W.J., Meier, B.: The concurrent validity of the N-back task as a working memory measure. Memory **18**(4), 394–412 (2010)
8. Flanagan, J.R., Johansson, R.S.: Action plans used in action observation. Nature **424**, 769 (2003)
9. Selye, H.: A syndrome produced by diverse nocuous agents. Nature **138**(July 4), 32 (1936)
10. Lazarus, R.S., Folkman, S.: Stress, Appraisal, and Coping. Springer, New York (1984)
11. Smith, E., Jonides, J.: Storage and executive processes in the frontal lobes. Science **283**, 1657–1661 (1999)

12. Dickerson, S.S., Kemeny, M.E.: Acute stressors and cortisol responses: a theoretical integration and synthesis of laboratory research. Psychol. Bull. **130**, 355–391 (2004). https://doi.org/10.1037/0033-2909.130.3.355

13. Diamond, A.: Executive functions. Annu. Rev. Psychol. **64**, 135–168 (2013). https://doi.org/10.1146/annurev-psych-113011-143750

14. Meyer, D.E., Evans, J.E., Lauber, E.J., Gmeindl, L., Rubinstein, J., Junck, L., Koeppe, R.A.: The role of dorsolateral prefrontal cortex for executive cognitive processes in task switching. J. Cogn. Neurosci. **10** (1998)

15. Del Missier, F., Mäntyla, T., Bruine de Bruin, W.: Executive functions in decision making: an individual differences approach. Think. Reasoning **16**(2), 69–97 (2010)

16. Cannon, W.B.: The Wisdom of the Body. Norton, New York (1932)

17. Boucsein, W.: Electrodermal Activity. Plenum, New York (1992)

18. Setz, C., Arnrich, B., Schumm, J., La Marca, R., Tröster, G., Ehlert, U.: Discriminating stress from cognitive load using a wearable EDA device. IEEE Trans. Inf. Technol. Biomed. **14**(2), 410–417 (2010)

19. Taelman, J., Vandeput, S., Spaepen, A., Van Huffel, S.: Influence of mental stress on heart rate and heart rate variability. In: Proceedings Conferences for International Federation for Medical and Biological Engineering, pp. 1366–1369 (2008)

20. Sharma, N., Gedeon, T.: Objective measures, sensors and computational techniques for stress recognition and classification: a survey. Comput. Methods Programs Biomed. **108**, 1287–1301 (2012)

21. Sun, F., Kuo, C., Cheng, H.-T., Buthpitiya, S., Collins, P., Griss, M.: Activity-aware mental stress detection using physiological sensors. In: Mobile Computing, Applications, and Services, pp. 211–230. Springer, Heidelberg (2011)

22. Van Orden, K.F., Limbert, W., Makeig, S.: Eye activity correlates of workload during a visuospatial memory task. Hum. Factors: J. Hum. Factors Ergon. Soc. **43**, 111–121 (2001)

23. NaturalPoint, Inc.: DBA OptiTrack (2016). http://www.naturalpoint.com/optitrack/

The Effect of Anthropomorphization and Gender of a Robot on Human-Robot Interactions

Hongjun Ye[1(✉)], Haeyoung Jeong[1], Wenting Zhong[1],
Siddharth Bhatt[1], Kurtulus Izzetoglu[1], Hasan Ayaz[1,2],
and Rajneesh Suri[1,2]

[1] Drexel University, 3141 Chestnut Street, Philadelphia, PA 19104, USA
{hy368,hj325,wz326,shb56,ki25,ayaz,surir}@drexel.edu
[2] Drexel Business Solutions Institute, 3220 Market Street, Philadelphia,
PA 19104, USA

Abstract. The popularity of assistant robots has increased in the recent past. Past research has looked at the effect of anthropomorphization of robots and considered the consumers' gender as an important factor. However, research has not examined the interaction between a robot's gender and its level of anthropomorphization on human-robot interactions. Our results indicate that males and females perceive a service failure differently depending upon the level of anthropomorphization of the robots involved in the failure.

Keywords: Automation · Consumer behavior · Service failure ·
Human-robot interaction · Gender

1 Introduction

Anthropomorphization is to apply human characteristics and features to non-human objects [1], such as robots. The effect of anthropomorphization has been a focal topic in robot research. Roboticists have speculated that people are likely to work more naturally and easily with humanoid robots than with machine-like robots [2–4]. However, the theory of Uncanny Valley [5] and research based on this theory suggests that high level of anthropomorphization might induce negative effects [6, 7].

Gender of participants has been considered as a relevant variable in most social science research. Prior research on robots has investigated the effects of the robots' gender on human-robot interaction [8]. However, to our knowledge no research has examined differences in how men and women will respond to a robot when considering both the gender of the robot and the level of anthropomorphization of the robot at the same time. The present research examines these two factors in a retail setting involving a service error (failure) committed by a robot assistant.

© Springer Nature Switzerland AG 2020
H. Ayaz (Ed.): AHFE 2019, AISC 953, pp. 357–362, 2020.
https://doi.org/10.1007/978-3-030-20473-0_34

2 Method

Two pilot studies and one main study were conducted as part of the present research. The first pilot study investigated consumers' expectations toward sales personnel who was either a human or a robot. The second pilot study surveyed consumers' expectations of a smart speaker with a described gender. The main study examined the effects of anthropomorphization and the gender of a robot on human-robot interaction when a service failure had occurred.

2.1 Study 1: Pilot

We explored consumers' expectations of human and robot sales assistants in the first pilot study. Specifically, for what types of products or services, consumers are likely to take recommendations from a human versus a robot. Participants were presented with 20 different products and services that represented 6 major product categories – utilitarian, hedonic, material, experiential, functional and symbolic products\services.

Forty-one participants (22 females; M_{age} = 34.46) were recruited from an online panel for a modest compensation for their participation. In a between-subject design, participants were randomly assigned to one of the two conditions (human sales assistant vs. robot sales assistant). All participants presented with a role play of shopping at a large shopping mall where they were approached by a sales person. On the following screens, we asked the participants to infer the gender of the sales assistant and indicate to what extent they would like to take the recommendations from the sales assistant for the 20 products on the list (1 – extremely unlikely; 7 – extremely likely). The order of the products on the list was randomized.

2.2 Study 2: Pilot

This study manipulated the gender of a smart speaker by changing the pronouns associated with the speaker (he\him\his vs. she\her\hers).

Forty participants (18 females; M_{age} = 35.08) were recruited online for a small financial compensation. The participants were randomly assigned to one of the two gender conditions and were asked to list at least three functions that they expected from the smart speaker.

2.3 Study 3: Main

We recruited 298 participants (190 females; M_{age} = 37.67) online the main study. This study used a 2 (service failure vs. control) × 4 (female machine-like robot vs. male machine-like robot vs. female human-like robot vs. male human-like robot) between-subject design (Appendix A). Participants participated in a role play where they were visiting an office supply store to seek a refund. The customer service associate at the store was a robot assistant.

In the service failure condition, the participants were informed about a customer ahead of them in the cue who had received a wrong refund (10 dollars difference from the correct price). In the control condition, this customer received a correct amount of refund. Two hundred and three participants (133 females; M_{age} = 37.97) were assigned to the service failure condition and there were 95 participants (58 females; M_{age} = 37.13) in the control condition.

Next, all participants were told that they received a correct amount of refund from the robot assistant. On the following screen, the participants were asked to rate the likeability of the robot assistant that they interacted with (4 items, 7-point scale; Appendix B) [9]. In addition, we asked the participants to rate the level of anthropomorphization of the robot assistant (4 items, 7-point scale, Appendix C) [10].

3 Results

We used R Studio for the analyses and the default level of significance was set at $p < 0.05$.

3.1 Pilot Study 1

There was no difference in preference for the recommendations given by humans versus robots ($t = 0.016$, $df = 39$, $p = 0.98$). However, participants preferred to take recommendations from a robot sales assistant when the product or service was technological in nature. For experiential products or services requiring human-human interactions (e.g., restaurant for dinner), participants preferred to take recommendations from a human sales assistant (Table 1).

Table 1. Pilot study 1 results (top three preferred products\services).

	Rank = 1	Rank = 2	Rank = 3
Robot assistant	Scientific calculator	Business software	Headphones
Human assistant	Restaurant for dinner	Sneakers	Gift for friends

3.2 Pilot Study 2

Text analysis revealed that when a robot assistant was described as a female, it was perceived as passive and less active (e.g., I want her to be able to listen to me); when a robot assistant was described as male, participants expected it to be more active and have more advanced functions (e.g., I want him to find the optimal route for me).

3.3 Main Study

In the control condition, no difference was found in terms of the participants' preference for male robots versus female robots ($F = 0.09$, $p = 0.76$). The interaction effect (Subject's Gender × Robot Type) on likeability was marginally significant ($F = 2.16$, $p = 0.09$). This indicated that male participants like the female machine-looking robot more, while female participants like the male human-looking robot more (Table 2).

Table 2. Robot likeability (means).

	Female machine-like robot	Male machine-like robot	Female human-like robot	Male human-like robot
Male participants	4.79 ± 1.25 (N = 11)	2.65 ± 1.42 (N = 5)	4.02 ± 1.40 (N = 12)	4.63 ± 1.16 (N = 9)
Female participants	4.19 ± 1.33 (N = 13)	4.15 ± 1.24 (N = 18)	4.40 ± 1.41 (N = 13)	5.28 ± 1.23 N = 14

In terms of the level of anthropomorphization, we found that male participants tend to humanize female robots more, while female participants tend to humanize male robots more ($F_{subject\_gender \times robot\_gender} = 3.59$, $p = 0.06$) (Table 3).

Table 3. Level of anthropomorphization of the robots (means).

	Male robot	Female robot
Male participants	2.20 ± 0.88 (N = 14)	3.21 ± 1.37 (N = 23)
Female participants	2.79 ± 1.57 (N = 32)	2.69 ± 1.31 (N = 26)

In the service failure condition, our results showed that female participants rated the likeability of the robot assistant significantly lower than the male participants did to their robot assistants ($M_{Subject\_F} = 4.04$, $M_{Subject\_M} = 4.57$, $t = -2.73$, $df = 132.25$, $p = 0.007$). This suggest that female participants penalized the robot assistant more for the service failure.

However, when we looked at the effect by the gender of the robots, we found that overall participants penalized the male robots significantly more for the service failure ($M_{Robot\_M} = 4.01$, $M_{Robot\_F} = 4.45$, $t = 2.41$, $df = 200.63$, $p = 0.01$). Specifically, the difference between the two robots on the two extremes (i.e., female machine-like robot and male human-like robot) was significant – participants penalized the male human-like robot much more than the female machine-like robot ($M_{Robot4} = 3.86$, $M_{Robot1} = 4.43$, $F = 4.36$, $p = 0.03$).

4 Discussion

Autonomous mobile robots today can identify and track people and objects, understand and respond to spoken questions, and travel to a destination while avoiding obstacles [11]. Robots can be built to have abilities that complement human abilities and it is not surprising that many contend that the workplace including consumer retail will increasingly contain robots and people working together [12]. If professional service robots are to share the workplace with people, we need to understand what expectations consumers have about the form and function of robots that they will encounter in the market place.

Our results showed that male participants preferred machine-like female robots while female participants preferred human-like male robots more. When service failure occurred, both male and female participants penalized male robots more, but the likeability of female robots was not affected by the service failure they committed.

Such results suggest that gender assigned to robots may activate stereotypes that will impact human-robot interaction and such effect might also interact with the level of anthropomorphism of the robots. To further examine the effects, we designed a follow-up study using functional near infrared spectroscopy (fNIRS) and other devices to capture neurophysiological correlates in scenarios with three different themes (shopping, investing, driving). By analyzing the changes in neurophysiological correlates, we expect to gain more insight of the effects on consumers' decision-making process [13].

Appendix A: Types of Robots

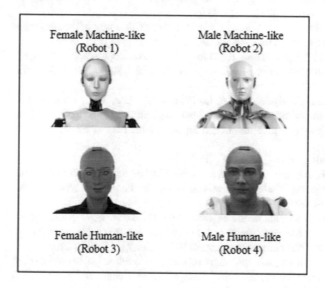

Female Machine-like (Robot 1) Male Machine-like (Robot 2)

Female Human-like (Robot 3) Male Human-like (Robot 4)

Appendix B: Likeability

Item no.	Description
1	I would enjoy knowing the robot assistant
2	The robot assistant is friendly
3	The robot assistant is likeable
4	I would enjoy interacting with the robot assistant

Note: items were rated on a seven-point scale (1 = strongly disagree; 7 = strongly agree)

Appendix C: Level of Anthropomorphism

Item no.	Description
1	It seems almost as if the robot assistant has its own beliefs and desires
2	It seems almost as if the robot assistant has a conscience (i.e., able to perceive what is right and what is wrong like a human)
3	It seems almost as if the robot assistant has a mind of its own
4	It seems the robot assistant has a consciousness (i.e., can receive and process information like a human)

Note: items were rated on a seven-point scale (1 = strongly disagree; 7 = strongly agree)

References

1. Epley, N., Waytz, A., Cacioppo, J.T.: On seeing human: a three-factor theory of anthropomorphism. Psychol. Rev. **114**(4), 864 (2007)
2. Hinds, P.J., Roberts, T.L., Jones, H.: Whose job is it anyway? A study of human-robot interaction in a collaborative task. Hum. Comput. Interact. **19**(1), 151–181 (2004)
3. Złotowski, J., Proudfoot, D., Yogeeswaran, K., Bartneck, C.: Anthropomorphism: opportunities and challenges in human–robot interaction. Int. J. Social Robot. **7**(3), 347–360 (2015)
4. de Visser, E.J., Monfort, S.S., Goodyear, K., Lu, L., O'Hara, M., Lee, M.R., Parasuraman, R., Krueger, F.: A little anthropomorphism goes a long way: effects of oxytocin on trust, compliance, and team performance with automated agents. Hum. Factors **59**(1), 116–133 (2017)
5. Mori, M.: The Uncanny Valley. Energy **7**(4), 33–35 (1970)
6. Saygin, A.P., Chaminade, T., Ishiguro, H., Driver, J., Frith, C.: The thing that should not be: predictive coding and the Uncanny Valley in perceiving human and humanoid robot actions. Soc. Cogn. Affect. Neurosci. **7**(4), 413–422 (2011)
7. Walters, M.L., Syrdal, D.S., Dautenhahn, K., Te Boekhorst, R., Koay, K.L.: Avoiding the Uncanny Valley: robot appearance, personality and consistency of behavior in an attention-seeking home scenario for a robot companion. Auton. Robots **24**(2), 159–178 (2008)
8. Tay, B., Jung, Y., Park, T.: When stereotypes meet robots: the double-edge sword of robot gender and personality in human–robot interaction. Comput. Hum. Behav. **38**, 75–84 (2014)
9. Parasuraman, A., Zeithaml, V.A., Berry, L.L.: Servqual: a multiple-item scale for measuring consumer perc. J. Retail. **64**(1), 12 (1988)
10. Hur, J.D., Koo, M., Hofmann, W.: When temptations come alive: how anthropomorphism undermines self-control. J. Consum. Res. **42**(2), 340–358 (2015)
11. Fong, T., Nourbakhsh, I., Dautenhahn, K.: A survey of socially interactive robots: concepts, design and applications: Technical Report CMU-RI-TR-02-29 (2002)
12. Thrun, S.: Toward a framework for human-robot interaction. Hum. Comput. Interact. **19**(1), 9–24 (2004)
13. Ayaz, H., Onaral, B., Izzetoglu, K., Shewokis, P.A., McKendrick, R., Parasuraman, R.: Continuous monitoring of brain dynamics with functional near infrared spectroscopy as a tool for neuroergonomic research: empirical examples and a technological development. Front. Hum. Neuros. **7**, 871 (2013)

Recognition of the Cognitive State in the Visual Search Task

Shuomo Zhang, Yanyu Lu$^{(\boxtimes)}$, and Shan Fu

School of Electronic Information and Electrical Engineering,
Shanghai Jiao Tong University, Shanghai, People's Republic of China
172810919@qq.com, {luyanyu, sfu}@sjtu.edu.cn

Abstract. Electroencephalogram (EEG) is a research subject that has been studied constantly. By the analysis of EEG signals, the mental state of the humans can be detected, so it would contribute to the design of human-machine interaction (HMI) systems. In this paper, we intend to study the cognitive state using EEG signals when the subject is performing a visual search task. We tried to obtain the different patterns of the EEG signals when the subject is performing differently in the task. Several pattern recognition algorithms on the signal are conducted to find the principal features in the EEG signals. We can see that the features of EEG signals can present the differences between cognitive states and this result will be beneficial to the recognition of the cognitive state of the operators in the complex systems.

Keywords: EEG · Signal processing · Cognitive state · Pattern recognition

1 Introduction

The human-machine interaction (HMI) has been one of the most important research fields in complex systems. The performance of complex systems is affected not only by the machine systems, but also by the human factors. With the improvement of technology, the effect of human factors has become more and more significant [1]. Of all the human factors considered, emotion and cognitive states of operators have been more and more important issues [2]. As a result, the study of the cognitive states of operators while interacting with the machine interface has been of greater importance.

Generally, there are two groups of methods to obtain the information about the operators' emotion or cognitive states. One way is based on the observation of the operators' overt behavior (e.g. operators' facial expression [3], speech [4] or the prosody [5]). The other way is about the processing of the physiological signals, such as the electroencephalogram (EEG) signals [6], the cardiovascular responses [7, 8] and the muscle activity [9, 10].

Among all the physiological signals, the electroencephalogram (EEG) has always been regarded as a crucial information source about human beings' mental states or behaviors, for the reason that brain activity is strongly associated with perceptive, experiential and expressive emotional processes [11]. Accordingly, numerous studies are concerned with the processing of the EEG signals, attempting to correlate the EEG signal patterns with the human beings' mental states (e.g. the emotions). In all these

© Springer Nature Switzerland AG 2020
H. Ayaz (Ed.): AHFE 2019, AISC 953, pp. 363–372, 2020.
https://doi.org/10.1007/978-3-030-20473-0_35

researches, considerations include the adjustment of the signal processing methods [12] and the different choices of the EEG signal features [13] and the reform of the pattern recognition algorithms [14].

The correlations between EEG signals and human states made by these researches can be instructive and meaningful in the design of HMI systems. But to ensure the generalization of the laboratory results to the real-world application, the experiments are best designed in a way that resembles the concrete environment, such that the applicability of the experimental results can be largely broaden.

In this paper, we have designed a visual search task, trying to analyze the cognitive states of the human beings when they are conducting the visual search task. The study is aimed to investigate whether cognitive states of the humans are different when the conditions of the experiment changes, based on the analysis of the EEG signals we have extracted. These results will be beneficial to the recognition of **the cognitive state of the operators** and be helpful for the design of the complex HMI systems.

The remainder of the paper can be divided into four parts. In part two, we will introduce the details of experiment procedure and the raw EEG signal data collected. In part three, we will process the EEG signals, transforming the signals into frequency domain, and select the features that can best represent the data collected. In part four, we will visualize the data processed in two-dimensional spaces and train a SVM model to complete the final classification task. Part five will be a conclusion and the review of all the work we've done.

2   Experiment Procedure

In this experiment, we are aiming at comparing the EEG signal patterns while humans are in different cognitive modes. In order to simulate the task-finishing environment, we have designed the visual searching tasks and extracted the EEG signals of different cognitive states of the subjects.

2.1   Experiment Design

In our experiment, the subjects would be presented with a table of numbers ranging from 1 to 50 on a touchscreen. The numbers on the screen will be arranged randomly. The subject is instructed to press the number with his/her finger in order (i.e. 1, 2, 3…). Obviously, the subject will spend some time searching on the screen for the right number before he/she presses it. Figure 1 shows an example of the arranged numbers. Once the right number has been pressed, it will disappear from the screen, but if the wrong number is pressed, it will make no change on the screen, until the right one has been pressed.

To record the EEG signal data, the subject is wearing an electrode cap (Cognionics, Cognionics, San Diego, CA, USA) on his/her head while he/she is performing the visual searching task. The sampling rate of the EEG signals is 500 Hz. Moreover, there are 32 electrode channels on the electrode cap, which can record the EEG signals from different locations of the hemisphere. Figure 2 gives the approximate locations of the 32 channels on the electrode cap.

9	12	45	47	1	15	49	17	19	46
28	36	13	50	20	48	35	14	34	3
5	21	30	8	6	26	43	40	7	23
11	32	39	22	18	4	44	38	37	24
42	33	29	16	31	41	2	27	25	10

Fig. 1. The table of numbers

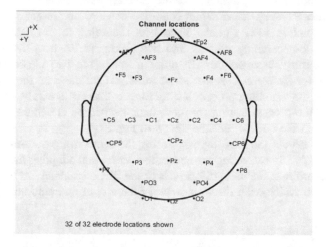

Fig. 2. The 32 channels and the corresponding locations

At the same time, the touchscreen also recorded the number that the subject truly pressed each time and the time he/she spent searching and pressing the number. Besides that, every time when the experiment begins, we make a special mark signal that can be received by the electrode cap, so that the EEG signals being recorded by the cap can synchronize with the touchscreen recording the time of the pressing action. Based on the start time of the whole experiment and the time intervals spent on each searching, the EEG signals can be split into pieces that correspond to the searching process of each number.

2.2 Subjects

We have invited five subjects to take part in this experiment. All of them are students or employees in Shanghai Jiao Tong University (SJTU). They all had perfect vision and stated healthy when they participated in this experiment. Before the experiment, all of them were well informed of the purpose and the procedure of the experiment and signed the experimental consent agreement. We adhered to the tenets of the Declaration of Helsinki in this research.

2.3   Data Collection

Figure 3 shows the details about the time that the five subjects spent on pressing the numbers. In Fig. 3, each subject is represented in a single subfigure. In each subfigure, the x-axis represents the numbers the subject was supposed to press (from 1 to 50) and the y-axis represents the time he/she spent searching for and pressing it. (Here, the time intervals corresponding to the pressing of wrong numbers have been ignored). From the subfigures, it can be seen that the distribution of time periods spans widely. The longest one is longer than 8 s, while the shortest one is about 0.1 s.

In this experiment, we assume that the different time duration of the searching process indicates the subject's different cognitive states, which might be verified based on the corresponding EEG signals. Therefore, according to the data collected, we roughly divide the searching process into two classes and extract the corresponding EEG signals during these periods. In order to make the two classes (namely two different cognitive states) as separate as possible, we set the time thresholds as 0.4 s and 4 s, that is, the searching process lasting for less than 0.4 s will be classified as the first class; and the process lasting for more than 4 s will be classified as another. (As can be seen from the dashed red line drawn in Fig. 3).

Finally, we extract the EEG signals during the searching processes that has been labelled differently. By extracting the adequate features and adopting the proper pattern recognition algorithms, the main differences of the EEG signals in different class could be differentiated, which, as a result, can verify our earlier assumptions.

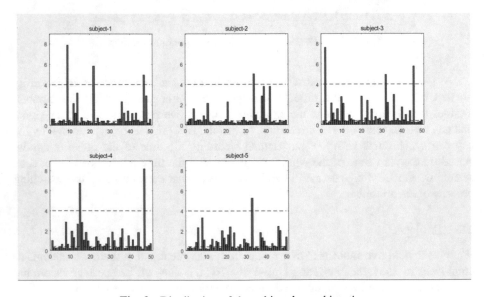

Fig. 3. Distribution of the subjects' searching time

3   Data Processing and Feature Selection

3.1   Raw Data Processing and Visualization

In the processing of EEG signals, spectral power in various frequency bands is often considered most [15]. So firstly, we analyze the frequency components that are contained in the EEG signals. Here, in order to gain a result that seems more robust, we first concatenate all the signals that have been classified in the same class into a long one. We assume that the signals that are classified in the same class will be of similar properties in the frequency domain.

After the concatenating, two long signals that represent the two separate class respectively are obtained. Then, we transform the two signals in the frequency domain to detect their differences. Here the Welch method has been used, by which the signal was split into 256 parts and every part is multiplied with the Hamming window function. Then, we make the FFT transformation on every part. The final results will be obtained by averaging over all the 256 FFT results.

Figure 4 shows the detailed results obtained after performing the procedure described above. Here we deal with the EEG signals in 32 channels separately, as the signals in different channels will be of different frequency properties. Each one of the subfigures in Fig. 4 depicts the power spectrum of the EEG signal in one channel and the two different class have been plotted in black and red lines respectively. According to Fig. 4, it is obvious that two main frequency components are contained in almost every one of the 32 channels, frequency components within the range from 8 to 15 Hz, which is α-band wave and frequency components within the range from 45 to 55 Hz, which is the γ-band wave. In addition, there exist obvious differences in the signal power between the two classes.

In order to give a more obvious presentation of the two main band power, Fig. 5 shows Brain Electrical Activity Mapping (BEAM) [16] of the band power results.

The upper two subfigures in Fig. 5 show the α-band power in all the 32 channels of the two types of searching behaviors. From the two subfigures, we can see that the power distribution of the two types of searching is different. In shorter time searching, the peak value of the power lays in the location of channel Fz, CP5 and PO4; the power on other regions of the hemisphere seem to be weaker comparatively. However, when in the case of longer time searching, the peak value of the α-wave power lays on the location of the channel Pz and Fp2. Moreover, the peak values in two types of searching behavior is different as well. The peak value of shorter time searching is larger than that of longer time searching.

The bottom two subfigures in Fig. 5 depict the γ-wave power of the two searching behaviors. Different from the α-wave, the γ power distribution is similar in the two types of searching. Both of the peak values lay on the location of the channel F3, but also, the absolute peak value of the two circumstances is different. The peak value in longer time searching is approximately three times larger than the value in the case of shorter time searching.

368   S. Zhang et al.

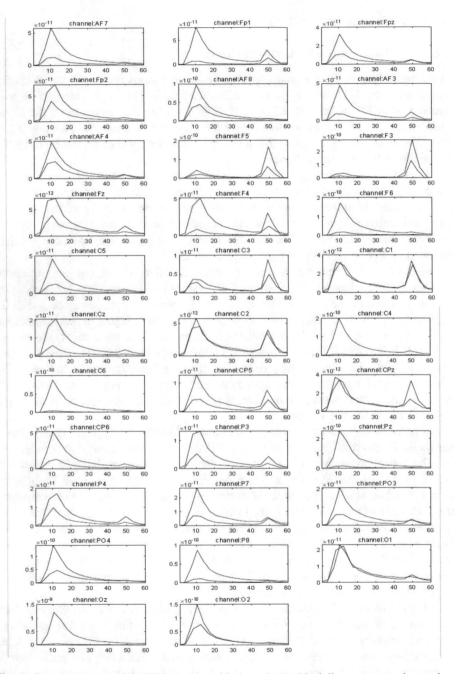

Fig. 4. Power spectrum of the EEG signals in 32 channels (the black line represents shorter time searching; the red line represents the longer time searching)

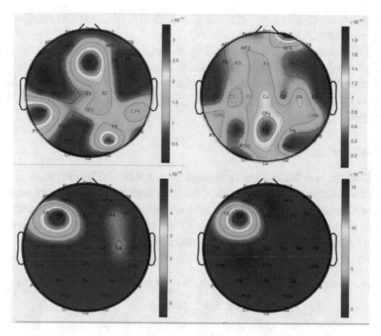

Fig. 5. Band power of two types of searching behavior (Upper left: α-band power of class 1; upper right: α-band power of class 2; bottom left: γ-band power of class 1; bottom right: γ-band power of class 2)

3.2 Feature Selection

In the light of the property that is showed in the power spectrum of the EEG signals, we decide to choose the band power, which ranges from 8 to 15 Hz, namely the α-band power, and the power ranging from 45 to 55 Hz, the γ-band power as the main features. We extract these two features from every one of the signals in 32 channels, and then concatenate them. So, all of the data collected is represented by a feature vector of 64 dimensions.

Here, all the samples needed to train the pattern recognition model should be of equal time length, such that the band power calculated will be under the same standard. Therefore, we cut the two long time signals obtained before into pieces of equal length (here we choose the length to be 500 data points, namely 1 s in time) to obtain the samples in two classes.

4   Dimension Reduction and Classification

4.1   Dimension Reduction

Before the training of the classification model, we will visualize all the samples to see the distribution of the data that have been collected. We conduct the PCA algorithm on all the sample data collected, reducing the dimensionality of each sample data from 64 dimensions to two dimensions. The left figure in Fig. 6 shows the sample data after being processed by PCA algorithm.

Here it is shown in the left figure of Fig. 6, that the sample data have been grouped in two separate classes obviously. The data in the first class, which represent subjects' cognitive state of shorter time searching, lay almost in the lower part of the data space, and the data in the second class, representing longer time searching state, lay almost in the upper part.

4.2   Classification

For classification, we use the linear support vector machine (SVM) algorithm to process the data collected. All the data used to train the SVM model is the ones that has been processed by the PCA algorithm. The right-hand side of Fig. 6 shown below depicts the model that has been trained and the classification results it attained. The final classification accuracy of the SVM model is 96.07%.

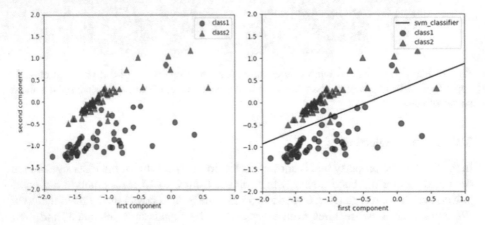

Fig. 6. Data distribution and the SVM model trained

This result can verify the statement that the two types of the searching behavior, the shorter time searching and longer time searching, can truly represent two very different cognitive states of the subjects. Moreover, the different cognitive state of the subjects can be seen from the power spectrum of the subjects' EEG signals and the classification results that the SVM model finally trained.

5   Review and Conclusion

In this paper, we try to find whether the two types of visual searching behavior (shorter time searching and longer time searching) represents different cognitive states of human beings. In order to do this, we collected the EEG signals of the subjects through a visual searching task.

As the different cognitive states can be shown in the EEG signal patterns, we extract the EEG signals corresponding to shorter time searching and longer time searching respectively to compare their pattern differences. Here, the features considered are the frequency features of the signals. The frequency components are mainly centered on 8–15 Hz and 45–55 Hz and the power of the two types of signals are truly different.

Next, we extracted the signal features – the frequency band power of the two ranges concatenated by the 32 channels – to represent each EEG signal. Through the PCA algorithm, the data shown in the two-dimensional space could be grouped in two separate regions. Finally, a SVM classification model was trained to complete the classification task and the model finally obtained had attained great accuracy of classification.

The result revealed that there exist differences in the EEG signal patterns when human beings are in different cognitive states. This result will be meaningful for the future study in the field of EEG signal and will be beneficial to the recognition of the cognitive state of the operators in the complex HMI systems.

References

1. Proctor, R.W., Van Zandt, T.: Human Factors in Simple and Complex Systems. Allyn & Bacon, Needham Heights (1994)
2. Wang, X.-W., Nie, D., Lu, B.-L.: EEG-based emotion recognition using frequency domain features and support vector machines (2011). Source: DBLP. https://doi.org/10.1007/978-3-642-24955-6_87
3. Black, M., Yacoob, Y.: Recognizing facial expressions in image sequences using local parameterized models of image motion. Int. J. Comput. Vision 25(1), 23–48 (1997)
4. Petrushin, V.: Emotion in speech: recognition and application to call centers. In: Proceedings of Artificial Neural Networks in Engineering, pp. 7–10 (1999)
5. Zeng, Z., Pantic, M., Roisman, G.I., Huang, T.S.: A survey of affect recognition methods: audio, visual and spontaneous expressions. In: Proceedings of the 9th International Conference on Multimodal Interfaces, ICMI 2007, pp. 126–133 (2007)
6. Musha, T., Terasaki, Y., Haque, H., Ivamitsky, G.: Feature extraction from EEGs associated with emotions. Artif. Life Robot. 1(1), 15–19 (1997)
7. Frazier, T.W., Strauss, M.E., Steinhauer, S.R.: Respiratory sinus arrhythmia as an index of emotional response in young adults. Psychophysiology 41(1), 75–83 (2004)
8. Lang, P.J., Greenwald, M.K., Bradley, M.M., Hamm, A.O.: Looking at pictures: affective, facial, visceral, and behavioral reactions. Psychophysiology 30(3), 261–273 (1993)
9. Cacioppo, J.T., Petty, R.E., Losch, M.E., Kim, H.S.: Electromyographic activity over facial muscle regions can differentiate the valence and intensity of affective reactions. J. Pers. Soc. Psychol. 50(2), 260–268 (1986)
10. Magnée, M.J., Stekelenburg, J.J., Kemner, C., de Gelder, B.: Similar facial electromyographic responses to faces, voices, and body expressions. NeuroReport 18(4), 369–372 (2007)
11. Demaree, H.A., Everhart, E.D., Youngstrom, E.A., Harrison, D.W.: Brain lateralization of emotional processing: historical roots and a future incorporating dominance. Behav. Cogn. Neurosci. Rev. 4(1), 3–20 (2005)
12. Sanei, S., Chambers, J.: EEG Signal Processing. Wiley, New York (2007)

13. Schaaff, K., Schultz, T.: Towards emotion recognition from electroencephalographic signals. In: Proceedings of International Conference on Affective Computing and Intelligent Interaction, pp. 175–180, September 2009

14. Kim, M.-K., Kim, M., Oh, E., Kim, S.-P.: A review on the computational methods for emotional state estimation from the human EEG. Comput. Math. Methods Med. **2013**, 1–13 (2013)

15. Jenke, R., Peer, A., Buss, M.: Feature extraction and selection for emotion recognition from EEG. IEEE Trans. Affect. Comput. **5**(3), 327–339 (2014)

16. Duffy, F.H.: Brain electric activity mapping, a method for extending the clinical utility of EEG and evoked potential data. Ann. Neurol. (2011). https://doi.org/10.1002/ana.410050402

Impact of Auditory Alert on Driving Behavior and Prefrontal Cortex Response in a Tunnel: An Actual Car Driving Study

Noriyuki Oka[1(✉)], Toshiyuki Sugimachi[1], Kouji Yamamoto[2],
Hideki Yazawa[3], Hideki Takahashi[4], Jongseong Gwak[5],
and Yoshihiro Suda[5]

[1] Department of Mechanical Engineering, Tokyo City University, Tokyo, Japan
`okanoriyuki0124@gmail.com, tsug@tcu.ac.jp`
[2] Yokohama Maintenance/Customer Service Center, Tokyo Branch,
Central Nippon Expressway Co., Ltd., Tokyo, Japan
`k.yamamoto@c-nexco.co.jp`
[3] Department of Corporate Strategy,
Central Nippon Highway Engineering TOKYO Co., Ltd., Tokyo, Japan
`h.yazawa.aa@c-nexco.co.jp`
[4] Department of Corporate Strategy, Central Nippon Highway Engineering,
NAGOYA Co., Ltd., Nagoya, Japan
`h.takahashi.a@c-nexco.co.jp`
[5] Institute of Industrial Science, The University of Tokyo, Tokyo, Japan
`{js-gwak, suda}@iis.u-tokyo.ac.jp`

Abstract. For traffic safety purposes, appropriate preparation time for emergent events is important. This study investigated the effect of an auditory (verbal) alert on a driver in a tunnel, specifically analyzing prefrontal cortical activation that related to attention and driving behavior. Eighteen healthy adults (35.2 ± 8.9 years old) participated by driving two times at about 60 km/h in the right lane of a two-lane road, once with an auditory alert of an oncoming construction event in the left lane, and once without the alert. The auditory alert increased prefrontal cortex activation and mitigated the sudden decrease in accelerator pedal opening related to the driver seeing the event. Our findings suggest that auditory alerts may promote smooth driving behavior and suggest they may be possible traffic safety measures.

Keywords: Car driving · Traffic safety measures · Acceleration change · Prefrontal cortex activation · Functional near-infrared spectroscopy (fNIRS)

1 Introduction

For safe driving, higher brain functions including cognition, decision-making, and other behaviors are required [1]. The foundation of brain executive function is attention, according to the neuropsychology pyramid [2]. Therefore, traffic safety measures that increase attention can enhance the cognitive function necessary for driving a vehicle and contribute to the avoidance of traffic accidents caused by human error.

© Springer Nature Switzerland AG 2020
H. Ayaz (Ed.): AHFE 2019, AISC 953, pp. 373–382, 2020.
https://doi.org/10.1007/978-3-030-20473-0_36

Driving is a complex activity that acquires data through vision, including more than 90% of the necessary information for input and execution [3]. As a traffic safety measure, it is important the driver identifies the events that should be avoided as soon as possible and be prepared to avoid them. However, in a situation where speed is high, such as highway driving, or when the field of vision of the driver is narrowed, the presentation of visual information (e.g., in a sign) provides limited safety benefit. For this reason, we focused on using auditory alerts. Providing information auditorily rather than visually can deliver information even when the driver is in a passive state and can trigger the driver to become active and visually attentive without looking away. Active visual attention is correlated with the neural network between the intraparietal sulcus and frontal eye field [4]. In addition, activity in the right dorsolateral prefrontal cortex (PFC) is reportedly increased when driving at an appropriate speed, as opposed to driving excessively fast [5]. Right dorsolateral PFC was also reported to be activated when choosing to reduce risk-taking behavior [6]. From the above, we hypothesize that an auditory alert would promote visual attention to the threat event and increase PFC activation, reducing risky behavior and thus prompting safe behavior.

In this study, we analyzed the influence of an auditory warning on changes in PFC activation and driving behavior. For measurement of PFC activity, we employed functional near-infrared spectroscopy (fNIRS), which has been found to be useful for automobile driving research [7] because the brain activity of a driver can be measured in real time [8–10].

2   Methods

2.1   Participants

Eighteen healthy subjects (35.2 ± 8.9 years old; 9 male and 9 female) participated in this experiment; all were right-handed according to the Edinburgh Handedness Inventory. Before the experiment, the participants received written and verbal explanations of the purpose of the experiment in accordance with documents approved by the Office for Life Science Research Ethics and Safety of the University of Tokyo, which approved all experimental procedures. Written informed consent was obtained from all participants.

2.2   Vehicle and Driving Behavior Data Acquisition

A Toyota Estima hybrid (Toyota Motor Corporation, Japan) was used as the test car (Fig. 1). Measures of driving behavior included the steering wheel angle, the amount of accelerator stroke and the amount of brake stroke; all were measured via the Controller Area Network (CAN) device. In addition, vehicle position was measured using GPS. For locations in the tunnel and low-precision GPS, position information was obtained by integrating the velocity. Vehicle speed was measured using a vehicle speed sensor. Synchronization between driving behavior data obtained by the CAN and brain activity data obtained by fNIRS was performed by inputting a trigger with a laser displacement meter attached to the front of the vehicle based on a conventional study [8, 9].

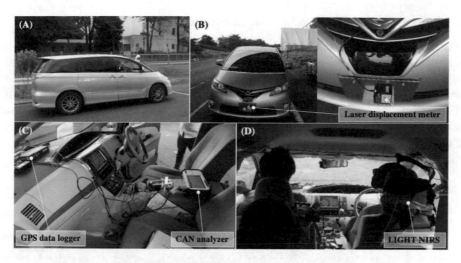

Fig. 1. Test vehicle and driver analysis equipment. (A) Overview of the test vehicle, and locations of (B) the laser displacement meter, (C) GPS data logger and Controller Area Network (CAN) analyzer, and (D) functional near-infrared spectroscopy (LIGHTNIRS) equipment.

2.3 Brain Activity Measurement

Brain activity was measured using a portable multi-channel fNIRS device (LIGHT-NIRS, Shimadzu Co., Japan) (Fig. 2). A handmade probe attachment with 8 light-emitter probes and 8 detector probes arranged in a lattice pattern was used. The distance between each probe was 3 cm. The number of channels, defined as any combination of adjacent, paired emitter, and detector probes, was 22 (Fig. 2A). The probe was applied such that the center channel closest to the face (channel 19) was placed 6 cm upward from the nasion (Fig. 2A); the prefrontal cortex (PFC), including Brodmann areas (BA) 8, 9, and 46, was analyzed (Fig. 2B). Concentration changes of oxyhemoglobin (ΔoxyHb) and deoxyhemoglobin (ΔdeoxyHb) were measured in continuous mode. Measurements were sampled once per 75 ms.

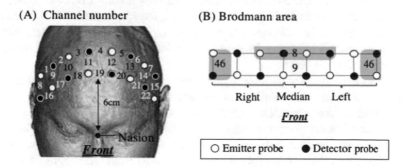

Fig. 2. Functional spectroscopic probe placement. (A) Numbers between probe pairs indicate the channel number. Probes were 3 cm apart, and the unit was placed using channel 19 and the nasion as the reference points. (B) Numbers (and gray areas) indicate the assumed Brodmann area based on probe placement.

2.4   Experimental Design and Task Environment

The experiment was conducted on a two-lane straight road with a real tunnel (total length 695 m) provided by the National Institute for Land and Infrastructure Management (Fig. 3). The distance to the entrance of the tunnel was 506 m from the start point. Speakers for auditory alerts were installed at six intervals of 50 m from the 631 m point. At 1248 m from the start, we set up a vehicle that stopped in the left lane and a signboard that urged drivers to move to the right lane due to construction. Participants were able to see the event at around the 776–806 m point.

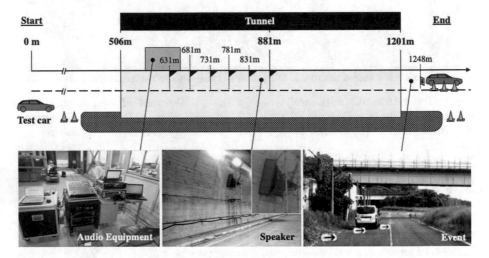

Fig. 3. Experimental course design. The auditory alert speakers were arranged so the voice could be heard two or more times by the driver within the 506 m to 881 m region. The driver was able to see the event at around the 776–806 m point.

At the start point, the experimenter instructed the participants to drive at 60 km/h in compliance with the Japanese Road Traffic laws, accelerating from the right lane to start. In order to avoid excessive attention to speed, the speedometer was hidden when the vehicle speed reached 60 km/h. Before entering the tunnel, all subjects had successfully exceeded 60 km/h and the speedometer was hidden, so the tasks were under a constant-speed driving condition.

Two tasks were set as experimental conditions: a no-alert condition ("without-alert" task); and an alert condition (auditory alerts used to notify the fact that the construction site is located at the destination ("with-alert" task). The alert comprised a voice announcement of "Construction is being carried out at running direction" with a noise level of 96 dB on average. Participants completed the experimental course once for each condition, and the task execution order was randomized. After finishing with the alert task, we confirmed whether the participants understood the contents of the voice announcement.

2.5 Data Analysis

In order to unify the analysis points of all acquired subjects' behavior and brain activity, we normalized the data from time series to distance-dependent points using increments of 1 m [8, 9]. The normalization of CAN and fNIRS data to mileage-dependent data was done in three steps. First, the same analog signal was used, and the CAN data was combined synchronously according to the NIRS sampling time (75 ms). Second, positional information was added to the time series data using location information obtained by integrating the vehicle speed to get the distance traveled in the tunnel (the GPS signal is blocked by the tunnel). Finally, data at the closest traveling position were extracted per meter and recreated using the traveling position as a reference. When the data according to the running speed of the vehicle were unobtainable, the values were determined by linear interpolation. All analyses were performed using these standardized data. Statistical significance was set to 0.05 for all analyses. The effect size was calculated using Eq. (1):

$$Effect\ size\ r = \sqrt{\frac{(t\ value)^2}{(t\ value)^2 + degree\ of\ freedom}} \qquad (1)$$

2.5.1 Brain Activation With- and Without-Alert

We attempted to detect the effect of the auditory alert on PFC activity based on NIRS-derived differences in the changes in oxy- and deoxyhemoglobin ([ΔHb difference (Hbdiff)] = [ΔoxyHb] − [ΔdeoxyHb]). Hbdiff is a validated indicator of blood oxygenation [11, 12] that is increasingly negative as cellular oxygen metabolism increases (i.e. with increased brain activity).

According to the criteria of the guidelines of the driving support system [13], the distances within which data are analyzed range from the 506 m point of the tunnel entrance (where the auditory alert can be heard) to the 731 m point (prior to the driver being able to see the event). To extract event-related brain activity, Hbdiff was referenced to its value at the entrance of the tunnel. The integrated value of Hbdiff for the range of 506 m - 731 m was calculated for each task (with/without-alert) and compared between tasks by paired t-test.

2.5.2 PFC Activation Related to Driving Behavior

Mapping analysis was carried out using the mean changes in Hbdiff for each meter after the 506 m point. In addition, for each channel in which PFC activation was significantly increased by the alert (see Sect. 2.5.1), paired t-tests were used to compare Hbdiff between tasks at each distance point. Also, the accelerator stroke value at each distance point was compared between tasks via paired t-tests; this value is a temporary reaction related to speed change that allows for an examination of the relationship between brain activity and driving behavior.

2.5.3   Effect of an Auditory Alert on the Smoothness of Driving Behavior

We evaluated the influence of the auditory alert on driving behavior by assessing the smoothness of the driving behavior; smoothness was represented by the absolute value of the amount of change in acceleration from meter to meter over the range of 506 m to 1201 m. The average between-meter change in the absolute values in all the sections of the tunnel was compared between the alert/no-alert tasks by paired t-tests.

3   Results

3.1   Greater PFC Activation with the Auditory Alert

PFC ΔHbdiff was significantly more negative (i.e. PFC was significantly more activated) in the "with-alert" task than in the "without-alert" task throughout the distances in which the auditory alert was given but the event could not be seen (Fig. 4, Table 1). No individual channels demonstrated significantly lower Hbdiff values in the without-alert task than in the with-alert task, indicating that the alert increased activity throughout the PFC.

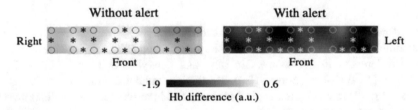

Fig. 4. Mapping analysis of differences in prefrontal cortex activation with and without the auditory alert. A map of the integrated values of the change in Hbdiff between 506 m and 731 m is shown. Asterisks indicate a significant difference depending on the presence or absence of the auditory alert.

Specific regions where brain activity was significantly increased included: bilateral BA9 (right hemisphere: channel 2, 9, 10, 11, 17, and 18, p value = 0.001–0.036, effect size r = 0.483–0.688; left hemisphere: channel 12, 14, and 21, p value = 0.015–0.035, effect size r = 0.0486–0.547); bilateral BA46 (right hemisphere: channel 8, p value = 0.039, effect size r = 0.478; left hemisphere: channel 22, p value = 0.007, effect size r = 0.594), and central BA8 (channel 4, p value = 0.002, effect size r = 0.665).

Of the thirteen channels that significantly increased activation but excluding the central two channels (channel 4 and channel 19), the number of channels that showed a decrease in Hbdiff was 1.75 times more in the right PFC than in the left PFC (left PFC:4 channels; right PFC: 7 channels). This suggests the possibility that PFC activation against the alert may increase more in the right hemisphere than left hemisphere.

Table 1. Differences in prefrontal cortex activation with and without the auditory alert. Only channels with significant difference by task comparison are shown.

Channel	Hemisphere	BA	Hb difference (a.u., Average ± SEM)		p value	Effect size
			With alert	Without alert		
2	Right	9	−0.898 ± 0.297	0.014 ± 0.371	0.026	0.510
9		9	−1.166 ± 0.506	0.318 ± 0.362	0.025	0.511
10		9	−1.381 ± 0.449	0.144 ± 0.351	0.010	0.577
11		9	−1.398 ± 0.386	0.530 ± 0.344	0.001	0.688
17		9	−1.175 ± 0.508	0.315 ± 0.286	0.004	0.628
18		9	−0.900 ± 0.461	0.389 ± 0.367	0.036	0.483
8		46	−1.581 ± 0.524	0.008 ± 0.417	0.039	0.478
12	Left	9	−0.881 ± 0.292	0.559 ± 0.418	0.015	0.547
14		9	−1.320 ± 0.334	−0.132 ± 0.570	0.031	0.496
21		9	−1.055 ± 0.467	0.191 ± 0.437	0.035	0.486
22		46	−1.937 ± 0.477	0.166 ± 0.742	0.007	0.594
4	Medial	8	−0.957 ± 0.268	0.287 ± 0.184	0.002	0.665
19		9	−0.918 ± 0.269	0.293 ± 0.303	0.019	0.533

3.2   Relationship Between PFC Changes and Driving Behavior

Figure 5A shows the mapping analysis of the distance-based data. Hbdiff decreased in the with-alert task. In the area of the most lateral channels (including BA46), ΔHbdiff decreased from the tunnel entrance to the exit, representing a gradual increase in activity, regardless of task. In the alert task, the activity of BA46 also increased in the speaker installation section. On the other hand, in the medial channel regions, mainly composed of BA8 and BA9, ΔHbdiff gradually decreased from the entrance of the tunnel and increased again toward the tunnel exit. Tracking of the Hbdiff changes for the channels with significant differences above (Sect. 3.1) identified a waveform showing a distinct unimodal negative peak only in channel 10 during the with-alert task (Fig. 5B). The most negative peak of Hbdiff was at 828 m, and was significantly lower in the with-alert task than in the without-alert task [t(17) = −2.197, p value = 0.042, effect size r = 0.470)]. In addition, no waveform showing a negative peak could be identified in the without-alert task.

The degree of accelerator pedal opening (i.e. accelerator stroke) showed a continuous and significant difference between 800 m and 857 m [t(17) = 2.145–4.984, p value = 0.046–0.0001; effect size r = 0.461–0.770]. The degree of accelerator pedal opening was significantly decreased in the without-alert task compared to the with-alert task (Fig. 5B). This decrease in accelerator pedal opening was identified from the 789 m point to the 839 m point in the without-alert task. The linear approximation for this range of distances in this task was represented by y = −0.233x + 17.04 (R^2 = 0.979). In the with-alert task, the linear approximation for the same distance range was represented by y = −0.067x + 19.11 (R^2 = 0.877). Regardless of task, the degree of accelerator opening decreased from the 789 m point. In summary, compared to the without-alert task, the increase in PFC activation in the with-alert task was correlated with a decline in the degree of accelerator opening and slower acceleration.

3.3   The Effect of an Auditory Alert in Promoting Smooth Driving Behavior

The average absolute value change in acceleration meter-over-meter (a smaller value indicates smoother driving behavior) was lower in the with-alert task than in the without-alert task, but not significantly different (with-alert: 0.030 ± 0.011; without-alert: 0.032 ± 0.012; t(17) = 1.367, p value = 0.189). A moderate effect size (r = 0.31) was detected. It is possible that the auditory alert reduced more drastic speed changes.

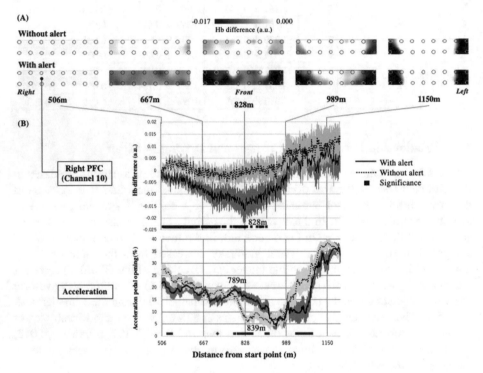

Fig. 5. Relationship between prefrontal cortex activation and driving behavior. (A) The result of mapping analyses of Hbdiff at each point. The blacker the decrease of Hbdiff, the higher the brain activity. (B) Average waveform data of Hbdiff and accelerator opening of channel 10 (right BA9) of the tunnel section at each point are shown. The solid line shows the with-alert task, and the dotted line shows the without-alert task. Bars indicate SEM. The plot shown at the lowest value on the vertical axis of each chart shows that there is a significant difference in the values between the tasks.

4 Discussion

4.1 Relationship Between PFC Activation and Driving Behavior

Higher PFC activities related to the auditory alert tended to differ between medial and lateral PFC. The brain activity of the medial PFC showed unimodal activity, especially in the right BA9, which clearly showed the peak at 828 m. We believe that the peak seen in brain activity at the 828 m point was not related to the auditory alert itself, despite being within the distances including the audio speakers. Rather, we hypothesize that it is the effect of the enhancement of attention and risk-reduction PFC activity related to safety considerations evoked by the alert. In fact, some reports [5, 6] have indicated that the activation of right dorsolateral PFC reduces the likelihood of choosing riskier behaviors.

The starting point of the decrease in the accelerator pedal opening (i.e. beginning the speed adjustment) occurred at about the same point regardless of the alert status. However, the reduction in acceleration and the degree of accelerator pedal opening were greater in the without-alert than the with-alert task. Although no eye-tracking data were obtained in this experiment, the 789 m distance puts the event within the range of the participant's sight. Thus, it was possible that the participant seeing the event caused the decrease in the accelerator pedal opening. Our data suggest that the PFC activity increased until the event was visually perceived and then decreased thereafter. Areas with increased PFC activity included BA8, which controls eye movements [14], and BA9, which plays roles in focusing attention on a human voice [15] and selecting actions to reduce dangerous driving [5]. In the with-alert task, we believe that driving continued without sudden or dramatic changes after visually confirming the event because the participant was already mentally prepared for the event.

Lateral PFC (BA46) activity increased as the event was approached. In a previous study [8], BA46 activity increased during acceleration. Thus, increased lateral PFC activity may be caused by accelerating back to regular speed following the deceleration needed to allow for visual confirmation of the event. The activation of the BA46 is compatible with the usage of working memory [16]. Therefore, we suspect BA46 activity increased in the speaker section due to the use of working memory to recognize and recall the meaning of the warning language.

5 Conclusion

The presentation of auditory information before visual information was obtainable suggests it is possible to enhance the driver's attention and promote smooth driving behavior. Depending on the content of the information provided, there may be individual differences in the possible behaviors, so it is necessary to consider the contents of the presentation with regard to individual differences between listeners.

References

1. Anstey, K.J., Wood, J., Lord, S., Walker, J.D.: Cognitive, sensory and physical factors enabling driving safety in older adults. Clin. Psychol. Rev. **25**, 45–65 (2005)
2. Tachigami, S.: Rusk institute of rehabilitation medicine, brain injury day treatment program. In: Ohashi, M., Ben-Yishay, Y. (Eds.) Igaku-Syoin Ltd., Tokyo (2010) (in Japanese)
3. Simms, B.: Perception and driving: theory and practice. Br. J. Occup. Ther. **48**, 363–366 (1985)
4. Corbetta, M., Shulman, G.L.: Control of goal-directed and stimulus-driven attention in the brain. Nat. Rev. Neurosci. **3**, 201–215 (2002)
5. Jancke, L., Brunner, B., Esslen, M.: Brain activation during fast driving in a driving simulator: the role of the lateral prefrontal cortex. NeuroReport **19**, 1127–1130 (2008)
6. Fecteau, S., Knoch, D., Fregni, F., Sultani, N., Boggio, P., Pascual-Leone, A.: Diminishing risk-taking behavior by modulating activity in the prefrontal cortex: a direct current stimulation study. J. Neurosci. **27**, 12500–12505 (2007)
7. Liu, T., Pelowski, M., Pang, C., Zhou, Y., Cai, J.: Near-infrared spectroscopy as a tool for driving research. Ergonomics **59**, 368–379 (2016)
8. Yoshino, K., Oka, N., Yamamoto, K., Takahashi, H., Kato, T.: Functional brain imaging using near-infrared spectroscopy during actual driving on an express-way. Front. Hum. Neurosci. **7**, 882 (2013)
9. Orino, Y., Yoshino, K., Oka, N., Yamamoto, K., Takahashi, H., Kato, T.: Brain activity involved in vehicle velocity changes in a sag vertical curve on an expressway: vector-based functional near-infrared spectroscopy study. J. Transp. Res. Board **2518**, 18–26 (2015)
10. Oka, N., Yoshino, K., Yamamoto, K., Takahashi, H., Li, S., Sugimachi, T., Nakano, K., Suda, Y., Kato, T.: Greater activity in the frontal cortex on left curves: a vector-based fNIRS study of left and right curve driving. PLoS ONE **10**, e0127594 (2015)
11. Wagner, B.P., Ammann, R.A., Bachmann, D.C., Born, S., Schibler, A.: Rapid assessment of cerebral autoregulation by near-infrared spectroscopy and a single dose of phenylephrine. Pediatr. Res. **69**, 436–441 (2011)
12. Tempest, G.D., Eston, R.G., Parfitt, G.: Prefrontal cortex haemodynamics and affective responses during exercise: a multi-channel near infrared spectroscopy study. PLoS ONE **9**, e95924 (2014)
13. Ministry of Land, Infrastructure, Transport and Tourism: Guidelines for communication assisted driving support system. (in Japanese). http://www.soumu.go.jp/main_content/000288894.pdf
14. Pierrot-Deseilligny, C., Milea, D., Muri, R.M.: Eye movement control by the cerebral cortex. Curr. Opin. Neurol. **17**, 17–25 (2004)
15. Nakai, T., Kato, C., Matsuo, K.: An fMRI study to investigate auditory attention: a model of the cocktail party phenomenon. Magn. Reson. Med. Sci. **4**, 75–82 (2005)
16. Petrides, M., Alivisatos, B., Meyer, E., Evans, A.C.: Functional activation of the human frontal cortex during the performance of verbal working memory tasks. Proc. Natl. Acad. Sci. U.S.A. **90**, 878–882 (1993)

A Research Agenda for Neuroactivities in Construction Safety Knowledge Sharing, Hazard Identification and Decision Making

Rita Yi Man Li[1(✉)], Kwong Wing Chau[2], Weisheng Lu[2], and Daniel Chi Wing Ho[3]

[1] HKSYU Real Estate and Economics Research Lab,
Shue Yan University, North Point, Hong Kong
ymli@hksyu.edu
[2] Department of Real Estate and Construction,
The University of Hong Kong, Pok Fu Lam, Hong Kong
[3] Faculty of Design and Environment, THEi, Chai Wan, Hong Kong

Abstract. Neuroscience is often associated with psychology or medical studies. In this research, we propose a research agenda for studying neuroactivities during construction knowledge sharing and hazard identification via a neuroscience tool such as fNIRS. It explores the different regions of the brain which may involve hazard decision making. Processing executive functions, for example, attention, memory and planning tasks are connected to the prefrontal cortex activation. As per the neural efficiency hypothesis, individuals with higher intelligence test scores exhibit less neural activity when they perform a complicated task. The temporal lobes contain many substructures, with functions such as perception, object recognition, memory acquisition, language understanding, and emotional reactions.

Keywords: Neuroscience · fNIRS · Memory · Construction safety

1 Introduction

Decision making is controlled by multiple interacting neural controllers [1]. Understanding the neural basis of human decision making has been the subject of numerous studies due to its impact on daily life activities [2]. To implement the primary brain-load-driven mechanism of task sharing, constraints applicable to different tasks were identified, for example, welding requires welder certification, and helpers are not allowed to assist with either position or roll welding. Similarly, an experienced certified welder or supervisor is merely authorised to perform quality control and inspection [3]. Likewise, the neuroscience-based construction safety knowledge sharing helps to understand the ways which may be helpful for knowledge retention, for example, if knowledge shared by the Internet of Things, Web 2.0, mobile apps or even chatbot contains some level of excitement, that may trigger a flashbulb memory.

Construction activities have long been regarded as hazardous activities [4], so identifying ways to reduce the hazard via knowledge sharing are essential to minimise

© Springer Nature Switzerland AG 2020
H. Ayaz (Ed.): AHFE 2019, AISC 953, pp. 383–389, 2020.
https://doi.org/10.1007/978-3-030-20473-0_37

risks. In this paper, we attempt to offer a conceptual idea and a new research agenda to study how people recognise dangers and enhance safety knowledge sharing through neuroscience viewpoints.

2 Neuroactivities and Decision

The various parts of the brain have different functions (Table 1). For example, the prefrontal cortex (PFC) is involved in high order cognitive functions and its activity can be assessed by measuring changes in the concentrations of oxyhemoglobin (O_2Hb) and deoxyhemoglobin (HHb) via functional near infrared spectroscopy (fNIRS). Processing of executive functions, such as attention, memory, task planning, and language production, is linked to activation in the prefrontal cortex. The efficiency with which individuals can perform these processes is reflected in the level of neural activity of the PFC, which is associated with glucose metabolism. According to the neural efficiency hypothesis, individuals who score better on intelligence tests and have more expertise in the task exhibit less neural activity when performing a complex task well. Conversely, those who play poorly exhibit more neural activity, that is associated with the recruitment of more neural circuits and higher rates of energy consumption. Nonetheless, factors such as age, sex, task type, task complexity and learning may also influence the observed neural activity. Regardless, an increase in neuronal activity is linked to an increase in regional blood flow to transport glucose and oxygen to meet the increased metabolic demands. This, in turn, alters the concentrations of O_2Hb, HHb and total haemoglobin (tHb) in the brain. Functional magnetic resonance imaging (fMRI) has been used since the 1990s as a neuroimaging technique to assess neural activity, however, although it is considered the gold standard of neuroimaging, it is costly and precludes measurements during movement [5].

The Research Group Ohio Kahiki presented dynamic, noisy, pictures of faces and cars performed at a perceptual categorisation task. Some subjects' dorsal dorsolateral prefrontal cortexes were inhibited by transcranial magnetic stimulation and showed an increase in the reaction time for decision making. The results showed that the Transcranial Magnetic Stimulation of the dorsolateral prefrontal cortex affected the drift rate of the diffusion process, suggesting that the dorsolateral and prefrontal cortex are linked to the accumulation of evidence for decision making. The diffusion model, the most powerful model in neuroeconomics, suggests that the brain monitors the difference for evidence of alternatives and when this difference reaches a threshold, the firing of neurons for decision making reaches the threshold, thereby finalizing the choice. It appears that like the neuronal system, the brain makes this calculation similar to the diffusion process, so, the diffusion model is only one model from a cluster of models called sequential sampling models [6].

Uncertainty decisions include decisions involving solely ambiguity and decisions affecting ambiguity plus risk. Compared to the former, in decisions affecting risk, safe choices usually have a higher probability of gaining a reward. Risky decisions are a particularly important topic for human action and thought. Recently, such risky decisions have attracted increasing attention in the field of social neuroscience regarding the question of how information is used, and a decision made at the level of interacting

Table 1. Roles of different parts of the brain [8]

Frontal lobe	Description	Associated functions
Frontal Lobe	• the largest of the brain structures, they are part of the cerebral cortex • primary site of cognitive functions • contain important substructures including the prefrontal cortex, orbitofrontal cortex, motor and premotor cortices, and Broca's area • involved in attention and thought, voluntary movement, decision making, and language	• executive voluntary behaviour such as decision making, planning, problem-solving, thinking, voluntary motor control, cognition, intelligence, attention, language processing and comprehension etc. • associated with many disorders, including ADHD, schizophrenia, and bipolar disorder (prefrontal cortex)
Parietal Lobe	• plays a vital role in integrating information from different senses to build a coherent picture • integrates information from the ventral visual pathways (which process what things are) and dorsal visual pathways (which method where things are) • coordinates movements in response to the environment; contains a reference map of the body, near and distant space, which are updated continuously as we move • parietal cortex processes attentional awareness of the environment is involved in manipulating objects and representing numbers	• perception and integration of sensory information (e.g. touch, pressure, temperature, and pain) • visuospatial processing • spatial attention • spatial mapping • number representation
Occipital Lobe	• occipital cortex is the primary visual area of the brain, receiving projections from the retina (via the thalamus) from where different groups of neurons separately encode different visual information, such as colour, orientation, and motion • dorsal stream projects to the parietal lobes and processes where objects are located • ventral stream projects to structures in the temporal lobes and processes what objects are	• vision

Frontal lobe	Description	Associated functions
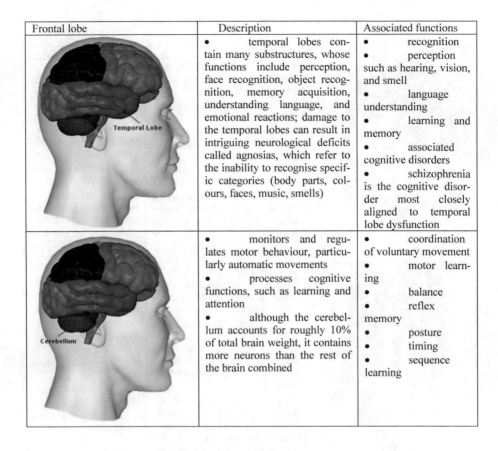	• temporal lobes contain many substructures, whose functions include perception, face recognition, object recognition, memory acquisition, understanding language, and emotional reactions; damage to the temporal lobes can result in intriguing neurological deficits called agnosias, which refer to the inability to recognise specific categories (body parts, colours, faces, music, smells)	• recognition • perception such as hearing, vision, and smell • language understanding • learning and memory • associated cognitive disorders • schizophrenia is the cognitive disorder most closely aligned to temporal lobe dysfunction
	• monitors and regulates motor behaviour, particularly automatic movements • processes cognitive functions, such as learning and attention • although the cerebellum accounts for roughly 10% of total brain weight, it contains more neurons than the rest of the brain combined	• coordination of voluntary movement • motor learning • balance • reflex memory • posture • timing • sequence learning

brains in natural social contexts. Embracing the multi-brain frame, this study aimed to examine the question of how our minds might integrate cognitive processes as well as risky decision behaviour across two persons [7].

All the above studies, however, did not study construction safety risks activities, which may be a new direction for future research.

3   Memory, Training Transfer and Knowledge Sharing

There is much research based on memory, for example, previous studies suggested that knowledge refers to a mixture of values within a social context, the construction of new experiences based on the past elaboration in memory [9]. The International Council on Monuments and Sites suggested that virtual heritage restores historic buildings' memory [10]. Interpretation of how things work is based on the elaboration of previous experiences in memory [11]. Moreover, corporate memory can be considered as a kind of institution which affects land reclamation [12]. In this paper, we propose to study flashbulb memory which may lead to a successful brain-based learning design.

Flashbulb memory is a unique memory mechanism triggered by an event exceeding criteria levels of surprise, creating a permanent record of the contents for the period immediately surrounding the shocking experience. Flashbulb memories have unique characteristics different from those of memories produced by "ordinary" memory mechanisms, implying that memories are products of a particular memory mechanism. The special flashbulb memory mechanism, when triggered by criteria levels of surprise and consequentiality, creates a detailed and permanent record in the memory of an individual's experience immediately before, during, and shortly after learning of the shocking event [13].

Training transfer refers to the extent to which concepts and practices learned via training are transferred or replicated in the workplace. High-engagement training requires that workers play an active role in the learning process, thus, it usually encompasses a high level of interaction between an expert facilitator and among workers. Training may be provided either on-the-job or off-site and feedback concerning performance in the field and during the exercise is offered regularly to inspire perfection [14].

The hippocampus, located in the front and centre of the brain joins the left and right sides, is part of the brain's limbic system which regulates people's emotions, receiving information from the five senses through the sensory nerves to the hippocampus. Neuroscientists discovered that our ability to remember and recall learned information is directly linked to the strength of activity in the hippocampus while we are learning. When we are emotionally aroused during a learning activity, the amygdala will signal to the hippocampus that the subject matter is worth paying attention to and remembering, so, emotional arousal helps learning to happen, that is, learning occurs when we are motivated; if a learner feels motivated to learn and anticipates and experiences reward as part of the learning process, dopamine will be released [15].

Thus, if construction safety knowledge sharing process via mobile apps, Internet of Things and chatbot involves emotional arousal or a certain level of surprise, that should be more effective theoretically speaking. However, the proof relies on the field or lab-based experimental studies, so utilisation of fNIRS may be able to indicate the effectiveness of various safety training (Fig. 3).

4 Functional Near Infrared Spectroscopy

Functional near infrared spectroscopy (fNIRS) is an emerging imaging technique and has distinct advantages because it is non-invasive, inexpensive, shows good temporal resolution, is portable and shows little artefacts with movement. Furthermore, O_2Hb data obtained using fNIRS is consistent with findings from fMRI [5]. As compared with the traditional MRI, fNIRS is less expensive and is even safer than EEG, as it may induce a seizure. The high level of safety means that even infants can be used as research subjects. The neuroimaging is useful for us to understand the how construction workers' neuro response when they see the construction hazard and construction safety knowledge sharing (Figs. 1 and 2).

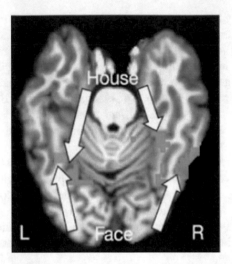

Fig. 1. Neural activities when we see different things [6]

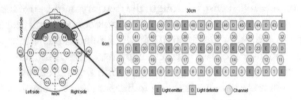

Fig. 2. Placement of optodes for fNIR [16].

Fig. 3. Turning the knowledge shared with long-term memory [15]

5 Conclusions

Construction activities are often high-risk. Different parts of the brain must function well to ensure that the individual is aware of the risks and appropriate responses are essential. Apart from showing the parts of the brain that are involved, we attempt to offer a research agenda to study how to enhance safety knowledge sharing via various means, such as Internet of Things, chatbot, Web 2.0, and mobile apps, from a flashbulb memory perspective.

Acknowledgement. The research is supported by:

Willingness to share construction safety knowledge via Web 2.0, mobile apps and IoT (UGC/FDS15/E01/17).

Ocular behaviour, construction hazard awareness and an AI chatbot (UGC/FDS15/E01/18).

References

1. Weygandt, M., et al.: Interactions between neural decision-making circuits predict long-term dietary treatment success in obesity. NeuroImage **184**, 520–534 (2019)
2. Dashtestani, H., et al.: Canonical correlation analysis of brain prefrontal activity measured by functional near infra-red spectroscopy (fNIRS) during a moral judgment task. Behav. Brain Res. **359**, 73–80 (2019)
3. Fini, A.A.F., et al.: Enhancing the safety of construction crew by accounting for brain resource requirements of activities in job assignment. Autom. Constr. **88**, 31–43 (2018)
4. Li, R.Y.M., Chau, K.W., Ho, D.C.W.: Dynamic panel analysis on construction accidents in Hong Kong. Asian J. Law Econ. **8**(3), 1–14 (2017)
5. Bonetti, L.V., et al.: Oxyhemoglobin changes in the prefrontal cortex in response to cognitive tasks: a systematic review. Int. J. Neurosci., 1–9 (2018)
6. Klucharev, V.: Introduction to Neuroeconomics: How the Brain Makes Decisions (2018). https://www.coursera.org/learn/neuroeconomics/home/welcome. Accessed 12 Dec 2018
7. Zhang, M., et al.: Social risky decision-making reveals gender differences in the TPJ: a hyperscanning study using functional near-infrared spectroscopy. Brain Cogn. **119**, 54–63 (2017)
8. The Memory Foundation.: What Do Areas of the Brain Do? (2013). https://memory. foundation/2013/11/06/brain-areas/. Accessed 28 Oct 2018
9. Li, R.Y.M., Pak, D.H.A.: Resistance and motivation to share sustainable development knowledge by Web 2.0. J. Inf. Knowl. Manage. **9**(3), 251–262 (2010)
10. Li, R.Y.M.: 5D GIS virtual heritage. Procedia Comput. Sci. **111**, 294–300 (2017)
11. Li, R.Y.M.: Knowledge sharing by Web 2.0 in real estate and construction discipline. In: Anandarajan, M., Anandarajan, A. (eds.) e-Research Collaboration: Theory, Techniques and Challenges, pp. 289–301. Springer, Heidelberg (2010)
12. Lai, L.W.C., Chau, K.W., Lorne, F.T.: "Forgetting by not doing": an institutional memory inquiry of forward planning for land production by reclamation. Land Use Policy **82**, 796–806 (2019)
13. McCloskey, M., Wible, C.G., Cohen, N.J.: Is there a special flashbulb-memory mechanism? J. Exp. Psychol. Gen. **117**(2), 171–181 (1988)
14. Namian, M., et al.: Improving hazard-recognition performance and safety training outcomes: integrating strategies for training transfer. J. Constr. Eng. Manage. **142**(10), 04016048 (2016)
15. Pulichino, J.: Brain-Based Elearning Design (2019). https://www.linkedin.com/learning/ brain-based-elearning-design/how-learning-happens-in-your-brain. Accessed 2 Feb 2019
16. Zhu, Y., et al.: Prefrontal activation during a working memory task differs between patients with unipolar and bipolar depression: a preliminary exploratory study. J. Affect. Disord. **225**, 64–70 (2018)

Research on the Influence of Dimension of Visual Grouping Cue for Guiding Sign Recognition

Zhongting Wang, Ling Luo, Chuanyu Zou, Linghua Ran[✉],
and Haimei Wu

SAMR Key Laboratory of Human Factors and Ergonomics, CNIS,
No.4 zhichun Road, Haidian District, Beijing, China
{wangzht, luoling, zouchy, ranlh, wuhm}@cnis.gov.cn

Abstract. As the advent of big data era, the structure of modern city is becoming bigger and more sophisticated. Many of us are likely to lose their way for fail to seek out the specific direction of their destination or waste time because they can't distinguish signage at once. So, the paper analyzes an experiment which about proportional relationship of information in the design of signage layout of Way-finding Signage System and attempts to find out the relationship of words, icons and marks in the design of signage layout and then offers further reference for the signage design. Moreover, the study is mainly assisted to figure out the influence of information distribution on sign recognition in signs design.

Keywords: Signage layout design · Way-finding signage system ·
Cognitive recognition · Information proportion · Guide signs

1 Introduction

As the expanse of urbanization, the number of floating populations in China is rapidly increasing, the content of city becoming bigger and the public transport system being more sophisticated. It may confuse quite a few individuals who cannot find out their location for the lack of guidance sign or unfamiliar with those signs. Thus, in order to cope with those urgent problems, the way-finding signage system, including a series of words, marks, signs, etc., occur. The expert, Kevin A Lynch, in his book–The Image of the City firstly put forth the word "Way-finding" and defined it as a continuously usable, accurate vision system organization designed for the external environment [1]. However, this conclusion is just an overall statement. The problem of layout of way-finding signs also belongs to a layout issue. Furthermore, it can effectively improve resource utilization by seeking out optimized solution of layout problem [2, 3]. Specifically, optimized solution of layout of guidance signs may be conducive to reasonably control the design of signage layout costs, help space users to identify their location easier and improve user's satisfaction about direction service, etc. In fact, layout situation and content as the core content of guidance sign layout are one of the research hotspots in this kind of research.

© Springer Nature Switzerland AG 2020
H. Ayaz (Ed.): AHFE 2019, AISC 953, pp. 390–396, 2020.
https://doi.org/10.1007/978-3-030-20473-0_38

Several researchers have studied maximizing induced information to find out the optimal choice of signage layout location and content [4–6]. This kind of research way indeed offers a conducive clue to the guidance sign layout optimized problem. Nevertheless, in fact, the objective of guidance sign layout is to provide users with the information they need to find their way. These researches only according to the subjective perspective of the designer to take "inducing information maximization" as the optimization target ignored signage user's demand of effective and induced information. Therefore, it is significantly important to find out a reasonable solution to design signs and arrange layout induced information based on users.

The role of public signs is based on the user's understanding of the layout information, especially the space location. Thus, after eliminating the effect of familiarity by selecting several signs which subjects easily recognize what these signs mean, the study focuses in testing the response time and accuracy of these signs to find out the influence of group gap and line width in the recognition of signs or icons. And it aims to provide further suggestions and bases for the design of public signs layout.

2   Methods

2.1   Subjects

Subjects in this experiment can be chosen without specific objective in that guidance signs in public place are used by public people. Thus, 30 subjects are chosen to involve in this study (males: 15, females: 15). And the average age of these subjects is M = 31.7 (SD = 10.5). All subjects have normal eyesight. Table 1 is the essential situation of subjects.

Table 1. The essential situation of subjects

Categories	Groups	Numbers
Gender	Male	15
	Female	15
Age	18–30	15
	31–50	12
	50+	3
Educational background	Associate degree and less	10
	Bachelor degree	14
	Beyond bachelor degree	6

2.2   Materials

Familiarity includes two kind of meanings: one is the frequency of the use of signs; second is the familiarity of the content of signs [7, 8]. Confronting with a certain sign more frequent and then will be more familiar with that sign, in contrast, strange. In other words, familiarity of guidance signs has a significant influence on study results. Hence, in order to reduce the impact of the test results caused by the subjects' familiarity with signs, the familiarity of signs is firstly tested.

20 icons of traffic signs were selected and then its familiarity was tested by using online questionnaire. Familiarity of signs was divided into five level and then asked subjects to rate the familiarity of sign subjectively. Specific process can be shown in the following Fig. 1: very strange and very familiar was respectively noted in the ends of icon. (1) stands for very strange, (2) for less strange, (3) for neither strange nor familiar, (4) for less familiar, (5) stands for very familiar.

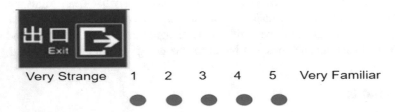

Fig. 1. The icon for testing familiarity

After collecting 200 valid questionnaires in total, and then repeated measurement analysis of variance is performed on the results of 20 symbols, we selected five icons at an average level of familiarity (statistical difference $P < .05$, familiarity is above 80%) as the elements of signs. According to the size of group gap and line width, 36 pictures of random combination were designed as experiment materials by using the five icons (as shown in Fig. 2).

Fig. 2. The materials

2.3   Environment and Instruments

The test was carried out in a laboratory with a stable illumination environment (as shown in Fig. 3). The test instruments include a laptop, an external keyboard and a large LED display that directly connected to the laptop. The experimental materials were put directly on the screen by used E-prime software through the laptop. Subjects looked at the experimental material presented on the screen and used the keyboard to respond. According to the proportional relationship between viewing distance and icon, the distance between subjects' eyes and LED display is about 1.8 m.

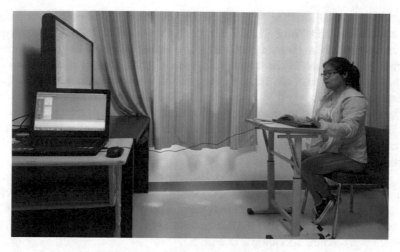

Fig. 3. The experiment environment

2.4   The Experiment Design and Procedure

2.4.1   Experiment Design

A two-factor within subject design was adopted, and the independent variable includes group gap (3 levels), line width (6 levels):

Group gap (3 levels): 0.4a, 0.6a, 0.8a;
Line width (6 levels): 1w, 1.2w, 1.4w, 1.6w, 1.8w, 2w.

The space between groups (group gap) and the line width are defined as the grouping cue to distinguish the information between groups of guiding sign system (see Fig. 4).

Fig. 4. The schematic diagram of visual grouping cue

2.4.2   Procedure

The experiment was conducted separately. Above all, after entering the laboratory, the subjects were firstly tested for version, and those with normal vision or corrected vision could participate in the experiment. And then, the subjects were asked to sit in the designated position of the experiment and began to read the instructions on the screen of the LED display silently. After reading, the experimenter summarized the instructions to ensure that the subjects correctly understood the process of the experiment and the reactions to be made during the process and highlighted that a question about location (such as "parking lot") which is needed subjects to look for would be presented on the screen in the course of the experiment. Next, subjects looked at the icon shown on the screen and fund out the position that was mentioned in the question and then used the keyboard to respond. (2) Conducting practice tests to ensure that subjects understand and master the experimental procedures. After that, subjects were asked to make a subjective report which is about the content of signs. The test will begin until all subjects were ensured to understand the meaning of all signs in the test. (3) Start the test.

36 pictures were randomly presented. Each picture fixed a test element and matched with a specific question, for instance, Fig. 2 for "Elevator", Fig. 4 for "Airport". The task is based on the question, and subjects were asked to respond "left" or "right" about the direction of a certain icon. The entire process of the experiment takes about 10 min. The e-prime procedure was shown as Fig. 5.

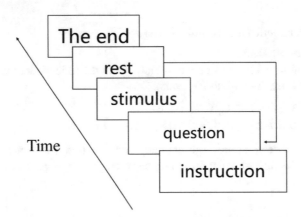

Picture 5. The e-prime procedure

3 The Result

Response time and accuracy were selected as the dependent variable. SPSS 16.0 was used to processing the experimental data.

Thirty subjects completed the entire experiment. However, one of them had a high error rate, thus, the data of this one was deleted. The average accuracy of the left 29 subjects is 99%. The average and standard deviation of response time of all subjects under various test conditions are shown in Table 2.

The index of response time was analyzed by two-factor repeated measurement analysis of variance. The results show that the response time has not main effect under the condition of group gap and line width: F group gap $(1,28) = .153$, P = .699 > .05, F line width $(1,28) = 1.036$, P = .317 > .05. The results show that the size of the line width and the space of group gap of signs do not affect the recognition efficiency.

Table 2. The result of response time (M ± SD, unit: MS)

Group gap	Line width					
	1w	1.2w	1.4 w	1.6w	1.8w	2w
0.4a	1749 ± 573	1786 ± 676	1568 ± 516	1882 ± 722	1746 ± 628	1750 ± 758
0.6a	1734 ± 579	1872 ± 1061	1844 ± 843	1873 ± 586	1820 ± 509	1830 ± 1358
0.8a	1897 ± 977	1843 ± 607	1815 ± 704	1757 ± 625	1564 ± 446	1695 ± 574

In addition, all subjects were divided into two groups (novice and expert) according to whether they could correctly understand the content of the signs at the beginning. And the response time and accuracy of the two groups were counted. Table 3 presents the subjects' understanding of signs at the beginning. As can be seen from Table 3, educational background is an essential factor and affects the understanding of signs.

Table 3. The result of subjective understanding

Categories	Understand correctly or not (number)		Accuracy (%)
	Yes	No	
Associate degree and less	4	11	36
Bachelor degree	11	14	79
Beyond bachelor degree	4	5	80

The response time of the two groups was analyzed by using the three-factor repeated measurement analysis of variance. The results demonstrate that the response time has not main effect under the condition of group gap and line width: F group gap $(2,54) = 2.268$, $P = .113 > .05$, $F_{line\ width}$ $(5,135) = 1.122$, $P = .352 > .05$. Nevertheless, there is a main effect in the different category of subjects: $F_{category}$ $(1,27) = 376.018$, $P = .000 < .001$. There is no difference in the accuracy of the two groups (novice 99.8%, expert 98%). The results reveal that the size of the line width and the space of group gap of signs do not affect the recognition efficiency. But the subjects' understanding of the content of the signs affects the efficiency.

4 Discussion and Conclusion

Guidance signs are set up to make it easy for travelers, especially those from other places, to choose the right direction to reach the destination successfully. It means that the design of public signs, especially guidance signs, may take a great important role to assist public seek out their way. Thus, this study selected five icons at an average level of familiarity (statistical difference $P < .05$, familiarity is above 80%) as the elements of the signs by using an online questionnaire. According to the group gap and line width, 36 pictures of random combination were designed as test materials by using the five icons. The results show that the size of the line width and the space of group gap of signs do not affect the recognition efficiency of signs. But the subjects' understanding of the content of the signs affects the efficiency. It means that the layout of guidance signs does not so much influence on the recognition of signs, especially the distribution of lines, gap of each part. In contrast, the availability of the content of sign is a significant factor. Hence, the study suggests that the design of guidance signs should be easily understood and fully takes different kind of people into account. And it is also crucial to spread the meaning of a certain sign.

When the spacing between groups is less than 1a, the size of the line width and the space of group gap of signs do not affect the recognition efficiency of signs. The subjects' understanding of the content of the signs affects the efficiency. Therefore, when the layout size of the guide sign is limited, it is necessary to ensure that the graphic symbols and words on the sign are easy to understand, the dimension of cues between groups does not affect the pathfinding efficiency.

Acknowledgments. We would like to thank the participants who took part in the experiment for their many efforts. We gratefully acknowledge the financial support from National Key R&D Program of China (2016YFF0201700) and National Science and Technology Basic Research (2013FY110200).

This work is supported by National Key R&D Program of China (2016YFF0201700) and National Science and Technology Basic Research (2013FY110200).

References

1. Kevin, L.: The Image of the City. Huaxia Press, Beijing (2001)
2. Liu, B.Y.: On the design of guiding signs in the field of visual symbols. J. Design Res. **8**(2), 91–95 (2018)
3. Zhang, H.J., Zhang, Y.: Research on packing problem based on simulated annealing algorithm. J. Comput. Eng. Design **27**(11), 1985–1988 (2006)
4. Zheng, Y.Q., Ding, K.X., Wang, Y.: Integrated cellular layout design based on cooperative particle swarm optimization. J. Comput. Integr. Manuf. Syst. **18**(5), 950–956 (2012)
5. Liang, Y.H., Zhang, X., Han, Y.X.: Optimization algorithm for sign-oriented identification system of railway terminal. J. Transp. Syst. Eng. Inf. Technol. **11**(6), 157–163 (2011)
6. Jiang, Y.: Research on the Guidance and Transfer System of Urban Rail Transit Hub. Nanjing University of Science & Technology, Nanjing (2012)
7. Lin, Y., Kang, L., Shi, Y.J.: The multi-objective optimization modeling and IFD-NSGA-II solving of layout of the pedestrian guidance sign. J. Syst. Manage. **22**(4), 553–559 (2013)
8. Ng, A.W.Y., Chan, A.H.S.: The variation of influence of icon features on icon usability. J. Ind. Eng. Res. **4**(1), 1–7 (2007)
9. Isherwood, S.J., McDougall, S.J.P., Curry, M.B.: Icon identification in context: the changing role of icon characteristics with user experience. J. Hum. Factors **49**(3), 465–476 (2007)
10. Wang, D.M., Hu, M., Ge, L.Z., Li, Y.J.: Navigation performance of typical traffic sign layout of Chinese cities. J. Chin. J. Ergon. **20**(3), 22–26 (2014)
11. Qi, E.S., Duan, Q., Bian, Y.Z., Shi, Y.J.: VI-based optimization model of layout of guidance sign in multilayer space. J. Comput. Integr. Manufact. Syst. **21**(9), 2528–2534 (2015)
12. Lin, C., Hsieh, M.C., Yu, H.C., et al.: Comparing the usability of the Icons and Functions between IE6.0 and IE7.0. J. Hum. Comput. Interact. **5610**, 465–473 (2009)
13. Everett, S.P., Byrne, M.D.: Unintended effects: varying icon spacing changes users' visual search strategy. In: Conference on Human Factors in Computing Systems, pp. 24–29 (2004)
14. Ni, Z.Y., Ge, L.Z.: A literature review of icon ergonomics research. J. Sci. Res. 578–581 (2012)

Cognitive Neuroscience and Health Psychology

Psycho Web: A Machine Learning Platform for the Diagnosis and Classification of Mental Disorders

Paulina Morillo[1](✉), Holger Ortega[1], Diana Chauca[2], Julio Proaño[1],
Diego Vallejo-Huanga[1,3,4], and María Cazares[5]

[1] IDEIAGEOCA Research Group, Universidad Politécnica Salesiana,
Quito, Ecuador
{pmorillo,hortega,jproanoo}@ups.edu.ec

[2] Department of Computer Science, Universidad Politécnica Salesiana,
Quito, Ecuador
dchauca@est.ups.edu.ec

[3] Department of Mathematics, Universidad San Francisco de Quito,
Quito, Ecuador

[4] Department of Physics and Mathematics, Universidad de las Américas,
Quito, Ecuador
diego.vallejo.huanga@udla.edu.ec

[5] Department of Psychology, Universidad Politécnica Salesiana, Quito, Ecuador
mcazares@ups.edu.ec

Abstract. In this paper, we present the development of a platform to collect data from cases diagnosed with mental disorders. It includes the use of a Machine Learning classification algorithm, k-NN with TF-IDF, to automatically identify the type of mental disorder suffered by a patient given his/her symptoms, when evaluated by a mental health professional. The platform called "Psycho Web" has a friendly web interface that will allow ergonomic interaction between the mental health professional and the system. The dataset used for the initial evaluation of our platform is composed of 114 instances in total, 56% of which were obtained from the taxonomy proposed by ICD-10. The rest of the instances correspond to real cases, whose symptoms and diagnoses were taken by professionals who voluntarily collaborated with the project. A raw application of the algorithm to the data available shows results with errors that go down to 5%.

Keywords: Machine learning · Diagnosis prediction · Mental disorders · ICD-10 · k-NN · TF-IDF

1 Introduction

A correct classification of mental disorders has important implications for a better diagnosis and treatment of them. Despite the Classification of Mental and Behavioral Disorders (ICD-10, Chapter 5) [1], provides a detailed taxonomy of mental and behavioral disorders, the process of classification is challenging due to the

© Springer Nature Switzerland AG 2020
H. Ayaz (Ed.): AHFE 2019, AISC 953, pp. 399–410, 2020.
https://doi.org/10.1007/978-3-030-20473-0_39

heterogeneity of the disorders, the ambiguity between the relationships and symptoms of some mental disorders and the lack of documentation of previous cases that could serve as a reference for new diagnoses.

Machine learning (ML) is a discipline of Artificial Intelligence (AI) that using algorithms has made possible the automation of processes and tasks in the field of business, engineering, and medicine [2]. In the case of psychology, the numerous disorders and symptoms, and their intrinsic relationship make it difficult to diagnose a specific disease. Therefore, the memory and calculation capacity of computers can be used to implement algorithms that assist the attending mental health professional for reliable diagnosis, reducing the risk of erroneous prescriptions.

The diagnosis of mental illnesses is usually supported by neuroimaging techniques, which allow us to obtain images of the brain to try to explain its functioning and the relationship between mental disorders and their symptoms, e.g., the fMRI scanner (functional Magnetic Resonance Imaging Scan) [3]. There are some approaches to the use of ML for the diagnosis of mental disorders. In [4] an electroencephalogram (EEG) was used to measure the electrical activity record of the brain together with classification algorithms to achieve early recognition of Alzheimer's. In [5] an advanced artificial vision algorithm is used to analyze magnetic resonance images of the brain and detect the incidence of schizophrenia. It also performs the prediction of the effectiveness of pharmacological treatment. Most of these works use Machine Learning algorithms to achieve an early and accurate diagnosis; however, although they are of great help, they do not consider the complete taxonomy of the disorders that are part of the ICD-10 classification. The main limitation is the great variety of disorders, added to the cost and the difficulties of acquiring images and/or data of diagnosed patients that allow the construction of a training dataset, a prerequisite for the application of any ML algorithm [6].

This paper presents the development of the Psycho Web platform, which will serve as a support tool in the diagnosis of mental disorders taking into account the main categories of ICD-10 taxonomy. The determination of the main category will be made using the information on the patient's symptoms.

The rest of the paper is organized in the following way: Sect. 2 shows the scheme and implementation details of the platform. The methodology for data collection and classification is defined in Sect. 3. Section 4 describes the experiments performed and the results. Finally, conclusions and prospects of this work are presented in Sect. 5.

2 Proposed Platform

In this section, we describe the Psycho Web platform that is used to collect data from cases diagnosed with mental disorders. You can find a web version of the tool in the url: http://35.229.66.36/psicoweb.

2.1 Scheme

The Psycho Web platform incorporates a web interface in charge of the interaction between the system and the attending mental health professionals. This interface also allows the management of users and manage databases by an administrator account. The core of the application consists of two modules: the Data Acquisition module and the Diagnosis Prediction module, as shown in Fig. 1.

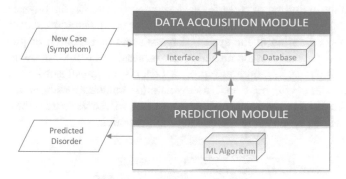

Fig. 1. Diagram of the Psycho Web platform

The Data Acquisition module was designed to receive each one of the cases diagnosed by the mental health professional. The process begins with an initial survey, which will serve as a filter for the subsequent choice of symptoms presented by the patient. Each new block of data (symptoms/disorder) will be part of the training set of the machine learning algorithm used for the diagnosis of future cases. On the other hand, the Diagnosis Prediction module incorporates a classification algorithm. This algorithm discriminates the main type of disorder presented by the patient, according to the symptoms that have been entered in the system. Then, the mental health professional, based on his/her knowledge and experience, can validate the prediction accepting or rejecting it. In case of not being satisfied with the prediction, the mental health professional can select the name of the category of disorder that he/she considers correct. Both categories, the one prescribed by the mental health professional and the one predicted by the system, are stored in the database, to check the performance of the classification algorithm.

2.2   Implementation Details

The methodology used for the development of the platform was UML-based Web Engineering (UWE). This methodology provides a guide to the process of creating a web application and standard diagrams for the design. It is an extension of the Unified Language Model (UML) [7]. UWE models such as requirements analysis, a model of contents, navigation models, presentation model, and process model were developed under the supervision of mental health professionals, to include important aspects that facilitate the use of the platform. The UWE methodology allows defining the basic functionalities of the system and the general scheme of the application.

The platform has a home page, which contains basic information about the project. For accessing the functionalities, it is necessary that users sign up, providing their data of contact and basic information, to be validated before granting access.

Although the main user profile of the platform is for the mental health professional, there are two other roles available: Administrator and Guest. It is worth to notice that the latter only has a view of basic and contact information of the web application. The Administrator profile allows the user the parameterization of the system, the algorithm training and the definition of roles for each registered user. On the other hand, the role of mental health professional enables the options for the registration of cases of patients diagnosed, the validation of the system diagnosis and consultation of previous cases.

The general process that the mental health professional would follow to enter a case and achieve diagnosis is shown in Fig. 2. All data (symptoms/disorder) selected in the diagnostic process, will be stored in a relational database, without registering the identity of the patient. The dataset of previous cases will serve to train the classification model that will be responsible for making the prediction.

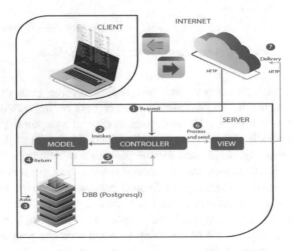

Fig. 2. MVC architecture of the Psycho Web platform

3   Methodology

3.1   Data Collection

The taxonomy of disorders considered in the development of the system was based on the diagnostic criteria proposed by the World Health Organization (WHO), in the handbook of international classification of diseases ICD-10 Chapter 5. This taxonomy groups disorders into eleven main categories (Table 1) and includes the descriptions disorders in each category and the symptoms associated to each disorder.

With the aim of having descriptors of the symptoms, which can be friendly to the users, two considerations were taken into account: (1) the symptoms of each disorder should have short names, maximum one or two words. (2) the symptoms were sub-classified by the area to which they belonged: cognitive, physiologic, behavioral or emotional. In this way, we organized 171 different symptoms (45, 44, 21 and 61 for each area respectively). The symptoms that did not belong to any of these categories

Table 1. Taxonomy ICD-10, Chapter V

Class	ICD disorders code	ICD main categories description
A	F00-F09	Organic, including symptomatic, mental disorders
B	F10-F19	Mental and behavioural disorders due to psychoactive substance use
C	F20-F29	Schizophrenia, schizotypal and delusional disorders
D	F30-F39	Mood [affective] disorders
E	F40-F48	Neurotic, stress-related and somatoform disorders
F	F50-F59	Behavioural syndromes associated with physiological disturbances and physical factors
G	F60-F69	Disorders of adult personality and behaviour
H	F70-F79	Mental retardation
I	F80-F89	Disorders of psychological development
J	F90-F98	Behavioural and emotional disorders with onset usually occurring in childhood and adolescence
K	F99-F99	Unspecified mental disorder

were classified as "initial questions" (14 in total). In this way, the information is filtered showing only the symptoms necessary to diagnosis.

The first training dataset was made up of 64 cases from the ICD-10. On the other hand, the test dataset consisted of 50 instances, which were collected from the clinical interviews made to patients of the psychological offices in the Faculty of Psychological Sciences of the Central University of Ecuador, from February to March 2018.

The data were taken retrospectively, i.e., they were uploaded to the application after the interview was completed, based on the information stored in the medical records, because, due to privacy issues, it is not possible to do so during the interview. We take a sample of 50 people who presented symptoms of some disorder and who agreed to participate in the research.

3.2   Model Training: k-NN & TF-IDF

The aim of ML algorithms is the automatic classification of an object according to its attributes (characteristics). This process consists of three faces such as training, evaluation, and testing. In this work, the ML algorithm must be able to determine the main category of disorder that a patient suffers (class) according to its symptoms (attributes).

The ML algorithm used is a particular case (when k = 1) of the algorithm named k-nearest neighbors (kNN) [8]. This algorithm is based on the calculation of distances between an object that will be classified and objects of the training set. In this way, the class to predict is chosen among the most frequent classes of the k-objects closest to the object to be classified. One of the most important characteristics of this algorithm is that in the development phase the instances are saved, but not built in a model, so the classification is given when the test instance arrives.

In our case, it is not easy to determine the true class using only the kNN algorithm since many disorders shared some symptoms and the symptoms array of each instance is sparse. So, to improve diagnosis, before calculating the distances, we have modified the training set adapting the measure TF-IDF used in Natural Language Processing (NLP) [9]. This measure is used to weight the symptoms according to their relevance to identify disorder category, thereby, greater weight is given to those symptoms that are less common with other disorders.

The formal definition of the algorithm applied to the automatic diagnosis of mental disorders takes into account the training dataset, formed by m instances. Each instance is represented by an array $o_j(j = 1, 2, \ldots, m)$ of coordinates $S_i(i = 1, 2, \ldots, N)$. The coordinates represent symptoms and can take values 1 or 0, depending on whether the patient presents the symptom or not, respectively. In general, a new case o_D is represented by an N-dimensional array of characteristics $(N = 185)$, of the form $o_D = [S_{1D}, S_{2D}, \ldots, S_{ND}]$, the sub-indices of the coordinates define the symptom and the instance, respectively.

Then, we calculated the TF-IDF, this measure is a combination of two measures $tf(s, o)$ (term frequency) which is the frequency (0 or 1) of the symptom in each instance and $idf(s, O)$ (inverse document frequency) which is defined by the Eq. 1:

$$idf(s, O) = \log \frac{|O|}{|\{o \in O : s \in o\}|} \tag{1}$$

Where: $|O|$ is the cardinality of the training set, that is, the total number of instances. $|\{o \in O : s \in o\}|$ number of instances where symptom appears.

Once tf and idf are obtained, we calculated TF-IDF by multiplying these two measures. In this way, the values of the coordinates of each array are recalculated as $o'_j = [S'_{1D}, S'_{2D}, \ldots, S'_{ND}]$. Then, we calculated Euclidean distance between each instance of the training dataset $\left(o'_j\right)$ and the instance to be diagnosed $\left(o'_D\right)$:

$$d_{o'_D, o'_j} = \sqrt{\sum_{i=1}^{N} \left(S'_{iD} - S'_{ij}\right)^2} \tag{2}$$

After that, we chose the class of the object o'_j closest to the object to be diagnosed o'_D (Eq. 3).

$$\text{Diagnosis} = \text{class}\left(o'_j | \min\left(d_{o'_D, o'_j}\right)\right) \tag{3}$$

In the Psycho Web application, each time a new instance enters the system, TF-IDF and distances are recalculated. After validating the disorder category, this test instance becomes a new example of the training dataset.

4   Experiments and Results

The official ICD-10 taxonomy provided us with 64 instances. On the other hand, the data provided by the professionals who collaborated with our project amounted to 50 instances, giving a total of 114 instances to work with. In this section, the two experiments performed to evaluate the system using this data are described. Before that, however, we take a glance at the data itself to assess its discriminatory power in the classification task we faced.

To begin with, let us analyze the cardinality of the data corresponding to each class, as shown in Table 2.

Table 2. Cardinality of the data corresponding to each class.

Class	Cardinality in the official ICD-10 set	Cardinality in the set generated by our collaborators
A	7	1
B	8	4
C	8	5
D	6	13
E	6	6
F	6	7
G	4	3
H	7	3
I	3	0
J	3	5
K	6	3
Total	64	50

These numbers clearly show that our dataset contains few instances in each class, which could hinder the classification task. Also, the classes are unbalanced. For example, class I is not represented in the second set, and the algorithm has only three instances to infer the peculiarities of the class from the first one. On the other hand, the algorithm has seven instances of class A in the first set, but only one for its evaluation in the second one.

Also to assess the discriminatory power of the data, this time in a graphical way, we used the technique known as multidimensional scaling (MDS) to observe if there are desirable relationships and clusters between them. MDS comprises performing principal component analysis (PCA) on the feature vectors and plotting the three first principal components in pairs [10]. The plots obtained are shown in Fig. 3.

As can be read from the plots, some classes are better discriminated by the features than others, e.g. classes B and C. Other classes do not show substantial differences between them, this may be due mainly to the fact that some symptoms are shared with more of a class and there are few relevant symptoms for each class. In addition, it should be mentioned that each instance contains only a few symptoms (1 to 23) of the 185 considered.

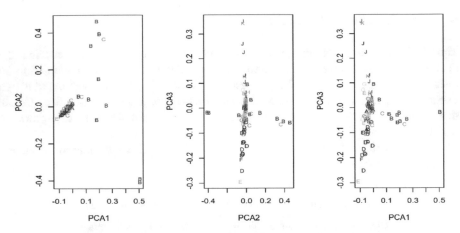

Fig. 3. Multidimensional scaling (MDS) of the dataset used.

With these conditions of the data, we performed the two experiments above mentioned, and which we describe as follows.

4.1   Experiment I

The first experiment concerned the usage of the 64 instances given by ICD-10 as training set and the following 50 as a test set (being added one by one to the training set). In this way, we used k-NN to predict the class of the test instance number one, taking as training set the 64 initial instances. Then, we added this instance to the training set, to form a 65-instance training set and compute the prediction for the test instance number two, and so on. In each case, and before the application of k-NN, TF-IDF was applied over the dataset conformed by the training set and the instance for which the prediction is computed.

The results of the classification made by k-NN following the procedure described are shown by means of a confusion matrix, Fig. 4. We also present in this figure the accuracy and the Kappa coefficient of the classification.

The value of accuracy means that less than 50% of the instances were classified correctly, which shows the impact of the dataset's size and the dimensionality of the vectors in the classification task. The Kappa coefficient [11] indicates a fair agreement level between mental health professionals and our system.

A positive fact in favor of the algorithm used is that the computation of the distances when applying it produced a solitary nearest neighbor: no other feature vector was so close to the new case, and so there was no ambiguity in the prediction. Figure 5 shows the evolution of the classification of the test instances as they were entered into the system.

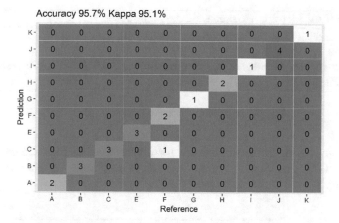

Fig. 4. Confusion Matrix of results for Experiment I

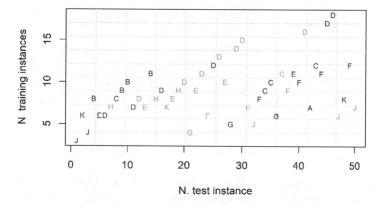

Fig. 5. Evolution of classification results

4.2   Experiment II

The second experiment was performed using the entire set of 114 instances as a training set. For testing, a synthetic dataset was generated in the following way: around 20% of the instances (23 feature vectors) were randomly taken from the training set and a "tweak" was performed upon them with probability p. The meaning of tweak in this procedure is as follows: we sweep each feature vector, component by component, and execute the pseudo-code showed in Fig. 6.

In this way, we end up with a different feature vector from the original one, slightly or utterly different depending on the value of p. For values of p close to 0, this would roughly correspond to patients with very similar symptomatology than that of the real ones. Therefore, we would expect them to have the same disorder category or a similar one.

Algorithm Algorithm to create synthetic instances
1: Extract u from the uniform distribution $U(0, 1)$
2: **if** $u < p$ **then**
3:      Switch the value of the component (0 instead of 1 or 1 instead of 0)
4: **end if**

Fig. 6. Algorithm to create synthetic instances

By running the algorithm 1000 times for different values of p and calculating the average of (1000) classification errors obtained in each case, we obtained the results shown in Fig. 7.

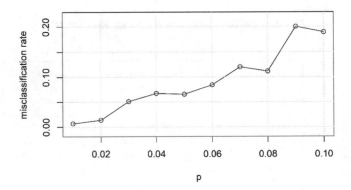

Fig. 7. Error average obtained for different values of p

As expected, the misclassification grows as p increases, which is coherent with the fact that patients with increasingly different symptomatologies should be classified in increasingly different categories of mental disorder. In Fig. 8 we show the confusion matrix for the classification made by k-NN using this procedure with $p = 0.025$.

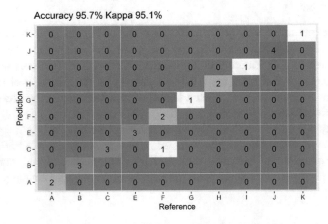

Fig. 8. Confusion Matrix of results for Experiment II, with $p = 0.025$

The accuracy, in this case, was 95.7%, this implies a misclassification error lower than 5%. The Kappa factor also increases to an almost perfect agreement level.

5 Conclusions and Future Work

The Psycho Web application allows data acquisition from patients diagnosed with a mental disorder, through a web platform with an ergonomic interface. The fact that the mental health professionals collaborating with this project were able to load new cases shows that our application can be used effectively to collect data from professionals not specialized in computer science.

As a proof of concept, the data collected was used to create a Machine Learning model which was able to generate predictions to be contrasted with the professional diagnosis. Thanks to the web interface the use of the platform can be extended to a greater number of mental health professionals, this will contribute to the construction of a more robust training dataset, and therefore of a more reliable model.

The relatively low accuracy obtained in the first experiment accounts for the impact of the dataset's size and the dimensionality of the vectors in the classification task. On the other hand, the low misclassification error obtained in the second experiment shows that the proposed model can achieve good performance when applied to patients with similar symptomatology to those represented in the training set. The use of the TF-IDF technique allows to evidence the most relevant symptoms in each category of disorder. This, in turn, can improve the accuracy of the classifier.

Certainly, the main tasks to improve the performance of the classification in future works are: first, the search of mechanisms to reduce the dimensionality of the symptomatology and the sparsity of the data, this is, to reduce the number of symptoms associated with each disorder; second, the increasing of the size of the training set; and third, the incorporation of alternative classification algorithms.

Acknowledgments. This work was supported by IDEIAGEOCA Research Group of Universidad Politécnica Salesiana in Quito, Ecuador.

References

1. World Health Organization: International statistical classification of diseases and related health problems, Chapter V, 10th revision (1992)
2. Bishop, C.M.: Pattern Recognition and Machine Learning. Springer, New York (2016)
3. David, A., Blamire, A., Breiter, H.: Functional magnetic resonance imaging: a new technique with implications for psychology and psychiatry. Br. J. Psychiatry **164**(1), 2–7 (1994)
4. Lehmann, C., Koenig, T., Jelic, V., Prichep, L., John, R.E., Wahlund, L.O., Dierks, T.: Application and comparison of classification algorithms for recognition of Alzheimer's disease in electrical brain activity (EEG). J. Neurosci. Methods **161**(2), 342–350 (2007)
5. Cao, B., Cho, R.Y., Chen, D., Xiu, M., Wang, L., Soares, J.C., Zhang, X.Y.: Treatment response prediction and individualized identification of first-episode drug-naïve schizophrenia using brain functional connectivity. Mol. Psychiatry, 1–8 (2018). https://doi.org/10.1038/s41380-018-0106-5

6. Gutiérrez Miras, M.G., Peñas Martínez, L., Santiuste de Pablos, M., García Ruipérez, D., Ochotorena Ramírez, M.M., San Eustaquio Tudanca, F., Cánovas Martínez, M.: Comparación de los sistemas de clasificación de los trastornos mentales CIE 10 y DSM IV. Atlas VPM **5**, 220–222 (2008)
7. Koch, N., Knapp, A., Zhang, G., Baumeister, H.: UML-based web engineering. In: Web Engineering: Modelling and Implementing Web Applications, pp. 157–191. Springer, London (2008)
8. Altman, N.S.: An introduction to kernel and nearest-neighbor nonparametric regression. Am. Stat. **46**(3), 175–185 (1992)
9. Trstenjak, B., Mikac, S., Donko, D.: KNN with TF-IDF based framework for text categorization. Procedia Eng **69**, 1356–1364 (2014)
10. Green, P.: Marketing applications of MDS: assessment and outlook. J. Mark. **39**, 24–31 (1975)
11. Carletta, J.: Assessing agreement on classification tasks: the kappa statistic. Comput. Linguist. **22**(2), 249–254 (1996)

Evaluation of Fatigue and Comfort of Blue Light Under General Condition and Low Blue Light Condition

Yunhong Zhang[1]([⊠]), Na Liu[2], and Hong Chen[3]

[1] National Key Laboratory of Human Factor and Ergonomics,
China National Institute of Standardization, Beijing, China
zhangyh@cnis.gov.cn
[2] School of Economics and Management,
Beijing University of Posts and Telecommunications, Beijing, China
liunal8@bupt.edu.cn
[3] Center of College Students' Mental Health Guidance,
Xi'an University of Architecture and Technology, Xi'an, China
psy3780683@126.com

Abstract. This study tested the visual comfort and visual fatigue of low blue light condition and the general condition on the same phone under the low illumination environment by critical fusion frequency (CFF), electroencephalogram (EEG) and subjective perception scale of visual fatigue. 17 participates (9 females, 8 males) participated in the experiment, and they were asked to conduct 60 min screen tasks including reading, watching the video and gaming, respectively. Their EEG indicators were recorded in the whole experiment. Before and after the screen tasks, they were measured by critical fusion frequency (CFF) and after the task they were required to finish a subjective perception scale. The result showed that the difference of CFF indicators under different conditions is remarkable, and there was remarkable difference of EEG attention index after completing the screen tasks between the two conditions. As a conclusion, the low blue light condition caused less visual fatigue and high level of attention than the general condition, and CFF and EEG were better indicators for evaluating the visual fatigue and comfort of the different phone screen.

Keywords: Low blue light · CFF · EEG · Visual fatigue

1 Introduction

The problem of visual fatigue due to the use of electronic products is increasingly prominent with the popularity of electronic products. In recent years, many manufacturers and researchers have paid close attention to the high-energy visible lights in display. The wavelength of this kind visible light is 400–500 nm and its light frequency is slightly lower than ultraviolet ray. The energy of high-energy visible light is in the highest part of the visible light and its color is blue and purple, which is known as "blue light". The photon energy of blue light is high, which can be used to excite

© Springer Nature Switzerland AG 2020
H. Ayaz (Ed.): AHFE 2019, AISC 953, pp. 411–416, 2020.
https://doi.org/10.1007/978-3-030-20473-0_40

fluorescence. Generally, the yellow light produced by LED turns into compound light mixed by blue and yellow after excited by blue light, which shown as white light. The amount of blue light needs to be increased if one wants the LED whiter. It needs to lower the amount of blue light if you like the LED becomes warmer. Some manufacturers directly increase the intensity of blue light in order to improve the brightness of white LED, which increases yellow light accordingly and finally increase the brightness of white light, causing the "blue light excess" problem.

As the gradually thorough research on light damage, the blue-ray excess is generally believed to cause the damage of retinal pigment epithelium (RPE) cells and photoreceptor cells, so the blue light needs to be filtered to protect the retina [1]. In 1966, Noell et al. showed that the blue light could cause the light damage of rhodopsin mediated, which will further cause the damage of rod cell. Meanwhile, the epidemiological investigation demonstrated that blue light exposure was associated with the incidence of age-related macular degeneration, but at present there is no large-scale clinical study verified that implanting blue-ray filtration intraocular lens can reduce the incidence of senile spot denaturation. It has also rarely been reported whether the implanting of blue-ray filtration intraocular lens will aggravate the damage of night vision for the patents existing retinal diseases like the age-related macular degeneration and diabetic retinopathy damaging the rod cells. But many clinical studies have confirmed that cataract extraction with or without the implantation of artificial lens will more likely to bring UV-light and blue light to the retina and cause the injury of RPE, which will lead to the increase of age-related macular degeneration incidence [2]. The damage to the eye of the short wavelength visible light, especially the 415–455 nm blue light was the maximum. Newman et al. proved that melatonin was functional short-wave (blue-ray 420–440 nm) photosensitive pigment by animal experiments [3]. Brainard et al. found that the normal exposed to the light of wavelength 446–477 nm could most regulate the secretion of melatonin [4]. Therefore, it is still controversial on whether filtering blue light.

Although the blue light may damage our retina, this mechanism may mainly exist in the patients with intraocular lens surgery or senile macular degeneration. Will the blue light be more likely to cause fatigue for the normal people? What kind of impact the blue light has on our visual comfort and fatigue in low illumination environment? There is little research on this problem. This study intends to investigate the influence of using ordinary mobile phones and blue light filtration mobile phones for a long time under low illumination environment on the visual fatigue and watching comfort of people, trying to experimentally verify the comfort of mobile phone display after filtering blue-ray.

2   Method

2.1   Design

This experiment tests visual comfort and visual fatigue differences of different blue light conditions with the same visual task on a same mobile phone. A within-subject factorial design was conducted to the experiment data.

2.2   Participants

Seventeen ordinary adults from 21 to 26 years old (8 male and 9 females, mean age = 24.59) were recruited and paid to participate in the experiment. All of them had 4.6 to 5.0 normal or corrected-to-normal visual acuities and healthy physical conditions, without ophthalmic diseases.

2.3   Materials

The size of sample phone is 5.0 inches. The screen resolution is 1280 * 720 and the brightness is 164 cd.

2.4   Apparatus

The research used Standard logarithmic visual acuity chart developed by the eye hospital of WMU [5], the BD-II-118 type critical fusion frequency, single electrode EEG equipment to record the EEG indicators during the experimental process. The visual fatigue scale developed by James E. Sheedy was used to test the visual fatigue after the task, including eye fatigue (such as eye burning sensation, eye pain, eye tightness, eye irritation, tearing eyes, blurred vision, ghosting and eye dryness, etc.) [6].

2.5   Procedures

Experiments were conducted in a quiet laboratory that low light indoor mobile phone could be used (experimental environment illumination value between 20 lx–80 lx). First, participants signed the informed consent and completed a general survey about their demographic information. The main contents of using mobile phone were reading text, watching video and playing games (when reading the phone's illumination was 100–140 lx or so; when playing game the phone's illumination was 0–50 lx or so; when watching video the phone's illumination was 0–60 lx or so). The order of three tasks was balanced by Latin square. The total length of the three tasks was about one hour. The content of the text was the article from the WeChat public no. The video was the classic documentary named "I am in the Imperial Palace repair artifacts". The game was two classic games named "Zuma" and "Eliminate". The fusion frequency of red, green and blue three colors before and after the mobile phone tasks were tested. The subjects need wear EEG to record EEG changes in the entire mobile phone task process. In order to avoid the visual fatigue influencing the experimental results, subjects were required to rest for at least 5 min between the two experiment conditions. In each condition, subjects need to complete the text reading, video viewing and visual games. In order to avoid the impact of learning effects, the test order of different experiment condition was balanced by ABBA method. After completing the experimental task, an interview was conducted to investigate the comfort, color, brightness, clarity and comprehensive evaluation of different experiment conditions. After completing the experiment, the subjects would got a certain reward.

2.6   Data Analysis

The changes value of visual search and visual fatigue data were analyzed by IBM SPSS 20 Statistics software (IBM-SPSS Inc. Chicago, IL). The method of repeated-measurement ANOVA analysis was applied to the experiment data to compare the differences of 3 factors from 3 quantitative indicators, namely: search time, response accuracy and subjective report data.

3   Results and Analysis

3.1   Comparison of Critical Fusion Frequency Data Results

A repeated-measurement ANOVA analysis was applied to the data of two conditions, the results showed that the difference between the two conditions was significant (F = 5.88, P = 0.021 < 0.05) and the difference between the test times was not significant. Further analysis showed that the critical fusion frequency of the low blue light condition (M = 29.49, SD = 0.74) was significantly higher than that of the general condition (M = 29.21, SD = 0.71, p < 0.05), which indicated that the visual acuity of time response in the low blue light condition was significantly better than that of the general condition. In this study, the difference between the test times was not significant, indicating the task of this experiment (for 1 h of mobile reading, games and watching video) did not induce the visual fatigue of subjects.

3.2   Comparison of EEG Data

A repeated-measurement ANOVA analysis was applied to EEG data, the results showed that the main effect was significant (F = 4.18, P = 0.051), and there was marginal significant interaction in blue-light control condition and test time (F = 3.86, P = 0.060). The simple effect analysis showed that the difference between the two conditions was not significant before the test, the difference between the two conditions was significant after the test (F = 6.70, P < 0.05), indicating that the two conditions focus on the same degree of consistency before the test. Through the use of mobile phones, the use of two mobile phone focus had undergone significant changes. Further analysis showed that the degree of focus in low blue light condition (M = 45.07, SD = 1.33) was significantly higher than that in general condition (M = 42.09, SD = 1.56, P < 0.05) (Fig. 1).

3.3   The Results of Subjective Perception Scale

From the results of the visual fatigue subjective perception scale, a repeated-measurement ANOVA analysis was applied to different conditions. The results showed that there was difference between the two conditions in the eye fatigue, mental fatigue and overall fatigue, but the difference was not significant. The overall value of the conditions was too small, indicating that the task did not induce fatigue.

Through the cumulative weighting analysis of the evaluation results between the two conditions, the comprehensive evaluation between two conditions had no

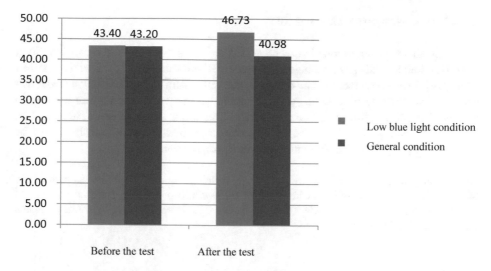

Fig. 1. Compare of EEG results under two kinds of conditions

difference. In terms of comfort, the low blue light condition had advantages over the general condition. In terms of color comfort, low blue light condition was slightly better than general condition. The subjects thought that general condition was better than low blue light condition in terms of brightness, the brightness of two conditions was consistent in fact, indicating that the subject's judgment may be affected by other factors, such as confusing the brightness and color. In terms of clarity, the subjects considered that the clarity of general condition was better than that of low blue condition, which was consistent with the actual situation of the test sample (Fig. 2).

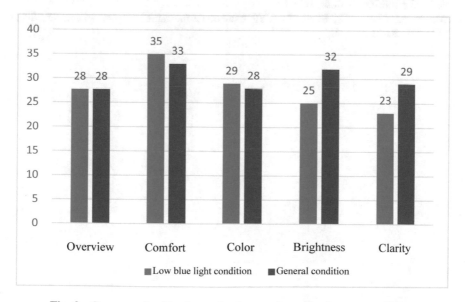

Fig. 2. Compare of subjective evaluation results under the two conditions

4   Conclusion and Discussion

Based on the above-mentioned subjective and objective evaluation, the low blue light condition had advantage of reducing the visual fatigue and could improve the comfort of the user to a certain extent. The comparison results from the critical fusion frequency showed that the low blue light condition could protect eyes. The current evaluation results also showed that compared to general condition, the low blue light condition had a short part in the clarity, color and other aspects, indicating that there may be contradictory aspects of clarity and comfort to some extent.

Acknowledgments. The authors would like to gratefully acknowledge the support of the National Key R&D Program of China (2016YFB0401203) and China National Institute of Standardization through the "special funds for the basic R&D undertakings by welfare research institutions" (522018Y-5984).

References

1. Li, Q., Lin, Z.: The advantages and disadvantages of blue light to visual function. Int. Rev. Ophthalmol. **30**(5), 336–340 (2006)
2. Mainster, M.A.: Spectral transmittance of intraocular lenses and retinal damage from intense light sources. Am. J. Ophthalmol. **85**, 167–170 (1978)
3. Newman, L.A., Walker, M.T., Brown, R.L., et al.: Melanopsin forms a functional short wave length photopigment. Biochemistry **42**, 12734–12738 (2003)
4. Brainard, G.C., Hanifin, J.P., Greeson, J.M., et al.: Action spectrum for melatonin regulation in humans: evidence for a novel circadian photoreceptor. J. Neurosci. **21**, 6405–6412 (2001)
5. Standard Logarithmic Visual Acuity Chart Developed by the Eye Hospital of WMU. People's Medical Publishing House, July 2012
6. Sheedy, J.E., Hayes, J., Engle, J.: Is all Asthenopia the same? Optom. Vision Sci. **80**(11), 732–739 (2003)

Applications in Human Interaction and Design

Embodied Interactions in Cognitive Manufacturing

Peter Thorvald[1,2(✉)], Jessica Lindblom[1], and Ari Kolbeinsson[1]

[1] University of Skövde, Skövde, Sweden
{peter.thorvald, jessica.lindblom,
ari.kolbeinsson}@his.se
[2] Chalmers University of Technology, Gothenburg, Sweden

Abstract. This paper presents a discussion on the role of embodied interaction with a basis in social embodiment effects and how they can be viewed in light of manufacturing ergonomics. The social embodiment effects are four statements, grounded in empirical findings, which highlight the interplay of social stimuli, embodied responses, and cognitive processing. They suggest and base an argument for how embodiment is central to cognitive processing, how bodily states interact extensively with cognitive states, and ultimately how embodied interaction is ubiquitous in human cognition. The paper further presents a view on how human based manufacturing can be studied in light of this argument, exploring other areas where social embodiment has been further researched, with an aim to suggest examples of where social embodiment effects might be found in manufacturing ergonomics and form a basis for future investigations.

Keywords: Manufacturing · Social embodiment·cognition ·
Human based manufacturing·ergonomics

1 Introduction

Since the early 80s, the view that human cognition must be understood within a social and material context has slowly gained traction. This is a view that is proposed by several different 'novel' approaches to cognition (i.e. situated cognition, embodied cognition, distributed cognition, 4E cognition, etc.,) that might differ in names but come together in the argument that the body, as well as the social and material context, are crucial to human cognition. This argumentative paper takes its starting point from there and aims to focus on the role of the body in human based manufacturing.

1.1 Embodied Cognition

Wilson and Golonka [1] explain that embodiment (or an embodied view on cognitive science) is simply the idea that the brain is not the only cognitive resource available to humans. This can be seen as a radical view, and can lead to the misunderstanding that a claim is being made about external elements *performing* cognitive tasks, while in fact the claim is merely that we use our action in the world and the changes we perceive in the world through our actions to support and supplant our cognitive abilities [1, 2].

© Springer Nature Switzerland AG 2020
H. Ayaz (Ed.): AHFE 2019, AISC 953, pp. 419–426, 2020.
https://doi.org/10.1007/978-3-030-20473-0_41

The central idea of the embodied view of human cognition is that cognition is evolutionarily rooted and exists to support action, replacing the notion that complex intra-mental representations are required with much simpler perception-action couplings and matching the information from the perception-action coupling to experience [1–4]. In the most radical form of embodiment, this is the closest to internal representations or simulations that is required [1]. The embodied view explains many complex actions performed by humans as relying on perception-action couplings, i.e. that humans use experience of the world to predict what happens next, and then continually update that prediction based on changes in the world. Wilson and Golonka [1] argue that the embodied view of cognition is best supported by how well it supports the observations of complex tasks in the real world while offering a theoretical explanation that can be explained using simple action based intra-mental cognitive processes that use the body and environment.

This reliance of embodied views of cognition on perception-action couplings and experience simplifies many otherwise complex tasks, such as catching a ball in flight, by offering a mechanism by which the observed actions of a ball catcher can be matched with theory. This relies completely on action and anticipation of action, using the world as the model with which to work instead of using complex models of the world within the mind. Another example of humans manipulating the world to solve cognitive problems instead of relying on intra-mental processing comes from Kirsh and Maglio's [5] experiment where skilled participants were observed to rotate Tetris pieces in the actual game instead of planning the rotation and movement of pieces internally and then executing the plan. Interacting in this way with the game is a prime example of using the world as its own best model to reduce the difficulty of the cognitive task, relying instead on the perception-action coupling offered by the instant feedback of the world to the actions performed.

The embodied view of cognition differs from a common view of cognition, referred to as the computer metaphor of mind, which presupposes that the mind is in the brain, and that the brain performs calculations using representations to create plans that are then executed to complete an activity. This view on cognition requires complex mental gymnastics involving detailed internal models of the world and the activity being performed, and is complex and sensitive to incomplete information about the world or the activity [4, 6].

1.2   Rationale

The role of the body, and indeed, even the social and material context, in cognition is gaining more and more traction as more and more evidence for it is produced. However, much of the research arguing for a situated view on cognition is, ironically, based on empirical evidence obtained in lab studies, thus being very non-applied. The social embodiment effects suggested by Barsalou et al. [7], discussed later, are excellent examples of this. They show an intricate connection between body and mind but the real world applications to this date, is quite far fetched, if present at all. The aim of this work is to discuss social embodiment effects and how they can be studied in a manufacturing environment. Ultimately offering us a new way of studying, understanding, and developing human based manufacturing.

2   Social Embodiment Effects

The social embodiment effects are four statements, grounded in empirical findings, which highlight the interplay of social stimuli, embodied responses, and cognitive processing. They suggest and base an argument for how embodiment is central to cognitive processing, how bodily states interact extensively with cognitive states, and ultimately how embodied interaction is ubiquitous in human cognition. Social embodiment effects have been investigated in a wide variety of areas outside of the original psychological experiments that formed the empirical basis of the statements in the early 2000s.

2.1   Embodied Interaction Is Ubiquitous in Human Cognition

Barsalou et al. [7] presented four statements, all grounded in empirical findings from psychology research during the past half-century that highlight the intricate interplay of physical and affective states. They can be crudely summarized to say that embodiment seems to be central to cognitive processing, that bodily states interact extensively with cognitive states, and that embodied interaction is ubiquitous in human cognition. This is shown and argued through a variety of empirical data from a variety of sources. The four statements and some of the research that they are built on are presented below.

Social Stimuli Elicit Embodied Responses in the Self. This effect shows how stimuli which is inherently cognitive, such as perceptual stimuli, has an effect on our physical behavior, affecting the way we move about in a material world. Weisfeld and Beresford (in [3, 7]), for example, reported that high school students who received good exam results adopted a more erect posture than those who received poor grades, who instead adopted a less erect position. Another example is results from Bargh, Chen and Burrows's [3, 7] who primed subjects with concepts related to the elderly (gray, bingo, and wrinkles), and found that the words primed embodiment effects accordingly such as moving more slowly when leaving the laboratory while this effect was not seen in the control group, having been primed with neutral words. Additional evidence show that perceived social stimuli can produce corresponding facial responses, a pleasant visual scene produced positive facial responses, and also that social stimuli can affect embodied aspects of communication. The latter being found in an experiment by Dijksterhuis and van Knippenberg [3, 7] where subjects primed with words being related to a stereotype showed behavior related to that particular stereotype.

Embodiment in Others Elicits Embodied Mimicry in the Self. This shows how the embodied responses mimic the perceived social stimuli. For instance, two persons engaged in interaction, tend to mimic each others body language or viewing a smiling person produces, to some degree, the same facial expression in the observer [8]. Barsalou et al. [7] identified three forms of mimicry; bodily, facial and communicative mimicry. With regard to bodily mimicry, bodily actions between interacting people often become synchronized [3, 7]. For instance, Chartrand and Bargh [3, 7] point out that subjects mimicked the experimenter's actual behavior, such as rubbing the nose or shaking a foot pretty often. Evidence of facial mimicry is widely documented in the literature [3, 7]. O'Toole and Dubin [3, 7], for example, point out that mothers tend to open their mouths after their infants have opened their own during feeding. Embodied mimicry is also present during communication, matching the speech rate, utterance duration, and emotional tone. For instance, Bavelas et al. [3, 7] report that listeners often mimic speakers' manual gestures, and it has been argued that different kinds of synchronization do assist interacting partners during conversation by establishing understanding, cooperation and empathy [3, 7].

Embodiment in the Self Elicits Affective Processing. Embodiment not only functions as a response to a social stimulus but also constitutes a tentative stimulus in itself. Hence, this is the reverse of the first identified social embodiment effect and basically can be said to mean that whatever physical state the body is in, the affective state of the mind will be influenced accordingly. I.e. an erect posture results in more positive feelings about your own performance which was shown by Stepper and Strack [3, 7], for instance. In their experiment, subjects were induced into either an upright or slump position during a cover story about measuring 'ergonomic positions'. During the experiment all the subjects performed an achievement test, and received false feedback that they had done well on the assignment. Afterwards the subjects were asked to rate their feelings of 'pride' at the time, and those that had been in an upright position experienced more pride than those who had been in a slump position. Hence, the subjects' posture actually influenced their affective states. As Barsalou et al., point out, facial elicitation, on the other hand, is well documented in the literature and is often referred to as facial feedback [3, 7]. Despite the fact that the accounts of these effects differ, it can be noted that many studies demonstrate that the shaping of the face into an emotional expression, for instance a smile, tends to produce the corresponding affective state, namely 'happiness'. Strack, Martin and Stepper [3, 7] demonstrated the famous effect that occurs when holding a pen with the lips or with the teeth, the former case inducing a frown-like activation of the facial muscles and the latter inducing a smile-like activation of the same. Results indicate that the subjects who held the pen with their teeth rated cartoons that they were asked to read as funnier than the others who held the pen between their lips.

The Compatibility of Embodiment and Cognition Modulates Performance Effectiveness. This final social embodiment effect concerns more complex relationships in cognitive behavior. It essentially means that when bodily and cognitive states are compatible, cognitive work will be more successful and require less effort. Consequently, when they are incompatible, cognitive performance is less efficient. Several examples of this compatibility effect are reported, for example, Chen and Bargh [3, 7] demonstrated that subjects respond faster to 'positive' words than 'negative' ones, when pulling a lever towards them instead of pushing it away. This suggests that motor performance is more optimal when compatible with cognitive processes. Similar results are reported for memory tasks, face recognition, facial categorization, word recognition, reasoning, and secondary task performance [3, 7]. Broadly speaking, Barsalou et al. stress that the compatibility between embodiment and cognition is also present for non-social stimuli, supporting a broader pattern of embodiment-cognition compatibility. They conclude that common mechanisms which produce this compatibility effect across different domains exist. Additionally, embodiment seems to be central to cognitive processing, in view of the fact that bodily states interact extensively with cognitive states. Thus, they conclude that embodied interaction is ubiquitous in human cognition.

2.2   Embodiment of Attitudes, Social Perception and Emotions

In more recent work, Niedenthal et al. [9] investigated and suggested embodiment effects in areas more traditionally associated with social psychology, such as:

- Embodiment of attitudes
- Embodiment of social perception
- Embodiment of emotions

Embodiment of Attitudes concerns how our bodily activity not only affects and modulates performance effectiveness, but also has bearing on our attitudes in terms of positive and negative feelings towards different concepts. For instance, Tom et al. [9] studied how head movements influenced attitudes towards a pen placed before the subjects during the testing of headphones. Participants were asked to nod or shake their head to see if the headphones would fall off as a cover story where the effect of head shaking or nodding on attitudes towards the pen before them was what was really being investigated. Afterwards, an experimenter offered the 'old' pen that hade been placed on the table during the experiment or a 'new' pen that the subjects had not seen before. Depending on the performed head movements, i.e., nodding in agreement or shaking in disagreement, the participants favored the pen that correlated to the developed attitude. In other words, the nodding participants chose the 'old' pen, whereas the head-shaking participant preferred the 'new' one.

Embodiment of Social Perception refers to the effects on embodiment that is a result of different social stimuli. With regards to the aforementioned experiment by Bargh, Chen and Burrows [7] where subjects were primed with concepts related to the elderly, these concepts were not associated with motoric functions whereby they can be argued to be socially and culturally bound. Contrary to a similar experiment where subjects who were primed with a fast animal (such as a cheetah) were observed to leave the lab more quickly than subjects primed with a slower moving animal.

Embodiment of Emotions highlight that the linking between embodiment and emotions can actually be traced back to William James who, generally speaking, argued that the basis of emotion is the bodily activity that occurs in response to an emotional stimulus [10]. Returning to Niedenthal's et al. [9] position, embodiment is essentially engaged in information processing about emotion, online, and most notably, also offline, in which humans represent the emotional meaning in abstract entities such as words.

Niedenthal et al. [9] stress that besides the ubiquity of embodied responses to what they call non-emotional actions and movements, there is accumulating evidence that humans also mimic others' *emotional* facial expressions [9]. Bavelas et al. [9], for instance, reported that when a co-researcher of theirs actually participated within their own experimental situation and deliberately faked an injury, grimacing in pain, the observing participants then also grimaced. The extent of the participants' grimaces correlated with how clearly they could see the confederate's face. This means, emotion imitation seems to occur automatically without any higher mediating conscious awareness.

3 Embodied Interactions in Cognitive Manufacturing

As mentioned earlier, while research on social embodiment effects sheds light on the intricate relationship between body and mind, understanding how they occur in real life and how to exploit them, requires more applied research taking place outside the lab, focusing on more naturalistic occurrences of such effects. In this section, we will attempt to draw some parallels and discuss social embodiment effects in manufacturing, a more naturalistic environment, where understanding and recognizing these effect might actually benefit the outcome of work.

There are several examples in the psychological literature that can be used to argue for how external social stimuli result in embodied responses. The aforementioned experiments where subjects were primed with concepts related to the elderly, or primed with fast or slow animals, are just the ones mentioned by Barsalou et al. [7]. Another, quite well known effect, is that listening to arousing or non-arousing music respectively, while driving, affects the driving speed and general driving behavior [11]. With respects to the application at hand in this paper, we might conclude that whatever social stimuli is present at the workstation, is likely to have an effect on the physical behavior of the worker. Considering the often chaotic environment that can be found in a manufacturing facility, there are many variables to keep track of here. In our experience, it is quite common with flashing lights and sirens to alert someone about malfunctions or to alert about an approaching forklift. Combined with the social aspects of work such as music, conversations, etc., and the chaotic environment becomes quite difficult to handle for management, and equally difficult to study for us as researchers.

Postural synchrony of two people interacting, or body mirroring, is a concept that has been known for a long while [8]. Mirroring is done subconsciously as opposed to intentional imitation and often goes unnoticed by both parties. Recent developments on mirror neurons have shed additional light on mirroring of postures and behavior, and studies have even shown mirror neuron activation in human subjects when interacting with a humanoid robot [12, 13]. This raises several interesting research questions with bearing towards manufacturing such as if this effect is exclusive to humanoid robots, i.e. requiring some kind of anthropomorphism, or if it might be found also in interaction with industrial robots. Previous ideas on this have been discussed in Fasth Berglund et al. [14], who discussed using different types of robot configurations to investigate various levels of anthropomorphism and how they might affect feelings of trust towards the robot. They suggested the use of traditional industrial cobots, industrial cobots equipped with a robotic head (Furhat), Sawyer or Baxter who are one and two armed cobots equipped with a screen where you can visualize information or in this case, a set of eyes or a face, and ultimately a humanoid robot (Pepper).

The aforementioned effect with a pen held in two distinctive ways between the lips or the teeth was first researched by Laird [15] but the general ideas go as far back as William James at the end of the 19th century and allegedly even as far back as to Dutch

17th century philosopher Baruch Spinoza (although this is not confirmed). As mentioned, several similar effects have been observed, most notably the effect of posture on self satisfaction. In some sense, this changes the playing field slightly because all of a sudden, physical ergonomics is not only about avoiding physical wear on the body, but has potential impact on emotional states which in turn has great impact on work outcome. On another note, one of the main challenges for the western world in the coming years is the change in demographics [16]. The population is getting older which can have many potential outcomes. Later retirement and a shortage of personnel being two outcomes that has relevance to this argument. Shortage of personnel forces each employer to focus on what their offer to a potential employee is. Work satisfaction is already viewed to have great future potential in hiring and retaining personnel [17].

The compatibility effect is perhaps the most complex of the four social embodiment effects as proposed by Barsalou et al. [3, 7], but it is also one that has some clear cut applications. Additionally, there are several related effects that have potential impact on work outcome. Stimulus-response compatibility for instance, is an effect that describes that the compatibility between stimulus and response, improves outcome of the response. So a stimulus coming in from the left hand side of the subject, is more efficiently responded to using the left side of the body [18]. Widening the scope to product design, the compatibility of the semantics of a product to the actual actions it offers, can also be said to be a compatibility effect. Consider what Norman [19, 20] initially called 'affordances' and later 'perceived affordances', the invitations to action that an object contains, should match the actual actions that can be performed with it. I. e. a knob invites turning but if it rather is meant to be pushed, the compatibility between the perceived affordance and the actual action capabilities of the object is lessened (for a more accurate account of affordances, see [21]).

4 Discussion

The aim of this paper was to discuss the intricate connection between body and mind through social interaction effects. As mentioned, these effects are largely based on off-line lab experiments with limited applications to the real world. As a result of this, we also wanted to suggest potential areas for future investigations based on existing work, such as the work on mirror neurons, or simply just areas that seem to show potential to the work at hand.

A lot of work on embodied and situated cognition is currently taking place and, no doubt, the future will see further similar evidence as the social embodiment effects towards the complex connection between the physical and the cognitive human.

References

1. Wilson, A.D., Golonka, S.: Embodied cognition is not what you think it is. Front. Psychol. 4, 58 (2013)
2. Wilson, M.: Six views of embodied cognition. Psychon. Bull. Rev. 9(4), 625–636 (2002)
3. Lindblom, J.: Embodied Social Cognition. Springer, Berlin (2015)

4. Vernon, D.: Artificial Cognitive Systems: A Primer. MIT Press, Cambridge (2014)
5. Kirsh, D., Maglio, P.: On distinguishing epistemic from pragmatic action. Cogn. Sci. **18**(4), 513–549 (1994)
6. Rogers, Y.: HCI Theory: Classical, Modern, and Contemporary, pp. 1–129. Morgan & Claypool, San Rafael (2012)
7. Barsalou, L.W., Niedenthal, P.M., Barbey, A.K., Ruppert, J.A.: Social embodiment. In: Ross, B.H. (ed.) The Psychology of Learning and Motivation, pp. 43–92. Academic Press, San Diego (2003)
8. Chartrand, T.L., Bargh, J.A.: The chameleon effect: the perception–behavior link and social interaction. J. Pers. Soc. Psychol. **76**(6), 893 (1999)
9. Niedenthal, P.M., Barsalou, L.M., Winkielman, P., Krath-Gruber, S., Ric, F.: Embodiment in attitudes, social perception, and emotion. Pers. Soc. Psychol. Rev. **9**(3), 184–211 (2005)
10. Damasio, A.R.: The Feeling of What Happens: Body and Emotion in the Making of Consciousness. Harcourt, New York (1999)
11. North, A.C., Hargreaves, D.J.: Music and driving game performance. Scand. J. Psychol. **40**(4), 285–292 (1999)
12. Gazzola, V., Rizzolatti, G., Wicker, B., Keysers, C.: The anthropomorphic brain: the mirror neuron system responds to human and robotic actions. Neuroimage **35**(4), 1674–1684 (2007)
13. Oberman, L.M., McCleery, J.P., Ramachandran, V.S., Pineda, J.A.: EEG evidence for mirror neuron activity during the observation of human and robot actions: toward an analysis of the human qualities of interactive robots. Neurocomputing **70**(13–15), 2194–2203 (2007)
14. Fasth Berglund, Å., Thorvald, P., Billing, E., Palmquist, A., Romero, D., Weichhart, G.: Conceptualizing embodied automation to increase transfer of tacit knowledge in the learning factory. In: The 16th IEEE International Conference on Intelligent Systems (IS), Funchal, Portugal, Sept 2018
15. Laird, J.D.: Self-attribution of emotion: the effects of expressive behavior on the quality of emotional experience. J. Pers. Soc. Psychol. **29**(4), 475–486 (1974)
16. EC: People in the EU - statistics on demographic changes: European Commission (2017). https://ec.europa.eu/eurostat/statistics-explained/index.php?title=People_in_the_EU_-_statistics_on_demographic_changes
17. Manuwork: Balancing human and automation levels for the manufacturing workplaces of the future. http://www.manuwork.eu
18. Proctor, R.W., Reeve, T.G.: Research on stimulus-response compatibility: toward a comprehensive account. Adv. Psychol. **65**, 483–494 (1990)
19. Norman, D.: The Psychology of Everyday Things. Basic Books, New York (1988)
20. Norman, D.: Affordance, conventions, and design. Interactions **6**(3), 38–42 (1999)
21. Gibson, J.J.: The Ecological Approach to Visual Perception. Lawrence Erlbaum Associates, Hillsdale (1986)

Modeling and Multi-objective Optimization of Insulating Lining Using Heuristic Technique "Exploration of Variable Codes (EVC)"

Umer Asgher[1(✉)], José Arzola-Ruiz[2], Riaz Ahmad[1,3],
Osmel Martínez-Valdés[4], Yasar Ayaz[1], and Sara Ali[1]

[1] School of Mechanical and Manufacturing Engineering (SMME), National
University of Sciences and Technology (NUST), Islamabad, Pakistan
{umer.asgher,yasar,sarababer}@smme.nust.edu.pk,
riazcae@yahoo.com
[2] Studies Center of Mathematics for Technical Sciences (CEMAT),
Technological University of Havana, Havana, Cuba
jarzola@cemat.cujae.edu.cu
[3] Quality Assurance Directorate,
National University of Sciences and Technology (NUST), Islamabad, Pakistan
[4] ACINOX Ingeniería, MINDUS, Cotorro, Cuba
osmelmv@gmail.com

Abstract. In this study, a conceptual mathematical model for the refractory and isolating lining of high temperature installations is deduced based on system constraints. Model's decomposition leads to a bi-level objective having discrete optimization task with higher number of layers of compositions. Solutions are generated by a series of lower layers and multiple-objective optimization in accordance with a hierarchical participative structure. Graphical modelling of case study is developed with AutoLisp and OpenDCL for AutoCAD, which allows, for each refractory lining's proposal generation. The user's application and the associated graphic information among other possibilities allows to evaluate factors that could not be included in the model used for the solution of the original task. A novel heuristic approach is developed in the frame of the Integration Variables Method called "Exploration of Variables Codes (EVC)". The results of EVC are numerically compared with the ones obtained using the elitist genetic algorithm approach "Non-dominated sorting genetic algorithm II (NSGA II)" for the same model's structure. The optimized results demonstrate the higher solutions diversity, that facilitates the human-machine interaction procedures associated to develop the optimized solution.

Keywords: Multiple criteria optimization · Isolating lining ·
Genetic algorithm (GA) · Thermo mechanical modeling ·
Human-machine interaction

© Springer Nature Switzerland AG 2020
H. Ayaz (Ed.): AHFE 2019, AISC 953, pp. 427–437, 2020.
https://doi.org/10.1007/978-3-030-20473-0_42

1   Introduction

The selection of lining materials and its thickness for thermal installations is commonly based on the empiric knowledge, while evaluating new lining options, that limits the possibility to estimate options to carry out an appropriate decision making. Although in the existing literature of materials and design [1, 2], Ashby developed a methodology for material's selection that has received a great general acceptance in field of materials engineering. This methodology is based on the contrast derived from combining the attributes of the processes with the properties of the materials [3, 4]. The general method avoids concepts decision of the thermal facilities design and used to reduce the number of alternatives to take into account in the selection of materials. The refractory and isolating lining design process for thermal facilities under high temperature settings is carried out, by various researchers [3–7]. These studies include the objective functions like minimization of the heat losses, total thickness, total weight of the refractory mass [5]. These indicators although are attributable to the installation in general, but could also be attributed to each individual area of the facility [3, 4, 6, 7]. Other indicators that have systemic character are related to reducing the thermal stress, to minimize the resulting tension between the lining and the metallic foil [3, 4], requirement of equality of thickness of the different areas, speed of the gases as a result the internal pressure of the installation and the available space for its movement in case like heating ovens [6]. To take into account the availability of materials, its characteristics, working areas and the required properties are mentioned in the existing literature [8]. These approaches, shows the thermal lining of the installations as an integral problem that is linked to the selection of the materials, its thicknesses with a set of different restrictions and objectives. The materials and its thickness selection are a complex process, for that the solution must be based on the analysis and synthesis of engineering systems [9].

Our research work in addressing this problem allowed us to formulate it in the frame of bi-level and multi-objective optimization. The result of the interaction among entities that make decisions distributed in parts of a system characterized by the existence of a higher and lower levels. Higher level is commonly denominated as leader and lower's as followers. The leader's decisions are implicitly affected by the reactions of its followers. In the last decade's various studies have been devoted to the modeling [10], solution methods [10–12], applications [13] and decomposition techniques of multi-level systems have been published for different classes of problem. Among these classes are: linear [14], non-linear [15], continuous [16], discrete and mixed [17–19] variables. For all these classes many solution outlines related to mono [20] and multi-objectives problems, principally linear [21, 22] for bi-level [22] and, in occasions for tri-level [23] tasks. In the present work a systemic analysis is carried out that allows us to formulate the conceptual mathematical model of an optimal multi-objective task for the refractory and insulating lining. This task breaks down into a bi-level, multi objective, discrete optimization sub-tasks and a solution outline, using the Exploration of Variables Codes Algorithm (ECV), and is developed in the frame of the Integration of Variables method (IVM). On the other hand, when solving a multi-objective task, the decision-making process expects to obtain its Pareto set. Decision making in multi-objective optimization problems select a single solution to make a selection. This decision-making process may rely on an additional criterion and make it a control function to find the single solution [27]. Based on this idea, we propose a methodology to transforms the multi-objective task with Exploration of Variables Codes Algorithm (ECV) that break down tasks into sub tasks with signal objective function to be optimized.

2 Modelling Objective Function: Installation Under High Temperature

The optimization of a refractory wall's thickness task for solving discrete optimization problems starting from its homologous continuous function [5]. Various properties like maximal internal temperature $T_{i,j}^{\max}$ without changing physical properties for each j material, a number of layers n_i for every i area. It is required to determine the materials $m_{i,j}$ and its thickness $x_{i,j}$ for each i layer of refractory and insulating wall of area j, made up of bricks corresponding to an area of an eventual installation assuring the best possible efficiency indicators such as: total thickness ($z_{1,j}$), material cost ($z_{2,j}$), density of the caloric flow q ($z_{3,j}$), external temperature ($z_{4,j}$). Other indicators, including those added ahead. The analysis and synthesis methodology [5, 24] is applied. In Fig. 1 the external analysis of one-area optimization task is illustrated:

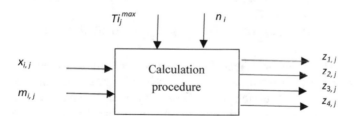

Fig. 1. External analysis representation of any j area task

Although the quantity of layers' n of the area is a decision variable it is considered, for simplification, as a coordination variable. The optimization task associated could be solved for different values of n. Mathematical formalization of the isolating and refractory optimization task, associated to the previous analysis [24]:

Minimize:

$$Z_j = \max\left\{\frac{\left|z_{1,j} - z_{1,j}^d\right|}{\left|z_{1,j}^d\right|}, \frac{\left|z_{2,j} - z_{2,j}^d\right|}{\left|z_{2,j}^d\right|}, \frac{\left|z_{3,j} - z_{3,j}^d\right|}{\left|z_{3,j}^d\right|}, \frac{\left|z_{4,j} - z_{4,j}^d\right|}{\left|z_{4,j}^d\right|}\right\} \tag{1}$$

Where:

z_{1j}^d, z_{2j}^d, z_{3j}^d, z_{4j}^d: whished values of the corresponding efficiency indicators. The possible values of every one of these parameters obtained starting from ideal solutions, solving the corresponding mono objective optimization tasks.

Must be assured the fulfillment of the following constraints:

$$T_{ij} \leq T_{ij}^{\max}(m_{ij}), \ i = 2, \ldots, n \tag{2}$$

$$0 \leq x_{ij} \leq x_{ij}^{\max}, \ i = 1, \ldots, n \tag{3}$$

$$x_{ij} \in \left(d_{ij}^1, \ldots, d_{ij}^{k_i}\right), \ i = 1, \ldots, n \tag{4}$$

$$m_{ij} \in M_{ij} \tag{5}$$

$T_{i,\ j}$:   Internal temperature of the layer i closer to the atmosphere.

M_{ij}:   Set of materials selected as candidates to be used in the layer i of the area j.

x_i^{max}, $T_i^{max}(m_i)$: m_i maximal value of the thickness and the temperature, respectively, of the material m_i of the wall, without affecting its physical properties.

d_i^k: Thickness of the layer i, in the k variant of the bricks disposition.

Restrictions (2) establish that the temperature of the layer cannot be over the maximal permissible without affecting the physical properties of the selected m_i material. Restrictions (3) establish that the thickness of the layers should be between the allowed limits values. Restrictions (4) establish that the thickness of each layer should adopt one of d_i possible values, according the allowed variants of the bricks disposition.

Restrictions (5) indicate that the materials of the layer i has to be selected from the set of materials.

The determination of the heat density q and T_i, $i = 1 \ldots, n + 1$ for well-known x_i values, is carried out by an implicit character calculation procedure. The minimum possible values of the layer's thickness and of the cost of the wall materials are, evidently, similar to zero, equivalent to the non-existence of refractory wall, for what the decision-making process can choose desirable minimum values of these magnitudes. Similarly, the minimum possible values of the heat density flow through the wall q and of the external temperature are also similar to zero and correspond to an infinitely length wall, for what the decision-making process always to choose desirable minimum values of these magnitudes with evidently subjective origin. Starting from the initially selected desirable values, the decision-making process begins sampling the set of efficient solutions, specifying in each new iteration, new desirable values for these magnitudes. Using heuristic methods, it is possible to generate a population of solution options for layers, those can select the one that better satisfies the requirements of the decision-making model of the whole installation. In studies [25, 26] it is exposed, step by step, solution algorithms for solving this problem and its quality indicators with regard to elitist genetic algorithm NSGAII linked to the optimal design under several multiple criteria of the tasks studied.

3 Bi-level Objective Optimization of the Refractory and Insulating Lining

A mathematical model for the refractory and insulating lining design has necessarily the following mathematical structure, given the systemic character of the installation:

$$Z = \sum_{j=1}^{m} Z_j + \varepsilon(x) \qquad (6)$$

Where:

Z_j is the vector of multiple objectives functions of the j area of the installation.

$\varepsilon(x)$ is a vector of systemic losses of the efficiency indicators obtained from the interaction between areas. That include the penalty functions for the non-fulfillment of the systemic restrictions that are added to Eqs. (2)–(5) for each one of the areas of the installation, that could be expressed as:

$$g_k(x) \geq 0, \ k = 1, \ldots, kf \qquad (7)$$

Where constraints $g_k(x)$ describe the interactions among the areas of the installation.

The model's structure among the areas correspond to a hierarchical participative decomposition structure [5, 8], according to the outline of the Fig. 2, where the decision-making system is composed of a higher-level task. That establishes $(Y_1, Y_2, \ldots Y_n)$ tasks to the n lower level systems and select the optimal combination of proposals (Or) from the close to optimal lining options $(Or_1, Or_2, \ldots Or_n)$, generated by the lower level systems with multiple objective optimization tasks, that minimize the generalized criterion z and assure all the restrictions settled down by the higher level system.

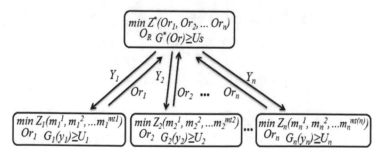

Fig. 2. General representation of the decomposition of the lining selection tasks for the areas

In the case of metallurgical ladles [3], shown in Fig. 3, the following restrictions are added to (2)–(5):

$$(\delta b_i - \delta s_i)^2 \leq \delta, i \in I \tag{8}$$

$$B \geq B^{\min} \tag{9}$$

$$hc \geq h^{\min} \tag{10}$$

Restrictions (8) indicates that the dispersion of the radial dilation of the lining δb_i that must be close enough to the radial dilation of the foil δs_i, for every area of the ladle.

Restriction (9) indicates that the quantity of metal calculated for the corresponding area that must be more than the minimum quantity of metal required according ladle nominal capacity B^{min}.

Restriction (10) indicates that the reserve of metallurgical height hc, which is opposite of metallurgical height that must be more than its permissible value, h^{min}.

In the foundry of metals furnaces, it is required that the mechanical resistance of the lining of the floor should be higher than the required for resisting the coalition derived from the deposition of the metal.

$$\sigma\left(m_{1,1}, \ldots, m_{1,r}; x_{1,1}, \ldots, x_{1,r}\right) \geq \sigma^{din} \tag{11}$$

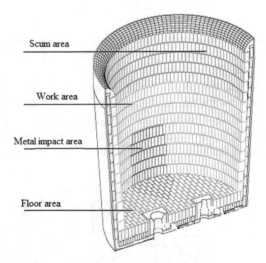

Scum area

Work area

Metal impact area

Floor area

Fig. 3. Internal view of a metallurgical ladle

Where σ^{din} is the dynamic resistance of the floor of the furnace. In most of the cases it is required that the total or a part, of the lining of the areas would be the same.

```
Start
  For i = 1 to cz // coding alternatives
    Inter0(i) = Cod (MaterialesOrd (i), nc(i))
  Next i
  Repeat
    For i = 1 to cz // creating initial search intervals
      If ((CodMax(i) - CodMin(i)) <= 2 Then
        CodMin(i) = 0, CodMax(i) = Inter0(i)
      Else // generating aleatory materials combinations
        Cod1(i) = Rand (CodMin(i), CodMax(i))
        SolEspMat1(i) = DET_ESPESOR (DeCod (Cod1(i)))
        Cod2(i) = Rand (CodMin(i), CodMax(i))
        SolEspMat2(i) = DET_ESPESOR (DeCod (Cod2(i)))
    Next i
    ListCombSol = Combina (SolEspMat1, SolEspMat2)
    For k = 1 to Length (ListCombSol)
      Z_Fitness = GetFuntionZ (ListCombSol (k)) // calculating objective function
      If Length (Poblacion) < Cind Then // creating population
        AddPoblacion (ListCombSol(k), Z_Fitness): UpPS = True
      Else UpPS = UpdatePoblacion (ListCombSol(k), Z_Fitness)
      If Z_Fitness < Z_FitnessBest Then
        Z_FitnessBest = Z_Fitness : ListBestComb = ListCombSol(k)
    Next k
    ListBestCod = GetCod (ListBestComb)
    For i = 1 to cz // verifying search intervals
    If ListBestCod(i) = Cod1(i) Then CodMax(i) = Cod2(i)
      Else CodMin(i) = Cod1(i)
    Next i
    If UpPS = True Then NoAct = 0 Else NoAct = NoAct + 1
  Until (NoAct > NoActMax) // evaluating stop criterion
End
```

Fig. 4. Pseudocode - exploration of the extremes function of variable codes algorithm applied to thermal refractory and insulating lining.

For the material selection, in various concrete applications, the Exploration of the Extremes of a Function of Variables Codes (ECV) algorithm, developed in the frame of the Integration of Variables method [15, 16, 25] is used. The Fig. 4, shows the pseudocode of this algorithm. For every solution code generated a discrete optimization task for finding optimal thickness is used in the algorithm [4].

In each iteration, search of the minimum of a function of j variable codes (that constitute the values, in the decimal system of numeration, of the variable-codes of the looked-for solution for every j area) is carried out. The initial variable code values, for every j area are generated stochastically inside the interval of possible values of the solution variable code $\left(0 - \text{MaxCode} = \prod_{i=1}^{n} Cod(i) - 1\right)$. Searching solutions is carried out by the operator of the Exploration of the minimum of the function of variable code by area. The quality function Z could be interpreted in the same way that Genetic Algorithms does, as fitness, and it could include the result of the calculation of a penalty function for non-fulfillment of the restrictions. This value is calculated by the generation and decoding of random values inside the current search interval (*CodMin, CodMax*) that decreases in each iteration by the elimination of the sub-interval that doesn't contain the best solution. Each localization iteration contains the population of solutions, while the population's size is smaller than the established one (*CInd*) or the population can be upgraded in case where it is already reached the established size. Once the foreseen precision is reached the generation of random values process is restarted to its initial values. This procedure continues until the iterations reached up to a level without the population of solutions upgrading (*Noact*) reach a pre-established value (*NoactMax*). Every current solution *x* is decoded by the universal algorithm shown in Fig. 2. Every generated code x is decoded by the algorithm shown in Fig. 5.

$$\text{For } i = 1 \text{ to } n$$
$$Cod\,(i) = x \bmod MaxCod\,(i) + 1$$
$$x = \text{Int}\,[\,x \,/\, MaxCod\,(i)]$$
$$\text{Next } i$$

Fig. 5. Decoding *x* codes algorithm

4 Numerical Results

The study is a continuation of experimentation and the results are based on experimental findings carried out in previous studies [3, 4, 6, 7, 24–26]. The experimental results [26], where solving the refractory and isolating lining of metallurgical ladles design are shown in Table 1. The results obtained using the Exploration of the Extremes of a Function of Variable Codes algorithm (EFVC) presents a better behavior during the solutions generation by areas stages (3 areas) which carries out a mean value of 500 iterations less than an elitist Genetic Algorithm (EGA) with better mean values of the objective function (VFOp) and higher diversity of individuals in the population (DCp). The same situation is observed in case of indicator where the behavior of the conciliation task for the overall installation is observed. These results depend on the programming details of both algorithms but demonstrate the viability of the successful

Table 1. Summary of the execution of implemented algorithms

Operator		Materials selection by areas		Conciliation task			Total
	CI	VMFp	DCp	CI	VMFp	DCp	CI
EGA	33828	1,327	0,401	1486	0,380	0,423	35314
EFVC	33328	0,394	0,497	1408	0,443	0,510	34736

application of the approach proposed in the present article, using different modern heuristics algorithm in the frame of the decomposition scheme that leads to the bi-level decision-making task. In this particular case the objectives were: total cost, heat and weight of the lining.

Where:

OG - Genetic Operator, EFVC - Exploration Function of the Extremes of Variables Codes algorithm, CI - Total quantity of calculations of the objective function, $VMFp$ - Main value of the objective function of solutions in the population, DCp - Diversity of solutions in the population. The solutions diversity obtained by EFVC algorithm facilitate the human-machine iteration and the possibility to choose between different characteristics solutions

5   Human-Machine Interaction

In both decision-making levels, the human-machine interaction plays a vital role. In the higher level, the decision-making process must select the refractory and isolating materials-candidates to be part of the corresponding walls of the different areas. This depends of many factors, including readiness of materials and the set of materials most convenient for different layers and areas. Formulated model facilitates that decision-making process could iterative vary the desirable values of the criteria. Select the most convenient solution. Using the possibilities of the CAD systems and the installations configurations factors not taken in consideration in the model, thus the possibility of charge more metal into de furnace (or the ladle), or to enlarge some areas. In the lower level, the possibilities of performance of the decision-making process are multiple: same selection of the lining design obeys to the adoption of some commitment among the efficiency indicators according the decider preference, same design could be modified as required, using the possibilities of the CAD system, taking into consideration factors no considered in the model, including technological considerations.

6   Findings and Conclusions

The Integration of Variables method enlarges the possibilities of construction evolutionary algorithms that not only could be effective for its execution speed and quality of the obtained solutions, but also obtaining a variety of different solutions for an easier selection of the most appropriate one by the developer.

In this research, scientific and technical principles that enable the application of bi-level and multi-objective optimization refractory and isolating lining design of high temperature installations are presented that allows to increase the efficiency of the solutions found and the automation level of the design, while maintaining the required flexibility in the decision-making process by the designer. The implementation of the proposed procedure using the CAD technology facilitate the introduction of any kind of modifications in the solutions proposed by the optimization process at both decision-making levels. The conception of the developed system presupposes the narrow interaction human-machine with significant influence on the pattern recognition in the complex decisions making processes.

References

1. Ashby, M.F., Shercliff, H., Cebon, D.: Materials: Engineering, Science, Processing and Design, 2nd edn. Editorials Elsevier Ltd., Oxford (2010)
2. Ashby, M.F.: Materials Selection in Mechanical Design, 4th edn. Elsevier Ltd., Oxford (2011)
3. Martinez-Valdés, O., Arzola-Ruiz, J.: Selección óptima bajo criterios múltiples de materiales refractarios y aislantes para cazuelas metalúrgicas. Rev. int. métodos numér. cálc. diseño ing (RIMNI) 32(4), 252–260 (2016)
4. Martinez-Valdés, O.: Optimización del revestimiento refractario para cazuelas metalúrgicas. MSc, Thesis, Holguín (2011)
5. Arzola-Ruiz, J.: Sistemas de Ingeniería, 2nd edn. Editorial Félix Varela, La Habana (2012)
6. Pinder, M.: Diseño de aislamiento térmico de hornos de cabina para el calentamiento de crudos. MSc, Thesis, El Tigre, Anzoátegui, Venezuela (2014)
7. Toledo, H.: Diseño óptimo de revestimiento para horno de reverbero de fusión de aluminio. MSc. Thesis, C. Bolívar, Venezuela (2014)
8. Arzola-Ruiz, J.: Análisis y Síntesis de Sistemas de ingeniería (2009). http://www.bibliomaster.com
9. DIDIER: Refractory Products and Services. Didier-Werke AG (1996)
10. Lu, J., Han, J., Hu, Y., Zhang, G.: Multi-level decision-making: a survey. Inf. Sci. 346, 463–487 (2016)
11. Lu, J., Shi, C., Zhang, G.: On bi-level multi-follower decision making: general framework and solutions. Inf. Sci. 176, 1607–1627 (2006)
12. Lu, J., Zhang, G., Montero, J., Garmendia, L.: Multi follower tri-level decision making models and system. IEEE Trans. Ind. Inf. 8, 974–985 (2006)
13. Kalashnikov, V.V., Dempe, S., Pérez-Valdés, G.A., Kalashnykova, N.I., Camacho-Vallejo, J.F.: Bi-level programming and applications. Math. Prob. Eng. 181, 423, 442 (2015)
14. Glackin, J., Ecker, J.G., Kupferschmid, M.: Solving bi-level linear programs using multiple objective linear programming. J. Optim. Theory Appl. 140, 197–212 (2009)
15. Wan, Z., Wang, G., Sun, B.: A hybrid intelligent algorithm by combining particle swarm optimization with chaos searching technique for solving non-linear bi-level programming problems. Swarm Evol. Comput. 8, 26–32 (2013)
16. Angulo, E., Castillo, E., García-Ródena, R., Sánchez-Vizcaíno, J.: A continuous bi-level model for the expansion of highway networks. Comput. Oper. Res. 41, 262–276 (2014)
17. Fontaine, P., Minner, S.: Benders decomposition for discrete–continuous linear bi-level problems with application to traffic network design. Transp. Res. Part B: Methodol. 70, 163–172 (2014)
18. Nie, P.Y.: Dynamic discrete-time multi-leader–follower games with leaders in turn. Comput. Math Appl. 61, 2039–2043 (2015)
19. Nishizaki, I., Sakawa, M.: Computational methods through genetic algorithms for obtaining Stackelberg solutions to two-level integer programming problems. Cybern. Syst. 36, 565–579 (2005)
20. Bard, J.F.: Practical Bi-level Optimization: Algorithms and Applications. Kluwer Academic Publishers, Dordrecht (1998)

21. Audet, C., Haddad, J., Savard, G.: Disjunctive cuts for continuous linear bi-level programming. Optim. Lett. **1**, 259–267 (2007)
22. Zhang, T., Hu, T., Guo, X., Chen, Z., Zheng, Y.: Solving high dimensional bi-level multiobjective programming problem using a hybrid particle swarm optimization algorithm with crossover operator. Knowl. Based Syst. **53**, 13–19 (2013)
23. Han, J., Lu, J., Hu, Y., Zhang, G.: Tri-level decision-making with multiple followers: model, algorithm and case study. Inf. Sci. **311**, 182–204 (2016)
24. Arzola-Ruiz, J.: Selección de Propuestas. Ed. Científico Técnica, La Habana (1989)
25. Martínez-Valdés, O., Arzola-Ruiz, J.: Operadores genéticos y de búsqueda aleatoria aplicados a tareas de selección de materiales de revestimiento para cucharas metalúrgicas. Revista de Metalurgia **53**(3) (2017). http://dx.doi.org/10.3989/revmetalm.099
26. Martínez-Valdés, O., Arzola-Ruiz, J.: Exploration of variable codes algorithm for linings materials and its thickness selection of steel casting ladles. IEEE Latin Am. Trans. **15**(8), 1528–1532 (2017)
27. Augusto, O., Bennis, F., Caro, S.: A new method for decision making in multi-objective optimization problems. Pesquisa Operacional, Sociedade Brasileira de Pesquisa Operacional **32**(3), 331–369 (2012)

Use of Active Test Objects in Security Systems

Marina Boronenko[✉], Yura Boronenko, Vladimir Zelenskiy,
and Elizaveta Kiseleva

Laboratory of Complex Engineering, Ugra State University,
628012 Khanty-Mansiysk, Russia
marinaboronenko@gmail.com, yurab7@yandex.ru,
w_selenski@ugrasu.ru, kyndryashka@mail.ru

Abstract. Some modern security systems use pupillometry to identify people in a state of drug or alcohol intoxication. Our work is aimed at developing an optoelectronic system to search for people who represent a hidden threat. We propose in security systems to use test objects that are significant only for a certain category of people. Other people will not have a reaction to the test object. Synchronous registration of pupillograms and oculograms allows to determine the deviation of the emotional state from the original. Oculograms specify the detail of the test object to which the reaction occurred. We experienced the idea in the older age group (45–50 years old) under ideal conditions (the pupil is clearly visible). In the future we plan to expand the age range and increase statistics. Experiments have shown that the use of thematic images as test objects allows the pupil to cause a reaction that is greater than that caused by other factors. Synchronization of pupillograms and oculograms indicates the centers of attention and indicates the presence (or absence) of emotional attachment.

Keywords: Pupillogram · Oculogram · Test object · Attention wave

1 Introduction

Modern studies have shown that the process of perceiving the visual nervous system is difficult due to the contribution of various parts of the brain that affect the pupil response. For this reason, pupil size is a rather indicative psychophysical parameter. The use of pupilography is quite wide. For example, a change in pupil size during speech is recorded [1, 2]. If at the same time a person is deceiving, it means that he is in a state of stress, the size of the pupil changes, which is registered by the equipment. Using pupometry to study the autonomic nervous system allows us to study various neurological disorders [3], to recognize Alzheimer's disease, neuropsychiatric disorders, sleep disorders, migraines, Parkinson's disease. By the change in the normal pupillary reaction, it is possible to judge the presence of diabetes mellitus in the early stages of the disease, amyloid and rheumatic diseases, and Chagas disease, one of the most common parasitic diseases in Latin America. Pupilograms are widely used to study the effect of printed text design on the visual system of people of different ages [4]. In 2008, Wilson and a group of researchers noted that the dynamics of pupil diameter can be used as an indirect measure of brain function. Scientists Fomenko V.N., Kupriyanov A.S. The fact

© Springer Nature Switzerland AG 2020
H. Ayaz (Ed.): AHFE 2019, AISC 953, pp. 438–448, 2020.
https://doi.org/10.1007/978-3-030-20473-0_43

of the influence of mental load on the size of students is unequivocally established and presented by a mathematical model of such reactions [5, 6]. Maddess T.L., James A.C. investigated the functional relationship of the reaction of pupils with simultaneous and sequential exposure of the eyes to visual stimuli. They built a map of visual functionality of the visual field of the eye, which will help all subsequent experiments on the reaction of the eye to carry out with greater accuracy. The reaction of pupils to pain in 2014 was studied by Ukrainian scientists O.Ya. Mokrik, V. O. Zaplatinsky.

The development and emergence of new information technologies led to the emergence of man-machine interaction (HCI) and the emergence of the concept of "emotional computing" (affective computing). Sensors embedded in the bracelet monitor several physiological signals, such as pulse, galvanic skin reactions, skin temperature, and system algorithms translate biological signals into the "language" of emotions. The machine is taught to recognize and respond accordingly to the user's feelings and emotions, determined by facial expression, posture, gestures, speech characteristics, and even body temperature. In marketing research, vision tracking [8–10] has become almost indispensable. All the above facts testify in favor of high reliability of pupillometry results. Thus, the combination of existing methods and high-tech equipment contributed to the creation of special optoelectronic security systems. In countries such as the United States, Israel, and Germany, biometric security systems have long been used in stadiums, subways, airports. In Russia, Akhmetvaleev A.M., Katasyev A.S. The presented concept of contactless identification of persons under the influence of alcohol and drugs is a threat to public safety. In aviation (transport) security systems, Vozzhenikova O.S. and Kuznetsova D.A. applied profiling technology. Specialists of the Center for Measuring Technologies and Automation of the Physics Faculty of Moscow State University. M.V. Lomonosov developed "VizioS-KAN" - a mobile system for tracking the direction of vision of a driver of a moving vehicle, which will improve safety on the roads [8].

Despite all the efforts, there are no security systems that can search for the evil intentions of people. By evil intentions, we mean, for example, the beliefs of terrorists. The standard method of identifying emotions, oriented to the face of the face is good, but if a person is good at hiding emotions, then it is difficult to recognize evil intentions. Our efforts are aimed at expanding the boundaries of the application of existing methods used in security systems. Here are described the results of applying test objects to determine the significance of the stimulus presented to an individual.

2   Experimental Methods

We propose to move from passive to active safety systems. In such systems you need to use test objects on the selected topic. Using synchronization of pupillograms and oculograms, it is possible to identify deviations of the emotional state. It is known that all people have their own preferences. You can even associate each person with a group in accordance with the selected attribute or idea. It is clear that not all people have the same emotional response to the same test object. However, people belonging to the same group will react approximately equally to a significant irritant. If a person hears or sees information relating to an essential topic, then at least he will be given involuntary attention. According to this principle, it is necessary to develop specific test objects, with the help of which it becomes possible to register the pupil size with the help of ECO to detect the instability of the psychophysical state of a person. The reaction of the pupils is proportional to the intensity of the emotions caused by the external test object. And this reaction will be noticeable only when the significance of all other internal and external stimuli is less than the significance of the test object $\sum \epsilon < \Psi$. It remains to make sure that the evoked emotional reaction is less than the threshold level activation $\Psi < \Psi_{activation}$, after which the person will begin to take any dangerous actions. The fulfillment of the condition $\Psi < \Psi$ activation corresponds to afferent synthesis, which does not lead to active actions. In a person for whom information is not valuable, the emotional state does not change, the size of the pupils is normal, fluctuations in the area of the pupil correspond to the norm (not more than the average ΔS). At the same time, fluctuations in the size of the pupil are caused by all internal and external stimuli, including both emotional and any other $\epsilon = \pounds_0 + \Omega_0 + \hbar_0 + \Psi_0$.

2.1 Experimental Setup

Slides with pictures shown from a laptop were used as test objects. The experimental setup is shown in Fig. 1.

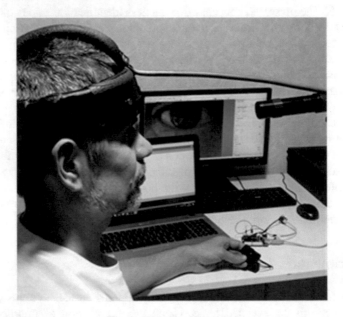

Fig. 1. Experimental setup

In order for the video camera to be used for recording pupillograms and oculograms, it is necessary to ensure its high spatial resolution and the same illumination of the eye surface, regardless of the brightness and color of test objects [9, 10]. In the process of implementing the tasks, an optoelectronic system (ECO) was assembled.

ECO hardware includes:

- Gavanic Skin Response Sensor (GSR - Sweating) sensor, measuring the resistance between two points;

Telescope Digital Lens for Guiding Astrophotographed T7 Video Camera T7 Astro Camera Astronomical Astronomy, 30 fps video mode, 1X-100X optical zoom microscope lens;

- Specially designed helmet, which creates a rigid coordinate connection between the video camera and the head;
- Hardware platform Arduino UNO R3;

Images were analyzed in the free FiJi software. Synchronization was carried out in the program OBS Studio.

Testing of the optoelectronic control system of psychophysical parameters was carried out on a group of students (18–24 years old) and older persons (45–50 years old), in the future it is planned to expand the age range and increase statistics. Eye diseases in humans were absent. The distance from the monitor to the eyes is 2 m, which made it possible to minimize fluctuations in the illumination of the surface of the eyes. Additional illumination control was carried out on the subject's skin tint (in the gray scale, normalized by the mean value). If necessary, you can enter a correction factor. To ensure correct interpretation of the results, the pupillograms were synchronized with oculograms. Experiments were performed with different test objects, gradually complicating them. At first there were circles and squares on the slides: one, two different sizes alternated, the empty poor surrendered. After the simplest test objects, we switched to thematic pictures shared by Internet users and their combinations. The intensity of emotions caused by the test objects was estimated by statistical survey. The intensity level of emotionally colored video files on a scale from 1 to 10 does not exceed 4, where 10 is the strongest emotion that can be. For testing, test objects were chosen in such a way that they were a carrier of information that is obviously significant for some and indifferent for others. The reaction of the pupils in response to an external stimulus is proportional to the significance of the information. If a person does not know the object, and he does not cause interest, then even a change in any part will not be noticed by the individual. At the final stage, participants in the experiment were shown complex images: nature, but in one of the corners was the emblem of a community.

2.2   Results

Subjects were asked to simply look at the pictures. The duration of the demonstration of one slide is 3 s. After processing video files, the results were presented in the form of time-synchronized graphs of oculograms and pupillograms (Fig. 2a, b). In the first part of the experiment, circles or squares were depicted on the slides. It was characteristic of all the results that on the same time sections of the oculograms [11, 13] and pupillograms [12, 14, 15] there were saccades and microsaccades. After linking the tracks (Fig. 2c) of the center of mass of the pupil to the image of the test object under consideration, it becomes logical to explain the observed changes in the pupil size by focusing on the considered element of the slide.

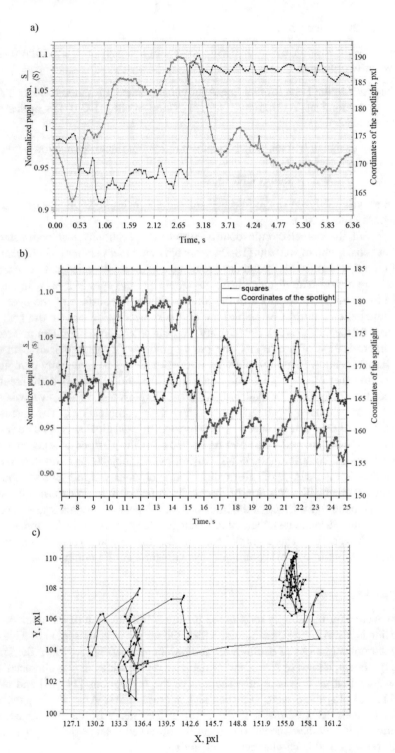

Fig. 2. Pupilogram and oculogram, obtained by viewing the simplest figures: (a) circles, (b) squares, (c) characteristic track with a consistent focus on two simplest figures.

At the same time, saccades in the oculograms correspond to a short-term V-shaped section in the pupilograms. Microscopes in the oculograms correspond to areas of smooth change in the pupillograms, and are relatively slow glance when looking at a test object. From graphs 2 a, b it can be seen that all changes in the size of the pupils have no other reason than focusing attention, since accompanied by a change in the coordinates of the radius vector. However, more careful research is needed to establish the functional dependence of the change in pupil sizes and the magnitude of the center of attention bias. Analysis of the pupillograms showed that their shape depends to some extent on the object considered by the individual. This can be explained by the fact that the average healthy eye views an object along an average statistical trajectory. If an object of a known shape is considered, then the average shape of the pupilogram must exist. Thus, it will be possible to calibrate the optoelectronic system in order to identify waves of attention with emotional coloring in the pupillograms. Basically, the pupillograms can be approximated by Gaussians.

$$y = y_0 + \frac{A}{w}\sqrt{\frac{2}{\pi}} exp\left(\frac{-2(x - xc)^2}{w^2}\right).$$

Conducted two series of experiments. First, the slides showed circles: one, two different sizes, then a repeat. Slides with images alternated with white blank slides. After the break, the experiment was repeated with images of squares. It can be assumed that the pupillograms, when examining identical test objects, will have a similar shape.

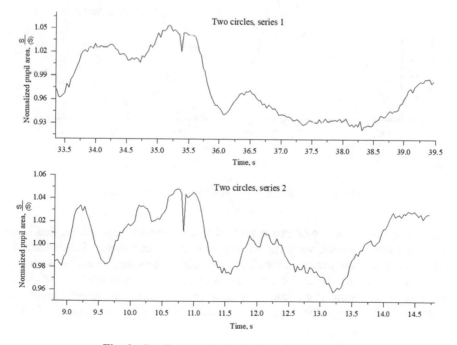

Fig. 3. Pupillorams obtained when viewing circles

To confirm the hypothesis put forward, we conducted a correlation analysis of the pupillograms obtained by looking at the simplest objects-circles and squares (Figs. 3 and 4.). Even visually, pupillograms have similar structures.

The peaks of the pupillograms obtained when viewing squares are sharper, which is explained by the presence of focuses of attention at the corners of the figures. Most often in psychological studies for the analysis of data using the linear correlation coefficient r - Pearson and the methods of Spearman and Kendal rank correlation.

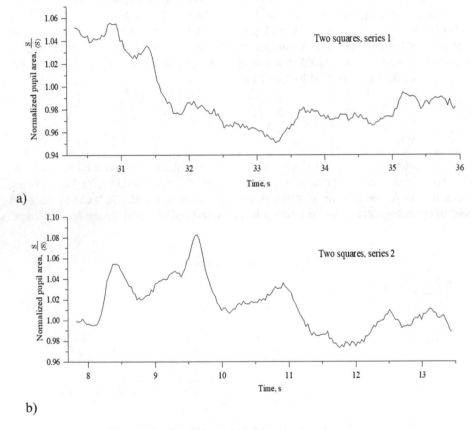

Fig. 4. Pupillograms obtained by viewing the squares

The Spearman's rank correlation method is a non-parametric method that is universal and works with data measured on any scale and is easy to use. The advantage of the method of rank correlation in the ability to compare the structure of data sequences. The coefficient of rank correlation made it possible to check the consistency of changes in the size of the pupils of the same test subject when reacting to certain simplest test objects and their groups. The results of the correlation study are presented in Tables 1 and 2.

Table 1. Correlation coefficients matrix.

Missing values removal	Pairwise deletion	Circle	Square
Circle	R	1.0000	
	R Standard Error		
	F		
	p-value		
	HO (5%)		
Square	R	0.0095	1.0000
	R Standard Error	0.0007	
	t	0.3553	
	p-value	0.7224	
HO (5%)		Accepted	
R			
Variable vs. Variable	R	No# of valid cases	
Square vs circle	0.0095	1396	

Denote: large square 1-1, small and big squares 1-2, small and big squares 2-3, big circle 1-4, small and big circles 1-5, small and big circles 2-6, large square 2-7, big circle 2-8.

From the fact that r measures only the linear relationship between X and Y, in some cases the values of r -are close to zero, if interpreted without taking the scatterplot into account. In the resulting scatterplots, the points are oriented - although they deviate randomly - relative to the curve, the relationship between X and Y is essentially nonlinear. Thus, the pupillograms obtained when looking at the simplest objects correlate. Therefore, it makes sense to conduct further research. With the complication of test objects to complex images, the interpretation of the pupillograms becomes more complicated. The fact that information relevant to the individual causes a change in the size of the pupil is known to all. Therefore, we will not dwell on this. We note only that even very old attachments cause a significant increase in the pupil (in our experiments 1.2 times). The hypothesis of focusing attention on significant elements for the subjects was tested on complex pictures with added emblems of any communities. Example of the results in the Fig. 5.

The reasoning is analogous to the previous one. The pupillogram tracks the spotlight, this allows you to take into account the effect on the size of the pupil, not only lighting but also focusing.

Table 2. Rank Correlations, Alpha (significance level) $\alpha = 0{,}1\%$

Missing values removal			Pairwise deletion		Inversion		
var vs. var	**Rho**	t	p-value	**Tau**	s	Z	p-value
1 /2	-0.3379	-3.0253	0.0017	-0.2097	-1 100.	2.6243	0.0043
1 /3	0.3538	3.1875	0.0011	0.1543	810.	1.9306	0.0268
1 /4	0.3286	2.9319	0.0023	0.2324	1 220.	2.9083	0.0018
1 /5	-0.6805	-7.825	1.7648E-11	-0.4765	-2 502.	5.9634	1.2355E-9
1 /6	-0.587	-6.1102	2.3982E-8	-0.3715	-1 950.	4.6495	1.6639E-6
7 /3	0.3297	4.7629	1.9152E-6	0.2237	7 862.	4.5589	2.5708E-6
7 /4	-0.3816	-3.5272	0.0004	-0.256	-1 420.	3.2495	0.0006
7 /5	-0.3904	-5.784	1.5189E-8	-0.2718	-9 552.	5.5387	1.5232E-8
2 /3	0.1612	2.227	0.0136	0.0815	2 864.	1.6611	0.0483
2 /8	0.2534	3.5724	0.0002	0.1699	5 968.	3.4613	0.0003
2 /5	0.4998	7.87	0.	0.3556	12 494.	7.2461	0.
2 /6	0.8004	18.1605	0.	0.5959	20 714.	12.1101	0.
3 /5	0.3053	4.3732	1.0178E-5	0.1733	6 090.	3.5316	0.0002
3 /6	0.4687	7.2176	6.6251E-12	0.2833	9 850.	5.758	4.2562E-9
4 /6	-0.3007	-2.6941	0.0044	-0.1811	-1 004.	2.2984	0.0108
8 /5	0.1322	1.8245	0.0348	0.0912	3 238.	1.863	0.0312
8 /6	0.2727	3.8554	7.9647E-5	0.181	6 294.	3.6793	0.0001
5 /6	0.6365	11.2235	0.	0.4553	15 828.	9.2523	0.

a)

b)

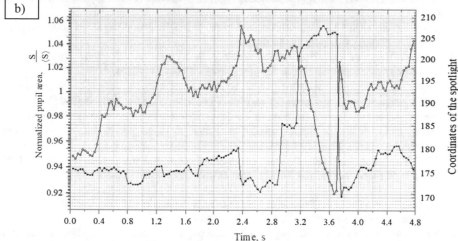

Fig. 5. Attention waves (a) and the spotlight track when viewing thematic test -objects (b)

2.3   Conclusions

In the course of the research it was found:

- ECO allows you to register changes in the amplitude, time and speed parameters of pupils under the influence of test objects.
- Analysis of the pupillograms showed that their shape depends to some extent on the object considered by the individual.
- The pupillogram tracks the spotlight, this allows you to take into account the effect on the size of the pupil, not only lighting but also focusing.
- Due to the fact that images containing elements of known sizes will be used in security systems, you can set a threshold value for changing pupil sizes, which can be explained by moving the center of attention. Thus, security systems can use a marketing research method to which individuals of a certain category will be sensitive. The rest will be as blind. Synchronization of the pupillogram and oculogram allows you to select areas that cannot be explained by a shift of view.

However, more careful research is needed to establish the functional dependence of the change in pupil sizes and the magnitude of the center of attention bias individual.

Thanks. The study was carried out with the financial support of the Russian Foundation for Basic Research in the framework of the research project 18-47-860018 p_a.

References

1. Cacioppo, J.T.: Feelings and emotions: role for electrophysioogical markers. Biol. Psychol. **67**, 235–243 (2004)
2. Stern, R.M., et al.: Psychophysiological Recording. Oxford University Press, New York (2001)
3. Meyberg, S., et al.: Microsaccade-related brain potentials signal the focus of visuospatial attention. NeuroImage **104**, 79–88 (2015)
4. Costela, F.M., et al.: Changes in visibility as a function of spatial frequency and microsaccade occurrence. Eur. J. Neurosci. **45**(3), 433–439 (2017)
5. Lowet, E., et al.: Microsaccade-rhythmic modulation of neural synchronization and coding within and across cortical areas V1 and V2. PLoS Biol. **16**(5), e2004132 (2018)
6. Perotti, L., et al.: Discrete structure of the brain rhythms. Sci. Rep. **9**(1), 1105 (2019)
7. Potter, M.C., et al.: Detecting meaning in RSVP at 13 ms per picture. Atten. Percept. Psychophys. **76**(2), 270–279 (2014)
8. Gao, Y., Huber, C., Sabel, B.A.: Stable microsaccades and microsaccade-induced global alpha band phase reset across the life span. Invest. Ophthalmol. Vis. Sci. **59**(5), 2032–2041 (2018)
9. Partala, T., Surakka, V.: Pupil size variation as an indication of affective processing. Int. J. Hum. Comput. Stud. **59**(1–2), 185–198 (2003)
10. Steidtmann, D., Ingram, R.E., Siegle, G.J.: Pupil response to negative emotional information in individuals at risk for depression. Cogn. Emot. **24**(3), 480–496 (2010)
11. Beatty, J., et al.: The pupillary system. Handb. Psychophysiol. **2**, 142–162 (2000)
12. Lanata, A., Armato, A., Valenza, G., Scilingo, E.P.: Eye tracking and pupil size variation as response to affective stimuli: a preliminary study. In: 2011 5th International Conference on Pervasive Computing Technologies for Healthcare (PervasiveHealth), 78–84. IEEE, Dublin (2011)
13. Onorati, F., et al.: Characterization of affective states by pupillary dynamics and autonomic correlates. Front. Neuroeng. **6**, 9 (2013)
14. de Gee, J.W., Knapen, T., Donner, T.H.: Decision-related pupil dilation reflects upcoming choice and individual bias. Proc. Natl. Acad. Sci. **111**(5), E618–E625 (2014)
15. Brambilla, R., et al.: A stimulus-response processing framework for pupil dynamics assessment during iso-luminant stimuli. In: 2018 40th Annual International Conference of the IEEE Engineering in Medicine and Biology Society (EMBC), 400–403. IEEE (2018)

Hand Gesture Based Control of NAO Robot Using Myo Armband

Sara Ali[1]([⊠]), Mohammad Samad[2], Faisal Mehmood[1], Yasar Ayaz[1],
Wajahat Mehmood Qazi[2], Muhammad Jawad Khan[1],
and Umer Asgher[1]

[1] School of Mechanical and Manufacturing Engineering (SMME),
National University of Sciences and Technology (NUST), Islamabad, Pakistan
{sarababer, faisal.mehmood, yasar, jawad.khan,
umer.asgher}@smme.nust.edu.pk
[2] Department of Computer Science,
COMSATS University Islamabad, Lahore Campus, Lahore, Pakistan

Abstract. Electromyography (EMG) has become an automation technique and has found its way in disciplines other than medical sciences. EMG based equipment is being used for automation of soft as well as hard robots. NAO is a humanoid robot developed by Aldebaran Robotics currently deployed in the health and education sector. In this paper we have attempted to develop an interface for controlling the NAO robot with an EMG sensor known as Myo Armband, we implemented a TCP client with the C++ Myo arm-band SDK and a corresponding sever with the Python NAO robot SDK. A Finite State Machine (FSM) architecture is incorporated at the client side to manage the network traffic rate as well as to increase efficiency of the system. The scope of this paper is limited to four different gestures instantiating four distinct actions performed by the robot using Myo Armband, this research is currently under development and in its testing phase. This technological aid supports the person who is waist down paralyzed or for an immobile patient on a hospital bed who wants to get the task done through a social robot.

Keywords: Human-robot interaction · Autism spectrum disorder ·
Client-server model · NAO robot · Myo arm band

1 Introduction

Electromyography (EMG) is a technique or procedure used to evaluate the condition of the motor neurons of the muscles [1, 2]. As all neurons, motor neurons also transmit little amount of electrical current that causes the muscles to contract or expand resulting in movements. The use of EMG enables us to record the values of these neurons and represent them as numeric values. EMG is both invasive and non-invasive. The fundamental difference between the two is that invasive EMG involves insertion of physical electrodes in the muscles while noninvasive EMG does not involve the insertion of electrodes inside the muscles [1]. Invasive techniques are known to cause discomfort and are not easy to use especially among patients [2]. Noninvasive or

© Springer Nature Switzerland AG 2020
H. Ayaz (Ed.): AHFE 2019, AISC 953, pp. 449–457, 2020.
https://doi.org/10.1007/978-3-030-20473-0_44

surface EMG is preferred nowadays to avoid discomfort among the patients or users. Many researchers have used EMG for human prosthetics especially arms [3], Isuru Ruhunage and his fellow researchers used EMG sensors to develop a prosthetic upper arm limb and later used EEG (electroencephalography) to make the prosthetic limb actions (opening and closing of hand) more efficient. Many other researchers have developed similar techniques using EMG sensors for prosthetics and medical experiments as well [4, 5].

Fig. 1. Thalmic Myo-Arm Band.

EMG based techniques are being used to improve the efficiency of exoskeletons and being used to control robots in some capacity as well. There are various devices available in the market that enables users to use EMG in their applications. One of such devices is Myo Armband by Thalmic labs. Figure 1 shows a typical Myo Armband.

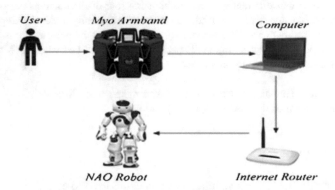

Fig. 2. General architecture of system.

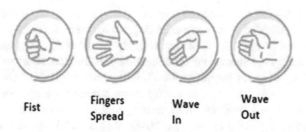

| Fist | Fingers Spread | Wave In | Wave Out |

Fig. 3. Gestures recognition

It is a gesture recognition device equipped with a set EMG sensor, a gyroscope, an accelerometer and a built-in lithium ion battery as well. It senses the muscle activity of the forearm and then predicts the gestures and the pose of the arm. It has built in application support for delivering presentations, keyboard support and mouse support. Myo Armband is currently being used for multiple research-based experiments. Malika Vachirapipop and his colleagues used the Myo Armband for communicating with Hearing-impaired people [6]. Kristian Nymoen and his team used Myo Armband for musical interactions, they concluded that the data (EMG values) provided by Myo Armband was sufficient enough for musical generation. Jonathan Cacace and team interfaced EEG and myo armband to control drones in rescue missions, where normal human-drone interaction is not favorable [7]. There is however one issues with EMG highlighted by [4] is that during the use of electromyography the muscle fatigue increases with the passing time, as this fatigue increases the signal strength gets weaker and it also gets noisy. [8] used deep learning approaches to recognize the human's arm motion patterns.

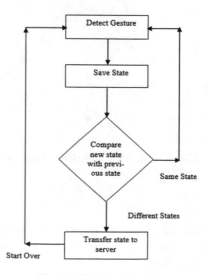

Fig. 4. FSM Architecture

Humanoid robots are becoming more common in human dominated workspaces. NAO robot is a humanoid robot having 25 degrees of freedom, equipped with cameras, sonar and other sensors [9, 12]. NAO has roughly the size of 3 years old toddler. NAO is the world's most sold humanoid robot specially in the education and health sector. NAO is also being used for the treatment of children under the ASD (Autism Spectrum Disorder) [10, 11]. NAO comes with multiple programming support for versatile usage and users, it comes with a visual programming tool Choregraphe enabling users with minimal programming skills to program NAO as to their liking, in addition to visual scripting NAO has its own framework known as NAOqi, programmers can program NAO using this framework on multiple programming languages.

The most prominent programming languages are C++ and Python, however support in Java and.Net is also available [9, 12]. More information on the NAO robot can be found at the Aldebaran robotics website.

In this paper, we have used the Myo Armband to get the EMG data and gestures and then transferred these gestures to instantiate certain actions on the NAO robot, humanoid robots are at the verge of forming a modern Man Machine society, such a society would include individuals incapable of speech and other normal communication mediums, by using this approach such individuals as well normal people would be able to have multiple actions performed with a humanoid robot. During the scope of this paper, we have used four simple gestures which come with the Myo armband, these gestures are recognized by a pre-trained model included in the Myo armband SDK, we used a TCP client server model with a Finite State Machine (FSM) to communicate efficiently. A detailed description and the experimental results of the entire project is explained in the further sections.

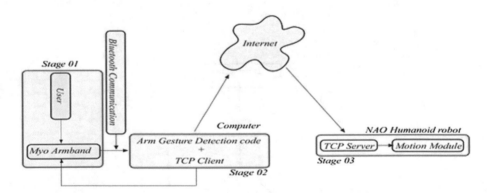

Fig. 5. Detailed architecture of system

2   Architecture

The goal of this paper is to make a gesture-controlled framework enabling the control of a humanoid robot. The actual program for this project was written in C++ and on Python. The current version of the project was tested on both Windows and Linux. A basic structure of the project t is illustrated in Fig. 2.

2.1 User Input

The system is designed in such a way that initially the user or operator has to wear the Myo Armband on the arm of his/her choice. The manager of Myo Armband allows to make custom profiles as well, but a default profile is loaded. The user inputs are detected in the C ++ Client side. As mentioned before within the scope of this paper we have limited inputs to 4 gestures only. The gestures are illustrated in Fig. 3 with their respect names.

2.2 Finite State Machine

The TCP client side works on the principle of FSM. It works in such a way that a request is transferred to the TCP server if and only if the last state send to the robot was different then the new requested state. The use of this state machine architecture, network latency issue was improved, and the overall system works much faster. Figure 4 illustrates the architecture of the finite state machine. It should be noted that this state machine was integrated at the Client or sender side, the rationale of this selection was the network latency and also this FSM serves as a verification module as to send only a valid request to the server. Actual hardware is involved in this project and it is very important that only valid commands are sent to the robot as to avoid any catastrophic damage to the robot as well as to environment.

2.3 Robot's Server

The FSM discussed above sends valid gestures to the robot server, in our case it is Python based TCP server. NAOqi which is mentioned above, is the framework on NAO robot. Figure 6 illustrates a high-level structure of the server side working. This server is rather very simple and is a smaller module as compared to the client module. One of the chief reasons to implement FSM at client side was to ensure that the server was light (low LOC count) so that the response at the robot could achieved immediately. The structure here is straight forward, when a socket is received, the server compares the incoming message with its preloaded command table. Table 1 presents the command table. After the verification from the command table, the system determines the action which needs to be performed and uses the NAOqi to send that respective command to NAO, the command is send remotely by using either an Ethernet cable or by Wi-Fi. During our study we used both Ethernet and wireless connection to the robot. Using an Ethernet cable has slightly better results but the overall difference with wireless connection is not a very decisive factor. The use of wireless network is encouraged as the project involves actual walking of NAO, Ethernet cable may become a cause for it fall down, or the connection might get interrupted.

Table 1. Commands Table

Gesture	Action
Fingers spread	Standup
Fist	Start walking
Wave in	Change arm poses
Wave out	Sit down and stop all processes

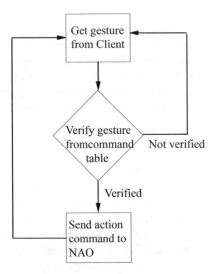

Fig. 6. Server flowchart

2.4   Network Setup

The network setup of this project is based on static IP's and Bluetooth (only for Myo Armband). Figure 5 illustrates a detailed system architecture with network representation. We setup a connection between our host computer and NAO robot using the SSH feature, it enables us to use secure network protocol services on unsecured networks, also providing a link between NAO and our computer.

3   Materials and Methods

3.1   Subjects

In this study three right handed healthy participants were involved to test the design system in real time. Two males and one female with an average age of 20 years approx. Each subject performed 10 gestures per day. Total number of tests by each person for each type of gesture was 50.

3.2   Experimental Setup and Design

The person was sitting on a comfortable chair in front of the robot such that he/she can easily see the robot. There was a distance of 1 m between the robot and person. Before initiating, person was instructed about the mapping of gestures and action that will be performed by NAO as shown in Table 1.

At the start of activity, the participant had to perform said gestures in a given order so that a cycle of robot's motion can be completed. During task performance, number of correct attempts were recorded along with the number of failures.

Each person had to perform at least 2 successive activities and 10 activities in a day. The states of the NAO robot were changing in a cyclic manner and were being recorded as well. In order to avoid slow network response, we implemented Finite State Machine (FSM) that sends the updated pose of human's arm only once to server, integrated in NAO robot's written algorithm. For this purpose, NAO's ALMotion API had been used.

4 Results

The experiment was simply conducted by having fifty trails with each gesture, the muscles fatigue factor [4] mentioned in introduction section of this paper plays a key role, as the more trails we had the more muscle fatigue increased. The results of the entire experiment were concluded with a commutative 80.5% success rate. Table 2 shows the results.

We had 3 volunteers testing the project, we collected all individual failed and succeed number of trails then calculated the average by taking the floor of the calculated average of each gesture, we computed the success percentage of each gesture as shown in Fig. 7.

Table 2. Results summary table

Gesture	Number of volunteers	Total trials each	Number of successful trials	Number of failed trials	Average success rate %age
Fingers spread	3	50	38	12	76
Fist	3	50	47	8	94
Wave in	3	50	39	11	78
Wave out	3	50	37	13	74

5 Discussion

Based on results, mentioned above in Table 2, the main contribution of this research is to observe the human robot interaction using biological sensors. The maximum accuracy achieved was 94% and minimum accuracy achieved was 74%. Moreover, the maximum number of correct attempts were 47/50 and minimum number of correct attempts were 37/50 as shown in Table 2 and Fig. 5. The observation is based on the experimentation done over 5 days. This helps disabled people who cannot speak and move to interact with the robots. The advantage of Myo integrated robot activity is that it does not require another human involvement.

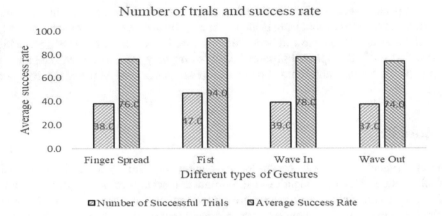

Fig. 7. Number of trials and success rate.

6 Conclusion

The main aim of this research is to exhibit a new way to control a humanoid robot for the upcoming massive man machine social society. We were able to identify another use of EMG and Myo Armband other than its applications in medical sciences. This project serves an interface for further development and as a potential framework for service robots targeted for individuals with severe medical conditions, unable to communicate with a robot using normal communication mediums. We were able achieve 80.5% success rate, the project is currently under development and the success rate including new features will be improved and updated in due amount of time.

References

1. Duiverman, M.L., van Eykern, L.A., Vennik, P.W., Koëter, G.H., Maarsingh, E.J.W., Wijkstra, P.J.: Reproducibility and responsiveness of a noninvasive EMG technique of the respiratory muscles in COPD patients and in healthy subjects. J. Appl. Physiol. **96**(5), 1723 (2004)
2. Aldrich, T.K., Sinderby, C., McKenzie, D.K., Estenne, M., Gandevia, S.C.: Electrophysiologic techniques for the assessment of respiratory muscle function. Am. J. Respir. Crit. Care Med. **166**, 548 (2002)
3. Ruhunage, I., Perera, C.J., Nisal, K., Subodha, J., Lalitharatne, T.D.: EMG signal controlled transhumerai prosthetic with EEG-SSVEP based approch for hand open/close. In: 2017 IEEE International Conference on Systems, Man, and Cybernetics, p. 3169 (2017)
4. Clarys, J.P., Cabri, J.M.H.: The use of surface electromyography in biomechanics. Presented at the (1997)
5. Lenzi, T., De Rossi, S.M.M., Vitiello, N., Carrozza, M.C.: Intention-based EMG control for powered exoskeletons. IEEE Trans. Biomed. Eng. **59**, 2180 (2012)
6. Vachirapipop, M., Soymat, S., Tiraronnakul, W., Hnoohom, N.: An integration of Myo Armbands and an android-based mobile application for communication with hearing-impaired persons. In: 2017 13th International Conference on Signal-Image Technology and Internet-Based Systems, p. 413 (2017)

7. Cacace, J., Finzi, A., Lippiello, V.: Multimodal interaction with multiple co-located drones in search and rescue missions. CoRR. abs/1605.07316 (2016)
8. Seok, W., Kim, Y., Park, C.: Pattern recognition of human arm movement using deep reinforcement learning. In: 2018 International Conference on Information Networking, p. 917 (2018)
9. Shamsuddin, S., Ismail, L.I., Yussof, H., Zahari, N.I., Bahari, S., Hashim, H., Jaffar, A.: Humanoid robot NAO: review of control and motion exploration. In: 2011 IEEE International Conference on Control System, Computing and Engineering (ICCSCE) (2011)
10. Tapus, A., Peca, A., Aly, A., Pop, C., Jisa, L., Pintea, S., Rusu, A.S., David, D.O.: Children with autism social engagement in interaction with NAO, an imitative robot: a series of single case experiments. Interact. Stud. **13**, 315 (2012)
11. Shamsuddin, S., Yussof, H., Ismail, L., Hanapiah, F.A., Mohamed, S., Piah, H.A., Zahari, N. I.: Initial response of autistic children in human-robot interaction therapy with humanoid robot NAO. In: 2012 IEEE 8th International Colloquium on Signal Processing and Its Applications (CSPA) (2012)
12. NAO - Technical Guide—Aldebaran 2.1.4.13 documentation. Doc.aldebaran.com (2019). http://doc.aldebaran.com/2-1/family/index.html. Accessed 1 Mar 2019

Research on Interactive Performance Optimization Strategy of Complex Systems Based on Information Coding

Tian Qin[1(✉)] and Siyi Wang[2]

[1] School of Design Art and Media,
Nanjing University of Science and Technology, 210094 Nanjing, China
qintian1996@126.com
[2] School of Mechanical Engineering,
Southeast University, 211189 Nanjing, China
220170380@seu.edu.cn

Abstract. In order to improve the interactive performance of the complex system, the usability evaluation is carried out. In this paper, the usability of the existing complex system interface was evaluated, and the optimization strategy was studied based on the information encoding. The results were analyzed through eye movement experiments. The performance of each task in the information interface after information coding optimization was better than that of the previous interface. It had different degrees of advantages in task completion time and task completion rate. The degree of information coding optimization of the interface was negatively correlated with the task completion time. The degree of optimization of interface information coding is positively correlated with click efficiency. The higher the degree of optimization, the higher the click efficiency.

Keywords: Information coding · Interactive interface · Usability assessment

1 Introduction

Vision is an important means for human beings to know nature and understand the objective world. It is also a breakthrough to understand human cognitive function. Human visual system is a complex information processing system composed of a large number of nerve cells through certain connections. When people start observing the outside world, the visual system will stimulate to the brain in the form of image, and through the brain's visual cortex area control the movement of the human eye to show interest in the image, a process called process of visual perception, visual perception is a very complex process, the user actually has experienced six mental process, including looking for, then, to identify and determine [1].

Visual search refers to the search target within the range of vision, and the specific search strategy is related to the search motivation. When it is found that the object detected by vision is consistent with the expected target, interference from other objects is excluded to fix the tracking object. Resolution refers to the in-depth detection of information about multiple similar objects. Identification refers to the difference

© Springer Nature Switzerland AG 2020
H. Ayaz (Ed.): AHFE 2019, AISC 953, pp. 458–466, 2020.
https://doi.org/10.1007/978-3-030-20473-0_45

between visual feature information and detail information and the identification of the meaning of the target. To determine is to determine the target by matching the locked object with the stored information in memory [2]. Memory search is the basis of the above visual process, and the information obtained by the above visual information should be compared with the memory information, so as to make a judgment.

2 Complex System Interaction Interface Requirements Analysis and Feature Extraction

Railway in China based on 12306 official website (https://www.12306.cn/index/) as an example, the demand analysis of complex system interactive interface, elements extraction, demand hierarchy, for later to optimize operation.

2.1 Demand Analysis

Users in order to complete the purchase of tickets as the purpose of the use of the website system, the website will be part of the operation to complete the purpose, in the current website from entering the website home page to complete the purchase process, need to complete the following operational requirements.

1. Determine the departure date and arrival date. The user needs to click on the "departure date" and click on the pop-up calendar window to select the date.
2. Determine the type of train (G/D/Z/T/K/Other). Users need to observe the available items, click on the desired option to determine the starting city and reach the city. The user needs to click on the "Departure" button to make a selection in the pop-up window. Or click the "Starting Point" button, enter the name of the departure place yourself, and confirm the entered information with the Enter key.
3. Check the train number. The user needs to directly present the "cars" information bar on the left side of the page.
4. Check the train departure time, arrival time, and duration. The user needs to be presented directly on the page "Dating Time, Arrival Time".
5. Check the departure station, route station, and arrival station. The user needs to click on the number of the train, view it in the pop-up list of the route station, and drag the wheel to browse.
6. Check the train fare. The user needs to click on any train seat and observe the price in the price bar that pops up below.
7. Check the number of train tickets. The user needs to directly observe the information of the train seat column, which is divided into three forms: "Yes", "Numbers ≤ 20" and "None".
8. Determine if the train is available for booking and direct booking. The user needs to directly observe the train remarks column, which can be blue button when booking, and gray button when not available.
9. Confirm that the train can check in to the station by ID card. The user needs to be presented directly next to the train information, and the ID card icon will be displayed when the ID card is checked in.

After the demand of ticket purchasing website is extracted, the above table is formed. On this basis, all user needs are still divided according to the degree of necessity.

2.2   System Function Requirements

According to the analysis of user behavior, the purchase behavior of users is divided into the following process [3]. The user's demand level can be divided into several levels. According to the design objectives and principles of the system, as well as the user's demand level, the ticket selling system needs to have the following main functions.

1. This level is the basis for all levels of requirements, and only after the completion of this level of requirements can subsequent levels be performed.

   (1) Book or not book:

   In the ticketing system, the availability of reservation is the primary requirement that surpasses all other functions. The demand elements at all levels below are all based on the availability of reservation. In another point of view, this feature can also be understood as at the end of all the other requirements, the user needs to confirm all information with its own demand was fully meet predetermined operations, will and if not scheduled, before all operations will only operate as useless, in this way, can you show is scheduled as heads of the other features of hierarchy, namely the first demand levels.

2. This level is the basic demand level, which is the information level that must be confirmed when purchasing tickets. When the demand of this level is met, the information of purchasing tickets in most cases has been understood and perfected, which is enough to support the completion of ticket purchase.

   (1) Departure station and arrival station:

   The user first needs to confirm whether the departure station and the arrival station meet their own needs. The criterion for this requirement is "correct" or "error." Subsequent operations can only be performed when the element is correct, and the behavior exists as a precondition for subsequent requirements.

   (2) Departure time and arrival time:

   The departure time and the arrival time are also one of the first information that the user needs to determine when purchasing the ticket. The judgment standard of the information has a certain fluctuation. The user needs to consider his own schedule, traffic situation, travel arrangements, climate factors, etc. In some scenarios, the user's needs can be met, but it is still the primary information that needs attention.

   (3) Residing votes:

   Users have different requirements according to factors such as price tolerance, reimbursement system, and comfort level. This factor may determine whether users purchase tickets or not, which is one of the first considerations.

3. When the basic ticket demand information is satisfied, users will start to pay attention to some secondary information. This information level has a low degree of demand, and even if it is not fully concerned, it can still be used to purchase tickets. However, this demand level has the function of assisting the second demand level.

   (1) Seat price:

   The user's consumption level and travel reasons are different, and the degree of attention to the price is different. Under certain circumstances, this factor is not a decisive factor affecting the ticket purchase process.

   (2) Train information:

   The information of the train does not affect the running behavior of the vehicle. It only encodes the vehicle. However, under certain circumstances, the user needs to discriminate the information of the train, which is not a decisive factor affecting the ticket purchase process.

   (3) Duration:

   The duration is only for the auxiliary user to judge. In the case where the departure time and the arrival time are displayed, the duration can already be calculated, and the information shows that the time is calculated only for the auxiliary user, so that the user can complete the ticket purchase operation.

4. This level is not important, and only a small number of users will pay attention to the following requirements when purchasing tickets. In most cases, it will not affect the ticket purchase operation.

   (1) Route station:

   The route station displays the information of all the stops from the departure station to the terminal, and most of the scenes do not affect the user's ticket purchase.

   (2) ID card check display:

   When the user enters the station, even if he does not go through the ID card to check in, he still needs to carry the ID card for the ticket collection operation, that is, the ID card must be carried. The demand is only for the user to understand the optimization information of the pit stop service and does not affect the ticket purchase process.

3   Research on Interactive Performance Optimization Strategy of Complex System Based on Information Coding

3.1   The Impact of Information Coding on Complex System Interfaces

With the human-computer interaction digital interface has more and more replaced the traditional human-machine hardware interface, and gradually become an important carrier for people to obtain information in complex information systems, the information coding of human-machine interface is called intra-thermal information perception [4]. An important basis for cognition, judgment, and decision making. The interface information enters the brain through the human eye and controls the behavior of the person through cognitive processing. However, due to the lack of design standards, more and more information is introduced into the display, which causes overload and confusion of information, which leads to the user's repeated cognitive calculations for comprehensive information processing, delaying the effective appointment and judgment of information. Time increases the cognitive load of the operator [5]. It can be seen that the complex information system has a large amount of information on the human-machine interface and a complex information coding structure, which easily causes imbalance between visual information coding and human cognitive mechanism and increases cognitive load.

Information coding is used in the field of communication by connecting information and carriers through a unique means to achieve signal recognition. Information has been encoded in the process of relying on the carrier for processing. By reasonably coding the information, the interaction between the human-machine interface can be smoother and the information can be better communicated, thereby improving the user's information recognition efficiency [6]. At present, the research on the display interface of complex systems in the society mainly improves the identification and reasonable division of layout from the dimension of information coding and viewport. The content of information coding covers character form, icon form, size, color, background color, brightness, contrast and so on.

3.2   Research on Interactive Performance Optimization Strategy of Complex System Based on Information Coding

1. Character form

   The main function of the website display interface is to convey information, and text is a major way of information transmission. You can improve your recognition efficiency by choosing the right font.

   The complexity and style of the font will all affect the recognition efficiency of the interface, but the style and complexity will not directly affect the cross. In the relevant analysis and analysis of the display and control interface of the motor car, it is found that in the commonly used Chinese fonts, the cognitive time of the black body and the Song body is the shortest, which is the preferred font for the display control interface. The response time of Hua Wen xing Wei, Fang Zheng Yao Ti, Hua Wen Li Shu, Qi Li and Li Shu was second.

2. Icon form

   In the interface information transfer, another important way besides text is the icon. The icon is an abstraction of the information to be delivered, and the use of graphics to communicate information. Good icon design enables users to identify information faster.

3. Application of color

   Color accounts for a large proportion in visual ergonomics. By planning a combination of colors, users can increase the cognitive efficiency of displaying information. Different color matching schemes have great differences in the efficiency of information recognition. The study found that under the black background, the correct efficiency of the human eye to identify different colors of information is red, yellow, light green, and blue, while the variance data shows that the recognition efficiency of red and yellow is significantly higher than blue. Therefore, in the complex system interface design of the website, reasonable matching color can improve the recognition efficiency [7].

4 Design and Evaluation of Complex System Interface Based on Information Coding

4.1 Experimental Purpose

The main purpose of this experiment is to study the influence of information coding changes on human visual cognition in complex system interface cognition. Through the display of two kinds of information interfaces, the cognitive effect of the optimization of information coding in the interface on the user's visual search processing is analyzed.

4.2 Test User

Ten college students participated in the experiment, including 5 male students and 5 female students. The age of the subjects was 22–24 years old, and the average age was 23 years old. All the subjects had naked eyesight or corrected visual acuity of 1.0 or above. Or weak color, normal hand operation ability, have not participated in this type of experiment before, the participants can get a certain amount of physical compensation after completing the experiment.

4.3 Experimental Equipment

The experimental materials come from the official website of China Railways 12306 (https://www.12306.cn/index/). The experimental equipment uses the eye movement recorder produced by Tobii to present the materials and record the eye movements of the subjects. The sampling frequency of the recorder is 1000 times/s. The experimental material is presented on a Hasee laptop with a display screen of 15.6 in. and a resolution of 1920×1080 pixels. The distance between the eyes of the test subject and the center of the display is about 54 cm.

4.4 Experimental Design

1. Experimental stimulus independent variables: This experiment uses the existing 12306 website interface, and the information interface based on information coding optimization, a total of two system interfaces as stimulus variables.
2. Referring to Rao Peilun's division of cognitive patterns, the experiment is divided into experimental tasks of different difficulty according to the operational tasks of ticket sales.

Tasks with low task difficulty are skill-based tasks, such as "Querying the ticket status from 'Nanjing' to 'Beijing'."

The task with difficulty in the task is a regular task, which requires the participants to think consciously and perform tasks according to some rules. For example, "you plan to take the high-speed rail or train to Beijing before 2 pm tomorrow, please arrange according to the schedule. You can purchase tickets for the trip, the amount and the number of trips are not limited."

A task with a high task difficulty is a knowledge-based task. The task requires an unfamiliar situation. A complex task that requires careful consideration and repeated scrutiny. The knowledge-based task used in the experiment assumes a scenario for the user: You are scheduled to travel to Beijing on a business trip. The schedule is to leave tomorrow morning and return the next afternoon. Since the time from Nanjing to Beijing takes four hours, considering the road traffic to and from the high-speed rail station and the need for normal food, you need to arrange your travel time. Between 11:00–19:00, the return time is between 7:00–12:30, the company is reimbursed only for the second-class seat of the high-speed train.", such a task has no clear operating instructions and needs to be tested. Choose your own round-trip time according to the schedule and operate with your own thoughts [8].

3. Experimental stimulus dependent variables:

In the experiment, the user is completely recorded: (1) The execution time of the completed task, (2) The number of errors, (3) The hot zone map of the eye movement experiment, (4) The gaze map of the eye [9].

The execution time of the task refers to the total time required to complete a group of tasks, which is automatically recorded by the computer to the accuracy of 1 s, and the time required for the three groups of tasks is recorded separately. The number of errors refers to the number of erroneous operations when each group of tasks is completed, and the experiment instructor observes the records when the experiment participants complete the tasks. The eye movement experimental hot zone map is used to reflect the user's browsing and gaze and shows the subject's attention distribution on the stimulating material. The annotation trajectory map can effectively display the visual attention overlap area of multiple subjects at the same time, and can judge the visual focus area, and the hot area map and the annotation trajectory map are automatically generated by the eye tracker and the equipped software.

4.5   Experimental Procedure

The experiment is divided into a preparation phase, an introduction phase, a testing phase, and an interview phase after the test.

1. Experimental preparation stage

During the experimental preparation phase, the laboratory personnel examine the experimental materials, equipment, and experimental environment to ensure that everything is ready.

2. Introduction stage

Before the official experiment begins, the experiment recorder will explain to the experiment participants the background, purpose, process, precautions, and remuneration of the experiment, answer the questions raised by the test user, and let the test user fill out the registration form.

3. Test phase

After ensuring that the subject understands the whole experimental procedure, the test recorder performs eye calibration on the eye tracker. After the calibration is completed, the test recorder shows the participant how to use the experiment interface through the display, and the test participant uses the interface to complete some practice tasks.

After the practice experiment is completed, the participants take a rest for 1 min and conduct a formal experiment. The experiment recorder issues the first low-level skill-based task. The time starts from the time the subject receives the task. The participant needs to complete in two different interfaces. For the same task, after the first task was completed, the subjects raised their hands and rested for 3 min. The experiment recorders paid the subjects in kind. The second regular task and the third knowledge task are identical to the first group. After the three tasks were completely completed, the experiment recorders conducted a retrospective interview with the subjects and recorded the interview contents.

During the experiment, the camera records all the operation processes, which is convenient for later statistical experiment time, total operation times, effective operation times, number of errors, and task completion rate.

4.6 Experimental Procedure

Table 1 shows the performance of the three tasks in the information interface optimized by information coding. Both the task completion time and the task completion rate have different degrees of advantages. It can be found that the degree of interface information coding optimization is negatively correlated with the task completion time ($r = -0.76403$, $P < 0.05$).

Table 1. .

The independent variables	1		2		3	
	The average time / s	Completion /%	The average time / s	Completion /%	The average time/s	Completion /%
12306 website interface	12.9	100	130.9	40	290.2	90
Optimized information interface	8.5	100	88.2	80	192.5	100

Table 2 shows that the degree of optimization of interface information coding is positively correlated with click efficiency (r = −0.527, P < 0.05).

Table 2. .

The independent variables	1			2			3		
	Click on the number of times	Valid hits	Effective click rate /%	Click on the number of times	Valid hits	Effective click rate /%	Click on the number of times	Valid hits	Effective click rate /%
12306 website interface	8	8	100	16.5	12	72.7	24.5	20	81.6
Optimized information interface	8	8	100	15	12	80	22.5	20	88.90%

5 Conclusion

The interaction performance of complex systems directly affects system energy efficiency and user perception, and the requirements for interactive performance are increasing in the new era. This paper combines the 12306 website for case analysis, through detailed user needs analysis and website interaction design, optimizes the website, and proposes a complex system interaction performance optimization strategy based on information coding, which can provide reference for the design and development of other complex systems and reference.

References

1. Sadaghiani, M.H., Dadizadeh, S.: Study on the effect of a new construction method for a large span metro underground station in Tabriz-Iran. Tunn. Undergr. Space Technol. **25**, 63–69 (2010)
2. Sharifzadeh, M., Kolivand, F., Ghorbani, M., Yasrobi, S.: Design of sequential excavation method for large span urban tunnels in soft ground – niayesh tunnel. Tunn. Undergr. Space Technol. **35**, 178–188 (2013)
3. Kim, S.K., Suh, S.M., Jang, G.S., Hong, S.K., Park, J.C.: Empirical research on an ecological interface design for improving situation awareness of operators in an advanced control room. Nucl. Eng. Des. **253**(12), 226–237 (2012)
4. Intraub, H.: Detection theory: a user's guide. Cognition **94**, 19–37 (2004)
5. Burns, C.M., Skraaning, G., Jamieson, G.A., Lau, N., Kwok, J., Welch, R., Andresen, G.: Evaluation of ecological interface design for nuclear process control: situation awareness effects. Hum. Factors: J. Hum. Factors Ergonomics Soc. **50**, 663–679 (2008)
6. Borst, C., Sjer, F.A., Mulder, M., Van Paassen, M.M., Mulder, J.A.: Ecological approach to support pilot terrain awareness after total engine failure. J. Aircr. **45**, 159–171 (2008)
7. Sawaragi, T., Shiose, T., Akashi, G.: Foundations for designing an ecological interface for mobile robot teleoperation. Robot. Auton. Syst. **31**, 193–207 (2000)
8. Burns, C.M., Kuo, J., Ng, S.: Ecological interface design: a new approach for visualizing network management. Comput. Netw. **43**, 369–388 (2003)
9. Duez, P., Vicente, K.J.: Ecological interface design and computer network management: the effects of network size and fault frequency. Int. J. Hum. Comput. Stud. **63**, 565–586 (2005)

Novel Metaheuristic Approach: Integration of Variables Method (IVM) and Human-Machine Interaction for Subjective Evaluation

Umer Asgher[1(✉)], Rolando Simeón[2], Riaz Ahmad[1,3],
and José Arzola-Ruiz[4]

[1] School of Mechanical and Manufacturing Engineering (SMME),
National University of Sciences and Technology (NUST), Islamabad, Pakistan
umer.asgher@smme.nust.edu.pk, riazcae@yahoo.com
[2] CAD/CAM Study Center, Holguin University, Holguín, Cuba
rsimeon@uho.edu.cu
[3] Quality Assurance Directorate, National University of Sciences
and Technology (NUST), Islamabad, Pakistan
[4] Technological University of Havana, José Antonio Echeverría,
Cujae-CEMAT, Habana, Cuba
jarzola@cemat.cujae.edu.cu

Abstract. Metaheuristics is recognized as the most practical approach in simulation based global optimization and during investigation of the state-of-the-art optimization approaches from local searches over evolutionary computation methods to estimation of distribution algorithms. In this study a new evolutionary metaheuristic approach "The Integration of Variables Method (IVM)" is devised for global optimization, that methodology is characterized by making vide solutions codes in an objective population aided by any composition of operators. The Genetic Algorithms (GA) are incorporated in this methodology; where genetic operators are used to evolve populations classifiers in creation of new concrete heuristics solutions for decision making tasks. In this work, the main consideration is devoted to the exploration of Extremes value distribution of Functions with multi Variables Codes in brain decision making process and Brain Computer Interface (BCI) tasks. The properties of Extremes value distribution algorithm are characterized by the diversity of measurements in populations of available solutions. Additionally, while generating adequate solutions of the basic objectives, subjective indicators of human-machine interaction are used to differentiate the characteristics of the different solutions in a population. The results obtained from this metaheuristic approach are compared with results of Genetic Algorithms (GA) in case study related with Optimal Multiple Objective Design and Progressive Cutting Dies implemented in a CAD system.

Keywords: Decision making · Human–machine interaction ·
Genetic Algorithms (GA) · Product design · Metaheuristics ·
Brain computer interface (BCI) · CAD systems

© Springer Nature Switzerland AG 2020
H. Ayaz (Ed.): AHFE 2019, AISC 953, pp. 467–478, 2020.
https://doi.org/10.1007/978-3-030-20473-0_46

1 Introduction

The multi-level optimization techniques are applied on the multi-levels of the real systems and processes. To initiate the interaction among entities that make decisions distributed system characterized by the existence of a higher and lower levels or even lower to lower sub levels. Higher level is commonly denominated as leader and lowers as followers. Both make decisions in sequence with the purpose of optimizing their respective objectives. The higher level has the priority making own decisions and the followers react later on knowing the leader's decisions, making their own decisions. At the same time, the leader's decisions are implicitly affected by the reactions of its followers. In the last decades hundreds of works devoted to the modeling [1], solution methods [1–3] applications [4] and decomposition techniques of multi-level systems has already been addressed by various researchers, for different classes of problem. Among these researches, various classes like: linear [5], non-linear [6], continuous [7], discrete or mixed [8–10] variables. For all these classes many solution outlines have been published to mono [11] and multi-objectives problems, principally lineal [12, 13] for bi-level [13] and tri-level [14] tasks.

The concept of the Integration of Variables Method (IVM) is linked to the evolution of any quantity of codes assisted by any set of operators to upgrade corresponding solutions populations [15, 16]. Its general features and flow are illustrated in Fig. 1. The possible solution variants are coded in one or more variables – codes, and is generated, according to a procedure characteristic for each particular application of the method, a set of n solutions close to the optimal one. Particularly, different procedures which are characteristic for different classes of Mathematical Programming methods applied to the solution codes could be used, with selected search environments in a random, deterministic or combined way. Each particular procedure of generation and upgrading populations has to do with a concrete application variant of the method.

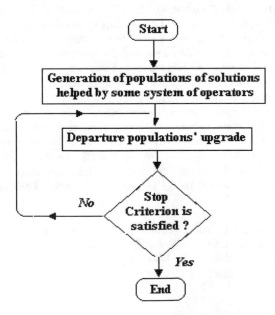

Fig. 1. Flow diagram of Integration of Varible Method (IVM)

So, the application of any heuristic technique derived from the Integration of Variables Method (IVM) requires the definition of the following problems:

a. A coding system for the possible solutions representation of the studied problem, a method for creating one or various initial population(s),
b. A quality (fitness) function that allows ordering the codes according the objective function values,
c. Operators that allow the altering of the composition of solutions codes in the successive populations,
d. Stop criteria and parameters values required by the algorithm used (population's size, probabilities associated with the application of operators).

Despite the general acceptance received in the last decades by other metaheuristic methods, difficulties have appeared in its application to optimization problems due to the following causes: there is no guarantee of obtaining the global optimal solution although generally have the tendency to do so. This tendency is frequently reduced when there is a loss of population diversity, which can lead to final populations of low quality and poor diversity [17–19], slowness of searching process in comparison with other methods [19, 20]. The authors had developed some algorithms in the frame of the IVM that could in occasions overcome mentioned difficulties [15, 16, 21], particularly, increasing the diversity of the solutions populations. This is close to the efficient ones, that facilitate the selection of the definitive solutions by deciders as a result of being able to choose between the objectives functions solutions with diverse characteristics [22–25]. Ahead the simpler of these algorithms is exposed and applied to the optimal multiple objective design of cutting and punching dies task.

2 Searching and Random Localization of the Functional Extremities - Variable Code Algorithm

In Algorithm 1 (Algo 1), the practical application of the Integration of Variables method (IVM), while searching the functional minima of variable codes among the possible solutions obtained by the discrete optimization with random variation of search intervals and upgrading the successive solutions populations [15, 16, 25] were proposed. In the pseudocode shown in Algorithm 1, each iteration is carried out to search the minimum of a function of variable code (that constitute the values, in the decimal system of numeration, of the variable-code of the looked-for solution). The initial variable code values are generated inside the interval of possible values of the solution variable code ($0 - \text{MaxCode} = \prod_{i=1}^{n} Cod(i) - 1$).

Searching solutions is carried out by the operator of the Localization of the minimum of the function of one independent variable method. The quality function Z could be interpreted in the same way that Genetic Algorithms (GA) does, as fitness, and it could include the result of the calculation of a penalty function with no fulfillment of the restrictions. This value is calculated by the generation and decoding of random values inside the current search interval (*CodMin, CodMax*) that decreases in each iteration by the elimination of the sub-interval that doesn't contain the best solution.

Each localization iteration, included in the population (solutions), while the population's size is smaller than the established one (*CInd*) or the population is upgraded in case it already reached the established size. Once the foreseen precision is reached the generation of random values process is restarted to its initial values. The procedure continues until the number of followed iterations without the population of solutions upgrading (*Noact*) reaches a pre-established value (*NoactMax*).

```
Start
Inter = MaxCod: CodMax =/ Inter: CodMin = 0
Repeat
If (CodMax - CodMin) ≤ 2
        Then CodMin = 0, CodMax =Inter
        Codd1 = Rand (CodMin, CodMax), Codd2= (Rand
(CodMin, CodMax))
If (Codd1<Codd2)
        Then Cod1=Codd1: Cod2=Codd2
Else
        Cod2=Codd1: Cod1= Codd2
S1=DeCod (Cod1 + CodMin )), S2 =  (DeCod (Cod2+ CodMax))
Z_Fitness1 = GetFunctionZ (S1), Z_Fitness2 = GetFunctionZ
(S2),
If Length (Population) <CInd
        Then Add Population (S1, Z_Fitness;
S2, Z_Fitness): UpPS = True
Else
        UpPS = False: Update Population (S1, Z_Fitness; S2,
Z_Fitness)
If  ZFitness1 > ZFitness2
        Then CodMin = Cod1,
Else
        CodMax =Cod2
If UpPS = True
Then NoAct =0, Else NoAct = NoAct +1
Until (NoAct =NoActMax)
Output (Population)
End
```

Algo 1. Pseudocode - localization of the extremes of a function "Variable Code Algorithm"

Every current solution x is decoded by the universal algorithm shown in Algorithm 2. Every generated code x is decoded by the algorithm.

```
For i = 1 to n
Cod (i) - x mod MaxCod(i) + 1
x = [ x / MaxCod(i)]
Next i
```

Algo 2. Decoding x-codes algorihm

3 Optimal Multi-objective Product Design

3.1 Multilevel System Analysis

Product fabrication by cutting, punching and other conforming technologies, is subordinated to the general design task of fabrication technologies. The product (pieces) could be manufactured by a given sequence of technologies or for a specific technology [26, 27]. Indeed, a design of piece system may require, the technology generation for the production of a particular piece, specifically by cutting and punching of simple and progressive dies. As decisive elements of the technology generations are necessarily, the distribution of the pieces in the foil (higher level) and the design of the appropriate die to manufacture the required piece (lower level). The die design task has to be conciliated with the task of piece distribution on the plate, that determines the dimension of the intermediate product to be used. There are, also, other defined parameters of the technology, such as the productivity of the conforming installation, the number of useful careers, the work force to be used (depending on the decided automation level), the useful life of the tool.

This way, the design simple and progressive cutting and punching dies task is subordinated to the general tasks of piece fabrication (multi-level system). The last one includes the determination of optimal production sequences, assignment in a first stage of concrete technologies of particular systems conceived for this purpose (for machinery, foundry, lamination, extrusion). In the process of concrete technologies generation the lower level systems generate, in a second stage, technology options, which include proposals of the intermediate pieces for the continuation of production by other technologies. In a third stage, the upper level system (the fabrication one) selects the combination of options that better satisfies the general requirements for the fabrication of the finished piece. In the Fig. 2 the hierarchy of technological processes is shown.

The hierarchy of the decisions tasks associated with tools design for cold cutting and punching, includes the configuration and technical specifications of the product (piece). This determines the production sequence and the most appropriate manufacturing operations. The systems, which generate different tool design, elaborate the corresponding manufacturing phases. These cold cutting and punching processes, stands out as key tasks, indissoluble linked with the distribution of the piece and the die design.

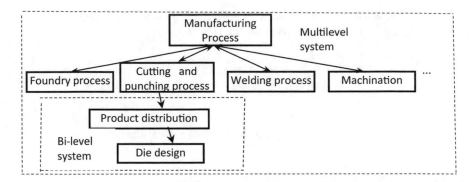

Fig. 2. Hierarchy of production process

3.2   System Analysis

The analysis of the pieces distribution task will be carried out starting from the study of the composition of entrance and exit variables and its decomposition by objectives (efficiency indicators) and decision variables. Other indicators: total production costs, cutting precision, energy consumption. Efficiency indicators and decision variables in the higher level are described as (Fig. 3 and Table 1).

Table 1. Decision variables

	Decision variables	Apch
a.	Axis X displacement between pieces	DespX
b.	Axis Y displacement between pieces	DespY
c.	Pieces positioning variant	VarN

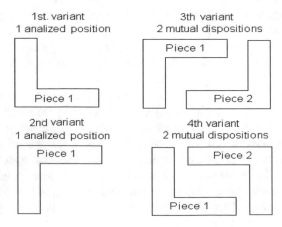

Fig. 3. Pieces mutual distribution paradigm

3.3   Analysis of the Die's Design Task

Cutting and punching dies are widely used tools in mechanical industry for making plane parts of various configurations. From the study carried out about the system of wide spans coordination variables stand out the following: the configuration of the finished piece, the technical specifications of the piece, and the size of the lot to be produced. Starting from surveys carried out with the participation of specialists and managers dedicated to the design and the production of dies, the following efficiency indicators for the specific case of the simple and progressive cutting and punching dies were defined. Efficiency indicators of the lower system are defined as (Table 2):

Table 2. Efficiency indicators

Efficiency indicators	Approach
Use of the plate	Apro
Press's productivity	Prod
Cutting forces	Fuer
Die cost	Cost
Die durability	Dura
Surface finish	Itac

For possible values of the efficiency indicators the decision making system (or the aiding decisions system) can modify the values of the following decision variables (Table 3).

Table 3. Product type, decision variables and approaches

Die type [3 possible solutions]	TTroq
Distribution of the pieces on the foil [1, 2, 3, 4]	DpCh
Side cutter [included or not]	ReLat
Cutting type [2 possibilities]	TiCor
Number of cutting attempts [1 or 2]	NuPas
Feeding system [2 possibilities].	SiAli
Centerpiece element [3 possibilities]	ElCen
Type of matrix and cutting edge [4 possibilities]	MaPar
Positioning Elements Guide Rule [included or not] Side pressurer [included or not]	ReGui PreLa
Material of the matrix [consecutive number of available materials]	MatMa
Material of the cutter [consecutive number of available materials]	MatPu
Type of the base [casted or laminated] Position of the columns [4 possibilities] Form of the columns [4 possibilities]	BasTip PoCol TiCol

That means, all variables are discrete and constitute general decisions that determine the designs die configuration. The internal analysis includes an entire methodology of engineering calculations and finite elements simulation procedures that complets the evaluation of the decision making system for different solution options generated later by the system.

3.4 Objective Function and Solution Outline

In a first stage, an orderly series of pieces distribution options is generated by an algoritm based in the Exploration method of Non-lineal Programing, that satisfies a local efficiency criterion. In a second stage, to the different distribution options indexes are attributed the character of possible values of a decision variable of the dies' design task. The preference criterion used in the second stage may have a more general character that in the first one. Indeed, during the piece distribution on the plate it can be influenced directly in the use of the foil and on the associated expenses, while in the second stage it is influenced directly In the solution of the optimal multi objective cutting and punching simple. Progressive dies task was used the Localization of funtional Extremities - Variable Code Algorithm (LE-VCA).

With the purpose of maintaining similar conditions as in case of using GA, the penalization function consists rejection of the non feasible solutions. Stop Criteria: The difference of the fitness values between the first and the last population solution overcomes 0.15. After 225 iterations the fitness value of the population's last element was not improved. For both algorithms the population's size is selected equal to 30. Each x value generated by LE-VCA algorithm is converted to binary code, for this code and all the indicators y_i, the value of the utility multiple objective function Z and of the fitness value $1/Z$ are calculated. if it is required, the graphical 3 D and 2 D images are generated.

3.5 Numerical Experiments

The TROQUEL System was developed for carrying out the necessary experiments that allowed comparing both methods. This system compared algorithms in the solution of the formulated task. Dies designs members of the populations were generated by 100 different executions of the GA and EAPSV algorithms described for a concrete piece. As indicators of the operation efficiency of both algorithms the following ones were considered:

a. Total quantity calculations of the objective function (QC) performed before reaching the final population.
b. Average minimum value reached by augmented Tchebycheff (AMTR) distance from the optimal solution for different designs.
c. Average mean Tchebycheff distance value (AMTD) reached by the final population.

Population's genetic diversity (GD) estimated as the sum of the normalized absolute deviations among the solution codes generated with regard to its half value, according to the expression (1). The obtained results are shown in the Table 4.

$$GD = \frac{1}{n} \sum_{i=1}^{n} \left| \frac{x_i - \bar{x}}{\bar{x}} \right| \tag{1}$$

Table 4. Comparative indicators obtained by algorithms

Indicators Methods	QC	AMTR	AMTD	GD
GA	4560	0.32	0.43	0.1
LE-VCA	4443	0.31	0.41	0.4

The number calculations for multiple objective function by LE-VCA is same order as by GA, reaching similar average objective function value, so for the best solution in the populations for its average values. Although the genetic diversity turned out to be much better for the case of LE-VCA, means that the decision making system could choose among more solutions with similar efficiency level. The results are more extensive in the exploration of search region, given by the same nature of the procedure used for the calculations of multi-objective function. The convergence speed for the LE-VCA algorithm is really higher, because the quantity and complexity of the operations carried out by each objective function calculation is lesser than by GA.

4 Human–Machine Interaction (HMI) in Design Processing

The Human machine interation (HMI) plays a decisive role right from the design section, fitness funtion till the final CAD design. In this multi-level settings the decisions are generation while interating with CAD systems as well as identifying some pattterns in the fitness funtion. In the higher level modelling, the decision making restricts the distributions generated by the system according to subjective preference and using the possibilities of the CAD systems. The LE-VCA algorithm could be used for fast convergence and near optimal search of the best design decision that may further lead to optimization interms on time and material. Decision making system modify some of the configurations, but not assisting to the factors that were not taken in consideration during model association. The possibility that some corner of a product (piece) could bend as a result of the cutting or by heat treatment are not eliminated in the solution space.

The input from the CAD system is directly translated as efficient solution selection and the fundamental step in decion making process. In the lower level the possibilities of performance of the decision making are multiple: the same selection of the design and the adoption of some commitment among the efficiency indicators according the decider preference. At each level the human-machine interation is considered in CAD designing process, selection of best design and design implementationn. On the other hand, the same design could be modified, using the possibilities of the CAD system used, taking into consideration factors that were not considered in the model, including technological considerations. The results of novel proposed alogritm LE-VCA have broad applications in decision making process, optimal solution solection of design process, HMI and Brain computer Interface (BCI) related design applications [28].

5   Findings and Conclusions

Integration of Variables Method (IVM) enlarges the possibilities of constructing evolutionary algorithms that not only be effective for its execution speed and quality of the obtained solutions, but also efficient and easier selection of most appropiate solution. In this research, Localization of funtional Extremities - Variable Code Algorithm (LE-VCA) that enable the application of bi-level and multi-objective optimization of cutting and punching processes are explored, that allows the decision support system to increase the efficiency of the solutions and the automation level of the design, while maintaining the required flexibility in the decision-making process. Comparative indicators obtained by Genetic Algorithms (GA) and Variable Code Algorithm (LE-VCA) shows a significant improvement in the design process in terms of fast convergence to the optimal decision making process. The implementation proposed Human-Machine Interaction using the CAD technology facilitates the introduction of broad modifications in the solutions anticipated by the optimization process at both decision making levels. The results of LE-VCA are superior to the GA. The conception of the developed system presupposes human-machine interaction with significant influence on the pattern recognition in the complex decisions making processes.

Conflicts of Interest: The authors declare no conflict of interest.

Acknowledgments. The authors would like to acknowledge School of Mechanical & Manufacturing Engineering (SMME)- National University of Sciences and Technology (NUST), Islamabad - Pakistan, CAD/CAM Study Center, Holguin University, Holguín - Cuba and the Technological University of Habana, José Antonio Echeverría, Cujae-CEMAT, Habana - Cuba for providing collaborative research support and opportunity to the authors in data collection, modelling and analysis of the Proposed Algorithms.

References

1. Jie, L., Jialin, H., Yaoguang, H., Guangquan, Z.: Multilevel decision-making: a survey. Inf. Sci. **346**, 463–487 (2016)
2. Lu, J., Shi, C., Zhang, G.: On bilevel multi-follower decision making: general framework and solutions. Inf. Sci. **176**, 1607–1627 (2006)
3. Lu, J., Zhang, G., Montero, J., Garmendia, L.: Multifollower trilevel decision making models and system. IEEE Trans. Ind. Inf. **8**, 974–985 (2006)
4. Kalashnikov, V.V., Dempe, S., Pérez-Valdés, G.A., Kalashnykova, N.I., Camacho-Vallejo, J.F.: Bilevel programming and applications. Math. Prob. Eng. **181**, 423, 442 (2015)
5. Glackin, J., Ecker, J.G., Kupferschmid, M.: Solving bilevel linear programs using multiple objective linear programming. J. Optim. Theory Appl. **140**, 197–212 (2009)

6. Wan, Z., Wang, G., Sun, B.: A hybrid intelligent algorithm by combining particle swarm optimization with chaos searching technique for solving non-linear bilevel programming problems. Swarm Evol. Comput. **8**, 26–32 (2013)
7. Angulo, E., Castillo, E., García-Ródenas, R., Sánchez-Vizcaíno, J.: A continuous bi-level model for the expansion of highway networks. Comput. Oper. Res. **41**, 262–276 (2014)
8. Fontaine, P., Minner, S.: Benders decomposition for discrete–continuous linear bilevel problems with application to traffic network design. Transp. Res. Part B: Methodol. **70**, 163–172 (2014)
9. NieP, Y.: Dynamic discrete-time multi-leader–follower games with leaders in turn. Comput. Math Appl. **61**, 2039–2043 (2015)
10. Nishizaki, I., Sakawa, M.: Computational methods through genetic algorithms for obtaining Stackelberg solutions to two-level integer programming problems. Cybern. Syst. **36**, 565–579 (2005)
11. Bard, J.F.: Practical Bilevel Optimization: Algorithms and Applications. Kluwer Academic Publishers, Dordrecht (1998)
12. Audet, C., Haddad, J., Savard, G.: Disjunctive cuts for continuous linear bilevel programming. Optim. Lett. **1**, 259–267 (2007)
13. Zhang, T., Hu, T., Guo, X., Chen, Z., Zheng, Y.: Solving high dimensional bilevel multiobjective programming problem using a hybrid particle swarm optimization algorithm with crossover operator. Knowl. Based Syst. **53**, 13–19 (2013)
14. Han, J., Lu, J., Hu, Y., Zhang, G.: Tri-level decision-making with multiple followers: model, algorithm and case study. Inf. Sci. **311**, 182–204 (2016)
15. Arzola, R.J., Simeón, R.E., Maceo, A.: El Método de Integración de Variables: una generalización de los Algoritmos Genéticos. In: Proceeding of Intensive Workshop: Optimal Design of Materials and Structures, París (2003)
16. Arzola-Ruiz, J.: Análisis y Síntesis de Sistemas de ingeniería (2009). http://www.bibliomaster.com
17. Díaz, A., et al.: Optimización heurística y redes neuronales. Paraninfo, Madrid (1996)
18. Man, K.F., Tang, K.S., Kwong, S.: Genetic algorithms: concepts and applications. IEEE Trans. Industr. Electron. **43**(5), 519–533 (1996)
19. Costa, M.T., Gomes, A.M., Oliveira, J.: Heuristic approaches to large-scale periodic packing of irregular shapes on a rectangular sheet. Eur. J. Oper. Res. **192**(1), 29–40 (2009)
20. Egeblad, J., Pisinger, D.: Heuristic approaches for the two- and three-dimensional knapsack packing problem. Comput. Oper. Res. **36**(4), 1026–1049 (2009)
21. Toranzo-Lorca, G., Arzola-Ruiz, J.: Un algoritmo del método de integración de variables para lasolución del problema Máximo Clique Ponderado. Rev. Inv. Operacional **35**(1), 27–34 (2014)
22. Cordovés-García, A., Arzola-Ruiz, J., Ashger, U.: Incorporating the cultural and decisions factors in multi-objective optimization of air conditioning conduit design process. In: Advances in Intelligent Systems and Computing. Springer, Cham (2017)
23. González, I., Arzola, J., Marrero, S., Legrá, A.: Operación con criterios múltiples para la compensación de potencia reactiva en redes industriales de suministro eléctrico. Parte II. Energética **XXVIII**(1), 45–52 (2007)
24. Hechavarría-Hernández, J.R., Arzola-Ruiz, J., Asgher, U.: Novel multi-objective optimization algorithm incorporating decisions factors in design modeling of hydraulic nets. In: Advances in Intelligent Systems and Computing. Springer, Cham (2019)
25. Martínez-Valdés, O., Arzola-Ruiz, Y.J.: Exploration of variable codes algorithm for linings materials and its thickness selection of steel casting ladles. IEEE Lat. Am. Trans. **15**(8), 1528–1535 (2017)

26. Huang, K., Ismail, H.S., Hon, K.B.: Automated design of progressive dies. Proc. Inst. Mech. Eng.: J. Eng. Manuf. **210**(4), 367–376 (1996)
27. Arzola-Ruiz, J., Símeón-Monet, R.E.: Aplicación de los algoritmos genéticos al diseño óptimomulti objetivo de troqueles de corte y punzonado simples y progresivos. Rev. Latinoam. Metal. Mater. **22**(2), 11–17 (2002)
28. Liang, Z., Jian, Z., Li-Nan, Z., Nan, L.: The application of human-computer interaction idea in computer aided industrial design. In: 2017 International Conference on Computer Network, Electronic and Automation (ICCNEA), pp. 160–164, Xi'an. https://doi.org/10.1109/iccnea.2017.71 (2017)

Using Adaptive Integration of Variables Algorithm for Analysis and Optimization of 2D Irregular Nesting Problem

José Arzola-Ruiz[1(✉)], Arlys Michel Lastre-Aleaga[2],
Alexis Cordovés[2], and Umer Asgher[3]

[1] Studies Center of Mathematics for Technical Sciences (CEMAT),
Technological University of Havana, Habana, Cuba
jarzola@cemat.cujae.edu.cu
[2] Equinoccial Technological University, Quito, Ecuador
arlysmichel@gmail.com, alexiscordoves60@gmail.com
[3] School of Mechanical and Manufacturing Engineering (SMME),
National University of Sciences and Technology (NUST), Islamabad, Pakistan
umer.asgher@smme.nust.edu.pk

Abstract. In this article a new evolutionary technique using 2D irregular-shaped nesting problem optimization algorithm applied on material cutting processes having processes like cutting of irregular foils pieces and complex geometrical shapes is proposed. This technique is adaptive and improves results of material cutting process and its optimization in comparison to other methods reflected in the earlier literature. A formalization and mathematical modelling are based on CAD/CAPP/CAM/CAP integration methodology for irregular two-dimensional pieces produced as a multilevel optimization with constraints of component part and the bi-level distribution and cutting process on foils. The decision of selection and the final solutions using adaptive algorithm based on dynamical modification of the objectives during the pieces' distribution and using the population obtained for the elaboration of cutting trajectories of the cutter device. This procedure defines a new approach for the bi-level optimization of geometric iterative increasing polygon pieces' configurations using the particular case of the Integration of Variables method called Exploration of a Function of Variable Codes algorithm. As final result a new optimal manufacturing integration scheme that include human-machine interaction as a component part is deducted.

Keywords: Irregular cutting · Stock problem ·
Nesting, distribution of irregular parts · Adaptive · Evolutionary methods ·
Graphic treatment · Boundary surfaces

1 Introduction

The multi-level optimization reflects the multi-level structure of most of the real systems. An interaction among entities that make decisions distributed in parts of a system characterized by the existence of higher and lower levels. Higher level is commonly

© Springer Nature Switzerland AG 2020
H. Ayaz (Ed.): AHFE 2019, AISC 953, pp. 479–489, 2020.
https://doi.org/10.1007/978-3-030-20473-0_47

denominated as leader and lowers as followers. Both make decisions in sequence with the purpose of optimizing their respective objectives. The higher level has the priority making own decisions and the followers react later on knowing the leader's decisions, making their own decisions. At the same time, the leader's decisions are implicitly affected by the reactions of its followers. In the last decade, various research works devoted to the modeling [1], solution methods [1–3], applications [4] and decomposition techniques of multi-level systems and different classes of problem have been developed. Among these classes, are linear [5], non-linear [6], continuous [7] and discrete or mixed [8–10] variables. For all these classes many solution outlines have been analyzed to mono [11] and multi-objectives problems, principally linear [12, 13] for bi-level [13] and, in occasions, tri-level [14] tasks.

The Integration of Variables Method was developed for searching solution to very complex systems, including multilevel ones, and is linked to the evolution of any quantity of codes by set of operators to upgrade corresponding solutions populations [15, 16]. It general, features are illustrated in Fig. 1. The possible solution variants are coded in one or more variables – codes. It is generated, according to a procedure characteristic for each application of the method. A set of n solutions close to the optimal one. Particularly, different procedures for different classes of Mathematical Programming methods applied to the solution codes could be used, with selected search environments, in a random, deterministic or in hybrid method. Each particular procedure of generation and upgrading populations has to do with a concrete application and variation of the method.

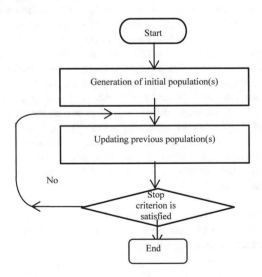

Fig. 1. General outline of the Integration of Variables method

On the other hand, the formulation of multi-level decision-making systems requires the application of methodological contributions to the analysis and synthesis of complex systems. Various researchers worked in solving theoretical and applications problems [15, 16]. These contributions include an analysis methodology that presupposes external and internal analysis. The output of the external analysis constitutes the conceptual mathematical model of the studied problem. Depending on its complexity, this model could be or could not be decomposed. The decomposition depends on the mathematical structure of the conceptual model. 2D nesting problem is analyzed using this methodology and the solution procedure for bi-level decision-making task will be organized using an algorithm developed in the frame of the Integration of Variables method.

2 Applying the Systems Analysis and Production Planning of Cutting Problems

These planning problems have been investigated in numerous works [17–19] and many other. The integration of CAD/CAPP/CAP/CAM systems tasks for the design (CAD), technologies generation, process planning (CAPP), production planning (CAP) and manufacturing (CAM) obey the same principles as that of the remaining systems [16]. In a first phase of more span task, the design of the piece, tool and installation be manufactured is subordinated, in quality of subtask. In the second stage the analysis of the design task is carried out. The process planning task is subordinated, necessarily, to the design of the product. The CAPP system responsible to generate different production options of the product and the production direction system of the shop. The aggregate of selected proposal for each production phase should be implemented in the moment settled down by the production direction system (CAP) of the shop. The execution of the productive tasks should be carried out by transferring the directive information from the CAPP to CAM system that governs the technological aggregate associated.

From the decomposition of the production direction task [20], the relationship between management of technical resources and production sequences generation is illustrated in Fig. 2. That requires the transfer of directive information from system of technical resources to the elaboration of production sequences. The decisions-making system generates options, used by the production system in the stage of sequences generation. Ones selected the appropriate option; the manufacturing system uses the corresponding information to manufacture the product. This way, the direction system of technical resources includes the design of the piece (CAD System), the generating options of production technologies (CAPP System), the transfer of the information to the CAM System for the fabrication and option selected by the production direction system (CAP System).

In Fig. 3, the general structure of the CAD/CAPP/CAM/CAP integration applied to the production of irregular cutting pieces is presented. This includes the relationship between the exchange of directive information, CAPP system and the production planning system CAP for the cases of manufacturing pieces by cutting plane pieces of irregular sheets configuration as shown. This relationship consists the CAPP system fabrication options, the programmed sequences of manufacturing, the selection by CAP system, the appropriate fabrication option while taking into consideration the operative complexities arisen in the shop. The most appropriate fabrication option, the CAPP

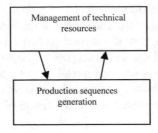

Fig. 2. Directive information flow

system sends is the numeric control codes and the necessary information to the CAM System while doing the production. External analysis of the manufacturing CAPP system and its analysis made it clear that the coordination variables are determined by the coordination of the CAPP, CAD systems and CAP, altogether. They are constituted by the planes of the pieces and the foils, as well as for the constructive restrictions. The requirements related with the production volumes and the raw materials readiness associated to the production lot. Thus, the coordination variables are: Geometric configuration and technical specifications of the pieces. Raw material (foils, or its polygonal residuals from previous processes) is used for guaranteeing the production volumes $DChaj$, $j = 1..., ju$.

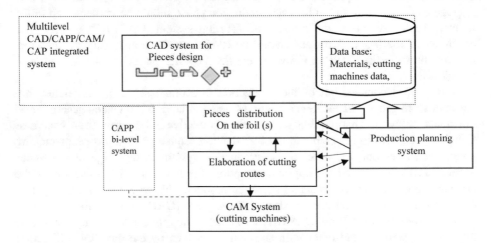

Fig. 3. CAD/CAPP/CAM/CAP integration applied to the distributions and cutting pieces

a. Lots of pieces to be produced by time unit Lpn.
b. Priority level of the piece in the lot Gpi, n.
c. Court technology to be used Tc.
d. Readiness of cutting machines according to the determined technology Dmt, m.

Efficiency indicators were searched from the literature, questioners and interviews with entrepreneurs and selected the following ones:

a. Maximal use of the processed raw material *Ap*.
b. Minimum trajectory of the cutter device used *Dhc*.
c. Maximum satisfaction by the user for the proposed solution, including the introduction of possible adjustments.
d. Maximum satisfaction of the restrictions imposed by the coordination variables.

The productivity of the cutting process, the consumption of resources such as gases, electric power, etc. and the minimal production time are directly related to the minimum displacement of the cutter device taken into consideration explicitly as efficiency indicators at the higher level, but at the lower one of the bi-level systems.

As decision variables are considered:

a. The coordinates of location of every point that describes the contour of every piece related to the other ones: *DespX, DespY*.
b. Among the most important intermediate variables to be calculated stand out:
c. Effective area of the foil *Ach*.
d. Codes of the pieces already distributed and/or cut Cod_i.
e. Dynamic values of the material use during the solution process *Ap*.
f. Time of the journey of the cutting device *Tr*.
g. Consumption of technological resource *Cct*.
h. Configuration of the reserve of free metal (polygonal foil) after cutting *Crd*.
i. Presence of curls in the corners of the pieces *Pbes*.
j. Perimeter of the cutting trajectory of the device: *Ptc*.

Decomposition of the production task into sub-tasks as it has been described in the analysis of the task under study, that stand out the sub-tasks of the distribution of pieces and the determination of the cutting trajectory for the distributions proposed in solutions population. In the first one, the different values of material used are calculated on the foil, that constitutes the first efficiency indicator. In the second one the trajectories are generated where it should pass the cutter device following the profile of the pieces that determines it displacement, as the second efficiency indicator.

The first sub-task, corresponding to the higher-level system, has to do with the determination of the mutual positioning of the pieces, determined by the values of the decision variables *DespX, DespY*. While the second sub-task, corresponding to the lower level system, defines the order sequence of the cutter device by each contour of the pieces, through the nodes x_i that define the changes in the distributed pieces.

This way, the conceptual mathematical model of the task could be expressed as expression (1), given disperse-conciliated nature of the model [16, 17]:

$$\text{Minimize } \{Z(Z_1(u), Z_2(u, x_n))/h(u) \geq 0; g_s(u, x_s) \geq 0; \forall s \in S\} \tag{1}$$

Where:

$Z_1(u) = 1 - Ap$ non profitable material on the foil, *Ap* is the use of the foil.
$u = (DespX, DespY)$: geometric coordinates of the piece positioning on the foil.
$Z_2(u, x_n) = Dhc$ minimal trajectory of the cutter device used.
$h(u) \geq 0$ geometric restrictions related to the possible mutual positioning of the pieces on the foil.

x_n: Nodes that define the changes contour - contour of the distributed pieces $g_s(u, x_s) \geq 0$ restrictions to the possible cutting trajectories.

In the research [16, 17], the general expression (1) could formulate those processes where the efficiency, the conditions of operation and its elements depend on common coordination variables vector. That vector could not be partitioned by lower level subsystems, as take place in the particular case of the studied task. The bi-level system task, shown in Fig. 3, used for solving the distribution and cutting tasks. That means our system consists of just one leader and one follower. The distribution of pieces on the foil (leader system), determines the component $Z_1(u)$, and the elaboration of the cutting trajectory for the distributions, the follower system determines the component $Z_2(u, x_s)$. The solution of expression (1), for the higher-level system, is reduced to the solution of (2)–(4):

$$Min\{Z(Q_1(u),\ldots,Q_n(u))/h(u) \geq 0\} \tag{2}$$

Where

$$Q(u) = Min\{Z_s(u, x_s)/g_s(u, x_s) \geq 0\}; u/h(u) \geq 0, \quad \forall s \in S\} \tag{3}$$

$$x_s = x_s(u) \tag{4}$$

The bi-level optimization task (1) could be formulated, according [16, 17], in terms of Tchebycheff Program, by model (5):

$$Z = max\left\{\left[w\frac{|(z_1(u) - z_1^{id})|}{|z_1^{id}|}, (1 - w)\left(\frac{|z_2(u, x_s) - z_2^{id}|}{|z_2^{id}|}\right)\right]/h(u) \geq 0, g(u, x_s) \geq 0\right\} \tag{5}$$

Where w reflects the importance attributed to the criterion $z_1(w)$, and (1- w) the importance attributed to the criterion $z_2(w)$.

3   Adaptive Algorithm

The search of pieces' positioning variants on the foil consists minimizing $z_1(u)$ in (5) assuring the set of restrictions $h(x) \geq 0$, associated to the dimensions of the foil and the cutting technology, that are taken in consideration by the establishment of parameters, such as piece to piece and piece to foil limit distances. The mathematical representation of the different pieces can be carried out only graphically using CAD systems technology. The task (6) is solved iteratively:

$$Minimize\, Z_1(u) = \{max[\lambda(1 - y_1), (1 - \lambda)(1 - y_2)]/h(x) \geq 0\} \tag{6}$$

Where

$\lambda =$ Area of the conglomerate/Minimal area polygon contained the conglomerate

y_1 is function that characterizes the effectiveness of the coincidence perimeter in couples of pieces. y_2 is function that describes the effectiveness of the adjustment area inserted in the minimal area possible polygon. In Fig. 4 the general algorithm for generating populations of pieces' distribution is illustrated. In every step of the algorithm, the conglomerate for each node generates m solutions and adding in each solution another piece from the set of non-positioned pieces corresponding to this node by the Exploration of Variables Codes algorithm by solving optimization problem (6). For the following node of the same step, it is done with the same procedure conserving the m more promising solutions for both nodes. When concluding the node number m, the population of nodes of the next step will be constituted by the m more promising conglomerates of two pieces. The procedure is repeated while pieces are still without positioning or the foil doesn't admit a new piece. In this last case the best solution obtained for that foil is selected and the same process for non-positioned pieces continue for a new foil. At the beginning of the positioning process, the coefficient λ is very large that acquires the highest importance the common perimeter. As the area of the conglomerate grows it acquires every time higher importance the use of the polygon (rectangle for the particular case of a rectangular foil or residual of a foil).

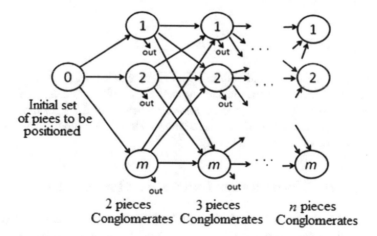

Fig. 4. General algorithm of conglomerates creation process

In Fig. 5 a pseudocode of the Exploration of Variables Codes algorithm applied for the particular studied case is shown. The code of pieces coincide with it. Initially, m conglomerations of 2 pieces are conformed by aleatory selection of 2 different pieces from the set of all the pieces to be distributed by exhaustive search minimizing objective (6). The indicators for the possible joining the concave parts of a piece with the convexities of the another piece for being able to evaluate are listed and referred bellow. Follows steps of close to optimal growing conglomerates formation. In each step the Exploration of Variables Method is applied by adding to each departure conglomerate a new piece not included in the set of pieces conforming it. For creating and upgrading the conglomerates population, variable code values are generated aleatorely inside the interval of possible values of the solution variable code. Searching solutions is carried out by the operators of the exploration of variables codes. These values are calculated by the generation and decoding random values inside the current search interval ($CodMin(i)$, $CodMax(i)$) for every i piece, considering the departure conglomerate as a piece, that decreases in each iteration by the elimination of the sub-interval of the codes that doesn't contain the best solution combination between the four generated solutions. In each exploration iteration it is included in the population the found solutions, while the population's size is smaller than the established one or the population is upgraded in case is already reached the established size ($Cind$). Once the foreseen precision is reached the generation of random values process is restarted to its initial values. The optimization procedure for any step of the procedure continues until the number of followed iterations without the population of solutions upgrading ($Noact$) reaches a pre-established value ($NoactMax$). The searching process of the overall pieces distribution is completed when the number of positioned pieces $Npieces$ is equal to the total it number $NpiecesT$ or become not possible to include one additional piece on the foil. Was obtained a mean value 0,5, of the diversity calculated by the expression (7) during the experimentation of 70 different pieces distribution of the same set of original pieces, enough high for taking into account subjective factors in the decision making process.

$$GD = \frac{1}{n} \sum_{i=1}^{n} \left| \frac{x_i - \bar{x}}{\bar{x}} \right| \tag{7}$$

Where: x_i codes of different solutions obtained; $\bar{x}$ mean value of codes. Different sets of pieces in different applications are associated to different values of the solutions population's diversity. Neverless, the exploratory character of the algorithm facilitates a sufficiently diverse sampling of the decisions making task domain, what allows to expect high values of solutions diversity.

4   Treatment of Surfaces for Exhaustive Optimization

The treatment of any irregular surfaces includes [21] concave and convex areas of the pieces and width of each concave or convex area. It is determined by the maximum distances among the extreme points belong to the area. Depth of the area Pr is determined by the maximum distance among the points of the area until the perpendicular direct line that defines the extreme points. Numeric approximation of the angle joining two piece (by the comparition of the convexities of one of pieces with the concavities of the second one). Carrying out the displacement by the frontier through the points of the contour of the piece, the angle Ang_i. Here for every value of i there possible two pieces contours coupling, and is calculated by each triad of serial vertexes, always taking the initial vertex according to the sequence of the contour. The sign of this angle and its perpetuity determines the extension and type of the defined area. The

```
01  Start
02  Npieces = 2
03  For i = 1 to m //coding first conglomerate formations solving task (6) for 2 aleatory pieces
04    Inter0(i) = Cod (Piece1 - Piece 2)
05  Next i
06  Repeat
07    Repeat
08      For i = 1 to m //creating initial search intervals
09        If (CodMax(i) - CodMin(i)) <= 2 Then
10          CodMin(i) = 0, CodMax(i) = Inter0(i)
11        Else // generating aleatory conglomerate-piece combinations
12          Cod1(i) = Rand (CodMin(i), CodMax(i))
13          SolEspMat1(i) = Det-Cong(DeCod (Cod1(i)))
14          Cod2(i) = Rand (CodMin(i), CodMax(i))
15          SolEspMat2(i) = Det-Cong(DeCod (Cod2(i)))
16      Next i
17      ListCombSol = Combina (SolCong1, ..., SolComg4)
18      For k = 1 to Length (ListCombSol)
19        Z_Fitness = GetFuntionZ (ListCombSol (k)) // calculating objective function
20        If Length (Poblacion) < Cind Then // creating population
21          AddPoblacion (ListCombSol(k), Z_Fitness): UpPS = True
22        Else UpPS = UpdatePoblacion (ListCombSol(k), Z_Fitness)
23        If Z_Fitness < Z_FitnessBest Then
24          Z_FitnessBest = Z_Fitness : ListBestComb = ListCombSol(k)
25      Next k
26      ListBestCod = GetCod (ListBestComb)
27      For i = 1 to m /.verifying search intervals
28        If ListBestCod(i) = Cod1(i) Then CodMax(i) = Cod2(i)
29        Else CodMin(i) = Cod1(i)
30      Next i
31      If UpPS = True Then NoAct = 0 Else NoAct = NoAct + 1
32    Until (NoAct > NoActMax)  // evaluating stop criterion
33    Npieces = Npieces + 1
34  Until (Npieces > Npieces:T)
35  End
```

Fig. 5. Pseudocode of conglomerates creation process

preliminary treatment of every piece allows to define the possible joining positions among any two pieces and, as a result, a the finite number of possible conglomerates and between them become possible to find the conglomerate with the less value of (6). Based on this possibility it is possible to face the general algorithm for the optimization.

5   Optimization of the Bi-Level Processes

For optimizing the bi-level distribution and cutting task is done by finding close to optimal cutting sequence for every distribution of pieces on the foil in the final population, according the model

$$\text{Minimize}$$
$$z_2(u, x_s), g(u, x_s) \geq 0 \tag{8}$$

An approximate solution of this task is presented in the work [22, 23, 24] for every distribution u, in the population of the pieces on the foil. The numerical values obtained from the solution of (7) are used for finding close to optimal solutions for the bi-level task (5). Human machine interactions take place during: Priority granted to each lot of pieces to be produced. The determination of the foils and foils wastes coming from previous cutting processes for being used in new faced distribution and cutting processes. Selection of cutting machine for every set of pieces distributions. Determination of distributions for being used for cutting taking into account non-desirable heat treatment of the pieces.

6 Conclusions

The analysis and synthesis of engineering systems methodology used in the distribution process and cutting process of irregular shaped pieces in the context of the multi-level integration CAD/CAPP/CAP/CAM processes is presented. The external analysis of the studied task allowed to elaborate the conceptual mathematical model of the task and its decomposition in the inter-related distribution and cutting irregular pieces tasks via disperse-conciliated structure. The multi-level function proposed with correcting weights demonstrated it effectiveness during the close to optimal distribution of pieces on the foil by means of an adaptive algorithm. The application of the exploration of variables codes algorithm allows to obtain a enough diverse solutions populations for an effective subjective evaluation.

References

1. Jie, L., Jialin, H., Yaoguang, H., Guangquan, Z.: Multilevel decision-making: a survey. Inf. Sci. **346**, 463–487 (2016)
2. Lu, J., Shi, C., Zhang, G.: On bi-level multi-follower decision making: general framework and solutions. Inf. Sci. **176**, 1607–1627 (2006)
3. Lu, J., Zhang, G., Montero, J., Garmendia, L.: Multifollower trilevel decision making models and system. IEEE Trans. Ind. Inf. **8**, 974–985 (2006)
4. Kalashnikov, V.V., Dempe, S., Pérez-Valdés, G.A., Kalashnykova, N.I., Camacho-Vallejo, J.F.: Bi-level programming and applications. Math. Prob. Eng. **181**, 423, 442 (2015)
5. Glackin, J., Ecker, J.G., Kupferschmid, M.: Solving bi-level linear programs using multiple objective linear programming. J. Optim. Theory Appl. **140**, 197–212 (2009)
6. Wan, Z., Wang, G., Sun, B.: A hybrid intelligent algorithm by combining particle swarm optimization with chaos searching technique for solving non-linear bi-level programming problems. Swarm Evol. Comput. **8**, 26–32 (2013)
7. Angulo, E., Castillo, E., García-Ródenas, R., Sánchez-Vizcaíno, J.: A continuous bi-level model for the expansion of highway networks. Comput. Oper. Res. **41**, 262–276 (2014)
8. Fontaine, P., Minner, S.: Benders decomposition for discrete–continuous linear bi-level problems with application to traffic network design. Transp. Res. Part B: Methodol. **70**, 163–172 (2014)
9. NieP, Y.: Dynamic discrete-time multi-leader–follower games with leaders in turn. Comput. Math Appl. **61**, 2039–2043 (2015)

10. Nishizaki, I., Sakawa, M.: Computational methods through genetic algorithms for obtaining Stackelberg solutions to two-level integer programming problems. Cybern. Syst. **36**, 565–579 (2005)
11. Bard, J.F.: Practical Bi-level Optimization: Algorithms and Applications. Kluwer Academic Publishers, Dordrecht (1998)
12. Audet, C., Haddad, J., Savard, G.: Disjunctive cuts for continuous linear bi-level programming. Optim. Lett. **1**, 259–267 (2007)
13. Zhang, T., Hu, T., Guo, X., Chen, Z., Zheng, Y.: Solving high dimensional bi-level multi-objective programming problem using a hybrid particle swarm optimization algorithm with crossover operator. Knowl. Based Syst. **53**, 13–19 (2013)
14. Han, J., Lu, J., Hu, Y., Zhang, G.: Tri-level decision-making with multiple followers: model, algorithm and case study. Inf. Sci. **311**, 182–204 (2016)
15. Arzola, R.J., Simeón, R.E., Maceo, A.: El Método de Integración de Variables: una generalización de los Algoritmos Genéticos. In: Proceedings of Intensive Workshop: Optimal Design of Materials and Structures, París (2003)
16. Arzola-Ruiz, J.: Análisis y Síntesis de Sistemas de ingeniería (2009). http://www.bibliomaster.com
17. Arzola-Ruiz, J.: Sistemas de Ingeniería. Editorial Felix Valera, La Habana (2000)
18. Arzola-Ruiz, J.: Selección de Propuestas. Editorial_Científico-Técnica, La Habana (1989)
19. Lastre-Aleaga, A.: Optimización de la distribución y corte de piezas irregulares en chapas. Ph.D. thesis, Holguín (2009)
20. Mitchell, F.H.: CIM Systems. An introduction to Computer Integrated Manufacturing. Prentice Hall, Upper Saddle River (1991)
21. Bennell, J.A., Oliveira, J.F.: The geometry of nesting problems: a tutorial. Eur. J. Oper. Res. **184**(2), 397–415 (2008)
22. Arzola-Ruiz, J.: Sistemas de Ingeniería, 2nd edn. Editorial Félix Valera, La Habana (2012)
23. Lastres–Aleaga, A.M., Arzola-Ruiz, J., Cordovés-García, A.: Optimización de la distribución de piezas irregulares en chapas. Ingeniería Mecánica **13**(2), 1–12 (2010)
24. Lastres–Aleaga, A.M. Optimización de la distribución y corte de piezas irregulares en chapas. Tesis en opción al grado científico de Doctor en Ciencias Técnicas, Holguín (2010)

Effect of Different Visual Stimuli on Joint Attention of ASD Children Using NAO Robot

Sara Ali[✉], Faisal Mehmood, Yasar Ayaz, Umer Asgher,
and Muhammad Jawad Khan

School of Mechanical and Manufacturing Engineering (SMME),
National University of Sciences and Technology (NUST),
Islamabad, Pakistan
{sarababer, faisal.mehmood, yasar, umer.asgher,
jawad.khan}@smme.nust.edu.pk

Abstract. Interaction of autistic children with socially assistive robots have shown improvements in their impairments. This research focuses on one of the core impairments i.e. joint attention of children with autism spectrum disorder (ASD) using robot-mediated therapy. The study aims to compare the effect of different visual stimuli on joint attention. Two different visual cues i.e. rasta and blink by NAO robot are introduced. The evaluation parameters are total time for eye contact and number of eye contacts in response to generated stimuli. The experiment is conducted on 12 ASD children, 8 sessions for each cue over the period of 2 months. Each session consists of 8 trials of each category. This approach achieves 80.3% of accuracy for rasta while 65.1% for blink cue. The average eye contact time for rasta is 38.8 s and for blink is 32.4 s, signifying the effect of a prominent visual cue in improvement of joint attention.

Keywords: Human-robot interaction · Autism spectrum disorder · Joint attention · NAO robot

1 Introduction

The characteristics of Autism Spectrum Disorder (ASD) is impairment in social interaction, atypical and repetitive behavior, and developmental language and communication issues [1]. 1 in 88 children are diagnosed with ASD in United States [2]. Literature suggests that early diagnosis and start of behavioral intervention of ASD children can improve their impairments [3]. Because of the resource limitation, evidence-based early intervention is limited [4]. Along with the mentioned impairments, an autistic child has a lack of focus issue called as joint attention (JA) towards any type of command that can be either verbal or non-verbal [5]. Further, in joint attention, it has two main divisions: (a) response to JA (RJA) and (b) initiation of joint attention (IJA) [6].

With the advancement in technology, socially assistive robots (SAR) are acting as a promising behavioral therapy tool for assessment of an autistic child [7]. The child interacts with the robots for specially designed interventions focusing on the

© Springer Nature Switzerland AG 2020
H. Ayaz (Ed.): AHFE 2019, AISC 953, pp. 490–499, 2020.
https://doi.org/10.1007/978-3-030-20473-0_48

improvement in social behavior of an autistic child. Several robots are used for the purpose of behavioral improvement in autism such as Paro [8], Keepon [9], robotic doll Robota [10, 11], KASPAR [12]. Lack of an ability to shift the gaze towards another object is a sign of autism. Research exists to improve the joint attention skills of an autistic child using these robotic interventions as the robots acts as a promising tool for improvement in ASD [13, 14]. Robotic interventions for joint attention focus their task towards improvement of eye contact only as joint attention is a pivot of effective social communication [15]. These interventions help the children to learn more effectively by sharing their attention with different people. In all these robotics interventions, appearance of robot plays an important role and significantly effects the interaction of child [16, 17].

Human-robot interaction is one of the potential intervention tools for children with autism spectrum disorder as according to the research robots can perform the arm and speech gestures better than humans [11]. Measure of its repeatability, preciseness and low cost the proofs effectiveness of robot-mediated therapy [18].

In this paper, we have presented a robot-mediated therapy focusing on the effect of different visual stimuli for improvement in joint attention of an autistic child when exposed to the robot NAO. The experimentation was conducted on 12 ASD children over a period of 2 months containing 8 sessions. Each session consists of 8 trials for each type of visual cue to improve the joint attention. Two cues were introduced i.e. rasta and blink to check the effectiveness and significance of visual cue based on prominent effect. The achieved results were 80.3% of accuracy for rasta while 65.1% for blink cue. The average eye contact time for rasta was 38.8 s and for blink was 32.4 s. The parameters used to measure the efficiency of results were number of eye contacts when a stimulus was given and time for which an eye contact was maintained. The result indicated that the prominent cue i.e. rasta is more effective in improving the joint attention of ASD children as they intend to develop more frequent eye contact when this visual cue is introduced in the intervention.

2 Architecture

2.1 Joint Attention Protocol

The robot prototype used for this research is NAO as shown in Fig. 1. The purpose of this research is to find the effect of different visual stimuli for robot mediated therapy in order to improve the joint attention for children with autism. The interaction protocol for joint attention module of this research uses two cues i.e. rasta and blink. Both cues differ in the level of prominence. The child was exposed to the cues to check the behavior in term of eye contact based on the significance of cue itself. The parameters used to evaluate are number of eye contacts and total time for which an eye contact was maintained. NAO gaze tracking module measured these parameters.

Fig. 1. Joint attention therapy involving different visual cues for an ASD child using NAO platform.

Joint attention therapies done previously involve least-to-most (LTM) based cues for screening and diagnosis of ASD [19]. In this therapy rasta (changing eye color of robot in cyclic manner) is the considered the prominent cue whereas blinking is considered as least intruding command. Previous researches have only used robot to check the improvement of joint attention without comparing the effect of different visual stimuli on child's behavior. Figure 2 shows the architecture diagram for our research. The designed architecture consists of three different states: start, execution, and stop. Execution state consists of two modules i.e. visual cue module and eye contact module which are running in parallel manner. Visual cues are further divided into two types i.e. blink and rasta as shown in Fig. 3.

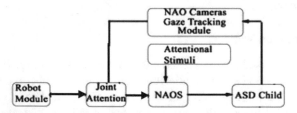

Fig. 2. Joint attention protocol architecture. In this module, NAO gives different visual attention stimuli to the child. The gaze-tracking module measures the parameters required to evaluate the JA of an ASD child based on cues.

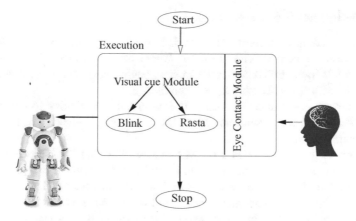

Fig. 3. Architecture with state machine diagram

3 Materials and Methods

3.1 Subjects

This therapy was tested on 12 ASD children including 11 males and 1 female. The subjects were recruited from Capitals's Autism Resource Center (ARC). The autism specialist and director board of ARC approved the study. The participant's parents also signed the consent form. There was no financial compensation for the participants. The clinical evaluation of participants was done by ARC based on Childhood Autism Rating Scale Score (CARS) criteria by the experts. The details of the participants is shown in Table 1 below.

Table 1. Details of ASD participants involved in experimentations

Subjects	Gender	Age
S1	M	4.5
S2	M	5
S3	M	5.8
S4	M	4.7
S5	M	3.5
S6	M	7.2
S7	M	6.9
S8	M	7
S9	M	5.8
S10	F	6.3
S11	M	6.7
S12	M	5

3.2   Experimental Setup and Design

The child was sitting on a comfortable chair in front of the robot such that he/she can easily make an eye contact with the robot. There was a distance of 1 m between the robot and child. For an initial brain reinforcement activity, the child was asked do count from 1 to 10 before the start of experimentation.

The robot starts the intervention by its first visual cue of blink to notice the joint attention of an ASD child. After that another cue i.e. rasta is given in order to measure the same parameters.

Eight sessions were conducted for this intervention over the period of 2 months. Each session involves 8 trials of each cue. In each session, number of eye contacts was noted against 8 trials. The eye contact for each session was recorded as an average eye contact time. The initial engagement of the child was noted as a baseline parameter. For this reason, the initial reading was taken several times until a stable baseline was achieved. This stable initial baseline parameter was compared to measure the success over the intervention.

Upper camera of NAO robot was used for measuring the eye contact of an ASD child. The color space was BGR and frame rate is 15 f/s. NAO's API, "ALGaze Analysis" associated with closing of eyes and opening of eyes.

4   Results

The results for each category was measured based on (1) number of eye contact for each visual cue and (2) total time for which an eye contact was established when a visual cue was given. The reading was observed over a period of 2 months and success was established based on a stable bassline reading. Based on the above parameters the improvement in eye contact for each visual cue as shown Fig. 4.

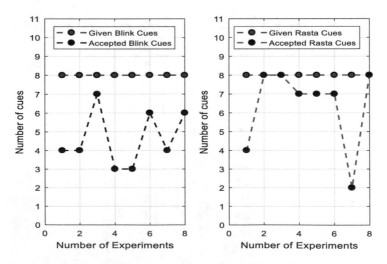

Fig. 4. This figure shows accepted and given cues for both blink and rasta of a single subject for 8 weeks of experimentation.

In Fig. 4 the performance of a subject was shown over eight weeks. The frequency was eights trial per experiment in a week. A clear comparison of an ASD child's inclination towards rasta is shown in Fig. 5. Figure 6 shows performance of a subject by comparing two visual cues in terms of total time for which an eye contact was established for each cue. This indicates the effectiveness of rasta over blink cue.

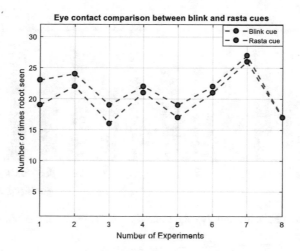

Fig. 5. This figure shows the comparison of blink by comparing the number of times robot was seen for each cue.

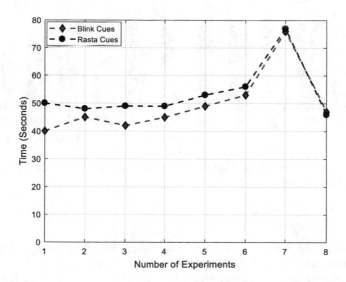

Fig. 6. This figure shows total time for which the eye contact was established for each visual cue i.e. blink and rasta.

The effective cue is towards which the child is more attentive can be seen in the results. From all of the above results, the improvement in joint attention of the child over the experiments can also be seen. The average results for all the subjects over 8 experiments for each cue using parameters i.e. number of times the eye contact is made for each cue and time for which an eye contact is made is shown in Figs. 7 and 8 respectively.

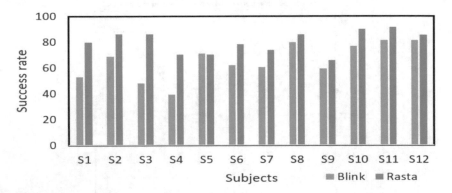

Fig. 7. This figure shows overall average accuracy rate for each cue i.e. blink and rasta of 12 ASD subjects.

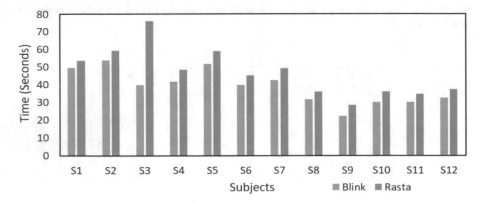

Fig. 8. This figure shows average time an eye contact is established for each cue i.e. blink and rasta of 12 ASD subjects

5   Discussion

Based on the results above, this main contribution of the therapy is the effectiveness of visual cue rasta over blink. This is observed based on the experimentation done over 8 weeks. This shows that the prominence of cue holds importance for establishing an eye contact o an ASD child. If the cue is more prominent, the child will show an effectiveness improvement of joint attention as compared to the non-prominent cue. The time and trials of each cue for establishing an eye contact is same. Therefore, the stimuli generated has not introduced any kind of biasness for an ASD child to establish an eye contact. The dominance of visual cue 'rasta' is shown in Fig. 9.

The advantage of robot-mediated therapy is that it does not require any human involvement. The reading for number of eye contacts and time for establishing the eye contact is done by NAO's camera. Therefore, the accuracy of the results is more as compared to a human therapist. However, these robot-mediated researches give rise to certain questions e.g., what if the robot has to control the child's behavior under certain situation? What will be the robot's reaction when child will be approaching it and how will it affect the therapy?

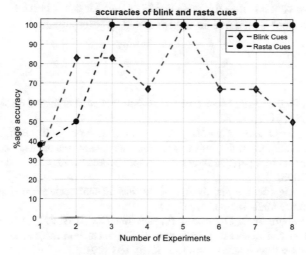

Fig. 9. This figure shows the dominance of visual cue rasta over blink in the form of percentage accuracy over the experiments.

Table 2. Average accuracies of all subjects over all experiments

Subjects	Blink stimulus	Rasta stimulus
S1	52.92	79.69
S2	68.75	86.00
S3	48.00	86.13
S4	39.29	70.29
S5	71.00	70.50
S6	62.13	78.38
S7	60.63	73.63
S8	79.75	85.88
S9	59.25	65.88
S10	76.75	90.13
S11	81.38	91.63
S12	81.38	85.38
Average	**65.10**	**80.29**

6 Conclusion

This research focuses on comparison of various visual stimuli given to an ASD child to observe the effectiveness of cue based on the prominence. The experiment was conducted on 12 ASD children over the period of 8 weeks. Each experiment was given 8 trials for a week. It is observed that each of the subject has shown improvement over the experiments for each cue. However, for rasta the improvement observed is better than blink. This shows that prominence of cue is important in contributing towards the improvement of joint attention with same number of trials and time given to each cue, achieving 80.3% of accuracy for rasta while 65.1% for blink cue as shown in Table 2. The average eye contact time for rasta is 38.8 s and for blink is 32.4 s, signifying the effect of a prominent visual cue in improvement of joint attention.

References

1. Mittal, V.A., Walker, E.F.: Diagnostic and statistical manual of mental disorders. Psychiatry. Res. **189**, 158 (2011)
2. Simonoff, E., Pickles, A., Charman, T., Chandler, S., Loucas, T., Baird, G.: Psychiatric disorders in children with autism spectrum disorders: prevalence, comorbidity, and associated factors in a population-derived sample. J. Am. Acad. Child Adolesc. Psychiatry. **47**(8), 921 (2008)
3. Ganz, M.L.: The lifetime distribution of the incremental societal costs of autism. Arch. Pediatr. Adolesc. Med. **161**(4), 343 (2007)
4. Warren, Z., Vehorn, A.C., Dohrmann, E., Newsom, C.R., Taylor, J.L.: Brief report: service implementation and maternal distress surrounding evaluation recommendations for young children diagnosed with autism. Autism. **17**(6), 693 (2013)
5. Damiano, L., Dumouchel, P.: Anthropomorphism in human-robot co-evolution. Front. Psychol. **9**, 468 (2018)
6. Dautenhahn, K., Nehaniv, C.L., Walters, M.L., Robins, B., Kose-Bagci, H., Mirza, N.A., Blow, M.: KASPAR – a minimally expressive humanoid robot for human-robot interaction research. Appl. Bionics Biomech. **6**, 369–397 (2009)
7. Dorsey, R., Howard, A.M.: Examining the effects of technology-based learning on children with autism: a case study. In: 2011 IEEE 11th International Conference on Advanced Learning Technologies, p. 260 (2011)
8. Hashemi, J., Spina, T.V., Tepper, M., Esler, A., Morellas, V., Papanikolopoulos, N., Sapiro, G.: A computer vision approach for the assessment of autism-related behavioral markers. In: Development and Learning and Epigenetic Robotics (ICDL), 2012 IEEE International Conference (2012)
9. Bryson, S.E., Zwaigenbaum, L., McDermott, C., Rombough, V., Brian, J.: The autism observation scale for infants: scale development and reliability data. J. Autism Dev. Disord. **38**, 731 (2008)
10. Sial, S.B., Sial, M.B., Ayaz, Y., Shah, S.I.A., Zivanovic, A.: Interaction of robot with humans by communicating simulated emotional states through expressive movements. Intell. Serv. Robot. **9**, 231 (2016)
11. Zheng, Z., Das, S., Young, E.M., Swanson, A., Warren, Z., Sarkar, N.: Autonomous robot-mediated imitation learning for children with autism. In: Robotics and Automation (ICRA), 2014 IEEE International Conference (2014)

12. Zheng, Z., Young, E.M., Swanson, A., Weitlauf, A., Warren, Z., Sarkar, N.: Robot-mediated mixed gesture imitation skill training for young children with ASD. In: Advanced Robotics (ICAR), 2015 International Conference (2015)

13. Goodwin, M.S.: Enhancing and accelerating the pace of autism research and treatment the promise of developing innovative technology. Focus Autism Other Dev. Disabil. 23(2), 125–128 (2008)

14. Kasari, C., Gulsrud, A., Wong, C., Kwon, S., Locke, J.: Randomized controlled caregiver mediated joint engagement intervention for toddlers with autism. J. Autism Dev. Disord. 40, 1045–1056 (2010)

15. Bekele, E., Lahiri, U., Davidson, J., Warren, Z., Sarkar, N.: Development of a novel robot-mediated adaptive response system for joint attention task for children with autism. In: RO-MAN, 2011 IEEE (2011)

16. Kumazaki, H., Warren, Z., Swanson, A., Yoshikawa, Y., Matsumoto, Y., Ishiguro, H., Sarkar, N., Minabe, Y., Kikuchi, M.: Impressions of humanness for android robot may represent an endophenotype for autism spectrum disorders. J. Autism Dev. Disord. 48(2), 632 (2018)

17. Robins, B., Dautenhahn, K., Dickerson, P.: From isolation to communication: a case study evaluation of robot assisted play for children with autism with a minimally expressive humanoid robot. In: 2009 Second International Conferences on Advances in Computer-Human Interactions, p. 205 (2009)

18. Warren, Z., Zheng, Z., Swanson, A., Bekele, E., Zhang, L., Crittendon, J.A., Weitlauf, A., Sarkar, N.: Can robotic interaction improve joint attention skills? J. Autism Dev. Disord. 45 (11), 3726 (2015)

19. Scassellati, B.: Quantitative metrics of social response for autism diagnosis. In: Rom. 2005. IEEE International Workshop on Robot and Human Interactive Communication, p. 585 (2005)

Assessment of User Interpretation on Various Vibration Signals in Mobile Phones

Nikko Marcelo C. Palomares, Giorgio B. Romero$^{(\boxtimes)}$,
and Jean Louie A. Victor

Department of Industrial Engineering and Operations Research,
College of Engineering, University of the Philippines Diliman,
Quezon City, Philippines
{ncpalomares, gbromerol, javictor}@up.edu.ph

Abstract. The study aims to identify the level of urgency of different vibration patterns in mobile phones and the commonly perceived phone notification (e.g. text message, phone call, emergency alert, etc.) associated with those patterns. After gathering information regarding the device usage of the user, the researchers collected the responses on perceived level of urgency and the most associated message through a five-point rating scale and a questionnaire, respectively. The gathered responses were summarized and analyzed in boxplot diagrams, tables, and pie charts and were justified using the Kruskal-Wallis Test and Fleiss' Kappa. The results showed that the different patterns are perceived differently, and respondents slightly agree to classify these patterns to different message categories. Vibration patterns were also grouped by most associated message type and level of urgency which can provide useful applications in mobile phone features and applications.

Keywords: Mobile phone · Vibration · Perception · Haptic

1 Background of the Study

The adoption of the mobile phone has been a global phenomenon in recent years and has dramatically changed people's ability to communicate. Mobile phones can now effectively interact with the three major human senses namely, vision, touch, and hearing [1]. Information (e.g. text message, phone call, etc.) is received in a form of a notification. The most commonly-used form of notification is audio notification but, in many situations, it may be inaudible due to external factors and unsuitable to some environments as it may disturb others. As an alternative to audio output, tactile output, such as vibration, is used to inform the user [2]. Providing alerts through vibration pulses has been a widely-used output mechanism and an important functionality in mobile phones. It became an essential information conveyor as it can notify the user privately [3]. It becomes more useful especially when the user's auditory and visual senses are overwhelmed or in some cases, absent. Vibration signals can represent different kinds of notification particularly in communication-related applications. However, most of the time, the same vibration signals are used in some incoming messages when the mobile phone is in silent mode [4].

© Springer Nature Switzerland AG 2020
H. Ayaz (Ed.): AHFE 2019, AISC 953, pp. 500–511, 2020.
https://doi.org/10.1007/978-3-030-20473-0_49

1.1   Rationale

The Philippines is the fastest-growing smartphone market in Southeast Asia to date. In addition to this, the smartphone users in the country will hit 90 million in 2021 from the current 40 million, as predicted by the Ericsson Mobility Report [5]. According to previous studies [4], 80% of mobile phone users are in silent and vibrate mode and at least 54% of the population with disabilities (e.g. blindness, deafness, etc.) in urban areas use a cell phone or a smartphone [6]. It is important to study about haptic technology, particularly vibrations, since the users mentioned above heavily rely on them when receiving messages and reminders. Misinterpretation of vibration signals in mobile phones can result to missed calls, unattended urgent matters, or wrong information perceived. However, when these signals are used properly, applications of vibration in mobile phones decrease mental workload of users and increase perceived usefulness of the device [7].

2   Objectives of the Study

The study aims to identify the level of urgency and the commonly perceived message associated with various vibration patterns in mobile phones.

3   Methodology

A total of 15 participants with ages 19–25 took part in the experiment. Studies have shown that there occurs a decline in the main sensory modalities such as touch sensation with advancing age [4, 8–10]. This age group was chosen due to its easy accessibility and better sensory modalities than higher age groups. The experiment was done inside a closed room and participants were asked to sit and to wear earphones to eliminate unwanted noise [3, 11].

Respondents answered a survey form that collects information regarding their name, age, year level, and course. Information regarding their mobile phone model in use and the frequentness (in hours) in a day their phone is in silent and vibrate mode was also gathered [12].

There were 10 vibration patterns used which are proven to be distinguishable by previous studies [8]. These patterns were felt in two scenarios: the first being the pattern played once, and the second being played over 4.5 s (looped). In total, there were 20 different vibration patterns felt by each respondent. A vibration pattern is a sequence of the on and off state of the phone's vibration motor, with specific lengths (short and long) assigned to each state. The vibration and gap length were limited to 200 ms for short (s) and 600 ms for long (l) [4]. The capital letter (S or L) signifies the length of the on state, while the small letter (s or l) signifies the length of the off state.

Participants were asked to feel the 20 vibration patterns through a mobile device with an application that allows custom vibration settings. The mobile phone was placed inside the trousers pocket because it is where mobile devices are commonly placed [13]. Only one mobile device was used for all respondents so that the device weight and vibration frequency are constant [3]. The order in which the vibration patterns are felt and the length at which it is played were randomized to satisfy the needed assumption that pattern perception by a respondent is independent from (or not affected by) his or her perception of the other patterns as shown in Table 1.

Table 1. Randomized pattern sequence

Pattern number	Sequence	Repetition
1	L-l-S-l	Over 4.5s
2	L-s-L-l	Over 4.5s
3	L-s-L-l	Once
4	S-s-S-l	Over 4.5s
5	L-s-S-s	Once
6	S-s-S-l	Once
7	S-s-L-l	Once
8	L-l-S-l	Once
9	L-s-S-l	Once
10	S-s	Over 4.5s
11	L-l	Once
12	S-l	Once
13	L-s-S-l	Over 4.5s
14	S-s	Once
15	L-s	Once
16	L-s	Over 4.5s
17	L-s-S-s	Over 4.5s
18	L-l	Over 4.5s
19	S-s-L-l	Over 4.5s
20	S-l	Over 4.5s

To identify the perceived level of urgency, a five-point rating scale was used where ordinal numbers are assigned to each level of urgency as enumerated below.

0 - Not Detected	3	-	High	Priority
1 - Insignificant	4	-	Emergency/Urgent	
2 - Low Priority				

A five-point rating scale was used because a study showed that users can distinguish at least four levels of urgency based on the design of the vibration patterns that was used [4]. In addition to the rating scale, a questionnaire was answered by the participants to identify what message was most associated with each vibration pattern - phone call, text message, app notification/update, alarm clock, low battery, emergency alert, and others as shown in Appendix A. The researchers recorded the responses first before playing the next vibration pattern.

Data Analysis. The descriptive statistics of the responses were summarized in tables, boxplot diagrams, and pie charts to identify any observable insights from these diagrams. For the statistical analysis, Kruskal-Wallis Test was used because the assumptions of a One-Way ANOVA were not met and the data gathered was ordinal.

It was done to determine first if there is a statistically significant difference between the independent variable (pattern) on an ordinal dependent variable (i.e. ratings).

Moreover, Fleiss' Kappa was used to analyze the data on perceived level of urgency and message association. This was done to determine the level of agreement the respondents give which indicates the consistency of the results. If the agreement is high, there is more confidence that the perceived level of urgency or the message associated with the vibration patterns reflects the actual circumstance. In grouping the patterns by rating and message, the mode was used to identify the most associated rating and message per pattern.

4   Results and Discussion

Among the 15 respondents (7 male and 8 female), 10 of them have silent mode enabled the whole day, while the average hours that their silent mode is enabled is 22.36 h. Moreover, seven of them have vibrations enabled when silent mode is enabled. The summary of the descriptive statistics per pattern is shown in Table 2.

Table 2. Summary of descriptive statistics per vibration pattern

Pattern number (randomized)	Pattern	Median	Spread of variability	Min	Max
1	L-l-S-l (4.5)	3	1	2	4
2	L-s-L-l (4.5)	3	2	1	4
3	L-s-L-l (1)	2	2	1	3
4	S-s-S-l (4.5)	3	2	1	4
S	L-s-S-s (1)	2	1	0	2
6	S-s-S-l (1)	2	1	0	2
7	S-s-L-l (1)	2	0	2	2
8	L-l-S-l (1)	2	0	2	2
9	L-s-S-l (1)	2	1	1	3
10	S-s (4.5)	4	1	3	4
11	L-l (1)	2	1	0	2
12	SJ(1)	1	0	0	1
13	L-s-S-l (4.5)	3	2	1	4
14	S-s (1)	1	1	0	2
16	L-s (1)	2	1	0	2
16	L-s (4.5)	3	1	2	4
17	L-s-S-s (4.5)	4	1	2	4
18	L-l (4.5)	3	1	2	4
19	S-s-L-l (4.5)	3	1	2	4
20	S-l (4.5)	2	2	0	4

Boxplots were used to show the shape of the distribution, central value, and its variability. There were various trends that were observed based from the Boxplot Diagram of Vibration Patterns found in Fig. 1.

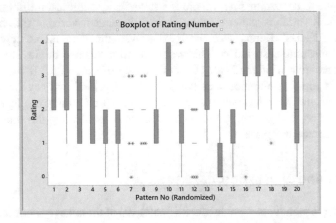

Fig. 1. Boxplot diagram of vibration pattern

In Fig. 1, it is notable that Pattern 10 (S-s Loop) has the greatest number of ratings of 3 or 4. This signifies that the pattern is perceived as the most urgent. Pattern 12 (S-l Once) and 14 (S-s Once) were almost undetectable since their ratings only range from 0 to 1. Pattern 20 (S-l Loop), on the other hand, has the largest variability with whiskers ranging from 0 to 4. This means that the perception of this pattern greatly varies, thus may not be an ideal pattern for a specific purpose. Figure 2 shows the common messages associated per rating using a boxplot.

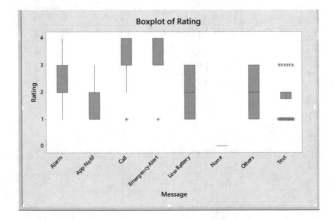

Fig. 2. Boxplot diagram of messages associated

In Fig. 2, it is apparent that emergency alert message has the greatest number of ratings of 3 or 4. This means that this message is perceived as the most urgent. However, it can be observed that alarm and low battery have a high range. This means that respondents have a varying perceived level of urgency towards those messages.

Kruskal-Wallis Test. The null hypothesis in the Kruskal-Wallis test states that all the means are equal, meaning that patterns were not perceived differently. In analyzing the data using Kruskal-Wallis Test, it was found that the p-value is 0.003678. Since it is lower than 0.05, we reject the null hypothesis. Therefore, the patterns are, indeed, perceived differently. The vibration patterns have to be perceived differently because if the user perception is the same for a number of vibration patterns, there would be no further analysis.

To identify the most associated rating and message per pattern, the mode for each pattern was determined and is summarized in Table 3. However, the variability of these ratings, as shown in the boxplot in Fig. 1, must be taken into consideration as well. Patterns with a high interquartile range means that the perception of this pattern varies and would not be an ideal pattern to use for a specific purpose.

Table 3. Most associated rating and message per pattern

Pattern number (randomized)	Most associated rating	Most associated message
1	High Priority	Phone Call and text
2	Low or High Priority	Phone Call
3	Low Priority	Text
4	High Priority	Alarm Clock
5	Low Priority	Text & App Notification
6	Low Priority	Text
7	Low Priority	Text
8	Low Priority	Text
9	Low Priority	App Notification
10	Emergency/Urgent	Emergency Alert
11	Low Priority	App Notification
12	Insignificant	App Notification
13	Emergency/Urgent	Phone Call
14	Insignificant	App Notification
15	Low Priority	Text
16	High Priority	Phone Call
17	Emergency/Urgent	Alarm Clock
18	Emergency/Urgent	Phone Call
19	High Priority	Phone Call
20	Low Priority	Alarm Clock

Fleiss' Kappa Test. For the Fleiss' Kappa Analysis for rating, the computed Kappa statistic is 0.171. Therefore, there is evidence towards a slight agreement in classifying patterns under the given ratings. It is likely that a pattern is assigned to a specific level of urgency than by chance. Thus, the grouping of the patterns by rating is justified. The test serves as a basis in grouping them by rating.

Meanwhile, in terms of the Fleiss' Kappa analysis for the message type, it was found out that the Kappa Statistic is 0.123. Therefore, there is evidence towards a slight agreement that respondents classify patterns under a given category, rather than over chance. Similar to the results of the first Fleiss' Kappa test, the patterns are likely to be assigned to a specific message. Thus, the grouping of of message by rating is justified. The test serves as a basis in grouping them by rating.

The grouping of vibration patterns and messages by level of urgency was determined by summarizing all the responses in a pivot table and getting its highest percentage per rating as summarized in Tables 4 and 5.

Table 4. Grouping of patterns by level of urgency

Level of urgency	Patterns
1 (Insignificant)	12 (S-l Once)
	14 (S-s Once)
2 (Low Priority)	2 (L-s-L-l Loop)
	3 (L-s-L-l Once)
	5 (L-s-S-s Once)
	6 (S-s-S-s Once)
	7 (S-s-L-l Once)
	8 (L-l-S-l Once)
	9 (L-s-S-l Once)
	11 (L-l Once)
	15 (L-s Once)
	20 (S-l Loop)
3 (High Priority)	1 (L-l-S-l Loop)
	2 (L-s-L-l Loop)
	4 (S-s-S-l Loop)
	16 (L-s Loop)
	19 (S-s-L-l Loop)
4 (Emergency/Urgent)	10 (S-s Loop)
	13 (L-s-S-l Loop)
	17 (L-s-S-s Loop)
	18 (L-l Loop)

In Table 4, it can be inferred that most vibration patterns played for over 4.5 s (looped) were categorized under High Priority or Emergency/Urgent while vibration patterns played once were categorized under Insignificant and Low Priority. Therefore, as shown in Table 5, specific vibration patterns under High Priority or Emergency/Urgent could be used for Alarm, Call and Emergency Alert and specific vibration patterns under Insignificant and Low Priority could be used for App Notification and Text. The complete breakdown of the components can be found in Appendix B.

Table 5. Grouping of messages by level of urgency

Level of urgency	Messages
1 (Insignificant)	App Notification (47.46%)
2 (Low Priority)	Text (44.76%)
3 (High Priority)	Alarm (27.94%) and Call (32.35%)
4 (Emergency/Urgent)	Call (50.94%) and Emergency Alert (37.74%)

The *Low Battery* message was not included in the grouping of messages since it was not associated most to any level of urgency. Pie charts were used to show the percentage of messages that were associated to each level of urgency as shown in Appendix B. The frequency of perceived ratings per message in percentage is also summarized in Fig. 3.

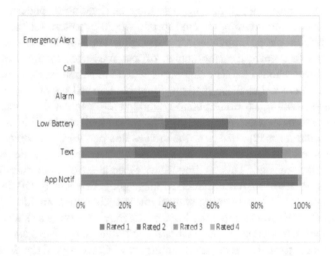

Fig. 3. Frequency of ratings per message in percentage

Figure 3 shows the percentage of the perceived level of urgency per message. It is observed that 97% of the responses perceived emergency alert as either high priority (3) or urgent (4). Meanwhile, similar to the boxplot in Fig. 2, Alarms and Low Battery has a more varying response in terms of level of urgency, which is divided among three or four levels.

5 Conclusion

The study investigated 10 basic vibration patterns that were played in two instances - the first being played once and the second being played over 4.5 s (looped). It was found out that the vibration patterns are interpreted differently. Pattern 10 (S-s Loop) was interpreted as the most urgent. In addition, the repetition of the pattern played a factor in the perceived message of the vibration pattern. Vibrations played only once were found to belong in the lowest rating, thus being the least urgent while looped vibration patterns were most likely to be perceived as urgent. Some patterns, like Pattern 20 (S-l loop), were found to have a high variety in rating and message interpretations, thus these patterns are not recommended to be used in specific purposes.

The kind of message and the level of urgency can be slightly classified to certain vibration patterns. The message perceived was also related to the level of urgency perceived by the user. The emergency alert message had the highest ratings of 3 or 4 which makes it the most urgent perceived message. However, the low battery message was not associated most to any level of urgency due to its high variability.

6   Recommendation

In general, mobile phone developers can use the classification of patterns by rating or message as a basis in determining the appropriate vibration pattern for their application. In terms of mobile phone applications, messaging applications should use longer vibration lengths to signify the importance of text messages to the user. Moreover, mobile applications should avoid using short repetitive vibrations on short bursts or patterns similar to L-s-S-l as these patterns could be reserved for emergency situations to avoid misinterpretation. Mobile phone manufacturers can set their emergency alert vibration pattern to have short, fast, and repetitive on and off states. This can easily alert the user that a disaster is reported near the area as this pattern is perceived as the most urgent by most of the respondents, regardless whether the respondents have their vibrations enabled or not. This is important especially in the Philippines, as the National Disaster Risk Reduction and Management Council (NDRRMC) utilizes emergency phone alerts in informing natural disaster alerts nationwide.

Groupings of vibration patterns per level of urgency was made as shown in Table 4 so people who wish to customize their vibration stimulation can have specific vibration patterns to choose from for each level of urgency. Currently, there is no available custom vibration application in iOS, while there are only limited options available on Android. Since a user can have different priorities in their mobile phone applications, having a customized vibration signal can greatly help the user in interpreting messages through haptic feedback, thus reducing mental workload.

Further studies can also be explored that consider other factors such as varying intensity, environment, and special populations. Particularly, for the special population such as those visually impaired since they primarily rely on hearing or touching. Since the study was done in a static environment, working on user perception on mobile environments (i.e. while moving) or on open environments can be considered. Further studies in assessing the difference in perception between people who have vibrations enabled in their phone and people who have it disabled can also be explored.

Acknowledgement. We would like to extend our utmost gratitude to our adviser, Ma'am Alyssa Jean Portus, who guided us through the entire duration of the study and gave us insights regarding our research topic. To Jurel Yap and Charmine Cramales from the UP School of Statistics, for providing us statistical assistance. To the developers of the Android application, *Good Vibrations*, whose application has made this study possible by providing custom vibration settings. Lastly, we would like to express our sincerest gratitude to our respondents, who made our data analysis possible through the raw data that they have provided for us.

Appendix

Appendix A: Pattern Perception Questionnaire

Likert Scale

Put a check [✓] mark on the space provided.

Describe the level of urgency of the Vibration Pattern	Not Detected (0)	Insignificant (1)	Low Priority (2)	High Priority (3)	Urgent (4)
Vibration Pattern 1					
Vibration Pattern 2					
Vibration Pattern 3					
Vibration Pattern 4					
Vibration Pattern 5					
Vibration Pattern 6					
Vibration Pattern 7					
Vibration Pattern 8					
Vibration Pattern 9					
Vibration Pattern 10					

Questionnaire

Put one check [✓] mark on the space provided.

What message do you most associate the vibration pattern with	Phone Call	Text Message	App Notification / Update	Alarm Clock	Low Battery	Emergency Alert	Others (please specify)
Vibration Pattern 1							
Vibration Pattern 2							
Vibration Pattern 3							
Vibration Pattern 4							
Vibration Pattern 5							
Vibration Pattern 6							
Vibration Pattern 7							
Vibration Pattern 8							
Vibration Pattern 9							
Vibration Pattern 10							

Appendix B: Pie Chart of the Percentage of Message Associated Per Level of Urgency

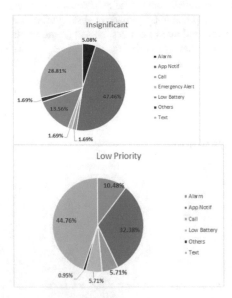

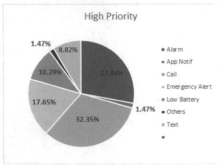

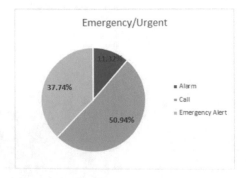

References

1. Ur Réhman, S., Liu, L.: Vibrotactile emotions on a mobile phone. In: 2008 IEEE International Conference on Signal Image Technology and Internet Based Systems (2008)
2. Sahami, A., Holleis, P., Schmidt, A., Häkkilä, J.: Rich tactile output on mobile devices. In: Aarts, E., et al. (eds.) Ambient Intelligence. AmI 2008. Lecture Notes in Computer Science, vol. 5355. Springer, Berlin, Heidelberg (2008)
3. Yao, H., Grant, D., Cruz, M.: Perceived vibration strength in mobile devices: the effect of weight and frequency. IEEE Trans. Haptics 3(1), 56–62 (2009)
4. Saket, B., Prasojo, C., Huang, Y., Zhao, S.: Notification Interface for Mobile Phones. NUS-HCI Lab, School of Computing, National University of Singapore 117418, Singapore (2013)
5. Abadilla, E.: Smartphone users up 25% to 32.5 M. Manila Bulletin (2016). https://business. mb.com.ph/2016/11/24/smartphone-users-up-25-to-32-5-m/. Accessed 14 Nov 2018
6. Reyes, C.: Persons with Disability (PWDs) in Rural Philippines: Results from the 2010 Field Survey in Rosario, Batangas. Philippine Institute for Development Studies (2011)
7. Rasche, P., Mertens, A., Schlick, C., Choe, P.: The effect of tactile feedback on mental workload during the interaction with a smartphone. In: Lecture Notes in Computer Science, vol. 9180 (2015)
8. Torcolini, N., Oh, J.: Effect of Vibration Motor Speed and Rhythm on Perception of Phone Call Urgency (2011)
9. Baek, Y., Myung, R., Yim, J.: Have you ever missed a call while moving? The optimal vibration frequency for perception in mobile environments. In: 6th WSEAS International Conference on Applied Informatics and Communications, pp. 241–245, Elounda, Greece, August 18–20, 2006. https://www.semanticscholar.org/paper/Have-you-ever-missed-a-call-while-moving%3A-the-for-Baek-Myung/7935f184e9b4408f5ac51773f3073d75f4620a46#paper-header
10. Wickremaratchi, M., Llewelyn, J.: Effects of ageing on touch. https://www.ncbi.nlm.nih.gov/pmc/articles/PMC2563781/. Accessed 16 Nov 2019
11. Ryu, J., Jung, J., Choi, S.: Perceived magnitudes of vibrations transmitted through mobile device. In: 2008 Symposium on Haptic Interfaces for Virtual Environment and Teleoperator Systems, Reno, NE, pp. 139–140 (2008)
12. Harkins, J., Tucker, P., Williams, N., Sauro, J.: Vibration signaling in mobile devices for emergency alerting: a study with deaf evaluators. J. Deaf Stud. Deaf Educ. 15(4), 438–445 (2010)
13. Ichikawa, F., Chipchase, J., Grignani, R.: Where's the phone? A study of mobile phone location in public spaces. In: 2005 2nd Asia Pacific Conference on Mobile Technology, Applications and Systems, Guangzhou, pp. 1–8 (2005)

Author Index

© Springer Nature Switzerland AG 2020
H. Ayaz (Ed.): AHFE 2019, AISC 953, pp. 513–515, 2020.
https://doi.org/10.1007/978-3-030-20473-0

Printed in the United States
By Bookmasters